# Microwave Photonics

# OPTICAL SCIENCE AND ENGINEERING



1. Electron and Ion Microscopy and Microanalysis: Principles and Applications, *Lawrence E. Murr*
2. Acousto-Optic Signal Processing: Theory and Implementation, *edited by Norman J. Berg and John N. Lee*
3. Electro-Optic and Acousto-Optic Scanning and Deflection, *Milton Gottlieb, Clive L. M. Ireland, and John Martin Ley*
4. Single-Mode Fiber Optics: Principles and Applications, *Luc B. Jeunhomme*
5. Pulse Code Formats for Fiber Optical Data Communication: Basic Principles and Applications, *David J. Morris*
6. Optical Materials: An Introduction to Selection and Application, *Solomon Musikant*
7. Infrared Methods for Gaseous Measurements: Theory and Practice, *edited by Joda Wormhoudt*
8. Laser Beam Scanning: Opto-Mechanical Devices, Systems, and Data Storage Optics, *edited by Gerald F. Marshall*
9. Opto-Mechanical Systems Design, *Paul R. Yoder, Jr.*
10. Optical Fiber Splices and Connectors: Theory and Methods, *Calvin M. Miller with Stephen C. Mettler and Ian A. White*
11. Laser Spectroscopy and Its Applications, *edited by Leon J. Radziemski, Richard W. Solarz, and Jeffrey A. Paisner*
12. Infrared Optoelectronics: Devices and Applications, *William Nunley and J. Scott Bechtel*
13. Integrated Optical Circuits and Components: Design and Applications, *edited by Lynn D. Hutcheson*
14. Handbook of Molecular Lasers, *edited by Peter K. Cheo*
15. Handbook of Optical Fibers and Cables, *Hiroshi Murata*
16. Acousto-Optics, *Adrian Korpel*
17. Procedures in Applied Optics, *John Strong*
18. Handbook of Solid-State Lasers, *edited by Peter K. Cheo*
19. Optical Computing: Digital and Symbolic, *edited by Raymond Arrathoon*
20. Laser Applications in Physical Chemistry, *edited by D. K. Evans*

21. Laser-Induced Plasmas and Applications, *edited by Leon J. Radziemski and David A. Cremers*
22. Infrared Technology Fundamentals, *Irving J. Spiro and Monroe Schlessinger*
23. Single-Mode Fiber Optics: Principles and Applications, Second Edition, Revised and Expanded, *Luc B. Jeunhomme*
24. Image Analysis Applications, *edited by Rangachar Kasturi and Mohan M. Trivedi*
25. Photoconductivity: Art, Science, and Technology, *N. V. Joshi*
26. Principles of Optical Circuit Engineering, *Mark A. Mentzer*
27. Lens Design, *Milton Laikin*
28. Optical Components, Systems, and Measurement Techniques, *Rajpal S. Sirohi and M. P. Kothiyal*
29. Electron and Ion Microscopy and Microanalysis: Principles and Applications, Second Edition, Revised and Expanded, *Lawrence E. Murr*
30. Handbook of Infrared Optical Materials, *edited by Paul Klocek*
31. Optical Scanning, *edited by Gerald F. Marshall*
32. Polymers for Lightwave and Integrated Optics: Technology and Applications, *edited by Lawrence A. Hornak*
33. Electro-Optical Displays, *edited by Mohammad A. Karim*
34. Mathematical Morphology in Image Processing, *edited by Edward R. Dougherty*
35. Opto-Mechanical Systems Design: Second Edition, Revised and Expanded, *Paul R. Yoder, Jr.*
36. Polarized Light: Fundamentals and Applications, *Edward Collett*
37. Rare Earth Doped Fiber Lasers and Amplifiers, *edited by Michel J. F. Digonnet*
38. Speckle Metrology, *edited by Rajpal S. Sirohi*
39. Organic Photoreceptors for Imaging Systems, *Paul M. Borsenberger and David S. Weiss*
40. Photonic Switching and Interconnects, *edited by Abdellatif Marrakchi*
41. Design and Fabrication of Acousto-Optic Devices, *edited by Akis P. Goutzoulis and Dennis R. Pape*
42. Digital Image Processing Methods, *edited by Edward R. Dougherty*
43. Visual Science and Engineering: Models and Applications, *edited by D. H. Kelly*
44. Handbook of Lens Design, *Daniel Malacara and Zacarias Malacara*
45. Photonic Devices and Systems, *edited by Robert G. Hunsberger*
46. Infrared Technology Fundamentals: Second Edition, Revised and Expanded, *edited by Monroe Schlessinger*
47. Spatial Light Modulator Technology: Materials, Devices, and Applications, *edited by Uzi Efron*
48. Lens Design: Second Edition, Revised and Expanded, *Milton Laikin*

49. Thin Films for Optical Systems, *edited by Francoise R. Flory*
50. Tunable Laser Applications, *edited by F. J. Duarte*
51. Acousto-Optic Signal Processing: Theory and Implementation, Second Edition, *edited by Norman J. Berg and John M. Pellegrino*
52. Handbook of Nonlinear Optics, *Richard L. Sutherland*
53. Handbook of Optical Fibers and Cables: Second Edition, *Hiroshi Murata*
54. Optical Storage and Retrieval: Memory, Neural Networks, and Fractals, *edited by Francis T. S. Yu and Suganda Jutamulia*
55. Devices for Optoelectronics, *Wallace B. Leigh*
56. Practical Design and Production of Optical Thin Films, *Ronald R. Willey*
57. Acousto-Optics: Second Edition, *Adrian Korpel*
58. Diffraction Gratings and Applications, *Erwin G. Loewen and Evgeny Popov*
59. Organic Photoreceptors for Xerography, *Paul M. Borsenberger and David S. Weiss*
60. Characterization Techniques and Tabulations for Organic Nonlinear Optical Materials, *edited by Mark G. Kuzyk and Carl W. Dirk*
61. Interferogram Analysis for Optical Testing, *Daniel Malacara, Manuel Servin, and Zacarias Malacara*
62. Computational Modeling of Vision: The Role of Combination, *William R. Uttal, Ramakrishna Kakarala, Spiram Dayanand, Thomas Shepherd, Jagadeesh Kalki, Charles F. Lunskis, Jr., and Ning Liu*
63. Microoptics Technology: Fabrication and Applications of Lens Arrays and Devices, *Nicholas Borrelli*
64. Visual Information Representation, Communication, and Image Processing, *edited by Chang Wen Chen and Ya-Qin Zhang*
65. Optical Methods of Measurement, *Rajpal S. Sirohi and F. S. Chau*
66. Integrated Optical Circuits and Components: Design and Applications, *edited by Edmond J. Murphy*
67. Adaptive Optics Engineering Handbook, *edited by Robert K. Tyson*
68. Entropy and Information Optics, *Francis T. S. Yu*
69. Computational Methods for Electromagnetic and Optical Systems, *John M. Jarem and Partha P. Banerjee*
70. Laser Beam Shaping, *Fred M. Dickey and Scott C. Holswade*
71. Rare-Earth-Doped Fiber Lasers and Amplifiers: Second Edition, Revised and Expanded, *edited by Michel J. F. Digonnet*
72. Lens Design: Third Edition, Revised and Expanded, *Milton Laikin*
73. Handbook of Optical Engineering, *edited by Daniel Malacara and Brian J. Thompson*

74. Handbook of Imaging Materials: Second Edition, Revised and Expanded, *edited by Arthur S. Diamond and David S. Weiss*
75. Handbook of Image Quality: Characterization and Prediction, *Brian W. Keelan*
76. Fiber Optic Sensors, *edited by Francis T. S. Yu and Shizhuo Yin*
77. Optical Switching/Networking and Computing for Multimedia Systems, *edited by Mohsen Guizani and Abdella Battou*
78. Image Recognition and Classification: Algorithms, Systems, and Applications, *edited by Bahram Javidi*
79. Practical Design and Production of Optical Thin Films: Second Edition, Revised and Expanded, *Ronald R. Willey*
80. Ultrafast Lasers: Technology and Applications, *edited by Martin E. Fermann, Almantas Galvanauskas, and Gregg Sucha*
81. Light Propagation in Periodic Media: Differential Theory and Design, *Michel Nevière and Evgeny Popov*
82. Handbook of Nonlinear Optics, Second Edition, Revised and Expanded, *Richard L. Sutherland*
83. Polarized Light: Second Edition, Revised and Expanded, *Dennis Goldstein*
84. Optical Remote Sensing: Science and Technology, *Walter Egan*
85. Handbook of Optical Design: Second Edition, *Daniel Malacara and Zacarias Malacara*
86. Nonlinear Optics: Theory, Numerical Modeling, and Applications, *Partha P. Banerjee*
87. Semiconductor and Metal Nanocrystals: Synthesis and Electronic and Optical Properties, edited by *Victor I. Klimov*
88. High-Performance Backbone Network Technology, *edited by Naoaki Yamanaka*
89. Semiconductor Laser Fundamentals, *Toshiaki Suhara*
90. Handbook of Optical and Laser Scanning, *edited by Gerald F. Marshall*
91. Organic Light-Emitting Diodes: Principles, Characteristics, and Processes, *Jan Kalinowski*
92. Micro-Optomechatronics, *Hiroshi Hosaka, Yoshitada Katagiri, Terunao Hirota, and Kiyoshi Itao*
93. Microoptics Technology: Second Edition, *Nicholas F. Borrelli*
94. Organic Electroluminescence, *edited by Zakya Kafafi*
95. Engineering Thin Films and Nanostructures with Ion Beams, *Emile Knystautas*
96. Interferogram Analysis for Optical Testing, Second Edition, *Daniel Malacara, Manuel Sercin, and Zacarias Malacara*
97. Laser Remote Sensing, *edited by Takashi Fujii and Tetsuo Fukuchi*
98. Passive Micro-Optical Alignment Methods, *edited by Robert A. Boudreau and Sharon M. Boudreau*

99. Organic Photovoltaics: Mechanism, Materials, and Devices, *edited by Sam-Shajing Sun and Niyazi Serdar Saracftci*
100. Handbook of Optical Interconnects, *edited by Shigeru Kawai*
101. GMPLS Technologies: Broadband Backbone Networks and Systems, *Naoaki Yamanaka, Kohei Shiomoto, and Eiji Oki*
102. Laser Beam Shaping Applications, *edited by Fred M. Dickey, Scott C. Holswade and David L. Shealy*
103. Electromagnetic Theory and Applications for Photonic Crystals, *Kiyotoshi Yasumoto*
104. Physics of Optoelectronics, *Michael A. Parker*
105. Opto-Mechanical Systems Design: Third Edition, *Paul R. Yoder, Jr.*
106. Color Desktop Printer Technology, *edited by Mitchell Rosen and Noboru Ohta*
107. Laser Safety Management, *Ken Barat*
108. Optics in Magnetic Multilayers and Nanostructures, *Štefan Višňovský*
109. Optical Inspection of Microsystems, *edited by Wolfgang Osten*
110. Applied Microphotonics, *edited by Wes R. Jamroz, Roman Kruzelecky, and Emile I. Haddad*
111. Organic Light-Emitting Materials and Devices, *edited by Zhigang Li and Hong Meng*
112. Silicon Nanoelectronics, *edited by Shunri Oda and David Ferry*
113. Image Sensors and Signal Processor for Digital Still Cameras, *Junichi Nakamura*
114. Encyclopedic Handbook of Integrated Circuits, *edited by Kenichi Iga and Yasuo Kokubun*
115. Quantum Communications and Cryptography, *edited by Alexander V. Sergienko*
116. Optical Code Division Multiple Access: Fundamentals and Applications, *edited by Paul R. Prucnal*
117. Polymer Fiber Optics: Materials, Physics, and Applications, *Mark G. Kuzyk*
118. Smart Biosensor Technology, *edited by George K. Knopf and Amarjeet S. Bassi*
119. Solid-State Lasers and Applications, *edited by Alphan Sennaroglu*
120. Optical Waveguides: From Theory to Applied Technologies, *edited by Maria L. Calvo and Vasudevan Lakshiminarayanan*
121. Gas Lasers, *edited by Masamori Endo and Robert F. Walker*
122. Lens Design, Fourth Edition, *Milton Laikin*
123. Photonics: Principles and Practices, *Abdul Al-Azzawi*
124. Microwave Photonics, *edited by Chi H. Lee*

# Microwave Photonics

Edited by

Chi H. Lee

CRC Press
Taylor & Francis Group
Boca Raton London New York

CRC Press is an imprint of the
Taylor & Francis Group, an informa business

CRC Press
Taylor & Francis Group
6000 Broken Sound Parkway NW, Suite 300
Boca Raton, FL 33487-2742


Printed in the United States of America on acid-free paper
10 9 8 7 6 5 4 3 2 1

International Standard Book Number-10: 0-8493-3924-3 (Hardcover)
International Standard Book Number-13: 978-0-8493-3924-0 (Hardcover)



**Library of Congress Cataloging-in-Publication Data**

Microwave photonics / edited by Chi H. Lee.
p. cm. -- (Optical science and engineering series ; no. 124)
Includes bibliographical references and index.
ISBN-13: 978-0-8493-3924-0
ISBN-10: 0-8493-3924-3
1. Optical communications--Equipment and supplies. 2. Optoelectronics. 3. Microwave communication systems. 4. Photonics. I. Lee, Chi H. II. Title. III. Series.

TK5103.59.M495 2007
621.382'7--dc22 2006036016

**Visit the Taylor & Francis Web site at**
**http://www.taylorandfrancis.com**

**and the CRC Press Web site at**
**http://www.crcpress.com**

# Preface

Rapid advance in microwave electronics, photonics, and ultrafast optics has engendered a new field: microwave photonics. Microwave photonics can be defined as the study of photonic devices operating at microwave or millimeterwave and even terahertz frequencies and their use in microwave or photonic systems. From pioneering works during the late 1970s, the field has expanded to produce numerous applications of commercial importance. In this multidisciplinary field at the interface between microwaves, ultrafast optics, and photonic technologies, typical investigations include, for example, measurements of ultrafast signals, high-speed and microwave signal generation, signal processing, and conversion as well as the distribution and transmission of microwave signals via broadband optical links. Microwave photonics, which experienced a slow start in the 1980s, has recently witnessed rapid growth. This volume reviews the major advances in the field of microwave photonics over the past decade. An in-depth coverage of various aspects by experienced workers in the field may be of timely help to new researchers and should be useful to scientists and engineers alike.

Because microwave photonics is still an expanding field, this book could not be all-inclusive. Some overlap among chapters could not be avoided, but the reader may benefit from presentations with different points of view.

The editor wishes to thank all contributors for their cooperation and effort.

# Editor

**Chi H. Lee** received a BS in electrical engineering from the National Taiwan University in 1959 and a PhD in applied physics from Harvard University in 1968. He has been a faculty member of the electrical and computer engineering department, University of Maryland since 1968, where he is now a professor emeritus.

Professor Lee's areas of research include microwave photonics, ultrafast optoelectronics, lasers, and electro-optic devices. He is a fellow of IEEE, the Optical Society of America, and the Photonic Society of Chinese-Americans. He was the chairman of the technical committee on lightwave technology in the IEEE Microwave Theory and Techniques (MTT) Society. He was the general co-chair of the International Meeting on Microwave Photonics in 1998. Professor Lee served as the chairman of the Steering Committee of the International Microwave Photonics Meeting for 1999. He was the chairman of the IEEE/LEOS technical committee on microwave photonics from 1997 to 2003, and a guest editor of the special issue on microwave photonics for the *Journal of Lightwave Technology* published in December 2003.

# Contributors

**Masuduzzaman Bakaul**
Department of Electrical and Electronic Engineering
Melbourne, Australia

**Frank Bucholtz**
Naval Research Laboratory
Washington, D.C.

**Eric E. Funk**
Red Mountain Radio, LLC
Ouray, Colorado

**Yan Han**
Department of Electrical Engineering
University of California
Los Angeles, California

**Dieter Jäger**
Zentrum für Halbleitertechnik und Optoelektronik
Universität Duisburg-Essen
Duisburg, Germany

**Bahram Jalali**
Department of Electrical Engineering
University of California
Los Angeles, California

**Teddy Kurniawan**
Department of Electrical and Electronic Engineering
Melbourne, Australia

**Erwin K. Lau**
Department of Electrical Engineering and Computer Sciences
University of California
Berkeley, California

**Chi H. Lee**
Department of Electrical and Computer Engineering
University of Maryland
College Park, Maryland

**Christina Lim**
Department of Electrical and Electronic Engineering
Melbourne, Australia

**Jason D. McKinney**
SFA, Inc.
Crofton, Maryland

**Tadao Nagatsuma**
NTT Microsystem Integration Laboratories
Kanagawa Prefecture, Japan

**Ampalavanapillai Nirmalathas**
Department of Electrical and Electronic Engineering
Melbourne, Australia

**Dalma Novak**
Pharad, LLC
Glen Burnie, Maryland
and
Department of Electrical and Electronic Engineering
Melbourne, Australia

**Albert Redo-Sanchez**
Rensselaer Polytechnic Institute
Troy, New York

**Hyuk-Kee Sung**
Department of Electrical Engineering and Computer Sciences
University of California
Berkeley, California

**Masahiro Tsuchiya**
National Institute of Information and Communication Technology
Tokyo, Japan

**Vincent J. Urick**
Naval Research Laboratory
Washington, D.C.

**Osamu Wada**
Department of Electrical and Electronics Engineering
Kobe University
Kobe, Japan

**Rodney Waterhouse**
Pharad, LLC
Glen Burnie, Maryland
and
Department of Electrical and Electronic Engineering
Melbourne, Australia

**Andrew M. Weiner**
School of Electrical and Computer Engineering
Purdue University
West Lafayette, Indiana

**Ming C. Wu**
Department of Electrical Engineering and Computer Sciences
University of California
Berkeley, California

**Xi-Cheng Zhang**
Department of Physics
Rensselaer Polytechnic Institute
Troy, New York

# Table of Contents

**Chapter 1**
Microwave Photonics—From Concepts to Devices and Applications .......... 1
*Dieter Jäger*

**Chapter 2**
Femtosecond All-Optical Devices for Ultrafast Communication and Signal Processing .......... 31
*Osamu Wada*

**Chapter 3**
Direct Modulation of Injection-Locked Semiconductor Lasers .......... 77
*Erwin K. Lau, Hyuk-Kee Sung, and Ming C. Wu*

**Chapter 4**
Optical Measurement Technologies for High-Frequency Electronics .......... 111
*Tadao Nagatsuma and Masahiro Tsuchiya*

**Chapter 5**
Hybrid Fiber Radio—Concepts and Prospects .......... 157
*Dalma Novak, Ampalavanapillai Nirmalathas, Christina Lim, Rodney Waterhouse, Masuduzzaman Bakaul, and Teddy Kurniawan*

**Chapter 6**
High Dynamic Range, 100 km Digital Radio-over-Fiber Links .......... 185
*Eric E. Funk, Vincent J. Urick, and Frank Bucholtz*

**Chapter 7**
Photonic Synthesis of Ultrabroadband Arbitrary Electromagnetic Waveforms .......... 213
*Jason D. McKinney and Andrew M. Weiner*

**Chapter 8**
Ultrafast Optoelectronic Switching Devices and
Applications Based on Optically Injected Free Carriers ............................ 259
*Chi H. Lee*

**Chapter 9**
Tera Sample-per-Second Time-Stretched
Analog-to-Digital Conversion ...................................................................... 307
*Bahram Jalali and Yan Han*

**Chapter 10**
Terahertz Photonics ...................................................................................... 363
*Albert Redo-Sanchez and Xi-Cheng Zhang*

**Index** ............................................................................................................ 417

# 1 Microwave Photonics—From Concepts to Devices and Applications

*Dieter Jäger*

CONTENTS

1.1 Introduction .......... 1
1.2 Concepts of Microwave Photonic Devices .......... 3
1.3 Microwave Photonic Components .......... 6
1.3.1 Electroabsorption Modulator .......... 6
1.3.2 Photodetector and Mixer .......... 7
1.3.3 Electroabsorption Transceiver .......... 10
1.3.4 Microwave Laser Diodes .......... 12
1.4 Photonic Integration Technologies .......... 13
1.5 Photonic Microwave Signal Processing .......... 14
1.5.1 Photonic Microwave Phase Shifter/Time Delay Control .......... 15
1.5.2 Photonic Ultrawide Band (UWB) Pulse Generation .......... 16
1.5.3 RF Spectrum Analyzer .......... 17
1.6 Broadband Fiber Optical Links .......... 17
1.7 Microwave Photonic Systems .......... 19
1.7.1 Photonic Local Oscillators .......... 20
1.7.2 EMC Sensor .......... 20
1.7.3 Hybrid Fiber-Coax (HFC) Systems .......... 22
1.7.4 Fiber-Radio Systems .......... 23
1.8 Conclusion .......... 24
Acknowledgments .......... 26
References .......... 27

## 1.1 INTRODUCTION

Within the last decade, the field of optic-microwave interactions has attracted growing interest worldwide. The specific term of microwave photonics was first introduced in 1991 and used to describe novel optoelectronic components based on the interaction of traveling optical and microwaves [1,2]. In the following years, it was foreseen that the merging of microwave and photonic technologies would develop and become a new approach to future fiber-radio communication systems, for example, in the transmission of radio frequency (RF) signal over optical carriers [3]. Since then the field of RF optoelectronics and microwave photonics rapidly broadened. Since 1996, International Topical Meetings on Microwave Photonics (MWP) are held annually [4]. The IEEE MTT Special Issue on Microwave Photonics that was first published in the year 1995 is now published regularly [5].

Microwave photonics is today an innovative multi- and interdisciplinary field combining and transferring different technologies [6–8]. In particular, microwave technologies are used and employed in photonics and vice versa. Moreover, in specific areas, novel synergistic concepts have been developed by merging both technologies, which particularly holds for the field of optoelectronics as their interface. Even more and as a result of ever-increasing frequencies, the term microwave stands here not only for GHz but also for THz frequencies in the frequency domain and, equivalently, for picosecond or femtosecond timescales in the time domain [8].

The general field of optic-microwave interactions, initially called RF optoelectronics, involves the study of high-speed photonic devices operating at micro- or millimeter-wave frequencies [2] and their corresponding use in microwave or photonic systems. This multidisciplinary field is located at the interface between microwave techniques and ultrafast electronics and photonic technologies, and typical investigations include, for example, high-speed and microwave signal generation, signal processing and conversion, as well as the distribution and transmission of microwave signals via broadband optical links such as glass fibers. From pioneering ideas and experiments in the 1970s, the field of microwave photonics has dramatically broadened and paved the way for an enabling novel technology with a number of innovative and, nowadays, even commercially important applications.

This chapter is intended to give an exemplary overview together with recent results in this multidisciplinary field of microwave photonics ranging from devices and technologies to systems under investigation and of worldwide interest. In particular, the following topics will be addressed by way of characteristic examples showing the synergetic mixture of the two technologies involved:

- Ultrafast photonic components such as optical modulators, photodetectors (PD), and photonic mixers and transceivers with special emphasis on traveling wave (TW) devices using the distributed interaction between traveling microwave and optical signals to avoid any RC time constant
- Integration technologies and packaging aspects
- Concepts and examples of microwave signal processing by way of using photonic technologies
- Broadband and analog optical links for high-speed interconnects
- Microwave photonic systems based on the merging of microwave and optical technologies, for example, wireless communication techniques

such as cellular radio systems or field sensor applications and optically controlled phased array antenna systems

Recent results, especially in this field of high-speed optical interconnects, clearly demonstrate the synergetic mixture and advantages of the different technologies involved in microwave photonics.

Section 1.1 provides an introductory overview of the field of microwave photonics and its significance for different microwave techniques. Section 1.2 is devoted to the principles of microwave–optical interaction devices starting with the basics of optically vertical and optical waveguide structures. Moreover, lumped elements as well as traveling wave devices are discussed. The so-called family of microwave photonic components includes photodetectors, laser diodes, modulators, mixers, etc., where different physical interactions are employed and utilized. In particular, recent advances in electroabsorption (EA) devices such as electroabsorption modulators (EAMs), EA transceivers (EAT), and EA mixers (EAX) as well as high-speed photodetectors and mixers are discussed in Section 1.3, including some results from experiments. Section 1.4 addresses optoelectronic integration techniques such as monolithic integration of different functions and photonic integrated circuits (PICs) and fiber-chip coupling techniques. In Section 1.5, a few most interesting examples of photonic microwave signal processing components and modules are introduced. Section 1.6 is particularly devoted to basic characteristics of microwave–optical links and in Section 1.7, several examples of optical microwave system applications are presented.

## 1.2 CONCEPTS OF MICROWAVE PHOTONIC DEVICES

Recent advances in the development of high-speed optoelectronic devices together with the broadband and low-loss transmission capability of optical fibers have, in large part, been responsible for the global growth of broadband communications. In particular, rapid progress has been made in the development of photodetectors and modulators that are already the key elements in our current ubiquitous high-speed optical internet where the optical fiber is now becoming available to the customer. In the following, an overview of the state of the art of ultrafast photonic components is provided with special emphasis on novel electroabsorption and traveling wave devices. Future trends in this emerging field of technology are also addressed.

In Figure 1.1 the four basic optoelectronic interaction devices are sketched. Here two types can be recognized: (i) Two-port devices, the photodetector and the laser or light emitting diode (LD or LED), as converters between optical and electrical signals, (see Figure 1.1a and Figure 1.1b, respectively); (ii) Three-port devices such as the electro-optical modulator (MOD) or the optically controlled electrical modulator shown in Figure 1.1c and Figure 1.1d, respectively, with wave like for optical and straight arrows for electrical signals. Please note that

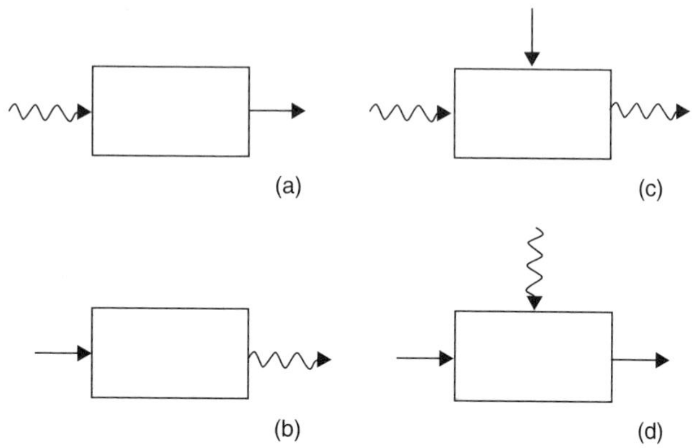


**FIGURE 1.1** Optoelectronic device family: basic microwave–optical interaction devices: (a) photodetector, (b) light emitting or laser diode, (c) optical modulator with electrical control, and (d) electrical modulator/device with optical control.

the two elements in Figure 1.1c and Figure 1.1d are a kind of optoelectronic transistor, which means that these devices, in contrast to those in Figure 1.1a and Figure 1.1b, can provide small-signal optoelectronic gain.

Two basic concepts of microwave–optical interactions may further be distinguished (Figure 1.2). As can be seen, the difference is due to different propagation directions of the optical wave in relation to the traveling charge carriers between the metallic contacts. In Figure 1.2a, the optical wave is traveling normally to the semiconductor surface, that is, parallel to the flow of the electrical charges and current. This leads to a design where the optical interaction length, in particular the absorption depth, is coupled and related to the distance the charge carriers are traveling. Hence, in a photodetector, for example, the absorption length cannot be designed independently from the

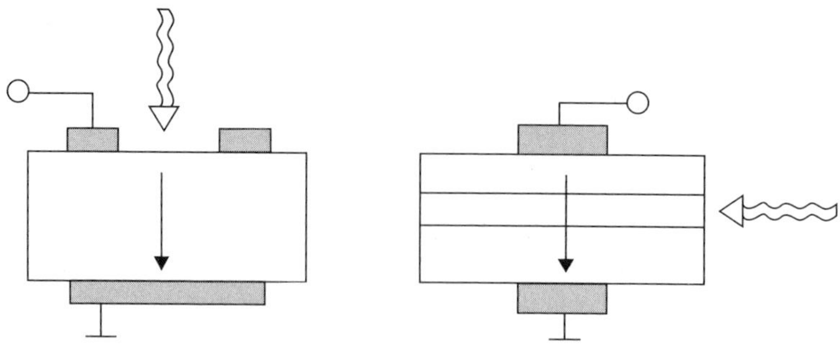

**FIGURE 1.2** Microwave–optical interaction devices with vertical (a) and horizontal (b) propagation of the light wave. (From Jäger, D., Stöhr, A., *Microwave Photonics—From Concepts to Applications*, Institut fur Mikrowellen- und Antennentechnik C.V. (IMA), 136–138, 2005. With permission.)

transit time. A large bandwidth is therefore connected to a short transit time, which can only be achieved by a small distance. This, however, will lead to less absorption. In Figure 1.2b, the optical wave is traveling parallel to the surface so the penetration depth can be made large, which is independent from the vertical design determining the transit time, in the example of the photodetector.

For high-speed operation, electronic devices, as shown in Figure 1.2, are usually scaled down with respect to the lateral dimensions of the electrical contacts to decrease the device capacitance and, correspondingly, the RC time constant thus increasing the cutoff frequency. This procedure, however, is usually limited by power considerations, because smaller devices will lead to smaller operating powers. An interesting and commonly accepted solution to this problem is to include electrical wave propagation phenomena also in the design of the electronic device as has frequently been described in the past [1,8–13] and already used in the development of high-speed integrated circuits such as radio frequency ICs (RFICs) and monolithic microwave ICs (MMICs) (cf. the well-known traveling wave amplifiers). According to the design of such traveling wave devices (Figure 1.3), the electrical contacts are realized in the form of well-known microstrips or coplanar transmission lines (CPWs). As can be seen, in these traveling wave (TW) devices, wave propagation effects are utilized not only in the optical but also in the electrical domain. Obviously, the concept is based on the fundamentals and physics of nonlinear optics where interaction always takes place during wave propagation. As a major result, the bandwidths of these elements are not limited by "RC time constants", if output and input impedances are identical with the characteristic impedance of the transmission line. In other cases, reflection

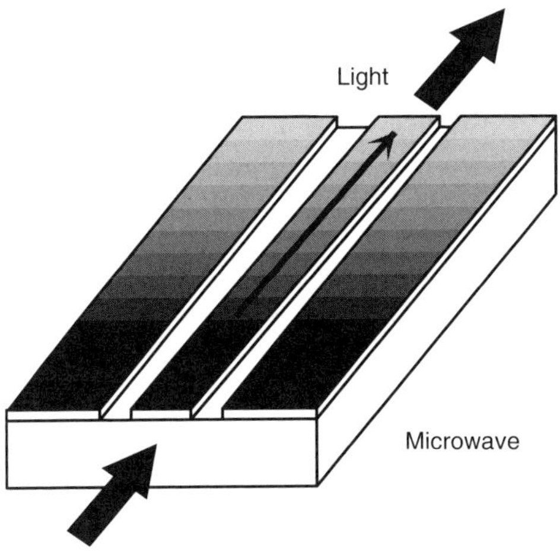


**FIGURE 1.3** Traveling wave concept. (From Jäger, D., Stöhr, A., *Microwave Photonics—From Concepts to Applications*, Institut fur Mikrowellen- und Antennentechnik C.V. (IMA), 136–138, 2005. With permission.)

occurs leading to a large increase in the microwave amplitude due to resonances and a corresponding enhancement of the effective interaction length. It should be mentioned at this point that the electrical waves are usually slow waves because of a high capacitance per unit length arising from pn- or pin diodes in the cross section.

It becomes obvious that further improvements of this TW concept can be foreseen. As an example, the transit time and wave propagation effects of space charges can also be taken into account and the interaction between electrical and optical waves can be oblique or in different directions (backward wave propagation), which may lead to even more interesting designs.

In summary, in modern microwave photonic components wavelike interaction phenomena between traveling electrons, holes, electrical fields, and photons take place and are summarized under "microwave–optical interactions." Clearly and consequently, different technologies meet and can—in a synergetic mixture—be merged to develop novel concepts with great advantages. Until now, just a few technologies and developments have been used and corresponding devices have been realized. However it can be foreseen that the emerging fields of nanotechnology and nano-optics as well as photonic band gap (PBG) and electromagnetic band gap (EBG) structures will lead to major progress in the years to come.

## 1.3 MICROWAVE PHOTONIC COMPONENTS

In the following, the concepts and realizations of some key microwave–optical interaction devices are discussed.

### 1.3.1 Electroabsorption Modulator

Electroabsorption modulators (EAMs) provide a great potential for low-voltage operation, large bandwidth, and monolithic integration with other components such as laser diodes [14–17]. In Figure 1.4, a broadband and high-speed EAM is sketched, which has been designed for a wavelength of 1.55 μm using InP-based semiconductor materials [15,16]. The EAM resembles electrically a well-known pin diode under reverse bias. Optically a ridge waveguide is realized where the core is made of a multiple quantum well (MQW) material sandwiched between InP cladding layers designed here for 1.55 μm. The component in Figure 1.4 is a lumped element with a coplanar input contact.

Experimental results of different digital and analogue devices are shown in Figure 1.5 and Figure 1.6. In Figure 1.5, the characteristics of two digital devices exhibiting a bandwidth of approximately 10 and 40 GHz, respectively, are shown. Alternatively, a traveling wave design leads to a cutoff frequency of >70 GHz [17]. In Figure 1.6, the frequency dependence of a lumped EAM for sensor application and the results of spurious free dynamic

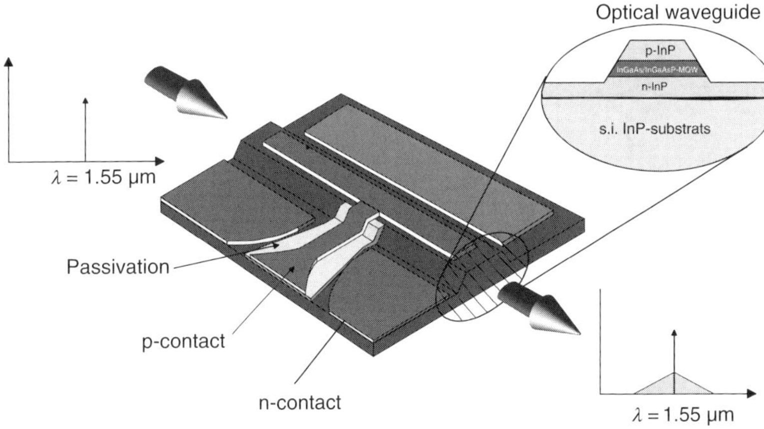


**FIGURE 1.4** Electroabsorption modulator (EAM) with cross section.

range (SFDR) measurements are plotted. As can be seen, the experimental results of the bandwidth agree very well with simulations based on an equivalent circuit model.

### 1.3.2 Photodetector and Mixer

In Figure 1.7, a traveling-wave photodetector is sketched. It consists of an absorptive optical waveguide and a coplanar waveguide (CPW) [18–23]. The device is designed for a wavelength of 1.55 μm using InP-based

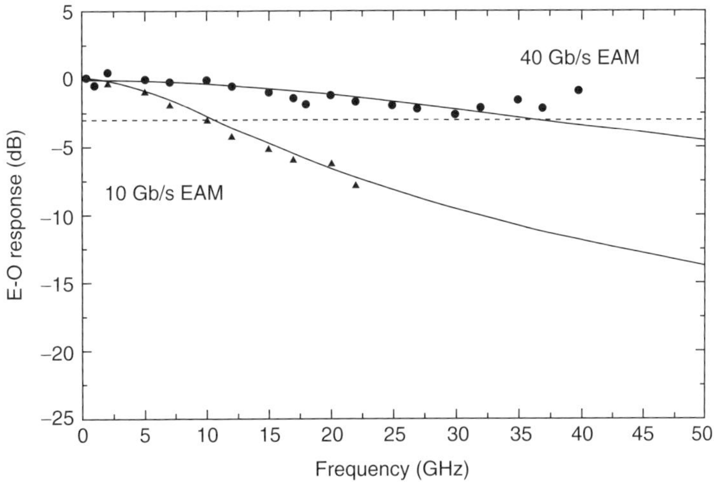


**FIGURE 1.5** Modulation of two digital EAMs.

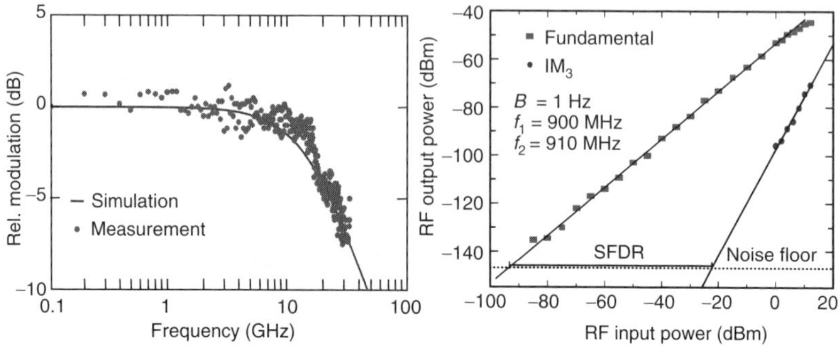


**FIGURE 1.6** Frequency-dependent modulation of EAMs, (a) lumped element fabricated for electric field sensor applications, and (b) measurement of SFDR.

semiconductor structures. As a result of optical absorption the optical power decreases exponentially along the waveguide while the generated microwave signal travels to the output with increasing amplitude. At the output, microwave reflection is prevented if the load is identical to the characteristic impedance of the CPW. As can be seen and as it is clear from nonlinear optics, an optimum power transfer can only be achieved for phase-matching conditions, that is, matching the underlying velocities of the optical and the electrical wave. It is further obvious that there is an optimum length of the device or, correspondingly, an optimum interaction distance where the gain due to optical absorption is less than the microwave losses of the CPW.

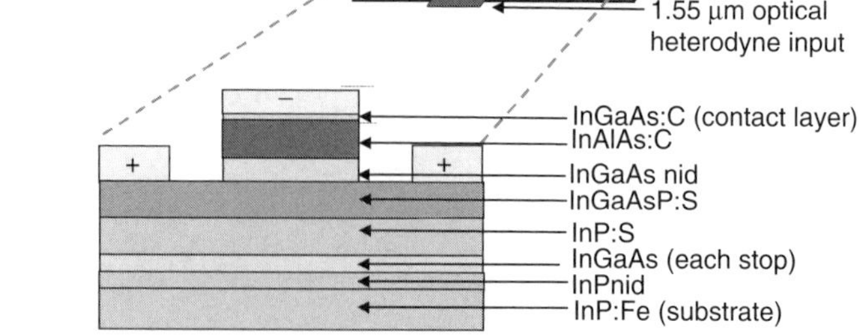


**FIGURE 1.7** Traveling wave photodetector.

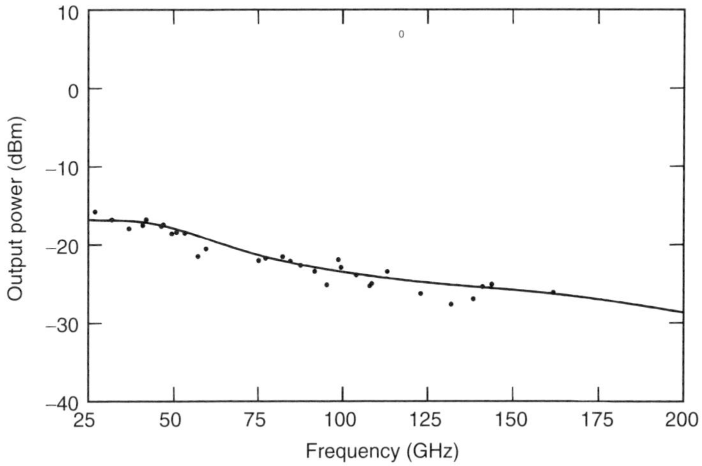


**FIGURE 1.8** Frequency response of a traveling wave 1.55 μm photodiode; here using electroabsorption in the MQW active material of the optical waveguide.

High-speed photodetectors are used in the time domain to generate ultrashort electrical pulses from ultrashort optical input pulses or in the frequency domain to generate high microwave power at a given frequency. This technique is called the heterodyne technique where two optical input signals at different optical wavelengths are used to generate the difference frequency, which is identical to the frequency of the optical envelope. Figure 1.8 shows results of such a heterodyne experiment using a TW photodetector, the design of which includes an optical waveguide and an electrical coplanar transmission line along the optical signal path, according to Figure 1.7. In this case, there is no optical output and no electrical input. Experimentally, a heterodyne setup has been used where two optical input signals at 1.55 μm with a corresponding difference frequency are fed into the optical waveguide. The photodetection process leads to a photocurrent at the difference frequency measured after down-conversion with a spectrum analyzer. In Figure 1.9, the results of an improved TW photodetector for operation at 110 GHz are shown. In recent years, major progress has been achieved concerning the micro- or millimeter-wave output power and the bandwidth.

Recently, a traveling wave photodetector has been shown to work up to THz frequencies [22–25] and, in particular, 460 GHz has been measured using an SIS mixer (a superconducting Josephson junction) in a THz-antenna system for radio astronomy. Here the photodetector together with the two laser diodes work as a photonic local oscillator (see Section 1.7).

It should be noted at this point that as a result of the (electro-) absorption mechanism, which is basically connected with the generation of charge carriers, the EAM can also be used as an electroabsorptive photodetector

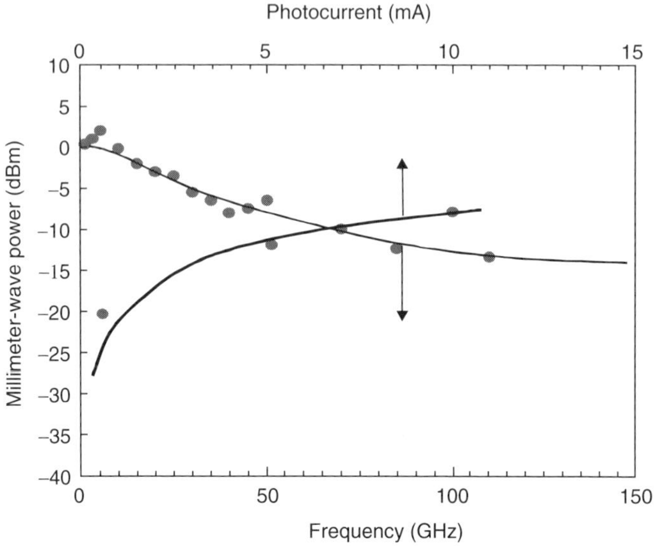


**FIGURE 1.9** Heterodyne RF generation using a TW photodetector and generated power versus photocurrent at 110 GHz. (From Jäger, D., Stöhr, A., *Microwave Photonics—From Concepts to Applications*, Institut fur Mikrowellen- und Antennentechnik C.V. (IMA), 136–138, 2005. With permission.)

(EAD). This further means that such an EAD can further be used for electrical tuning of the detector.

It is worth mentioning that due to the inherent nonlinearity of the photodetector where the photocurrent is proportional to the square of the electric field of the optical wave, such an absorptive device can further directly be employed as a photomixer. Experimental results using an EAM for up-conversion at 60 GHz are shown in Figure 1.10. In this experiment, again a heterodyne setup has been used where the two laser diode signals are separated by 60 GHz and one laser diode is additionally modulated at 2.6 GHz. Obviously, the electrical output signal exhibits the RF carrier frequency as well as the two side bands due to modulation. Such a device has been called an electroabsorption mixer (EAX).

### 1.3.3 Electroabsorption Transceiver

From the physics of electroabsorption it becomes clear that the EAM can simultaneously be used as a modulator and a photodetector with interesting applications to communication techniques where the electrical up- and downlink signals of the device have to be frequency multiplexed (see Section 1.7). Such a dual function EAM is called electroabsorption transceiver (EAT) [26] (Figure 1.11). As can be seen, a high contrast modulation can be obtained at 1.55 μm and an efficient photodetection characteristic is achieved at the

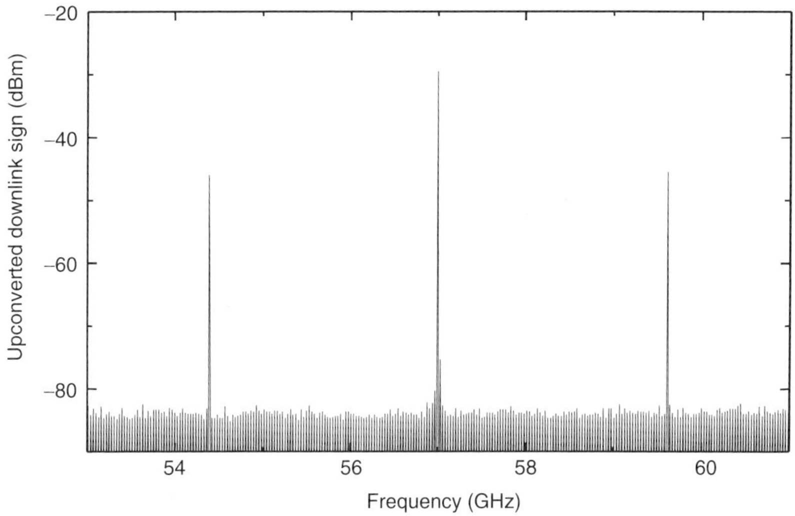


**FIGURE 1.10** Optoelectronic mixing in a photodetector, here an electroabsorption photomixer (EAX).

shorter wavelength of 1.51 μm in the present device. The multifunctional characteristics of an EAT have been demonstrated in different fiber wireless systems for broadband communications (see Section 1.7.4). For several

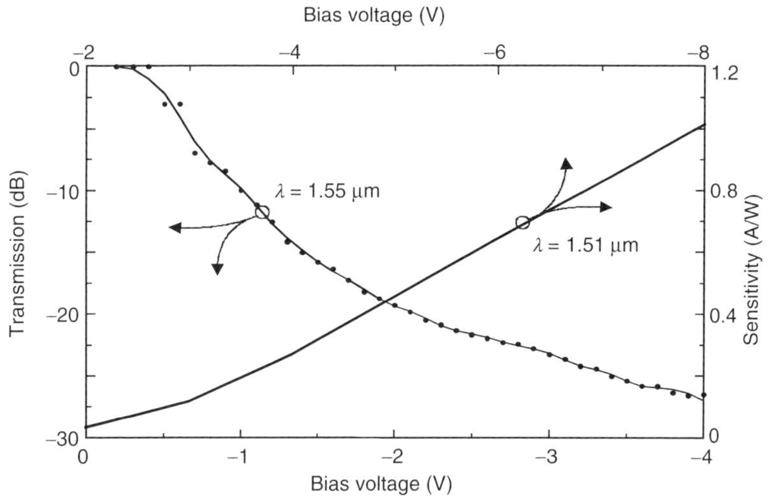


**FIGURE 1.11** Photodetection and modulation properties of an EAM. The device is called an electroabsorption transceiver (EAT).

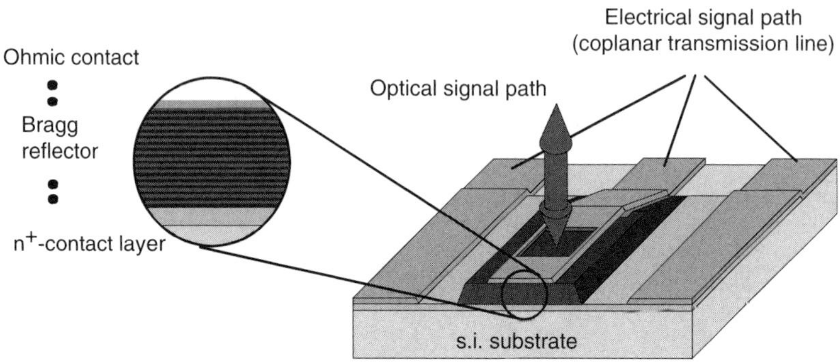


**FIGURE 1.12** Vertical EAT.

applications, for example, bidirectional communication, it is advantageous that just one fiber is used for the up- and downlink. As a result the modulator has to work in reflection, which has led to the development of a reflection-type EAM called R-EAT [27].

The multifunctional EAT, which behaves similar to the well-known self electro-optic device (SEED) element, can further be used to generate artificial optical nonlinearities such as optical bistability [28], which gives rise to switching, logic, and memory effects. From the physical point of view, this nonlinearity is caused by the internal positive feedback between the modulator and the photodetector properties of the device. It has been shown that such a bistable characteristic may be useful for multiple-GHz A/D conversion. The vertical EAT shown in Figure 1.12 is a device with a world record of switching energy and speed. The vertical EAT consists of an optical multilayer Bragg filter using a set of quarter wavelength semiconductor heterostructures such as GaAs/AlAs films. Please note that such a device is an R-EAT.

### 1.3.4 Microwave Laser Diodes

Until now laser diodes are usually not made using a special design for microwave applications [29]. In particular, no traveling wave structures are applied although the electrical contact size is already close to the electrical wavelength for frequencies of about 10 GHz. As a result, since several years the cutoff frequency is still in the range of 30 GHz and there is no breakthrough at higher frequencies. Technically, the monolithic integration of a laser diode with a high-speed EAM is a good solution and much development is devoted to the realization of such components.

## 1.4 PHOTONIC INTEGRATION TECHNOLOGIES

In an optoelectronic TW device, an optical waveguide (e.g., a strip loaded or a ridge waveguide) is used for optical wave propagation and an electrical transmission line (e.g., microstrip or coplanar waveguide) is used for guiding the microwave, usually in the same direction. In the region where the electrical fields overlap, the optoelectronic interaction occurs. Note that a dc bias may additionally be applied to control the operating point. From a physical point of view, the interaction is a nonlinear or active process. As discussed above, the photodetector and the laser diode are basic examples of two-port devices where optical power is converted into electrical power and vice versa. Typical three-port devices are electrically controlled optical modulators and switches or optically controlled microwave modulators and switches. Owing to the inherent nonlinearity, these devices can further be used for optoelectronic mixing of input electrical and optical signals where the output signal can be electrical or optical. As a result, such microwave–optical interaction devices show a variety of optoelectronic functions where in special integrated devices different functions may be achieved simultaneously.

A novel multifunctional device has recently been presented—the EAM integrated into the structure of a hetero-bipolar transistor (HBT) [30,31]. Such an HBT-EAM (Figure 1.13) includes the common transistor characteristics as well as the modulator behaviour leading to a novel approach to optoelectronic integrated devices and photonic integrated circuits. Because of the

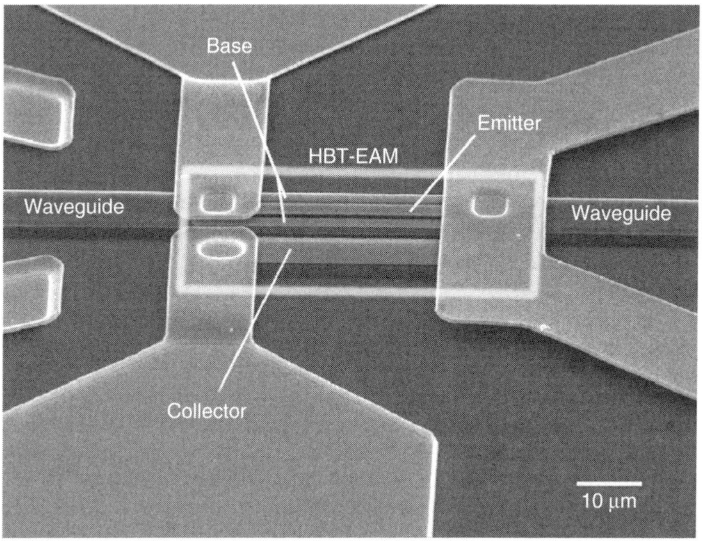


**FIGURE 1.13** SEM picture of an HBT-EAM.

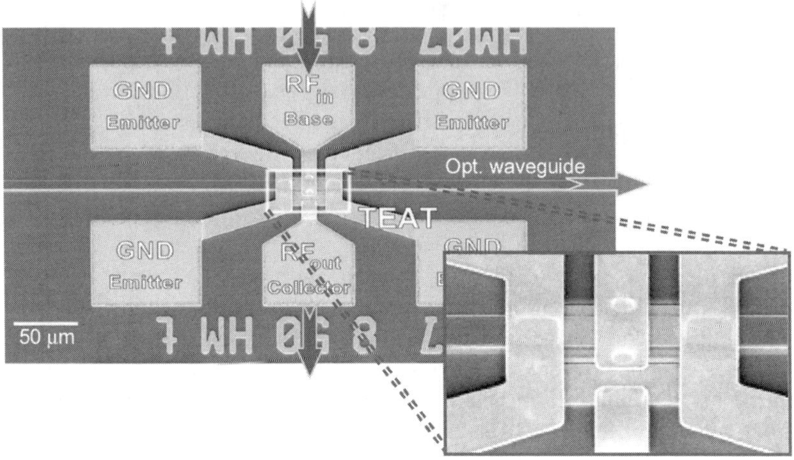


**FIGURE 1.14** Photo of a TEAT.

multifunctional characteristics this device is called a transistor-EAT, or briefly TEAT. In Figure 1.14, a photo of a TEAT is shown. The optical waveguide with input and output as well as the electrical contacts for input and output are obvious. This device combines the functions of the following devices: transistor, modulator with gain, detector with gain, and mixer with gain.

In recent years, first PICs and optical MMICs have been realized using the layer structure of the HBT-EAM. Figure 1.15 shows the layout of an EAM with a multistage HBT preamplifier for differential electrical inputs.

A further key integration technology is the optical coupling of the light between the optical waveguide and the external world, which is usually the optical fiber. This technique of fiber-chip coupling includes the processing of V-grooves into the substrate or the motherboard and the tapering of the optical fiber [32]. A further important technology is finally related to the adjustment and fixing of the fiber with respect to the optical waveguide. A result of measurements is shown in Figure 1.16 where a coupling loss of less than 1 dB is achieved for a mode field diameter (MFD) of about 1.5 μm. It should finally be mentioned that electrical coupling at micro- and millimeterwave, for example, of coplanar lines with coaxial cables and connectors, also requires careful design.

## 1.5 PHOTONIC MICROWAVE SIGNAL PROCESSING

In the following, a few examples of optically controlled microwave devices and modules are discussed. The physical mechanisms are due to the optical generation of free charge carriers in a semiconductor, leading to a control of some key electrical parameters determining the microwave properties.

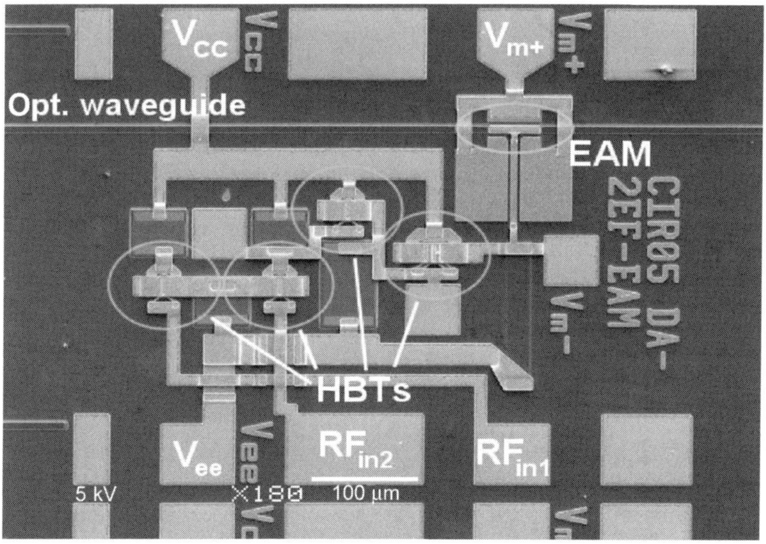


**FIGURE 1.15** EAM with multistage HBT preamplifier; the differential input amplifier and two power amplifier stages for an EAM provide differential electrical inputs and an optical output proportional to the difference of the inputs. (Courtesy of Tegude, F.J., Universität Duisburg-Essen.)

### 1.5.1 Photonic Microwave Phase Shifter/Time Delay Control

In Figure 1.17, electrical transmission lines—microstrip and coplanar waveguides—are sketched where the strip or the center conductor metal forms a Schottky contact with the semiconductor [8,10–12]. It is noteworthy to mention that the Schottky contact could be replaced by a pn-junction in the

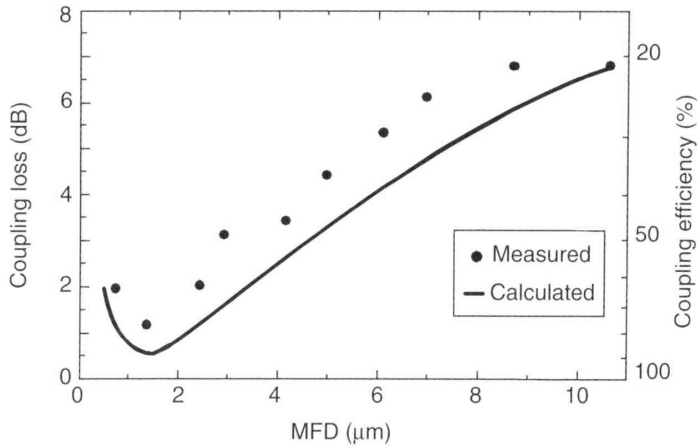


**FIGURE 1.16** Fiber-chip coupling: coupling loss versus mode field diameter (MFD).

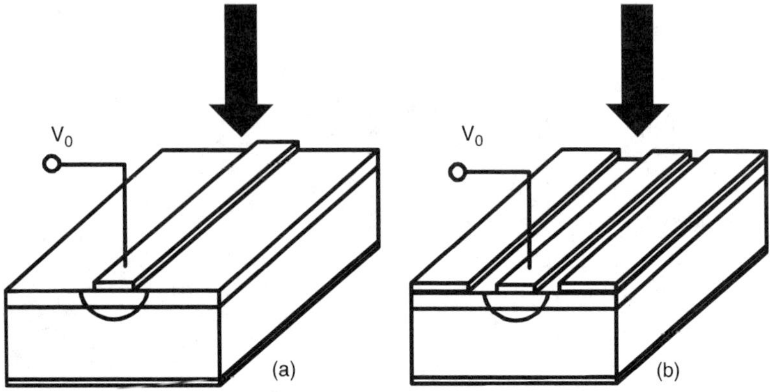


**FIGURE 1.17** Optically controlled phase shifter and time delay control using a Schottky contact microstrip line.

cross section. As mentioned before, the high capacitance per unit length gives rise to a slow mode behavior and the electrical wavelength or the phase velocity can be controlled by the capacitance per unit length of the transmission line, that is, by the width of the depletion region of the contact. The thickness of the space charge region, however, depends critically on the applied reverse bias voltage, which has been used frequently as an electrical control of the phase velocity and the delay time. On the other hand, when a fixed bias voltage is applied and an external series resistor is used, the voltage drop across the depletion layer can also be changed by absorption of an optical signal in the space charge region. As a result, the time delay is controlled optically. Such a device has been called an optically controlled electrical phase shifter (Figure 1.1d), but it should be noted that because the device is a usual transmission line, a true time delay shift occurs and not just a phase shift, if dispersion can be neglected.

### 1.5.2 Photonic Ultrawide Band (UWB) Pulse Generation

Figure 1.18 elucidates an optically controlled generation of microwave pulses. The structure consists of a stripline resonator in the center of the figure with two gaps on each side. On the left hand side a dc or RF source is connected to charge the resonator. On the right hand side an optoelectronic switch (OES), that is, a high-speed photoconductor—the second gap—is provided, which can optically be switched on to connect the output port of the resonator directly to an integrated antenna [33]. Two operating scenarios can be distinguished: (i) Using a dc source the resonator is charged to a maximum voltage and by closing the OES a pulse traveling to the antenna is generated. The temporal width of the pulse is given by the length of the resonator (frozen wave generator). Please note that specific pulse forms are

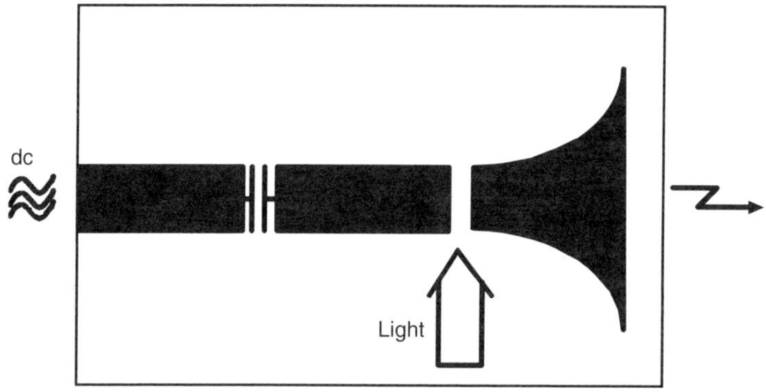


**FIGURE 1.18** Optoelectronic switching of a microwave resonator to obtain UWB signals. (From Jäger, D., Stöhr, A., *Microwave Photonics—From Concepts to Applications*, Institut fur Mikrowellen- und Antennentechnik C.V. (IMA), 136–138, 2005. With permission.)

generated, given by the structure of the charged line. (ii) Using a RF source at the input the resonator is also charged and a standing wave of large amplitude, depending on the quality factor of the transmission line, is established. When the OES is optically closed, the standing wave will become a propagating wave and a short ultrawide band (UWB) pulse is generated and radiated using, for example, a broadband slot antenna [34]. This technique that corresponds to the device concept in Figure 1.1d is a kind of pulse compression [35] well known from nonlinear optics and soliton theory.

### 1.5.3 RF Spectrum Analyzer

By using the techniques of Section 1.5.1 and Section 1.5.2 in a two-dimensional or even a three-dimensional array, parallel microwave signal processing becomes possible. Figure 1.19 shows a microwave circuit that has been studied in the past. Here the input signal is divided into several channels and each channel carries a time delay (T) and an amplitude (a) control using optical techniques as mentioned above. When the output signals are finally combined to a common connection a transversal filter results. In contrast, when the individual channels are connected to antennas, a phased array antenna system results. It should be noted that these concepts have been transferred to the optical domain by replacing the devices in Figure 1.19—here with the functions of those in Figure 1.1d—by the equivalent devices of Figure 1.1c or even all-optical ones.

## 1.6 BROADBAND FIBER OPTICAL LINKS

An analog or digital optical link consists of an optical transmission medium (preferably a fiber) and optoelectronic converters on both sides (Figure 1.20).

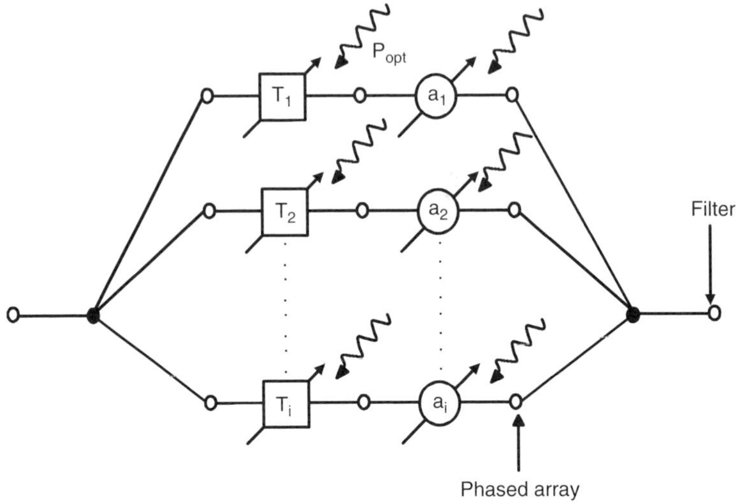


**FIGURE 1.19** Microwave signal processor.

One great advantage of such an optical link is that due to the broadband and extremely low-loss transmission capability of the optical fiber, such an optical interconnection can ideally be used to transmit microwave signals over large distances and therefore replace other lossy metallic waveguides (Figure 1.21). Depending on the specific and ever-increasing number of applications, a huge number of different techniques of optical links have been explored. There are analog and digital links as well as uni- or bidirectional and half- or full-duplex interconnections where different wavelengths, multiplexing techniques, and sources with internal or external modulation have been realized. For example, on the transmitting side a direct modulated laser diode or LED can be used or, alternatively, a cw laser diode and an external modulator (electro-optic or EAM). On the receiving side, an optoelectronic photodetector is the usual component. In addition to the bandwidth, a key parameter of such a link is the link loss, which depends on the conversion efficiencies of the optoelectronic elements, the optical coupling efficiencies, and the attenuation and dispersion of the transmission medium [36,37–38]. Note that a link gain can easily be

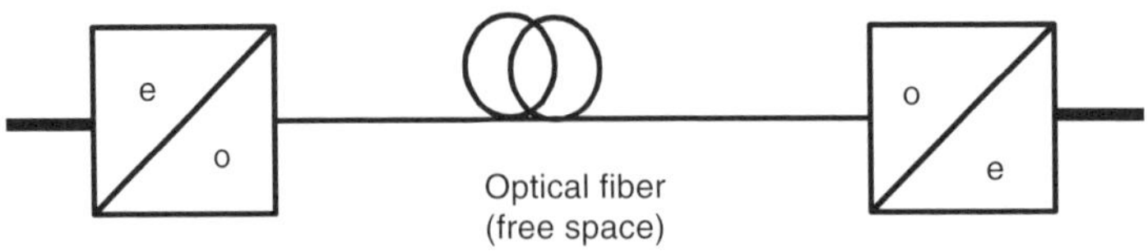


**FIGURE 1.20** Optical interconnection replacing an electrical cable.

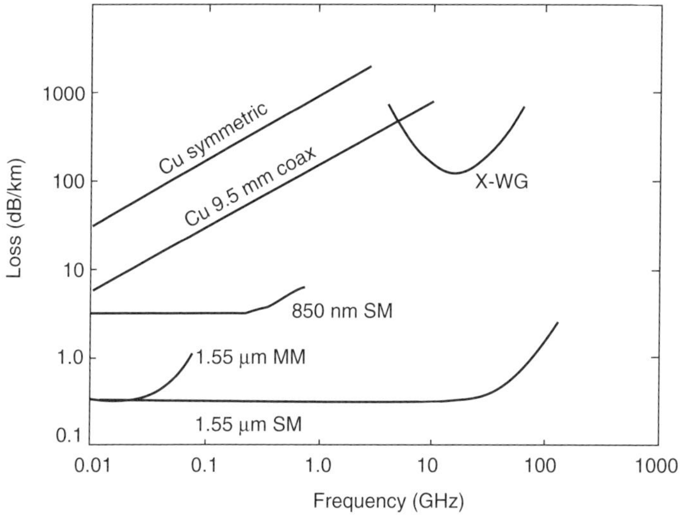


**FIGURE 1.21** Transmission loss of different transmission media. (From Jäger, D., Stöhr, A., *Microwave Photonics—From Concepts to Applications*, Institut fur Mikrowellen- und Antennentechnik C.V. (IMA), 136–138, 2005. With permission.)

achieved when an optical amplifier (EDFA) or externally controlled modulators according to the devices of Figure 1.1d, preferably on both sides, are being used [37]. For high-speed and broadband operation, the a.m. TW microwave photonic devices have successfully been employed.

For bidirectional fiber wireless communications, as shown in Figure 1.22, the link would require a duplication of the elements on both sides to provide up- and downlink transmission. This is the conventional system architecture. However there is another most interesting solution using elements as mentioned above. Figure 1.22 shows an example of an advanced system where the base station (BS) contains only one optoelectronic element, here the transceiver, as realized by an EAT. Optically, the transceiver receives two signals, one for downlink, which is absorbed, and one for the uplink, which is modulated by the received electrical signal to be transmitted to the central station (CS). Electrically the transceiver uses electrical frequency multiplexing, that is, the up- and downlink RF signals are at different frequencies. Given that due to the basic physical mechanisms the electrical bandwidth of the EAT is the same for the detection and the modulation processes, a bandwidth of more than 100 GHz can be achieved using a TW EAT (Figure 1.8).

## 1.7 MICROWAVE PHOTONIC SYSTEMS

Broadband fiber optic links are regarded as basic building blocks for different microwave systems where specific advantages of the optical interconnections

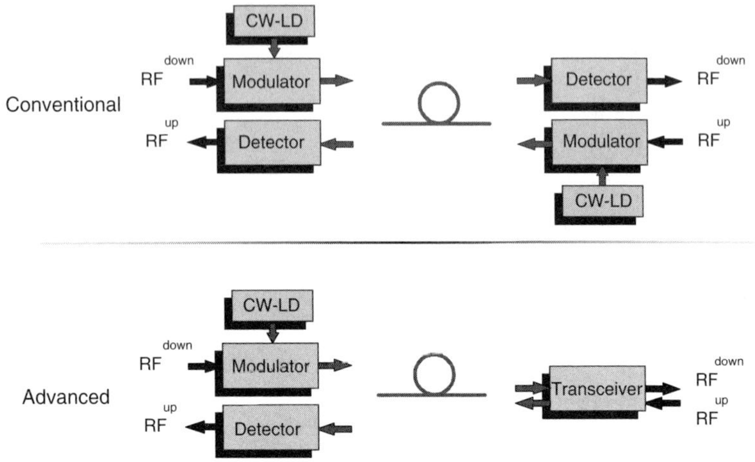


**FIGURE 1.22** Conventional and advanced IMDD scheme (IMDD is intensity modulation and direct detection). RF down and RF up are the RF signals of a down- and uplink communication between a central station (left) and a base station (right).

and optical signal processing capabilities are utilized. A few areas of significant applications are reviewed in the following.

## 1.7.1 Photonic Local Oscillators

Transmitting two optical wavelengths by using two frequency-locked lasers or a two-mode laser and mixing the two optical signals in a photodetector or mixer emulates a microwave local oscillator where the difference frequency is photonically generated by heterodyne techniques and where wavelength tuning provides a bandwidth of several THz depending on the bandwidth of the detector [39–43]. Note also that an optically induced phase shift in the optical path is directly transferred to the electrical domain. Figure 1.23 shows a module fabricated for radio-astronomical antennas. The module consists of a TW photomixer connected to a dc bias circuit with appropriate filters, a slot line antenna, and a quasioptical lens to focus the radiated beam at 460 GHz into a He-cooled SIS mixer.

## 1.7.2 EMC Sensor

When the modulator at the end of a fiber is driven by an electrical input signal at the given position, the received optical signal can be used to measure the electrical signal quantitatively at the transmitter side and a field sensor results [44,45]. Figure 1.24 shows the circuit diagram of such an E-field sensor for EMC applications. The sensor head consists of a dipole antenna. The output

**FIGURE 1.23** Photograph of a photonic local oscillator module.

signal is fed into a transimpedance amplifier driving an EAM. The optical input signal at 1.55 μm is delivered by a laser diode in the remote unit, which also contains the photodiode to measure the output signal from the EAM in the sensor head. The dc bias for the amplifier as well as for the operating point

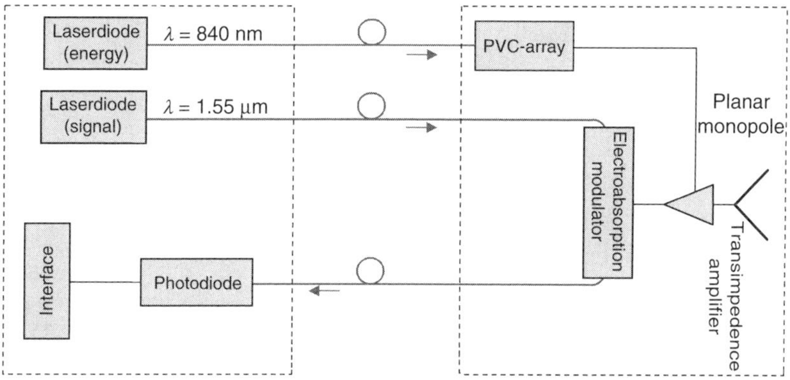


**FIGURE 1.24** EMC sensor.

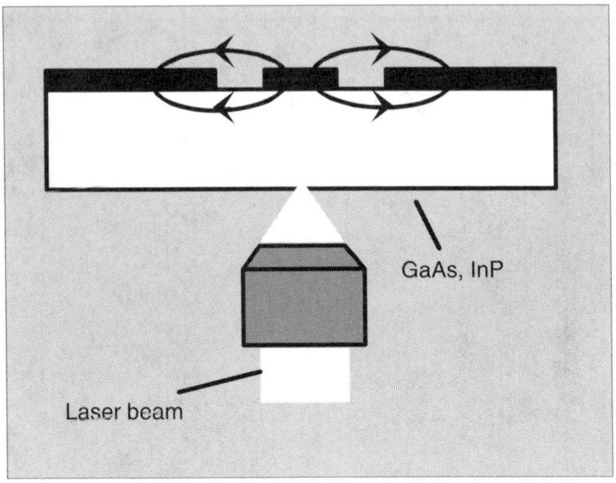


**FIGURE 1.25** Principle of electro-optic sampling. (From Jäger, D., Stöhr, A., *Microwave Photonics—From Concepts to Applications*, Institut fur Mikrowellen- und Antennentechnik C.V. (IMA), 136–138, 2005. With permission.)

of the EAM is provided by a photovoltaic cell (PVC) in the sensor head. A second laser diode in the read-out unit generates the necessary optical power.

Figure 1.25 elucidates the principle of electro-optic sampling. A laser beam is incident from behind into the substrate of a GaAs or InP integrated circuit (MMIC). As a result of the electro-optic properties of these materials, the light is modulated by the spatial distribution of the internal electric field. Hence, the reflected beam carries the information about the field distribution inside the circuit, and by scanning the beam across the circuit important information about the function of the circuit can immediately be obtained. Figure 1.26 shows the experimental setup that has been used in several laboratories over the world. Here a mode-locked Nd:YAG laser and an optical pulse compressor for picosecond resolution are used and a photodiode detects the received optical signal, which exhibits a polarization chance introduced by the electrical field. An example of the experimental results [46,47] is presented in Figure 1.27. The electrical field distribution at a coplanar T-junction clearly demonstrates the asymmetric behavior of the electrical field coming from a CPW to be split into two CPWs. The spatial resolution can go down to the sub-μm range as shown in Ref. [47].

### 1.7.3 Hybrid Fiber-Coax (HFC) Systems

In cable TV (CATV) techniques, the RF signals received from TV satellites are usually converted into the optical domain by directly modulating a laser diode or by using an external modulator together with a cw laser diode and fed into a fiber to be transmitted over long distances with only small attenuation. The received optical signals are converted back into the electrical

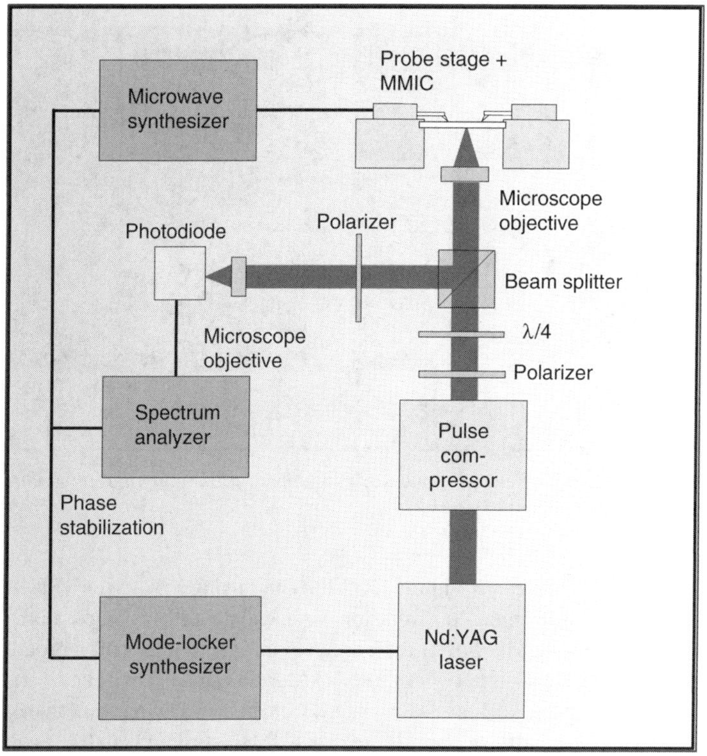


**FIGURE 1.26** Experimental setup for electro-optic sampling using a mode-locked Nd:YAG laser.

domain and guided to the customer via coaxial cable. This HFC technique, using again the advantages of the optical fiber transmission, has already been installed in major cities around the world where the unidirectional technique is replaced increasingly by bidirectional links to establish a back channel, for example, in multimedia applications.

### 1.7.4 Fiber-Radio Systems

It is agreed that fiber-radio access will provide a solution to the demands for a wireless connection to the customer ("last or first mile problem"), in particular for mobile communications [48]. For broadband services the frequencies are in the millimeter-wave range, for example at 40 or 60 GHz. Such a concept is based on an optical link between the central office and the base stations in a picocellular structure, which results from the high-atmospheric attenuation of the millimeter-wave signal and the corresponding reuse of frequencies at short distances. As an example, 60 GHz fiber-radio links

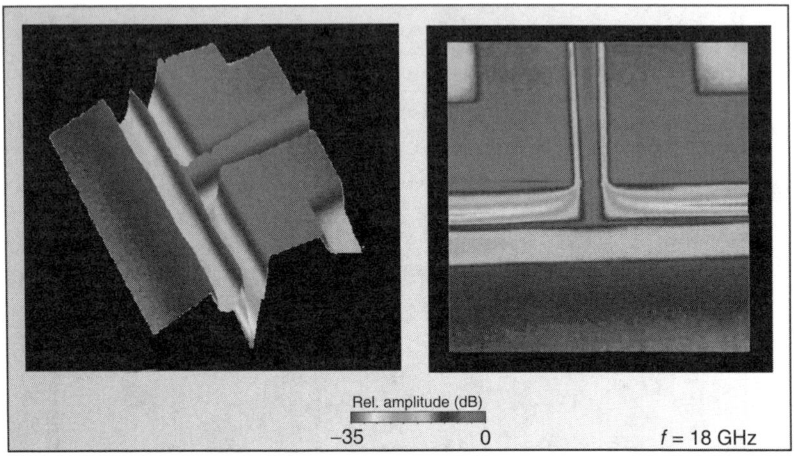


**FIGURE 1.27** Result of electro-optic sampling: electric field distribution at coplanar T-junctions measured at 18 GHz.

have been demonstrated providing 155 Mbps and using EAMs for half-duplex and EATs for full-duplex transmission in a wavelength division multiplex (WDM) ring network [49–56]. It is foreseen that for next generation broadband photonic communication networks, the EAM, and the EAT will be key elements. Experimentally, full-duplex transmission in a WDM ring network and electrical frequency division multiplex (EFDM), as sketched for two base stations in Figure 1.28, have been studied. Here two cw LDs at different wavelengths are used to drive the modulator function of the EAT in each base station addressed by the optical wavelength WDM technique. The other two LDs are externally modulated in the central station to provide the downlink transmission. These signals are absorbed in the EATs of the base stations to feed the wireless link. The electrical FDM technique is obvious from Figure 1.28. Downlink and uplink RF frequencies are 59.6 and 60 GHz, respectively. The transmission quality at a data rate of 156 Mbps is finally determined and evaluated by a bit error rate tester (BERT). An advanced fiber-radio system [49] is sketched in Figure 1.29. Here two laser diodes LD1 and LD1′ are used in the central station to provide the RF frequency by heterodyning the two optical beams in the EATX of the base station. One of the laser diodes, LD1, is externally modulated at 2.6 GHz subcarrier frequency to be mixed with the 57 GHz of the heterodyne signal (Figure 1.10).

## 1.8 CONCLUSION

In the past decade, the field of micro- and millimeter-wave photonics has become a key technology extending from components and modules to

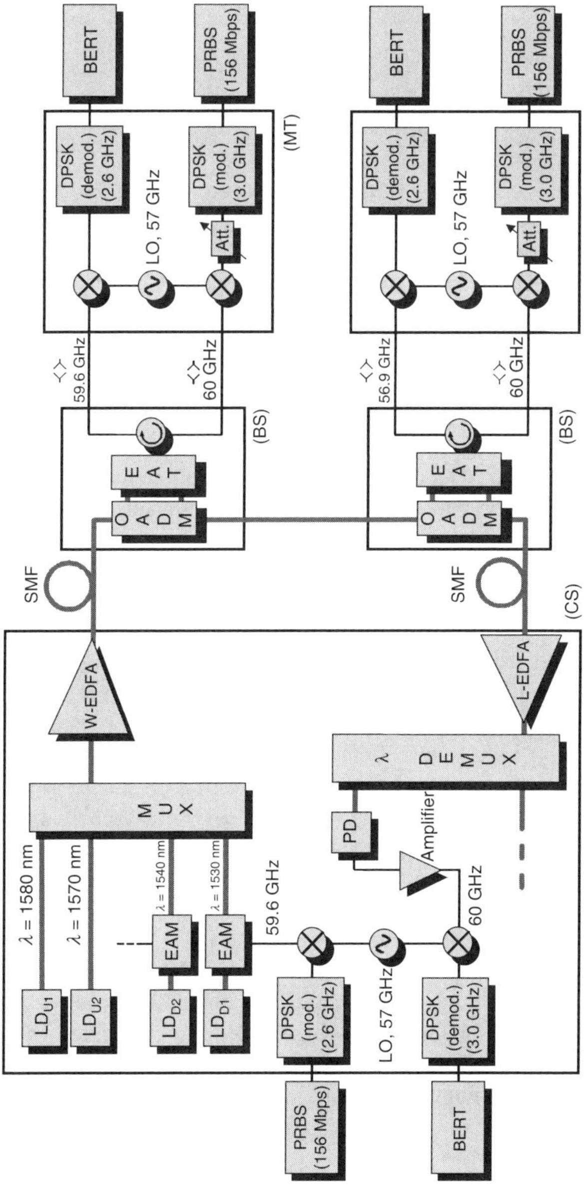


**FIGURE 1.28** Full-duplex 60 GHz millimeter-wave fiber wireless network employing an electro-absorption transceiver (EAT) as a key photonic component in the base station (BS).

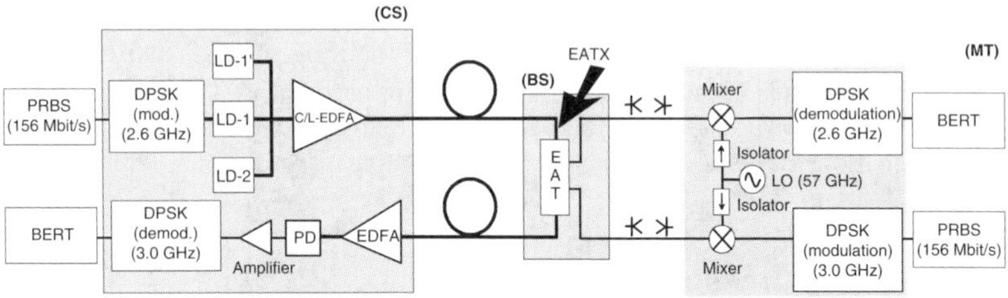


**FIGURE 1.29** Architecture for a 60 GHz fiber-radio access system. LD denotes a laser diode, PD a photodetector, and EAT-X an EAT mixer. MT is the mobile terminal. (From Jäger, D., Stöhr, A., *Microwave Photonics—From Concepts to Applications*, Institut fur Mikrowellen- und Antennentechnik C.V. (IMA), 136–138, 2005. With permission.)

systems with important applications. The driver has been twofold: on one hand the broadband low-loss and high-speed transmission capability of optical fibers has led to a considerable interest in their use for distributing and controlling micro- and millimeter-wave signals. On the other hand the breakthrough in the design and demonstration of several ultrabroadband photonic components has paved the way for wideband and high-efficiency optoelectronic converters used as important building blocks for microwave–optical links. As a result, it can be foreseen that this multidisciplinary field of microwave photonics will continuously expand and lead to further novel concepts due to the synergetic merging of different technologies.

In this contribution, specific high-speed and broadband photonic components have been studied. Special emphasis has been laid on electroabsorption and traveling wave devices. In particular, EAMs photodetectors, and mixers as well as transceivers have been presented. The multifunctional properties of the transceiver elements have been proven to be especially interesting. The performance of these elements has been demonstrated, for example, in a fiber wireless WDM ring network by BER measurements; however, the advantages will generally hold for other optical links with bidirectional transmission properties as well. The great advantage of this concept is that in the base station (in the sensor head or at the antenna side) just a single photonic component serves simultaneously as a converter for downlink and uplink communication and information transport. It is therefore concluded that these elements may play a significant and key role in fiber-coupled wireless micro- and millimeter-wave networks and for high data rate transmission services such as in multimedia techniques.

## ACKNOWLEDGMENTS

The author wishes to thank Professor K. Kitayama, Osaka University, and T. Kuri, Communications Research Laboratory, Tokyo, for their substantial cooperation in researching fiber-radio links. Special thanks, however, go to all the numerous members of the Zentrum für Halbleitertechnik und Optoelektronik, Universität Duisburg-Essen, and the Institute of Optoelectronics of the former Universität Duisburg, who have contributed to the results of this chapter for more than 10 years. The author is in particular indebted to Dr.-Ing Andreas Stöhr who has been the technical manager and the scientific assistant of the microwave photonics team in Duisburg for more than 10 years.

## REFERENCES

1. Jäger, D., Traveling-wave optoelectronic devices for microwave and optical applications, *Proc. Prog. Electromagn. Res. Symp. (PIERS)*, 327, 1991.
2. Jäger, D., Microwave photonics, in *Optical information technology*, Smith, S.D. and Neale, R.F., Eds., Springer-Verlag, New York, 328, 1993.
3. Polifko, D. and Ogawa, H., The merging of photonic and microwave technologies, *Microw. J.*, 35, 75, 1992.
4. *International Topical Meeting on Microwave Photonics*, Techn. Digest, Kyoto 1996, Duisburg 1997, Princeton 1998, Melbourne 1999, Oxford 2000, Long Beach 2001, Awaji 2002, Budapest 2003, Ogunquit 2004, Seoul 2005, Grenoble 2006.
5. *IEEE Trans. Microw. Theory Techn.*, Special Issue on Microwave Photonics, September 1995, September 1997, July 1999, June 2001, February 2006 and *IEEE J. Lightwave Techn.*, Special Issue on Microwave Photonics, Vol. 21, 2003.
6. Seeds, A.J., Microwave photonics, *IEEE Trans. Microw. Theory Techn.*, 50, 877, 2002.
7. Vilcot, A., Cabon, B., and Chazelas, J., *Microwave photonics*, Kluwer Academic Publishers, Boston, 2003.
8. Jäger, D. and Stöhr, A., Microwave photonics—From concepts to applications, in *Proc. German Micron. Conf.* (GeMiC 2005), Menzel, W., Ed., Ulm, 136, 2005.
9. Jäger, D., Characteristics of traveling waves along nonlinear transmission lines for monolithic integrated circuits, *Int. J. Electron.*, 58, 649, 1985.
10. Jäger, D., Kremer, R., and Stöhr, A., Traveling wave optoelectronic devices for microwave applications, in *IEEE Int. Microw. Symp. Digest*, Orlando, Kirby, L., Ed., 163, 1995.
11. Jäger, D., Kremer, R., and Stöhr, A., High-speed traveling-wave photodetectors, modulators and switches, in *IEEE/LEOS Summer Topical Meeting*, Keystone, Cox, Ch. and Wu, M.C., Eds., 23, 1995.
12. Jäger, D. and Kremer, R., Traveling-wave optoelectronic devices for microwave applications, in *Topical Meeting on Optical Microwave Interactions*, Paris, 11, 1994.
13. Jäger, D., Kremer, R., Humbach, O., and Stöhr, A., Traveling-wave optoelectronic devices for microwave photonics, in *Proc. Emerging Optoelectronic Technol.*, Selvarajay, A., Sonde, B.S., Shenai, K., and Tripathi, V.K., Eds., McGraw Hill, New Delhi, 409, 1994. Bangalore, 409, 1994.
14. Wood, T.H., Multiple quantum wells (MQW) waveguide modulators, *IEEE J. Lightwave Technol.*, 6, 743, 1988.
15. Humbach, O., Stöhr, A., Auer, U., Larkins, E.C., Ralston, J.D., and Jäger, D., Strained-layer multiple quantum well InGaAs/GaAs waveguide modulators operating at 1 μm, *IEEE Photon. Techn. Lett.*, 5, 49, 1993.

16. Stöhr, A., Humbach, O., Zumkley, S., Wingen, G., David, G., Bollig, B., Larkins, E.C., Ralston, J.D., and Jäger, D., InGaAs/GaAs multiple quantum well modulators and switches, *Opt. Quantum Electron.*, 25, 865, 1993.
17. Mineo, N., Nagai, K., and Ushikubo, T., Ultra-wideband electro-absorption modulator modules for DC to millimeter-wave band, in Proc. *Int. Top. Meeting Microw. Photon.* (MWP 2002), Furuhama, Y. and Izutsu, M., Eds., IEICE, Awaji, 9, 2002.
18. Stöhr, A., Heinzelmann, R., Malcoci, A., and Jäger, D., Optical heterodyne millimeter-wave generation using 1.55 μm traveling-wave photodetectors, *IEEE Trans. Microw. Theory Techn.*, 49, 1926, 2001.
19. Stöhr, A., Malcoci, A., and Jäger, D., GHz and THz InGaAsP travelling wave photodetector, in *Proc. IEEE LEOS Annu. Meeting*, Glasgow, 422, 2002.
20. Stöhr, A. and Jäger, D., Ultra wideband traveling-wave photodetectors for THz signal generation, in *Proc. IEEE LEOS Annu. Meeting* (LEOS 2004), Puerto Rico, 200, 2004.
21. Stöhr A. and Jäger, D., High-frequency photodetectors in Europe, in *Proc. Union Radio-Scientifique Int. General Assembly (URSI GA 2002)*, Maastricht, 17, 2002.
22. Malcoci, A., Stöhr, A., Lill, K., Siebe, F., van der Wal, P., Sauerwald, A., Güsten, R., and Jäger, D., Optical submillimeter-wave generation employing antenna integrated ultra-fast traveling-wave 1.55 μm photodetectors, in *IEEE Int. Microw. Symp. Digest*, Philadelphia, 143, 2003.
23. Stöhr, A. and Jäger, D., Optical generation of mm-wave signals, in *Microwave photonics*, Iezekiel, S., Ed., Wiley & Sons Ltd. (to be published).
24. Stöhr, A., Malcoci, A., Siebe, F., Lill, K., van der Waal, P., Güsten, R., and Jäger, D., Integrated photonic THz transmitter employing ultra-broadband traveling-wave 1.55 μm photodetectors, in Proc. *Int. Top. Meeting Microw. Photon.* (MWP 2002), Furuhama, Y. and Izutsu, M., Eds., IEICE, Awaji, 69, 2002.
25. Malcoci, A., Stöhr, A., Schulz, S., and Jäger, D., Optical THz generation, in *Proc. Int. Top. Meeting Microw. Photon.* (MWP 2003), Budapest, 179, 2003.
26. Stöhr, A., Heinzelmann, R., Alder, T., and Jäger, D., Electroabsorption transceiver (EAT) for fiber-wireless networks, in *Proc. Int. Top. Workshop Contemporary Phot. Techn.* (*CPT 2000*), Tokyo, 46, 2000.
27. Stöhr, A., Hagedorn, K., Gindera, R., and Jäger, D., Reflection-type EAT for full duplex wireless/wireline access networks, *Proc. 3rd Joint Symp. Opto-Microelectron.* (SODC 2004), Wuhan, 56, 2004.
28. Jäger, D., Large optical nonlinearities in hybrid semiconductor devices, *J. Opt. Soc. Am.* B, 6, 588, 1989.
29. Tauber, D., The microstrip laser diode, *IEEE Photon. Techn. Lett.*, 10, 478, 1998.
30. Schneider, M., Reimann, T., Heinzelmann, R., Stöhr, A., Velling, P., Neumann, S., Bertenburg, R.M., Tegude, F.-J., and Jäger, D., A novel 1.55 μm HBT-electro-absorption modulator, in Proc. *Int. Top. Meeting Microw. Photon.* (MWP 2001), Long Beach, 21, 2002.
31. Schneider, M., Reimann, T., Stöhr, A., Neumann, S., Tegude, F.-J., and Jäger, D., Monolithically integrated optoelectronic circuits using HBT, EAM, and TEAT, in *Proc. Int. Top. Meeting Microw. Photon.* (MWP 2002), Awaji, 349, 2002.
32. Alder, T., Stöhr, A., Heinzelmann, R., and Jäger, D., High efficient fiber-to-chip coupling using low-loss tapered single mode fiber, *IEEE Photon. Techn. Lett.*, 12, 1016, 2000.

33. Heidemann, R., Pfeiffer, Th., and Jäger, D., Optoelectronically pulsed slot line antennas, *Electron. Lett.*, 19, 316, 1983.
34. Paulus, P., Brinker, W., and Jäger, D., Generation of microwave pulses by optoelectronically switched resonators, *IEEE J. Quantum Electron.*, 22, 108, 1986.
35. Paulus, P., Stoll, L., and Jäger, D., Optoelectronic pulse compression of microwave signals, *IEEE Trans. Microw. Theory Techn.*, 35, 1014, 1987.
36. Cox, C.H., *Analog optical links*, Cambridge University Press, Cambridge, 2004.
37. Jäger, D., Stöhr, A., and Heinzelmann, R., Advanced microwave photonic devices for analog optical links, in *Proc. Int. Top. Meeting Microw. Photon.* (MWP 1998), Princeton, 153, 1998.
38. Jäger D. and Stöhr, A., High-speed photonic components for optical communications, in *Proc. 2nd Joint Symp. Opto—Microelectron.* (*SODC 2002*), Stuttgart, 59, 2002.
39. Stöhr, A., Heinzelmann, R., Hagedorn, K., Güsten, R., Schäfer, F., Siebe, F., Stüer, H., van der Waal, P., Krozer, V., Feiginov, M., and Jäger, D., Integrated 460 GHz photonic local oscillator, *Electron. Lett.*, 37, 1347, 2001.
40. Stöhr, A., Malcoci, A., Siebe, F., Lill, K., van der Waal, P., Güsten, R., and Jäger, D., Integrated photonic THz transmitter employing ultra-broadband 1.55 μm travelling-wave photodetectors, in *Proc. Int. Top. Meeting Microw. Photon.* (MWP 2002), Awaji, 69, 2002.
41. Stöhr, A., Malcoci, A., Sauerwald, A., Cámara Mayorga, I., Güsten, R., and Jäger, D., Ultra-wideband traveling-wave photodetectors for photonic local oscillators, *IEEE J. Lightwave Technol.*, 21, 3062, 2003.
42. Malcoci, A., Stöhr, A., Heinzelmann, R., Hagedorn, K., Güsten, R., Schäfer, H., Stüer, H., Siebe, F., van der Waal, P., Krozer, V., Feiginov, M., and Jäger, D., Photonic (sub)millimeterwave local oscillators, in *Proc. 14th Int. Conf. Microw., Radar Wireless Commun.* (MIKON 2002), Gdansk, 722, 2002.
43. Malcoci, A., Stöhr, A., Sauerwald, A., Schulz, S., and Jäger, D., Waveguide and antenna coupled traveling-wave 1.55-μm photodetectors for optical (sub)millimeter-wave generation, in *Proc. SPIE Photonics Eur. Microw. Terahertz Photon.*, Strasbourg, 202, 2004.
44. Heinzelmann, R., Stöhr, A., Groa, M., Kalinowski, D., Alder, T., Schmidt, M., and Jäger, D., Optically powered remote optical field sensor system using an electrabsorption modulator, in *IEEE Int. Microw. Symp. Dig.*, Meixner, R.P. and Cruz, J.J., Eds., Baltimore, 1225, 1998.
45. Stöhr, A., Heinzelmann, R., Bua, R., and Jäger, D., Electroabsorption modulators for broadband fiber electro-optic field sensors, in *Applications of Photonic Technology 2*, Lampropoulos, G.A., and Lessard, R.A., Eds., Plenum Press, New York, 871, 1996.
46. David, G., Tempel, R., Wolff, I., and Jäger, D., Analysis of microwave propagation effects using 2D-electro-optic field mapping techniques, *Opt. Quantum Electron.*, Special Issue, 919, 1996.
47. David, G., Bussek, P., Auer, U., Tegude, F.-J., and Jäger, D., Electro-optic probing of RF-signals in sub-μm MMIC devices, *Electron. Lett.*, 31, 2188, 1995.
48. Al-Raweshidy, H. and Komaki, S., *Radio over fiber technologies for mobile communications networks*, Artech House Inc., Stadt, 2002.

49. Kitayama, K., Stöhr, A., Kuri, T., Heinzelmann, R., Jäger, D., and Takahashi, Y., An approach to single optical component antenna base stations for broadband millimeterwave fiber-radio access systems, *IEEE Trans. Microw. Theory. Techn.*, 48, 2588, 2000.
50. Kitayama, K., Kuri, T., Heinzelmann, R., Stöhr, A., Jäger, D., and Takahashi, Y., A good prospect for broadband millimeter-wave fiber radio access system—An approach to single optical component at antenna base station, in *Proc. IEEE MTT-S Int. Microw. Symp.*, Perkins, Th., Ed., Boston, 1745, 2000.
51. Stöhr, A., Kitayama, K., and Jäger, D., Full-duplex fiber-optic RF subcarrier transmission using a dual-function modulator/photodetector, *IEEE Trans. Microw. Theory Techn.*, 47, 1338, 1999.
52. Miesner, C., Stöhr, A., and Jäger, D., Full-duplex simultaneous transmission of Gigabit Ethernet on analog transparent channel for broadcast and radio over fiber applications, *IEEE LEOS German Chapter News*, 7, 10, 2005.
53. Stöhr, A., Miesner, C., and Jäger, D., High-speed, high-frequency fibre optic modules and their applications in fibre-radio systems, in *Proc. 9th Europ. Conf. Netw. Opt. Comm.* (NOC), Eindhoven, 34, 2004.
54. Gindera, R., Hagedorn, K., Stöhr, A., and Jäger, D., Simultaneous transmission of 2.5 Gb/s base band signal and microwave signals employing a 1.55 μm R-EAT for hybrid wireless/wireline networks, in *Proc. 2nd NEFERTITI Winter School*, York, 2004, CD-ROM.
55. Stöhr, A., Miesner, C., and Jäger, D., All-optical radio-independent millimeter-wave radio-on-fiber system with lean antenna base station, in *Proc. Int. Top. Meeting Microw. Photon.* (MWP 2002), Awaji, 213, 2002.
56. Stöhr, A., Kitayama, K., Kuri, T., and Jäger, D., Electroabsorption transceiver (EAT) for radio-over-fiber applications, in *Proc. Union Radio-Scientifique Int. General Assembly* (*URSI GA 2002*), Maastricht, 2002.

# 2 Femtosecond All-Optical Devices for Ultrafast Communication and Signal Processing*

*Osamu Wada*

## CONTENTS

2.1 Introduction .......... 32
2.2 Advantages and Requirements of Femtosecond Devices .......... 34
2.3 Recent Progress in Femtosecond Light Sources .......... 37
  2.3.1 Monolithic Mode-Locked Lasers .......... 37
  2.3.2 Pulse Waveform Control Techniques .......... 40
2.4 Femtosecond All-Optical Switches .......... 41
  2.4.1 A Symmetric Mach–Zehnder (SMZ) All-Optical Switch .......... 42
  2.4.2 All-Optical Spin Polarization Switch .......... 46
  2.4.3 ISBT All-Optical Switch .......... 51
    2.4.3.1 InGaAs/AlAsSb System .......... 52
    2.4.3.2 GaN/AlGaN System and Other Material Systems .......... 57
  2.4.4 Wavelength Conversion Switches Using Four-Wave Mixing in SOAs .......... 60
  2.4.5 All-Optical Switches Based on Novel Materials .......... 61
    2.4.5.1 Quantum Dot SOAs .......... 61
    2.4.5.2 AlGaAs/GaAs Multiplayer System with Enhanced Nonlinearity .......... 64
    2.4.5.3 Carbon Nanotubes .......... 67
    2.4.5.4 Organic Thin Films .......... 67
  2.4.6 Current Status of Ultrafast All-Optical Switches .......... 70
2.5 Conclusions .......... 72
Acknowledgments .......... 72
References .......... 73

---

*(This paper has been rewritten with permission of IOP on the basis of the author's paper with the same title published in *New J. Phys.*, 6, 183, 2004.)

## 2.1 INTRODUCTION

Optical fiber communication system technology has been developed extensively since the first system installation in the early 1980s, and the most recent trunk lines run at a bit rate of 10 Gb $s^{-1}$; even 40 Gb $s^{-1}$ systems have been already prototyped and are expected to be commercialized soon. In addition to such high-bit-rate time division multiplexing (TDM) systems, wavelength division multiplexing (WDM) systems, which can involve hundreds of different wavelength channels to support the overall throughput close to 1 Tb $s^{-1}$, are used in recent systems, as illustrated in Figure 2.1 [1] . On the other hand, the most recent trend in the information transmission volume does not show any sign of saturation, regardless of the recent worldwide recession in the communication market at the very beginning of the new millenium [2]. This is thought to be caused by the ever increasing, strong demands for transmitting multimedia information through global networks involving the Internet and cellular phone services. Judging from an extrapolation of the increasing rate of information volume, a breakthrough in optical communication technology is required to improve the throughput in the range between 1 and 10 Tb $s^{-1}$ within a decade.

To achieve such an extremely high throughput in communication systems, the WDM technique could be extensively used to further increase the system

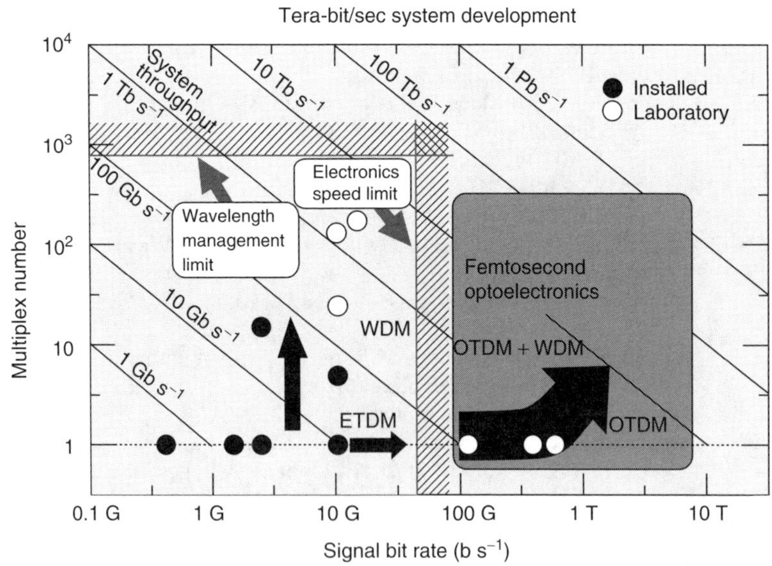


**FIGURE 2.1** Different approaches to improve the system throughput and their practical limits shown in a space of the signal channel multiplex number versus the signal bit rate for a single channel. O(E)TDM, optical (electronic) time division multiplexing.

throughput. However, this approach would suffer from a variety of problems: the use of many lasers, each of which must be readily tuned to a specific wavelength channel, becomes difficult or even impractical as the channel number increases. This limit in the wavelength management and handling may restrict the total system throughput. On the other hand, the operating bit rate of current TDM systems is basically limited by the speed of electronics used for signal processing and driving optical devices, and its improvement beyond a level of 100 Gb $s^{-1}$ seems to be rather difficult by solely relying on existing electronic technologies. In contrast to this electronic TDM technique, optical TDM (OTDM) will be able to break this limit, because all the necessary signal-processing functions are carried out all-optically, once the technological platform of ultrafast device design and fabrication is established [3]. A combination of ultrafast OTDM and ultrawideband WDM is expected to become a practically useful technique for supporting the throughput level exceeding 10 Tb $s^{-1}$ in future network systems.

Thus the development of ultrafast all-optical devices that can operate in the femtoseconds timescale is strongly desired for achieving very high-throughput OTDM systems. Technical feasibility of ultrafast signal transmission and processing at a bit rate in the range of 100 Gb $s^{-1}$–1.28 Tb $s^{-1}$ has been demonstrated in laboratories by employing primarily optical fiber–based lasers and nonlinear devices. A recent transmission experiment has shown 1.28 Tb $s^{-1}$ signal over 70 km [4]. The most recent development of OTDM device and system technology has been reviewed by Morioka et al. [5].

For establishing practical OTDM and OTDM–WDM combined systems, however, compact, stable and cost-effective, ultrafast all-optical devices are prerequisites. Based on this recognition, extensive research and development are being conducted in the area of ultrafast physics, materials, and devices by various research groups worldwide. Some of the latest results have been reported in a recently published book [6]. This paper discusses recent advances in femtosecond all-optical devices with a particular emphasis on ultrafast pulse generators and all-optical switches as basic devices for ultrafast systems [1,7]. The next section describes the advantages of femtosecond all-optical devices and explains basic signal-processing functions required for network systems. In the following sections, we review recent progresses in typical all-optical devices by considering some examples developed at the Femtosecond Technology Research Association (FESTA) Laboratories in the framework of Femtosecond Technology R&D Program funded by METI (Japan) [7,8]. They include ultrafast devices for mode-locked pulse generation, pulse compression, and dispersion compensation, and a few types of all-optical switches using various device principles and ultrafast materials. Some of these novel devices are already capable of basic signal-processing functions at the bit rate of 500 Gb $s^{-1}$–1 Tb $s^{-1}$.

## 2.2 ADVANTAGES AND REQUIREMENTS OF FEMTOSECOND DEVICES

The use of ultrafast pulses has a variety of potential advantages [9,10] as illustrated in Figure 2.2. They include the advantage of fully utilizing the material's nonlinearity by an extremely high peak intensity of field in ultrafast pulses. This is essential in the development of all-optical switching and modulation devices with high efficiency without increasing the average power consumption. An ultrashort optical pulse occupies an extremely short distance in space and propagates at the velocity of light, and this means a possibility of precisely controlling the delay time in a small dimension; the overall optical device and circuit can be very compact. An ultrashort pulse has a large spectral width due to the pulse shape–spectrum interdependence deduced directly from the Fourier transform relationship. This merits the use of various photonic functions in wavelength division, such as the extraction of multichannel wavelengths from an ultrashort pulse, and also in wavelength conversion and the shaping of pulse waveform by applying this property. The extraction of 200 wavelength channels has been shown using a femtosecond pulse [11]. These can contribute to combine both OTDM and WDM techniques to increase further the overall throughput of optical signal transmission and processing systems.

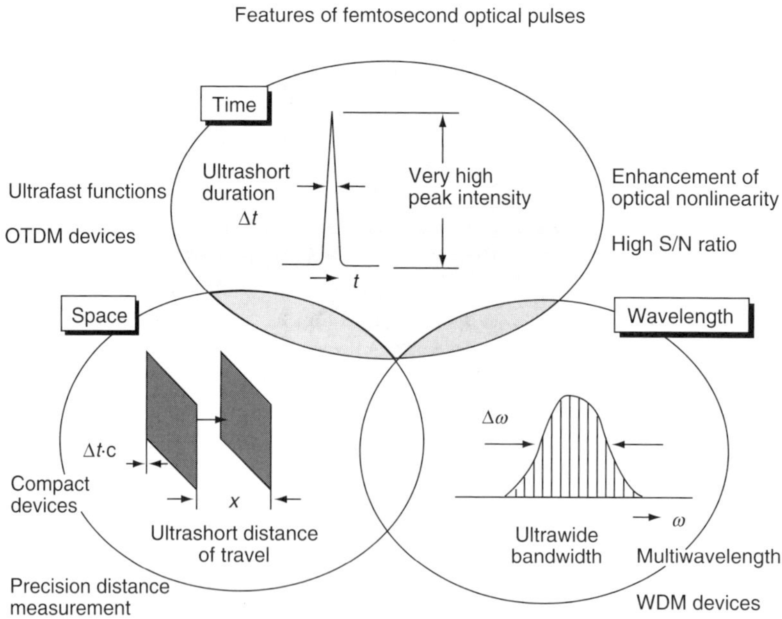


**FIGURE 2.2** Characteristic features and potential applications of ultrashort optical pulses indicated for three dimensions: time, space, and wavelength.

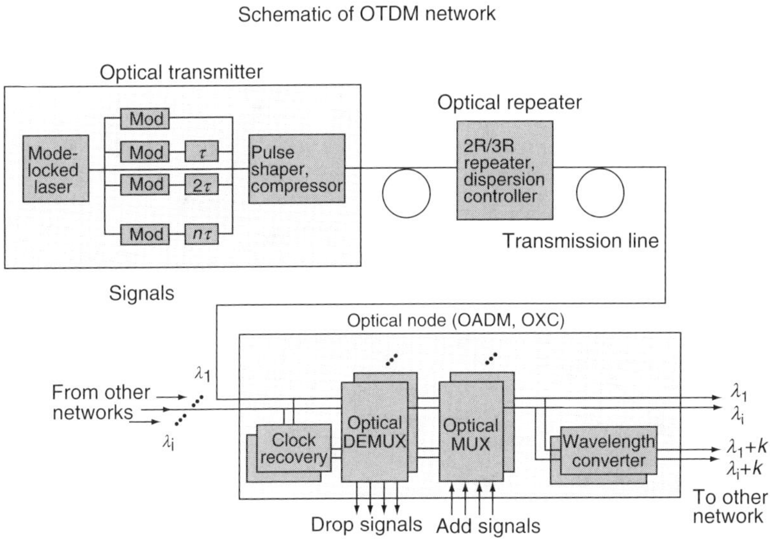


**FIGURE 2.3** Simplified schematic diagram of the OTDM network system. Optical data signals generated in the optical transmitter travel through the transmission lines and optical repeaters and are received by the optical node, in which optical add or drop multiplexing (OADM) and optical cross connect (OXC) functions are performed for networking.

The schematic diagram of a key part of the OTDM network system is given in Figure 2.3. In a transmitter, generation and coding functions of ultrafast pulses are required. Ultrafast pulses are generated by a mode-locked laser with a relatively low-repetition rate, and they are multiplied and coded by external control signals using an array of delay-lines and modulators, in which arms have different delay times. The pulse is compressed or shaped if necessary and then transmitted through a fiber or other transmission medium. The propagation loss and dispersion effect experienced in the transmission medium can be compensated by an optical repeater including either 2R function, which stands for reamplification and reshaping, or 3R function, which adds the retiming function to 2R. The function of optical nodes is to receive and drop necessary signals by using a demultiplexing circuit (DEMUX) and add slow rate signals prepared externally by using a multiplexing circuit. The clock recovery circuit is necessary to extract clock timing out of the incoming series of data signals, and wavelength conversion may be required for directing signals to other WDM network channels. This node function includes optical add or drop multiplexing function and sometimes covers even the exchange function for a large number of channels that are often called optical cross connect [12].

To illustrate a little more precisely the kind of optical functions and devices required to realize optical nodes, a photonic router circuit based on optical switches is shown in Figure 2.4. Routing function is achieved basically in a

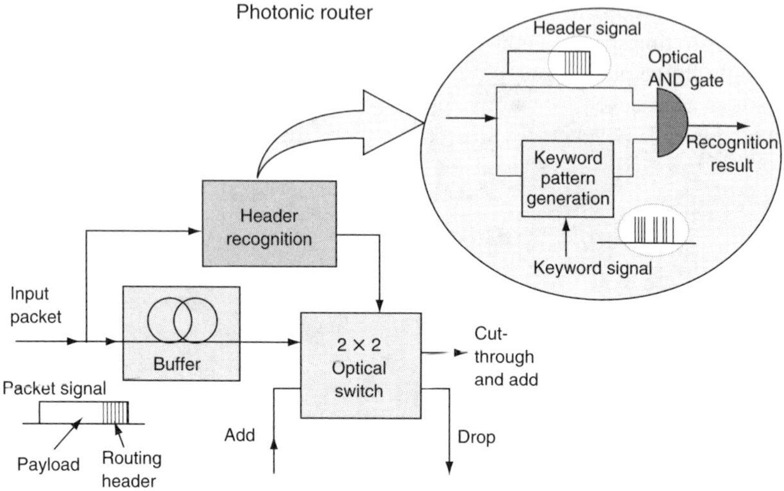


**FIGURE 2.4** Schematic diagram of an optical router used in a packet-switching network. Input signal packet is sent to either the cut-through port or the drop port depending on the header recognition result. In the header recognition circuit, the routing header signal pattern is compared with the reference keyword pattern at ultrafast speed.

header recognition circuit by recognizing the routing header pulses that are placed just before the pay load signal pulses within a signal packet, which is fed into the router. Recognition is carried out by comparing the header signal pattern and the reference keyword pattern by using an AND gate. In case the header signal pattern matches the keyword pattern, the input signal is dropped within this node through the cross state of a $2 \times 2$ optical switch, otherwise the signal is sent to the cut-through output port. Add signals are also allowed to transit to other nodes through this $2 \times 2$ switch.

A variety of such optical circuit-level functions are required for completing network functions, and they should be realized by combining basic device-level functions such as all-optical switch, modulator, low-level logic (AND, OR, EXOR, etc.), delay and memory, and also other optical device functions such as pulse waveform shaper, wavelength filter and converter, etc. Among these, the most basic, indispensable device function for ultrafast optical signal-processing systems should be the generation and switching of ultrafast optical signals. Ultrafast signal transmission and processing functions demonstrated so far have been achieved by using devices based on the nonlinearity of optical fibers. To establish a technical platform for ultrafast optical systems however, realization of compact, stable, and cost-effective femtosecond all-optical devices is a prerequisite. For realizing those ultrafast devices, new materials and device structures must be developed. For example, conventional optoelectronic devices consist of semiconductor materials and their operation speed is limited by the carrier recombination lifetime, which is usually in the nanosecond or subnanosecond range. To overcome this limit, various approaches are being exploited for developing new device concepts, which include novel device structures such as the Mach–Zehnder interferometer structure, new ultrafast phenomena such as spin relaxation and intersubband transition (ISBT) in quantum wells (QWs), or new materials (including quantum dots (QDs) and organic thin films), as will be discussed in the following sections.

## 2.3 RECENT PROGRESS IN FEMTOSECOND LIGHT SOURCES

Compact ultrafast lasers are indispensable for generating clock and data signals as well as control pulses for driving other all-optical devices. Ultrashort pulse width, sufficiently high optical power and repetition rate with low timing jitter, stable operation, and compact size are basic requirements of ultrafast light sources for practical applications in OTDM systems. Mode-locked lasers have been most extensively studied and promising results of ultrafast operation near 500 Gb $s^{-1}$ have been demonstrated. A combination of gain-switched laser and pulse compressor is another method of generating ultrafast pulses, and this technique is advantageous in flexibility of selecting the repetition rate. A novel, compact dispersion control device applicable to 1 ps class pulse compression has been developed based on matured III–V semiconductor device technology.

### 2.3.1 MONOLITHIC MODE-LOCKED LASERS

The semiconductor monolithic laser structures are advantageous in generating pulses with transform-limited waveforms and have a very high repetition rate owing to the very wide optical gain bandwidth and extremely short photon round-trip time, which are characteristic of microcavity semiconductor laser structures. Passive mode-locking, which requires no external electronic control circuits, is advantageous for high repetition rate operation and has been studied using discrete semiconductor lasers and then integrated laser structures. Monolithic integration of colliding pulse mode-locked (CPM) laser was first demonstrated in a GaAs-based laser [13], and improved performance, such as subpicosecond pulse width and the highest repetition frequency of 1.54 THz [14], was reported later in InP-based long-wavelength lasers. The greatest technical issues in semiconductor mode-locked lasers are the design of structures for generating ultrashort pulses with high extinction ratio at high repetition rate and stabilization of the timing and frequency of pulse generation.

A semiconductor mode-locked laser basically consists of a gain medium and a saturable absorber placed within a waveguide cavity. Saturable absorber is composed of a reverse-biased heterostructure waveguide, most often the same as that used for the active gain medium. To obtain ultrashort pulse width and high repetition rate, it is essential to have a very fast absorption recovery in the saturable absorber. These speedup effects have been shown by introducing a graded-index separate confinement heterostructure (GRIN-SCH) for the saturable absorber [15] and also by applying a high reverse bias voltage to a saturable absorber, which is designed intentionally for detuning to a shorter wavelength under no bias voltage [16]. Both these methods are effective in rapidly sweeping out photo-generated carriers in the absorber material. For further shortening the pulse width, the saturable absorber length must be sufficiently short for reducing the pulse propagation time inside it. It should be noted, however, that there is also a tradeoff between the propagation time and the absorption strength for minimizing the pulse width [17]. Another important factor for shortening the pulse width is to maximize the ratio of pulse compression, which is caused by the nonlinearity within the saturable absorber. To maximize the compression ratio, reduction in the background absorption in the laser cavity has been shown to be very effective. This has been confirmed by a laser structure involving a high-purity i-InP spacer layer inserted between the active and the p-doped cladding layer [17]. A monolithic structure involving a saturable absorber with optimized length (80 μm) and a low-loss waveguide structure incorporating an i-InP spacer layer has exhibited a pulse width well below 1 ps (790 fs) [17].

Pulse repetition rate, or the mode-locking frequency ($f_{ML}$), is determined by the cavity round-trip time or its harmonics: $f_{ML} = Nc/2n_gL$, where $c$ is the light velocity in vacuum, $n_g$ the group velocity index (typically 3.5 in semiconductor lasers), $L$ the cavity length, and $N$ the harmonic number for mode-locking. The fundamental mode-locking frequency for a conventional monolithic mode-locked laser having the saturable absorber at one end of the laser cavity is expressed by $N = 1$. In a CPM laser structure where a saturable absorber is located in the middle of two gain sections, the repetition frequency is increased by two times for the same cavity length ($N = 2$). To obtain a high repetition frequency, a short cavity length is effective.

Figure 2.5 shows the structure of a short cavity GRIN-SCH CPM laser [18]. The advantages of using the CPM configuration are both shorter round-trip time and lower absorption bleaching energy. The structure was grown and fabricated on an InP substrate by the metal organic chemical vapor deposition (MOCVD) technique. The active layer consists of a 0.8% compressively strained InGaAsP/InGaAsP multiple QW (MQW) structure. The total cavity length is 174 μm and the saturable absorber section has a length of 17 μm. Laser facets were coated with highly reflective (85%) $Al_2O_3$/Au thin films. Figure 2.6a shows the second harmonic generation (SHG) correlation trace and Figure 2.6b shows the optical spectrum of the laser output . The current biases for the two gain sections are 47 and 31 mA and the reverse bias for the saturable absorber is 0 V. The pulse repetition frequency observed is

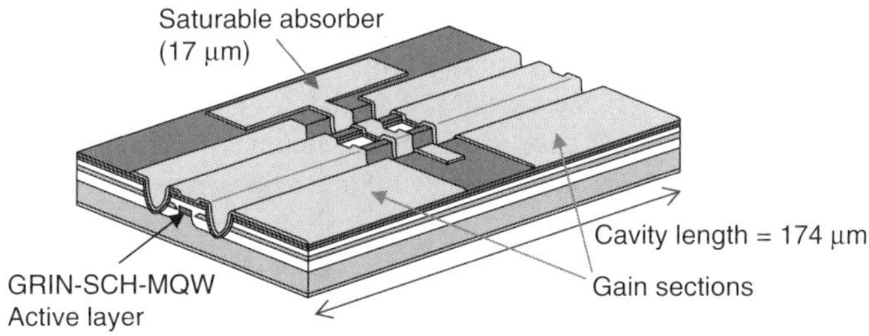


**FIGURE 2.5** Structure of the monolithic CPM laser fabricated on an InGaAsP/InP-based MQW wafer. A very short cavity length is used for ultrafast operation. (From Arahira S. et al., *IEEE Photon. Technol. Lett.*, 14, 537, 2002.)

480 GHz, agreeing with twice the cavity round-trip frequency. This is also in agreement with the spectrum in that the longitudinal modes are enhanced with every other two modes. The pulse width and spectral width are 0.58 ps and 8 nm, respectively, and the time-bandwidth product is 0.58, indicating a nearly transform-limited pulse waveform. It is noteworthy that the extinction ratio observed has been significantly improved when compared with previously reported results [14]. The averaged optical power is 1.6 mW.

Synchronization with external clock signals and low timing jitter characteristics is important in practical applications of mode-locked lasers. Various synchronization techniques including electrical and optical injection have been investigated. The most promising technique for ultrafast OTDM is the subharmonic optical synchronous mode-locking (SSML) using subharmonic frequency optical pulse injection. This technique can be applied by using low-repetition-rate optical pulses and no ultrafast driving source is required. SSML-stabilized operation over the range of 8.5–160 GHz has been demonstrated [19,20].

Since the timing jitter of ultrafast pulses cannot be measured by conventional electronic measurement techniques, a novel all-optical down-converting

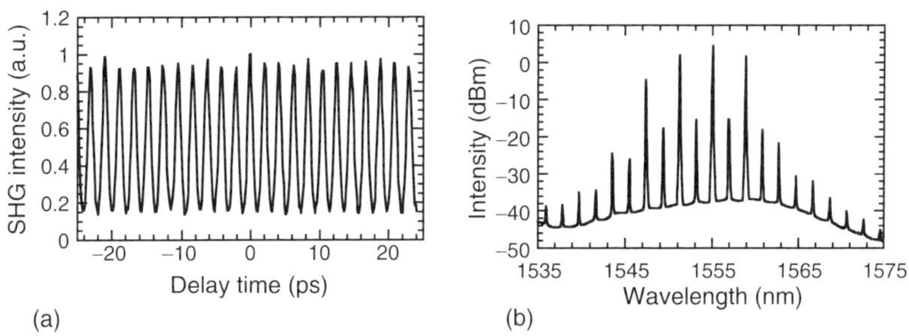


**FIGURE 2.6** (a) SHG correlation trace. (b) Optical spectrum of CPM laser, indicating 580 fs half-width pulse generation at a repetition rate of 480 GHz. (From Arahira S. et al., *IEEE Photon. Technol. Lett.*, 14, 537, 2002.)

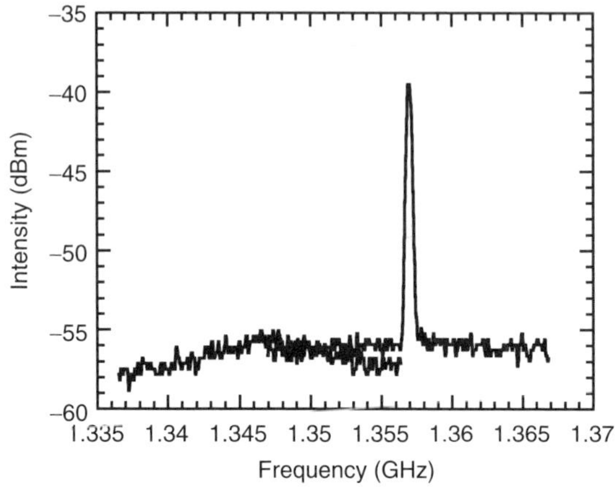


**FIGURE 2.7** Timing jitter characteristics measured using a beat signal generated between the CPM laser and a reference laser with known jitter property. Narrowing of the beat signal due to optical injection demonstrates optical synchronization. Jitter evaluated from these data is 0.14 ps. (From Arahira S. et al., *IEEE Photon. Technol. Lett.*, 14, 537, 2002.)

technique has been developed, in which a beat signal is generated between the output signal to be measured and a reference signal with known jitter property by using a nonlinear device [18]. Figure 2.7 shows the resultant electrical power spectra of the generated beat signal with and without master laser optical injection . In this case, the output signal frequency is approximately 480 GHz and the master laser and reference laser frequencies ($f_{master}$ and $f_{ref}$) are approximately 10 and 40 GHz, respectively. Without master laser excitation, the spectrum is broadened, indicating free running operation without long-term mode-locking. In contrast, a discrete frequency component ($48f_{master} - 12f_{ref}$) appears on optical injection, indicating that the 48th mode-locking is achieved. The timing jitter estimated from these data is 0.14 ps for an optical injection power of −12 dBm. This jitter is rather shorter than that of the master laser, which is interpreted to show superior short-term stability than the present laser [18,20].

### 2.3.2 Pulse Waveform Control Techniques

As described above, monolithic mode-locked lasers have been shown to be suitable for application to clock and data signal generation in OTDM systems because of their ultrafast operation with high stability. Extensive studies to further reduce the pulse width and increase the repetition frequency are in progress. For reducing the pulse width, the use of an external pulse compressor has been studied and a pulse width less than 20 fs has been achieved by a couple of groups using a well-designed pulse compressor composed of a series of fibers with different dispersion characteristics [21,22]. Also, the repetition frequency is increased by a novel technique called the repetition frequency multiplication technique based on the periodic interference of group velocity dispersion (GVD)–induced phase shifts of lasing modes in an external single-mode fiber [23]. This technique has been applied to a high-repetition-rate laser described above to generate 1 THz pulses [24].

A device element that enables the control of GVD within a compact chip of the solid state material such as a semiconductor would be very useful for practical ultrafast devices for pulse generation and waveform shaping. A novel GVD-controlling device based on the coupled-waveguide approach [25] has been developed. A coupled-waveguide device has been fabricated on an InP substrate, in which two core layers with different thicknesses and compositions are laminated using an intermediate layer. Two guiding layers have fast and slow group velocities and a very large GVD can be generated due to mode coupling between them. Successful compression of a chirped pulse with a pulse width from 2.2 ps down to 510 fs has been demonstrated in a device as short as 4 mm in length [26]. In addition, pedestal-free compression near the 1 ps range has been shown in an asymmetric-coupled waveguide made of $TiO_2$ and Si with a larger refractive index difference [27].

This result of compact, semiconductor-based devices is significant not only for providing compact dispersion compensators and pulse compressors but also for enabling future monolithic integration of light sources and pulse compressors. Another interesting approach for controlling the group delay, GVD, and pulse shapes is the application of photonic crystals (PCs) [28]. There are many active researches in this field. Among many new ideas reported, the optical delay device is interesting for ultrafast signal processing. A delay time as long as 600 fs has been demonstrated in a simple, planar, one-dimensional (1D) PC waveguide device with a length of 15 μm, which has been fabricated on a Si on an insulator substrate [29]. This result indicates four times reduction of the group velocity compared without the PC. Such an approach is considered useful in controlling the pulse timing, which will be a prerequisite in future all-optical signal-processing circuits.

Dispersion compensators, particularly adaptive dispersion compensators involving higher-order dispersion control, become increasingly important as the signal bit rate of the transmission system increases. A hybrid integration approach has been adopted to combine temperature-controlled fiber Bragg gratings to enable controlling the GVD and also the dispersion slope. The adaptive nature of the dispersion control has been demonstrated by a pulse shape control in the range of 8.5–1.8 ps for a starting pulse with 4.3 ps width [30].

## 2.4 FEMTOSECOND ALL-OPTICAL SWITCHES

Although performance improvement is being made continuously in optoelectronic and electronic switching devices, their operation at a bit rate higher than 100 Gb $s^{-1}$ seems to be still difficult because of various limiting response times including the carrier recombination lifetime, carrier transit time in the device, and also the external RC time constant. Ultrafast switching devices in OTDM systems must be all-optical; they are used to down-convert the repetition rate to a level at which signals can be processed by existing electronic devices. Requirements of all-optical switches include ultrafast response, high repetition rate, low switching energy, a high contrast ratio, and a low insertion loss. An early report on all-optical switching was based on nonresonant nonlinearity in the materials such as the optical Stark effect or virtual transition, but it required a impractically high excitation power to function. Instead, resonant nonlinearity is likely used in devices from the switching energy viewpoint, but a breakthrough is needed for avoiding the speed limit. A few different approaches examined to overcome this are given below:

- To develop a new device structure and operation mode relying on the existing nonlinear material sometimes at the sacrifice of device complexity. An example is the Mach–Zehnder interferometer-based switch incorporating a pair of optical nonlinear elements in both the interferometer arms.
- To use novel ultrafast phenomena in existing materials, represented by GaAs- and InP-based materials, which have been used in conventional optoelectronic devices. This includes spin relaxation in GaAs- and InP-based QW vertical structures, and four-wave mixing in semiconductor optical amplifiers (SOAs). Semiconductor QD materials can be useful in improving the performance of SOA-based devices. ISBT in QW is another important ultrafast phenomenon, and various semiconductors such as Sb-containing and nitride QWs have recently shown to be promising.
- Materials that have not been used in conventional optical communication devices may become important in developing novel ultrafast all-optical devices. Organic thin films exhibiting ultrafast nonlinearity have recently been developed and applied to two-dimensional (2D) all-optical switches. In addition, other novel nanostructure materials such as QDs and carbon nanotubes appear to be very interesting for future applications.

In the following sections, recent progresses in representative examples of these developments are discussed.

### 2.4.1 A Symmetric Mach–Zehnder (SMZ) All-Optical Switch

An Symmetric Mach–Zehnder (SMZ) switch first proposed by Tajima et al. [31] is a very powerful technique to realize ultrafast switching using conventional semiconductor materials. Its operation is based on differential

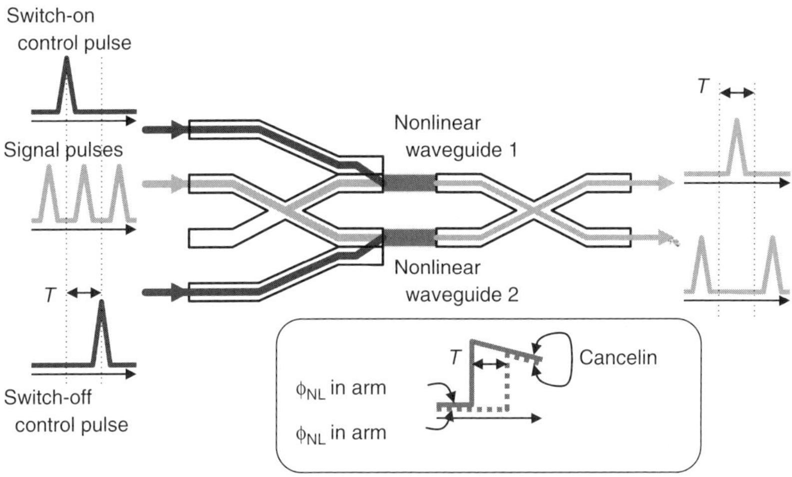


**FIGURE 2.8** Schematic representation showing the structure and operation of an SMZ all-optical switch. (From Tajima K. et al., *Jpn. J. Appl. Phys.*, 32, L1746, 1993.)

(push–pull) phase modulation of the nonlinear element pair incorporated into an interferometer configuration. The SMZ switch has two arms, including nonlinear waveguides (NLWGs), and two ports for control pulse injection as shown in Figure 2.8. In each NLWG, the control pulse produces a nonlinear refractive index change and causes phase shift in the propagating light pulse. When a control pulse is injected into one arm, the signal is switched from one output port to the other due to $\pi$ phase shift (switch-on). When another control pulse is injected into the other arm after a short interval, the same nonlinearity occurs in this arm and the phase mismatch is cancelled at the output port and the signal light is switched back (switch-off). Thus ultrafast switch-off and a square switching window are achieved regardless of the slow recovery in nonlinear elements. During its travel in an NLWG, the light pulse may exhibit frequency chirping, because chirp is in proportion to the rate of index change caused by the nonlinearity. It is interesting to note that the frequency chirp can also be cancelled in this configuration.

The hybrid-integrated SMZ device with a structure as shown in Figure 2.9 was fabricated using a silica-based planar lightwave circuit (PLC) and an InGaAsP/InP SOA pair array as the nonlinear elements. SOAs and optical fibers were self-alignedly mounted on a 20 mm long PLC platform [32]. To evaluate device performance, demultiplexing operation from 168 to 10.5 Gb $s^{-1}$ was examined. The wavelengths of the control and signal lights used were 1545 and 1560 nm, respectively. Signal and control pulses were generated using actively mode-locked fiber lasers with pulse widths of 1.4 and 2.5 ps. The input and output signal waveforms and the measured bit error

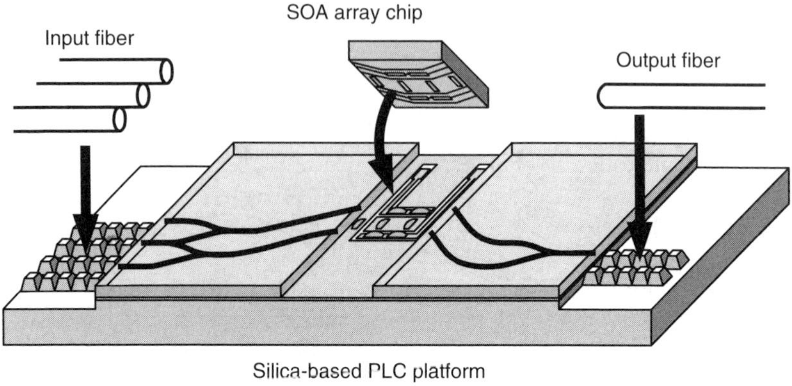


**FIGURE 2.9** Structure of the hybrid-integrated SMZ all-optical switch. (From Nakamura S. et al., *Abstr. Int. Workshop Femtosecond Technol.*, p. 61, 2002.)

rate (BER) versus received power characteristics are shown in Figure 2.10. An important advantage of using SOA is a marked reduction in the switching energy. The control pulse energy coupled into SOAs was estimated to be

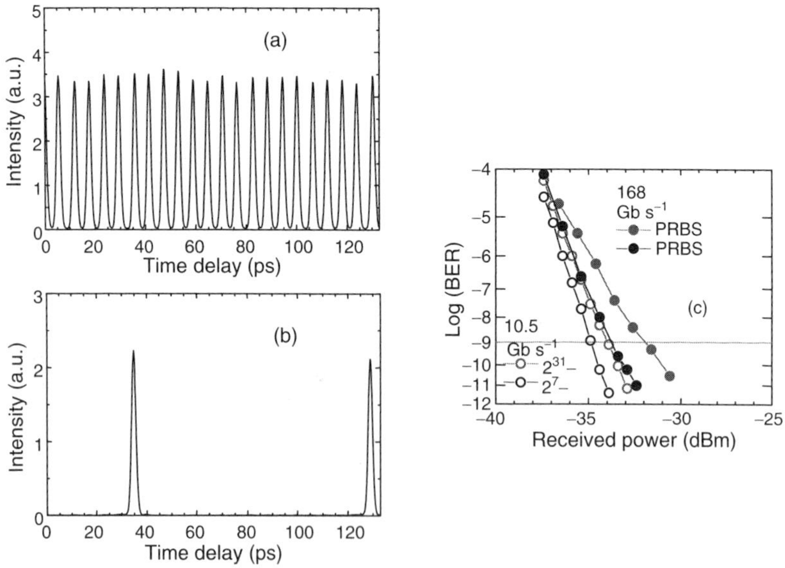


**FIGURE 2.10** The (a) input and (b) output signal pulses for the 168–10.5 Gb s$^{-1}$ demultiplexing experiment using an SMZ all-optical switch and (c) the BER versus the received optical power characteristics are shown. (From Nakamura S. et al., *Abstr. Int. Workshop Femtosecond Technol.*, p. 61, 2002.)

50–80 fJ in this experiment, indicating an extremely low energy at this ultrafast speed. The extinction ratio after the demultiplexing was better than 18 dB. The BER data show that error-free operation was achieved down to a BER of $10^{-11}$. A more recent study has shown successful demultiplexing operation even from 336 to 10.5 Gb $s^{-1}$ [33].

Another interesting and important operation of the SMZ device is found in a polarization-discriminating SMZ (PD-SMZ) switch [34]. As shown in Figure 2.11a, the PD-SMZ consists of an InGaAsP NLWG, birefringent crystals (BC1, BC2), and a polarizer. As the signal pulses pass through BC1, each pulse is split into a pair of orthogonally polarized signals with a time interval $T$, because of the group velocity difference between the ordinary and extraordinary lights in BC1. When a control pulse is injected to activate

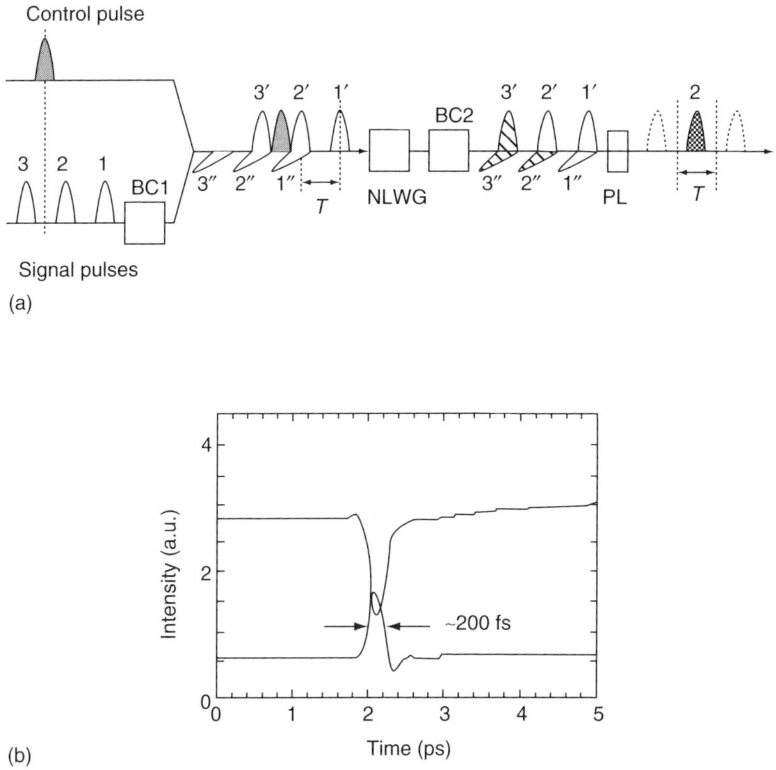


**FIGURE 2.11** (a) Schematic representation of a PD-SMZ all-optical switch. PL, polarizer. A control pulse is shown by shading. Hatching indicates that a signal component has experienced nonlinear phase shift. (b) Pump–probe trace of a PD-SMZ all-optical switch based on an InGaAsP NLWG. (From Nakamura S. et al., *IEEE Photon. Technol. Lett.*, 11, 1575, 1998.)

the nonlinearity, two orthogonally polarized pulses coming after the control signal experience phase shifts induced by the nonlinear change of the refractive index. Then, BC2 is used to recover the time difference and recombine the two orthogonal signals. The polarization of the output signal is determined by the phase difference between the two orthogonal components. Only those pulses for which the two orthogonally split components experience different phase shifts during the travel in NLWG, for example, the second pulse in Figure 2.11a, produce a polarization change in the output. Thus ultrafast switching similar to that in the original SMZ switch is achieved. Although the length of BC must be precisely determined in relation to the pulse repetition rate, the use of a single NLWG merits stable operation of the whole module. On the other hand, it may be worth noting that the same switching function can be realized by using a transverse-electric (TE)/transverse-magnetic (TM) beam splitter combined with Mach–Zehnder interferometer with different arm lengths (delays) instead of using BC, which is suitable for integrating the whole circuit.

Operation of the PD-SMZ switch has been demonstrated by a combination of InGaAsP bulk passive NLWG and calcite crystals, from which the shortest full width at half maximum (FWHM) pulse width of 200 fs has been measured at wavelengths of 1.73 and 1.52 μm for the control and signal, respectively, at the control pulse energy of 3 pJ, as shown in Figure 2.11b. Basic demultiplexing operation has also been demonstrated, as shown in Figure 2.12, for a series of pulses with 670 fs intervals, corresponding to 1.5 THz repetition signals [34]. Another experiment using PD-SMZ composed of SOA as the nonlinear element has shown successful error-free pulse regeneration operation, indicating squarelike modulation switching characteristics even at the bit rate of 84 Gb $s^{-1}$. Such an ultrafast operation speed is much faster than the carrier relaxation time in SOA (approximately 60 ps). In addition, the suitability of SMZ switches to integration and compact modules is very advantageous for achieving a much higher stability in comparison with the optical fiber–based counterparts using interferometer configuration. More recent studies have shown the applicability of SMZ switches for a variety of optical functional devices such as ultrafast 3R at 84 Gb $s^{-1}$ and wavelength conversion at 168 Gb $s^{-1}$ [35].

### 2.4.2 All-Optical Spin Polarization Switch

By using ultrafast relaxation of electron spin, an all-optical switch with a simple vertical structure can be designed. In semiconductor QW, since the degeneracy of heavy- and light-hole bands is split at the Γ point, it is possible to excite heavy-hole excitons selectively by illuminating light with a corresponding wavelength, as indicated in Figure 2.13. In addition, by using circularly polarized light, spin-polarized carriers can be excited because of the symmetry of the

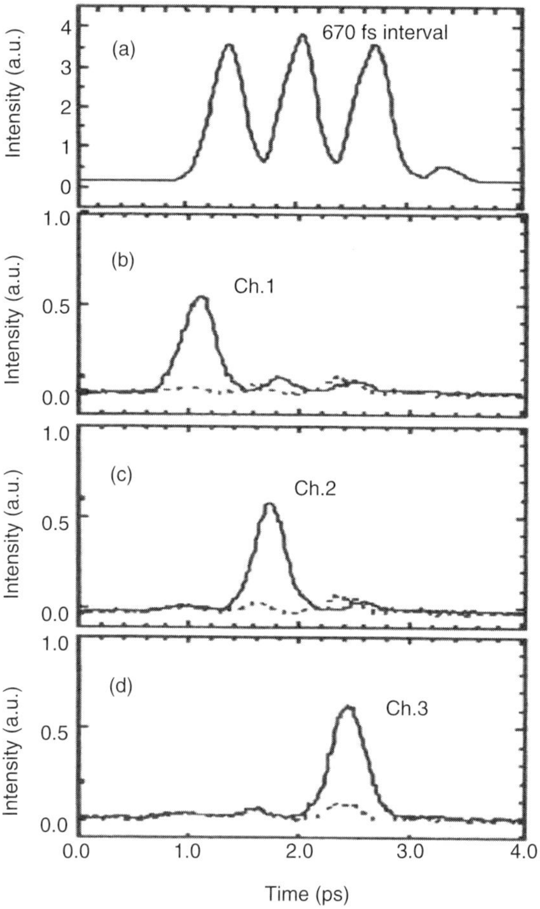


**FIGURE 2.12** Cross-correlation traces showing PD-SMZ-based demultiplexing operation. (a) A series of three signal pulses were input into the switch. (b)–(d) The demultiplexed outputs monitored by scanning the timing of the control pulse. Dotted lines are traces without control pulses indicating insignificant cross talk. (From Nakamura S. et al., *IEEE Photon. Technol. Lett.*, 11, 1575, 1998.)

crystal: the left (right) circularly polarized light excites the spin-up (spin-down) electrons. After the excitation terminates, electron spin begins to relax so as to reach an equilibrium state where spin polarization is 50% up and 50% down. The time for electron spin polarization to relax is in the range of picoseconds in GaAs MQWs at room temperature [36]. Electrons and holes generated in the structure eventually recombine in nanoseconds in usual semiconductors. Since the spin polarization causes circular dichroism, its relaxation produces a transient in the polarization sensitivity of the optical properties of the structure.

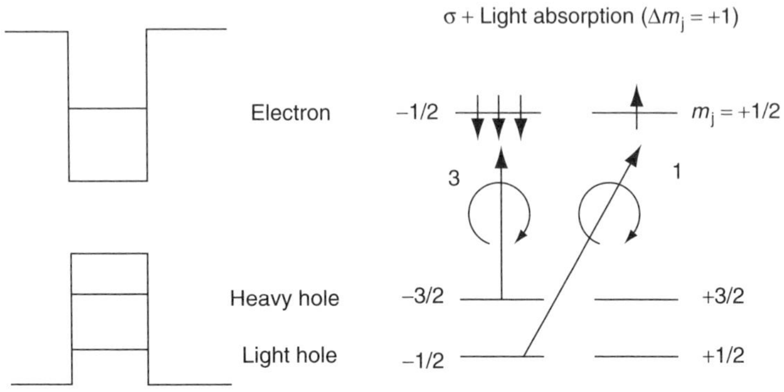


**FIGURE 2.13** Electronic states and selection rule of corresponding optical polarization transition in QW under excitation by right-circularly polarized light.

A typical example of the structure of a spin polarization all-optical switch is illustrated in Figure 2.14 [37] . This structure is essentially an etalon that is made by a pair of AlAs/(Al)GaAs distributed Bragg reflectors (DBRs) incorporating an AlGaAs/GaAs MQW. The configuration used for all-optical switching experiments is shown in Figure 2.15. Reflection (or transmission) from this etalon can be modulated by the excitation of MQW by a circularly polarized control (excitation) pulse at an appropriate wavelength. A signal (probe) pulse is linearly polarized—equivalent to a superposition of right and left circularly polarized lights—and fed into the etalon. Any polarization

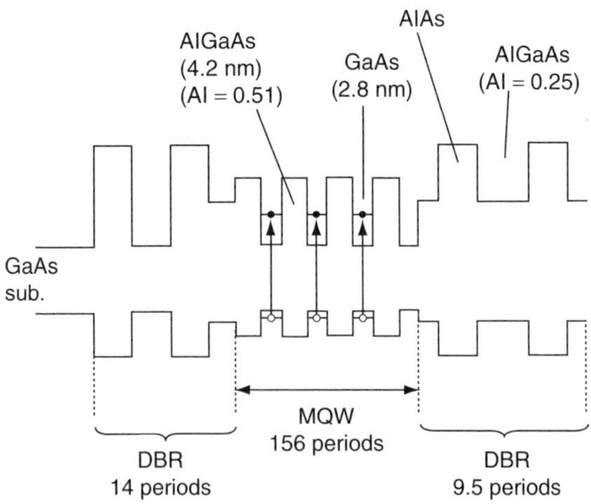


**FIGURE 2.14** Cross-sectional structure of an all-optical spin polarization switch.

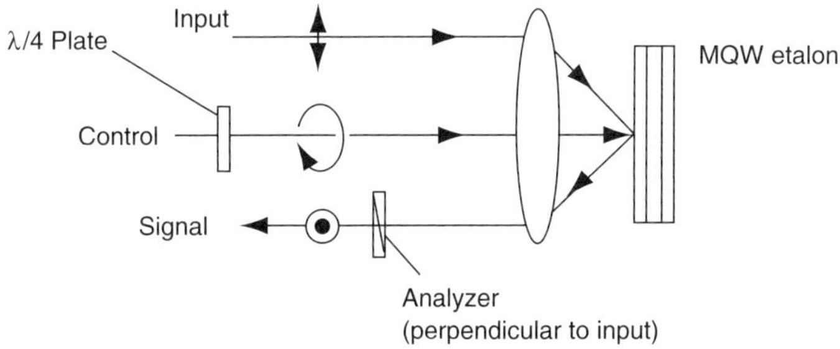


**FIGURE 2.15** Configuration used for all-optical switching by the MQW spin polarization switch.

change produced in the etalon by the control pulse causes elliptical polarization in the signal light, and this can be detected by an analyzer placed in the output signal path. The etalon structure is highly useful for enhancing the output signal intensity through the multiplication of absorption and phase changes during many reflections within the structure.

Figure 2.16 shows an example of the switching signal monitored in reflection configuration as shown in Figure 2.15. This indicates the switching pulse width to be 4 ps, which is consistent with an electron spin relaxation time of 28 ps determined from a separate pump–probe measurement. The switching energy estimated in this case is 50 fJ $\mu m^{-2}$, which is fairly low compared with other resonant nonlinearity-based devices. The total switching energy in this experiment was 9 pJ, but can be reduced by using a smaller beam size.

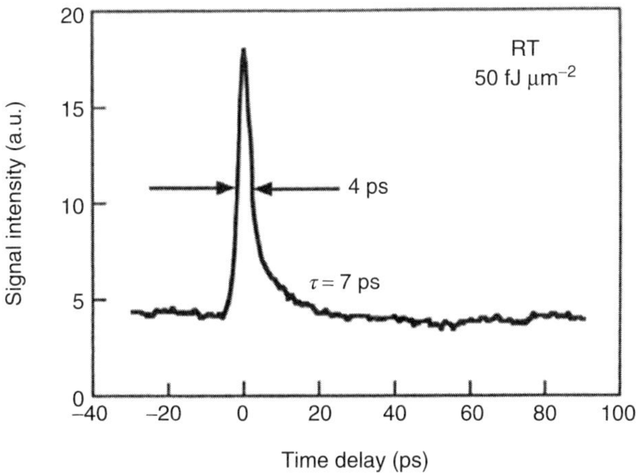


**FIGURE 2.16** Pump–probe trace of an AlGaAs/GaAs MQW all-optical spin polarization switch.

The contrast ratio obtained was 4:1, and this can be improved by increasing the total MQW thickness [37].

Knowledge of electron spin relaxation is important from the viewpoints of both low dimensional material physics and ultrafast device applications at different wavelengths. Figure 2.17 summarizes the spin relaxation time measured at room temperature as a function of the electron ground state confined energy measured from the conduction band edge for GaAs-based QWs and quantum wires (QWRs) [38] and InP-based QWs [39]. A slope of −2.2 obtained for GaAs-based MQWs is consistent with the D 'yakonov-Perel' (DP) mechanism in which spin–orbit interaction plays a dominant role. Our photoluminescence study on GaAs-based QWRs has shown an enhancement of the DP mechanism at lower dimensionality. It is interesting to note that InP-based MQWs exhibit shorter spin relaxation time. A smaller slope (−1) as found in Figure 2.17 suggests a different relaxation mechanism in this alloy system, probably caused by an enhanced contribution of the exchange interaction due to the Elliot–Yafet mechanism [36,40].

Such an ultrafast spin relaxation is independent of the usually long carrier lifetime (nanosecond or subnanosecond), and ultrafast switching is achieved. However, the repetition rate is still limited by the carrier lifetime.

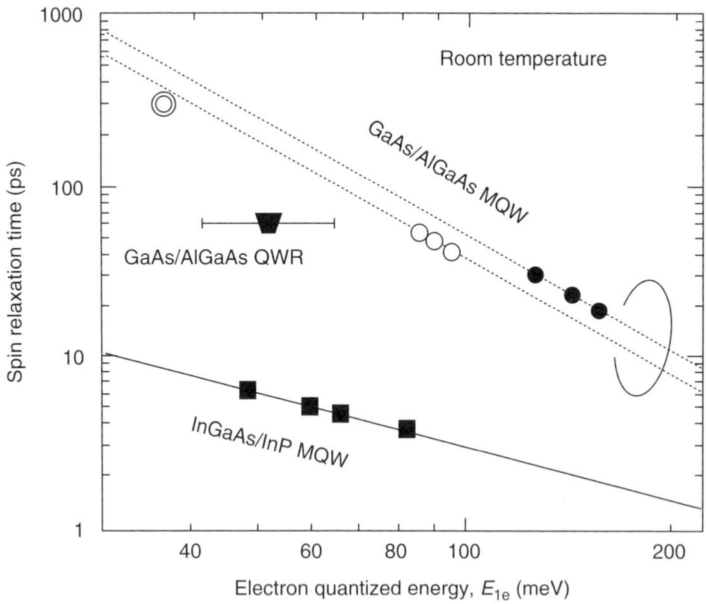


**FIGURE 2.17** Dependence of the spin relaxation time on the electron confined energy for the GaAs/AlGaAs and InGaAsP/InP MQWs and for GaAs/AlGaAs QWR, indicating the effects of the material and dimensionality.

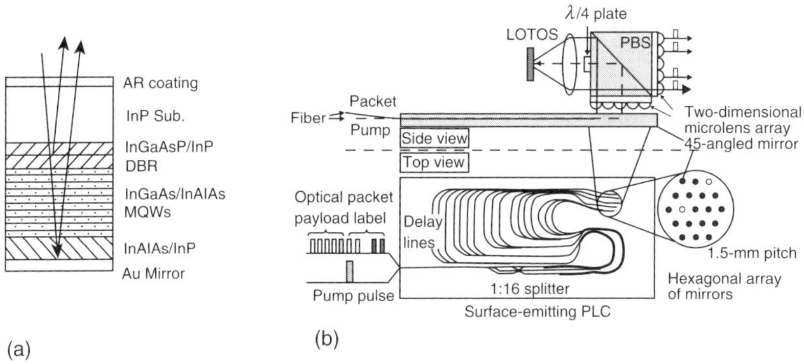


**FIGURE 2.18** (a) Cross-sectional structure of InGaAs/InAlAs MQW-based reflection-type spin polarization all-optical switch (LOTOS). (b) Side- and top-views of a 16-channel demultiplexing module using LOTOS. (From Takahashi, R. *Opt. Quantum Electron.* 33, 999, 2001; Takahashi R. et al., *Electron. Lett.* 39, 1135, 2003.)

The carrier lifetime can be reduced considerably by introducing defects in the crystal using proton bombardment, Be-doping, or low-temperature growth of the crystals [41,42]. Figure 2.18a shows the cross section of a low-temperature-grown InGaAs/InAlAs MQW etalon device, which is named low-temperature-grown total reflection optical switch (LOTOS). By introducing the polarization diversity configuration, polarization-insensitive switching operation together with an ultrashort pulse width of 300 fs and an extinction ratio greater than 30 dB has been demonstrated [41]. A signal repetition rate much greater than 100 GHz has been confirmed. Figure 2.18b shows the structure of a compact 16-channel demultiplexing module recently developed using the PLC technique [43].

### 2.4.3 ISBT All-Optical Switch

In conventional semiconductor optoelectronic devices, interband transition (IBT) is used and the response speed is determined by the electron hole recombination lifetime. Figure 2.19 illustrates a band diagram and possible transitions in a single QW (SQW) structure. ISBT in the conduction band can exhibit an ultrafast relaxation time unlimited by the carrier lifetime and also a large oscillator strength. A wide tunability is achieved by QW structure design using an appropriate material having a conduction band discontinuity large enough for the operation wavelength. Lasers [44] and photodiodes [45], which combine ISBT with the quantum cascade carrier transport in superlattice structures, have been demonstrated in the infrared region. Picosecond-class modulator and switch operation have also been shown in GaAs-based MQW in the 2.5 μm wavelength range [46]. Most of the previous works were conducted on conventional compound materials such as AlGaAs/GaAs and InGaAs/InAlAs systems.

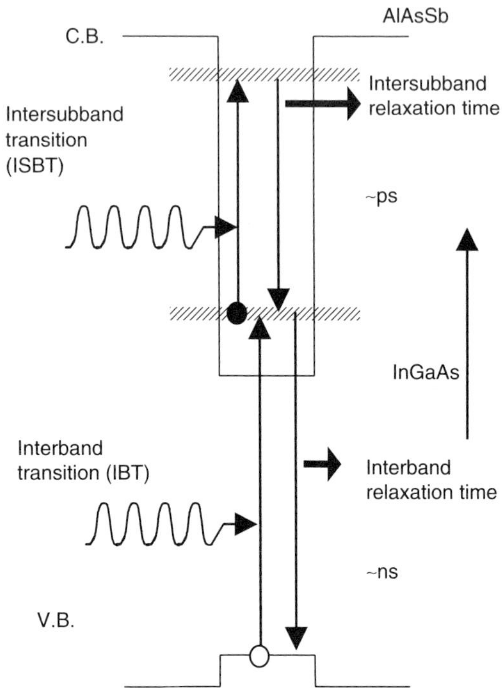


**FIGURE 2.19** Schematic diagram showing various transitions in an SQW. Ultrafast relaxation is expected for the ISBT in contrast with the IBT.

However, one of the most important issues in developing ISBT switches for practical applications would be the operation wavelength range covering 1.55 and 1.3 μm bands. In the following sections, we describe progresses in two different material systems that are being examined for 1.5 μm wavelength operation: the Sb-containing InP-based system and the gallium nitride–based system.

#### 2.4.3.1 InGaAs/AlAsSb System

The InGaAs/AlAsSb system is advantageous for application in long-wavelength devices because of its large conduction band discontinuity ($\Delta E_c = 1.6$–$1.75$ eV) [47] and small electron effective mass. In addition to this material property, the use of a coupled QW structure has been proposed to enhance the intersubband energy in QWs. Figure 2.20 illustrates the structure of a coupled-double-QW (C-DQW) structure consisting of two InGaAs wells separated by an AlAsSb barrier, which can be grown lattice-matched on an InP substrate [48,49] . The enhancement of intersubband energy is caused by the splitting of quantized electron states when the coupling between the wells becomes sufficiently strong as the barrier thickness is reduced. ISBT absorption spectra calculated for a doped ($n = 1 \times 10^{18}$ cm$^{-3}$) well using Bastard's

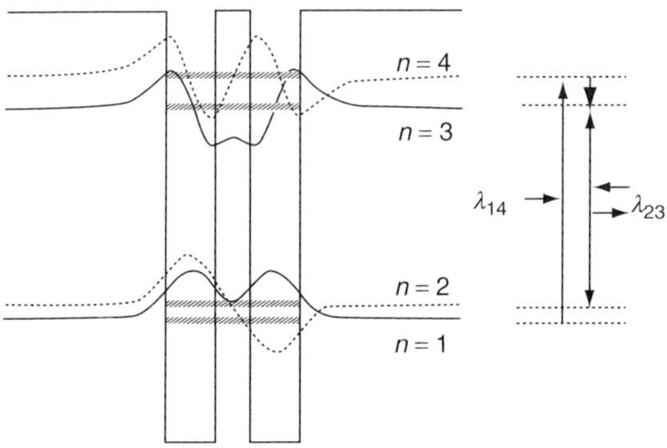


**FIGURE 2.20** Schematic band diagram of a C-DQW.

three-band model is shown in Figure 2.21 for different barrier widths. For appropriate barrier widths, two separate absorption peaks corresponding to the 1-4 and 2-3 transitions are observed. All-optical absorption modulation can be achieved by using the interaction between these two ISBTs. The population difference between the second and third subbands can be reduced by irradiating a light (control light) resonant to the 1-4 transition, assuming that the 4-3

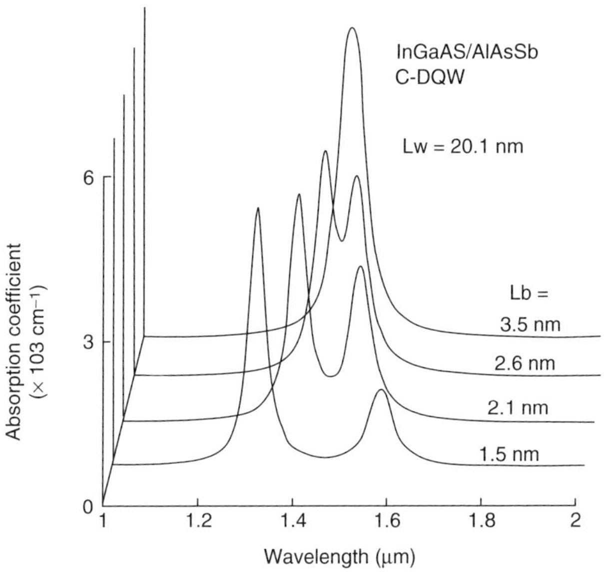


**FIGURE 2.21** Absorption spectra calculated for different barrier widths in InGaAs/AlAsSb C-DQW.

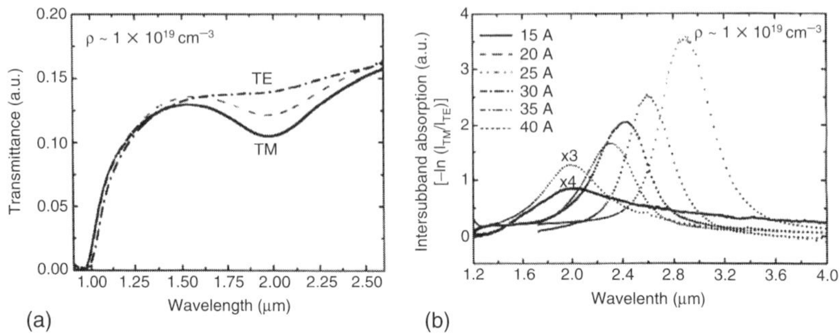


**FIGURE 2.22** (a) Absorption spectra for TE and TM polarization directions measured in an InGaAs/AlAsSb MQW. The well width is 2 nm and the barrier width is 8 nm. (b) Relative absorption spectra for various MQWs with different well widths in the range of 1.5–4 nm.

relaxation is sufficiently fast. This situation can be realized by designing the 4-3 energy difference larger than the longitudinal optical (LO) phonon energy. Thus the absorption for the 2-3 transition decreases and a modulation of the signal light is achieved. Although the absorption coefficient slightly decreases as the barrier thickness reduces, it is still larger than $10^3$ cm$^{-1}$, which is an order of magnitude larger than that of the 1-4 transition in a conventional uncoupled QW, and is sufficiently large for device operation. The design of the subband structure also has considerable influence on the response speed. The calculation of the rate equation has indicated that the response time is reduced to 300 fs (pulse half-width) by the optimization of the triple-coupled QW structure [50].

Experimental verification of all-optical switching in InGaAs/AlAsSb has been carried out by using SQW and C-DQW structures grown on (1 0 0) InP substrates by molecular beam epitaxy (MBE) technique. Figure 2.22a shows absorption spectra for TE and TM polarization directions measured in a 60-period InGaAs (2 nm)/AlAsSb (8 nm) MQW doped to $1 \times 10^{19}$ cm$^{-3}$. Absorption has been measured using the Fourier transform infrared on a sample with 45° facets to allow multiple internal reflections. This result indicates an absorption peak near 1.95 μm, and the observed polarization dependence confirms the origin of absorption to be ISBT. Figure 2.22b summarizes a series of absorption spectra for various MQWs with different well widths in the range of 1.5–4 nm [51].

Figure 2.23 shows a pump–probe measurement system used for evaluating the absorption recovery characteristics of ISBT. The pump pulse had a width of 150 fs and a wavelength of 1.95 μm to excite the intersubband energy, and the probe pulse with a spectrum broadened by a sapphire plate (about 1 μm) was used for monitoring the absorption change in the interband absorption in the InGaAs well, as indicated in the inset to Figure 2.23. Figure 2.24 shows an example of pump–probe experiments carried out on an 80-period

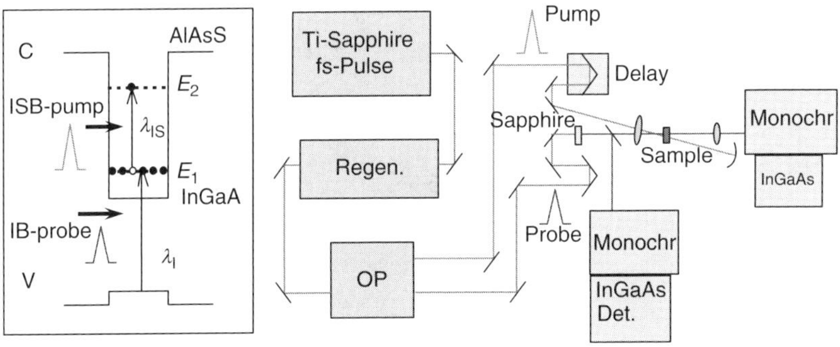


**FIGURE 2.23** Pump–probe measurement system used for evaluating the absorption recovery of ISBT. The inset indicates the functions of pump-and-probe beams. The optical parametric oscillator (OPA) output beam near 1.9 μm and the idler beam with wavelength broadened by a sapphire crystal around 1 μm are fed into an MQW sample.

InGaAs (2 nm thick, $1 \times 10^{19}$ cm$^{-3}$ doped)/AlAsSb MQW. The data show the interband (probe) absorption change at 1 μm as a function of delay after excitation with a 1.9 μm intersubband pump pulse, and the relaxation time was determined to be 2.1 ps [52].

To evaluate the nonlinearity, an absorption saturation measurement has been performed on the sample as described above [52]. An example of the results is shown in Figure 2.25, in which the absorption saturation is observed in the excitation power above 10 μW. Strong absorption observed at very high power is explained by the two-photon absorption in the material, predominantly in the InP substrate. The measured absorption

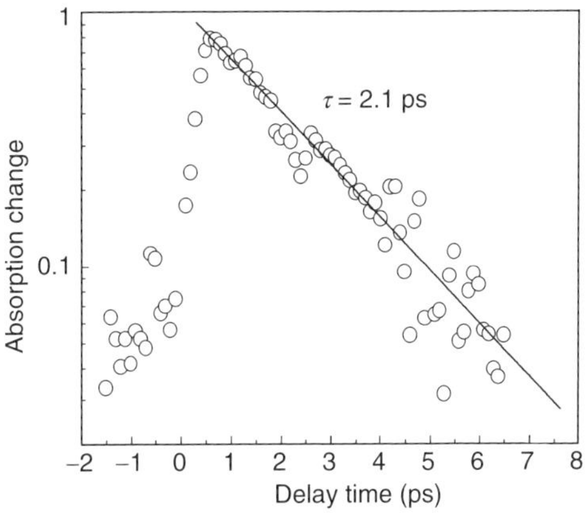


**FIGURE 2.24** Temporal change of intersubband absorption in InGaAs/AlAsSb MQW.

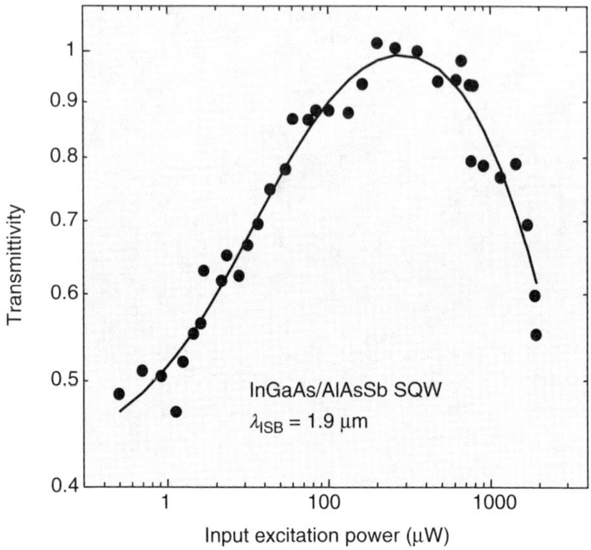


**FIGURE 2.25** Absorption saturation data using a femtosecond laser at 1.88 μm excitation. The solid line indicates a fit based on the thickness-dependent change of the incident pulse intensity after considering the substrate two-photon absorption.

characteristics have been fitted using the absorption coefficient as expressed by $\alpha(I) = \alpha(0)(1 + (I/I_s))^{-1} + \beta I$, where $I$ is the excitation energy, $I_s$ the saturation energy, and $\beta$ the two-photon absorption coefficient as previously reported. The absorption saturation power based on this relationship is about 50 MW $cm^{-2}$, corresponding approximately to 1 pJ $\mu m^{-2}$. This saturation energy has been found to be sensitive to the structure of QWs and the interface quality of the QWs, particularly the interface abruptness of composition, which would be easily degraded due to Sb interdiffusion during the MBE growth. A recent study [53] has shown that an SQW structure, grown by inserting thin AlAs interface layers for minimizing the interdiffusion effects, exhibits a very strong and sharp absorption peak with α larger than 10,000 $cm^{-1}$ and a half-width as small as 80 meV [53]. This QW structure has shown a record of lowest saturation energy of 3 fJ $\mu m^{-2}$, suggesting the suitability of ISBT for low-power all-optical switching devices [54].

All-optical switching demonstration has been carried out by using a device with ridge waveguide structure, in which a 80-period InGaAs/AlAsSb C-DQW active layer is sandwiched between InAlAs cladding layers as shown in Figure 2.26a. The pulse wavelength used is 1.55 μm, close to the 1-4 transition in this material. Figure 2.26b shows the result of the DEMUX operation, in which a series of four pulses with an equivalent repetition rate of 1 THz is demultiplexed into four different channels by changing the control

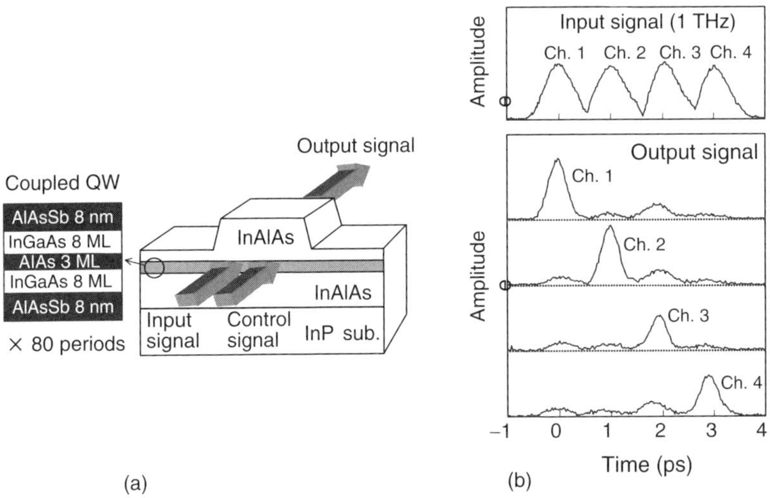


**FIGURE 2.26** (a) Structure of a ridge waveguide switch incorporating InGaAs/AlAsSb C-DQWs. (b) Input and output signal traces showing demultiplexing operation at an equivalent repetition rate of 1 THz.

pulse timing. The switching energy obtained is 27 pJ, and expected to be decreased by QW structure optimization [55].

### 2.4.3.2 GaN/AlGaN System and Other Material Systems

This material system has several advantages when applied to ISBT all-optical switches, including (1) extremely fast relaxation time down to 100 fs because of very strong electron–LO phonon interaction, (2) wide and continuous wavelength tunability because of direct transition nature over the entire composition range, (3) large absorption coefficient achievable by electron doping beyond $10^{19}$ $cm^{-3}$, and (4) the absence of two-photon absorption because of wide bandgap energy. On the other hand, piezoelectricity and spontaneous polarization effects are strong in this material system, and special attention should be drawn in designing the QW structures. Also various challenges have been examined to overcome the difficulty of crystal growth due to large lattice mismatch (2.5% in GaN/AlN). Both MBE and MOCVD have been used for growth on sapphire substrate [56,57].

Figure 2.27 shows the relationship between the ISBT wavelength versus the well thickness taking into account the built-in electric field at the interfaces. As shown in this figure, the wavelength is determined by the built-in field in thick QWs, because the quantized energy is determined by the slope of the potential in the QW, not by the well thickness. Very intense fields in the range of 1–3 MV $cm^{-1}$ in GaN/$Al_{0.65}Ga_{0.35}N$ and 5–6 MV $cm^{-1}$ in GaN/AlN have been found to exist at the interface. In the communication

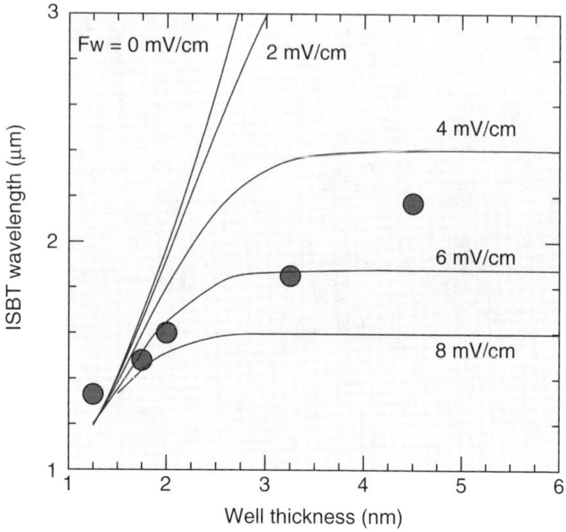


**FIGURE 2.27** Dependence of ISBT wavelength on the well thickness with built-in electric field in the well for GaN/AlGaN MQWs. Solid lines are the calculated data and circles are experimental data. (From Iizuka N. et al., *Optics Express*., 13, 3835, 2005.)

wavelength range, in contrast, the operation wavelength is determined almost solely by the well thickness, and the controllability and reproducibility of the growth technique in the nanometer range becomes critical. Transmission wavelength near 1.55 μm has been achieved by recent investigations [58].

Figure 2.28a shows the structure and Figure 2.28b shows the response time measurement result of a GaN/AlN QW waveguide switch most recently reported [58]. In the epitaxial growth, a 0.8 μm thick GaN layer was grown on a sapphire substrate by MOCVD, then a GaN(0.2 μm), two pairs of QW with $2 \times 10^{20}$ $cm^{-3}$ doping with Si, and GaN(1 μm) were successively grown by RF-MBE. A 1.5 μm wide ridge waveguide structure was formed by ECR-RIBE technique, and the waveguide length was 400 μm. The chip was moduled with a pair of polarization-maintaining fiber pigtails. The pump–probe experiment performed with a 1.7 μm, 150 pJ control pulse and a 1.55 μm, 5 pJ signal pulse has shown an FWHM pulse width as short as 230 fs [58]. The insertion loss has been estimated to be around 12 dB when the switch is ON state. Another absorption saturation measurement made on a chip not moduled with fibers has shown the saturation energy around 0.5 pJ $\mu m^{-2}$ at the wavelength of 1.48 μm [59]. Figure 2.29 summarizes the absorption recovery time as a function of ISBT energy including theoretical curves for both InGaAs/AlGaAs and GaN/AlGaN systems. It is noteworthy that the relaxation is more than one order of magnitude faster in nitride material compared with the GaAs-based system, although a sacrifice would be required in the switching energy.

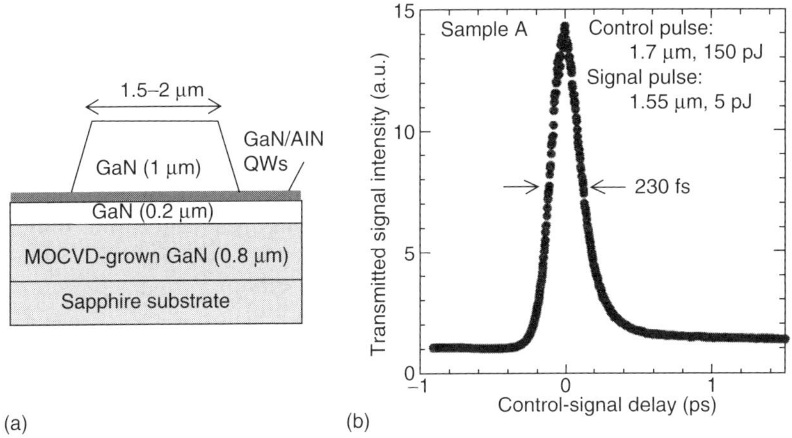


**FIGURE 2.28** (a) Structure of a ridge waveguide GaN/AlN QW ISBT switch. (b) Pump–probe trace of ISBT switch at signal wavelength of 1.55 μm, exhibiting a pulse width of 230 fs. (From Iizuka N. et al., *Optics Express*, 13, 3835, 2005.)

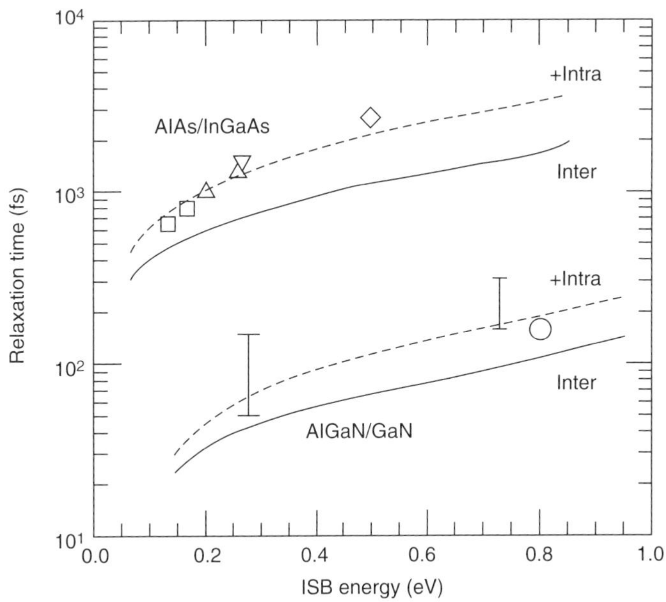


**FIGURE 2.29** Comparison of the intersubband relaxation time versus intersubband energy characteristics for InGaAs and GaN well structures, indicating faster relaxation in the GaN-based system by approximately 20 times. (From Iizuka N. et al., *Abstr. Int. Workshop Femtosecond Technol.*, p. 22, 2003.)

Another material system important for ISBT switch applications is II–VI compound alloys such as ZnSe/BeTe and CdS/BeTe, primarily due to very large conduction band discontinuity—2.3 eV for ZnSe and 3.1 eV for CdS. This material system can be grown on GaAs substrates by MBE. Recent research has shown that the introduction of a thin CdS layer in ZnSe/BeTe heterostructure can eliminate the slow carrier relaxation process of Γ(ZnSe)–X(BeTe) transfer that has been observed in ZnSe/BeTe QW [60]. Strong ISBT absorption at the wavelength of 1.55 μm has been demonstrated by a 40-period CdS(2.5 ML)/ZnSe(1 ML)/BeTe(10 ML) QW. Relaxation time monitored on a variety of CdS/ZnSe/BeTe QWs has indicated that ultrafast recovery in 1 ps without exhibiting slow tail is achieved at the wavelength range of 1.8–2.2 μm [60].

### 2.4.4 Wavelength Conversion Switches Using Four-Wave Mixing in SOAs

Wavelength conversion can be achieved by modulating an external light source Continuous Wave (CW) operating at a different wavelength by using cross-gain modulation and cross phase modulation in SOAs within a limit of the SOA modulation speed, which is usually below 100 Gb $s^{-1}$ [35]. Wavelength conversion based on coherent interactions such as SHG and difference frequency generation by using second-order nonlinear susceptibility $\chi^{(2)}$ in quasiphase matched $LiNbO_3$ devices [61,62] and nondegenerate four-wave mixing (NDFWM) using $\chi^{(3)}$ in InP-based SOAs [63,64] are more important for ultrafast operation beyond 100 Gb $s^{-1}$ in the communication wavelengths. We focus on the latter approach in this section.

Figure 2.30 shows the structure of a monolithically integrated distributed feedback (DFB) laser/SOA device enabling NDFWM-based wavelength conversion [65]. In this device structure, laser emission is used as a pump light source and injection of the signal pulse produces the conjugate output light at the converted wavelength, which is defined by $2\omega_p-\omega_s$, where $\omega_p$ and $\omega_s$ are pump (DFB laser) and input signal wavelengths, respectively. Although the converted signal intensity is 20 dB less than the amplified input, the SOA provides gain to recover this loss. This enables cascading such devices without additional amplifiers in a practical system. The pulse measurement has shown wavelength conversion of the subpicosecond pulse without any severe pulse broadening, confirming the advantage of coherent nature of the present device principle [66].

It is expected that the conversion efficiency is further improved over a wide range of detunings and also made independent of the detuning direction. QD material is promising in this respect due to its large oscillator strength and negligible line width enhancement factor. An experiment conducted to examine the capability of QD-SOA has shown promising results. Figure 2.31a illustrates the structure of the ridge waveguide QD-SOA device used and

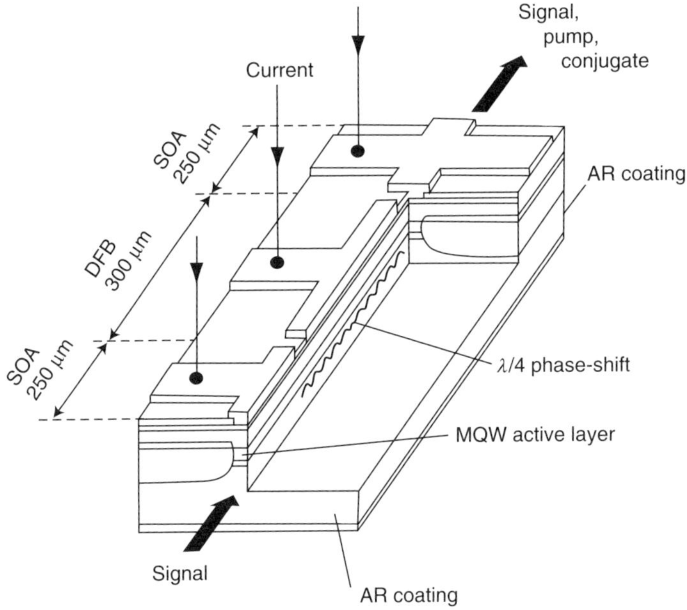


**FIGURE 2.30** Structure of a wavelength switch consisting of a (λ/4)-phase-shift DFB laser and SOAs integrated on an InP substrate. (From Shimoyama T. et al., *IEEE Photon. Technol. Lett.*, 12, 31, 2000.)

Figure 2.31b illustrates the spectra showing wavelength conversion in the wavelength region near 1.3 μm. Figure 2.32 depicts the wavelength conversion efficiency as a function of the detuning frequency between the pump and signal lights. A symmetric conversion efficiency, independent of the detuning direction, is achieved in QD-SOA in contrast with conventional bulk SOAs. In bulk SOAs where the line width enhancement factor is as large as 4, different nonlinear processes, namely carrier density pulsation and spectral hole burning, produce destructive interference, resulting in a decrease in the conversion efficiency, which is proportional to the square of $\chi^{(3)}$. In QD-SOAs, in contrast, the interference occurs constructively due to negligible line width enhancement factor and detuning direction-independent conversion is achieved [67,68].

## 2.4.5 All-Optical Switches Based on Novel Materials

### 2.4.5.1 Quantum Dot SOAs

Novel materials are required to be developed to fulfill the requirements of ultrafast device applications. In semiconductors, QD material has been extensively developed for application to uncooled lasers as well as ultrafast

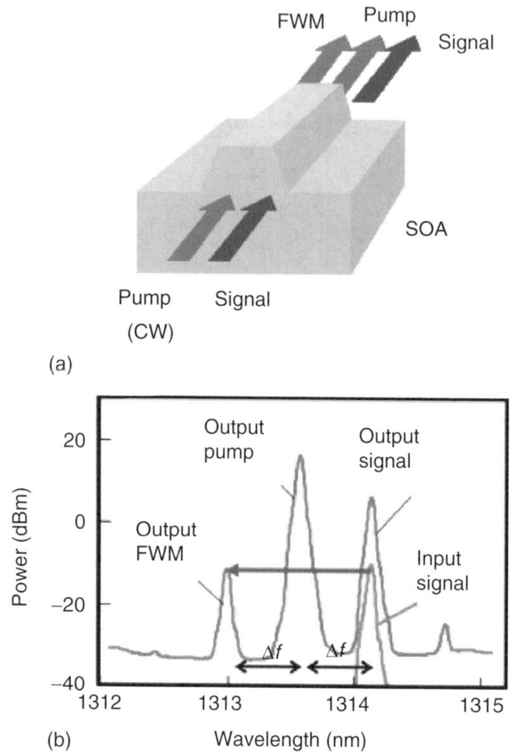


**FIGURE 2.31** (a) Schematic representation showing a QD-SOA wavelength conversion device. (b) Spectra showing wavelength conversion operation. (From Akiyama T. et al., *IEEE Photon. Technol. Lett.*, 14, 1139, 2002.)

SOAs at communication wavelength near 1.3 μm. Particularly, recent studies on QD-SOAs have shown various advantages, including ultrafast recovery time in the 1–3 ps range without exhibiting any pattern effect and also higher gain-saturation power compared to bulk SOAs [69]. Figure 2.33 shows output eye patterns of QD-SOA for different output power levels at the bit rate of 40 Gb $s^{-1}$. A clear eye opening and very little shift of the crossing point are found even in the nonlinear gain regime under high excitation, which is in a strong contrast with the case of bulk and quantum well SOAs as shown in the inset [70]. This is advantageous for the application of QD-SOAs for ultrafast switching. For example, wavelength conversion switching operation was achieved through cross-gain modulation by introducing a CW light and an ultrafast input light at a slightly different wavelength into a QD-SOA. An operation bit rate was expected to be at least 160 Gb $s^{-1}$ [70]. This high modulation speed is determined by the hole-burning effect in the QD system, where carriers are

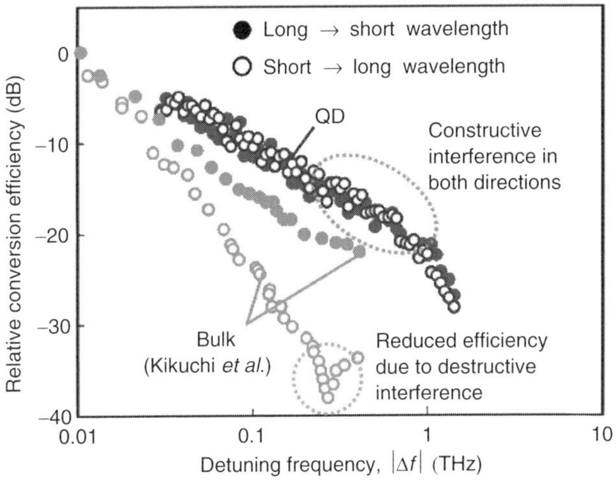


**FIGURE 2.32** Relative conversion efficiency as a function of the detuning frequency for a QD-SOA-based wavelength conversion device, indicating conversion direction-independent efficiency in QD-SOA. (From Akiyama T. et al., *IEEE Photon. Technol. Lett.*, 14, 1139, 2002.)

spatially separated from each other and carrier refilling is performed through direct transfer from the overlying cladding layer into the QDs, rather than the carrier-heating effect as often observed in bulk and quantum well SOAs.

More recent development of 1.55 μm QD-SOAs fabricated on InP substrates (cavity length = 6.5 mm, current density = 20–30 kA/cm$^2$) has shown a chip optical gain of 20 dB over a very wide wavelength range of 120–1460 nm owing to inhomogeneous size distribution of QDs, together with a noise figure less than 7 dB and a saturation gain over 19 dB. No pattern effect was observed up to 23.1 dBm as shown in Figure 2.34 [71].

In addition to such basic performance of SOA, whether the polarization insensitivity is established or not is critical in the application of devices to real systems. We have shown recently the possibility of adjusting the polarization sensitivity either by controlling the QD shape during the multilayer stacked growth process [72] or by controlling the strain through selecting the capping layer composition [73]. Figure 2.35 shows a result of the measurement on photoluminescence wavelength and polarization (defined by the emission intensity ratio in a form of TE/(TE + TM)) for columnar-shaped dots made of different stacking layers. As the stacking number increases and the dot shape varies, the polarization changes from TE-dominant to TM-dominant, and polarization-insensitive optical characteristics would be achieved by a spherical dot shape near the stack number around 7. This is a promising feature for designing polarization-insensitive SOAs for practical application.

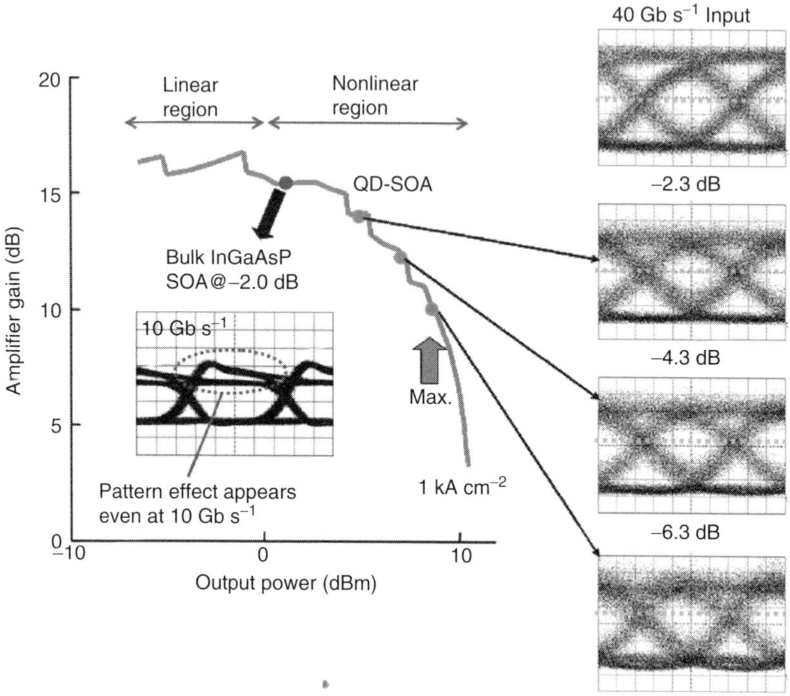


**FIGURE 2.33** Input signal and corresponding output signal eye patterns of QD-SOA for different power levels at the bit rate of 40 Gb s$^{-1}$. Clear eye openings are observed in both linear and nonlinear regions. (From Sugawara, M. et al., Proc., SPIE, 4905, 259, 2002.)

### 2.4.5.2 AlGaAs/GaAs Multiplayer System with Enhanced Nonlinearity

Another approach being developed focuses on nonlocal interaction in the confined exciton system. Based on a theory of Ishihara et al. [74], the third-order nonlinear optical response of excitons in semiconductor thin-film structure can exhibit a remarkable size dependence, resulting in a strong enhancement of the internal electric field and therefore the nonlinear response at particular film thickness, as has been shown by their early demonstrations [75,76]. Recently, they have combined this size-dependent electric field enhancement effect with the use of ultrafast nonlinearity based on two-photon absorption [77,78]. To avoid real carrier generation due to two-photon resonance, the bandgap energy of AlGaAs has been selected to be not more than twice the signal energy corresponding to 1.55 μm. An optical Kerr gate device structure similar to that used for spin polarization switches described earlier has also been introduced to enhance the refractive index variation induced by two-photon absorption. The device fabricated consists of 50 pairs of $Al_{0.18}Al_{0.72}As/AlAs$

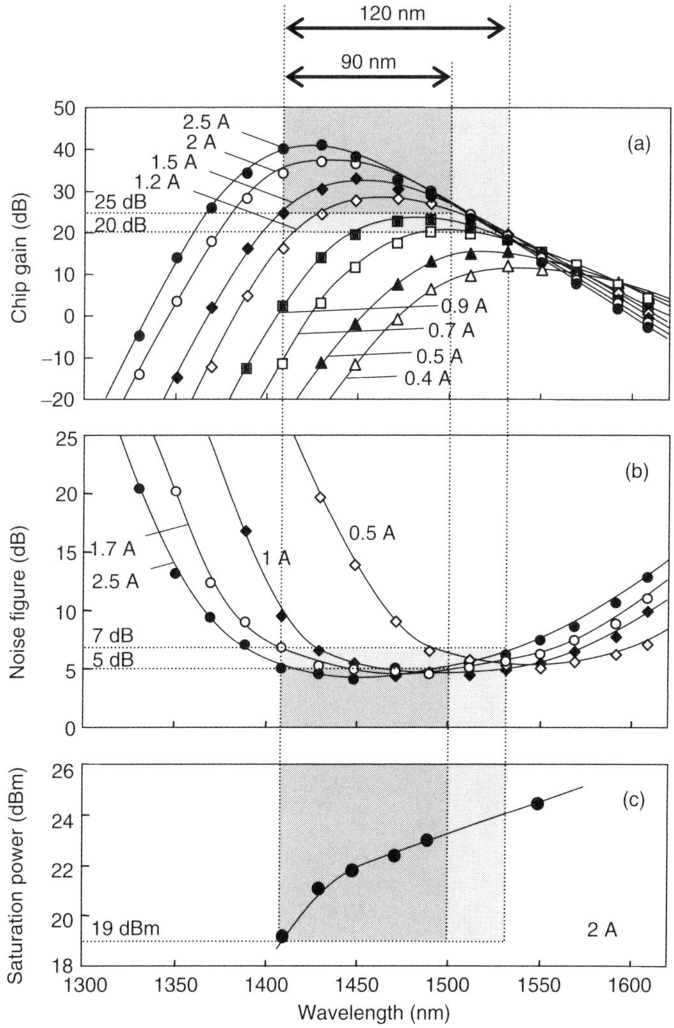


**FIGURE 2.34** Gain, noise, and saturation power as functions of operation wavelength for 1.55 μm QD-SOA. (From Akiyama T. et al., *IEEE Photon. Technol. Lett.*, 17, 1614, 2005.)

layers (each of the layer has a thickness of 100–120 nm), and the Kerr gate response has been measured by using two sequential control pulses in a cross-Nicol configuration using an analyzer to detect the normal polarization to the initial signal polarization.

Figure 2.36a shows transmission characteristics measured on this device, which demonstrates the enhancement of the electric field in the AlGaAs layer

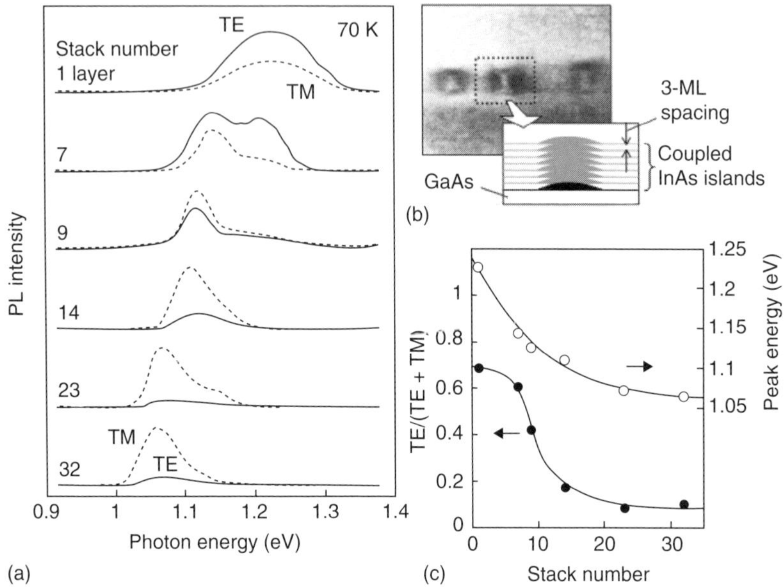


**FIGURE 2.35** (a) Dependence of TE and TM polarized photoluminescence spectra for a series of columnar-shaped quantum dots (QDs) with different stacking layer numbers. (b) SEM image of columnar dot. (c) A plot of emission wavelength and polarization determined for columnar QDs as functions of stacking layer number. (From Kita T. et al., *Jpn. J. Appl. Phys.*, 41, L1143, 2002.)

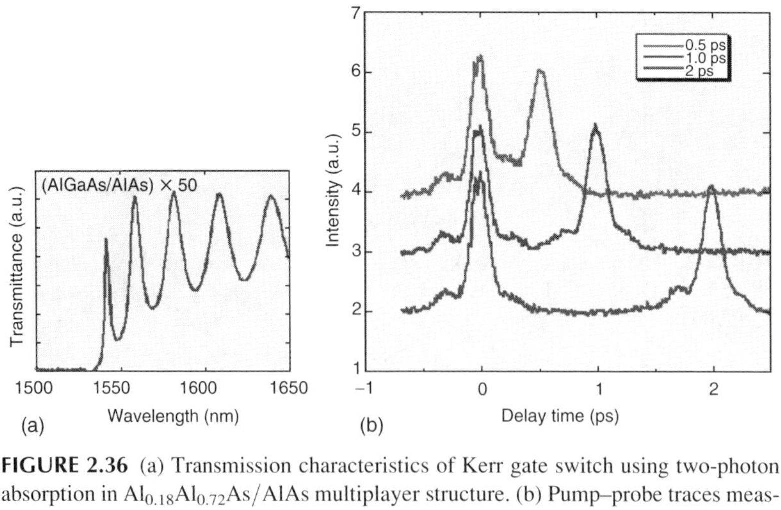


**FIGURE 2.36** (a) Transmission characteristics of Kerr gate switch using two-photon absorption in $Al_{0.18}Al_{0.72}As/AlAs$ multiplayer structure. (b) Pump–probe traces measured for two sequential control signals with different intervals at the wavelength near 1.55 μm. (From Isu T. et al., *Phys. Stat. Sol.(c)* 3, 671, 2006.)

and correspondingly the index change is expected to be enhanced at this wavelength. Figure 2.36b shows pump–probe traces of the Kerr gate signals for three different intervals between two control pulses near 1.55 μm. The signal FWHM width is as short as 200 fs and also no variation in the signal intensity is found, indicating no sign of pattern effect in the switching operation over 1 THz repetition rate. The switching energy required in a separate experiment was 15 pJ $\mu m^{-2}$ for an extinction ratio of 15 dB.

#### 2.4.5.3 Carbon Nanotubes

Recent researches on carbon nanotubes have shown their possibility of ultrafast optical switching. Single-walled nanotubes (SWNT) having semiconductor nature can exhibit an energy band structure with a bandgap energy determined primarily by the nanotube radius. Because the IBT is extremely fast in the range of $<1$ ps, SWNT can play an important role as an ultrafast all-optical switching material [79]. Initial application of carbon nanotubes as saturable absorbers in mode-locked lasers has recently been reported. Generation of pulses with FWHM of 315 fs at the wavelength of 1.56 μm has been demonstrated [80,81]. Further development of nanotubes both in material synthesis and optical characterization is required for the generation of useful ultrafast optical nonlinear materials.

#### 2.4.5.4 Organic Thin Films

Organic materials can be another attractive option due to their high nonlinearity. Also, they can be advantageous in forming large area films at low cost and are suitable for 2D array of all-optical switches. Molecular J aggregates such as pseudoisocyanine have been extensively studied for their large third-order nonlinear susceptibility $\chi^{(3)}$ and fast decay in bleached absorption via many exciton states and exciton–exciton annihilation. The structure of a J aggregate has been proposed to be a 1D molecular chain, and multiple molecules are excited coherently by a single photon. For developing practical J-aggregate films, an extensive study has been conducted recently on squarylium (SQ) dye derivates and the formation of stable J-ggregate films has been demonstrated by the simple spin-coating technique using two kinds of SQ derivatives [82].

Figure 2.37a shows the chemical formula of SQ dye-substituted with four butyl groups. SQ films with a typical thickness of 100 nm have been formed on glass substrates by spin-coating from 1 wt% SQ in 1,2-dichloroethane. Absorption spectra show a sharp J-band absorption centered typically at 776 nm, as shown in Figure 2.37b, whereas an SQ solution exhibits an absorption band centered at 640 nm for single molecules. The absorption coefficient determined at 776 nm is as high as $3.5\times10^5$ $cm^{-1}$, sufficiently high for designing thin-film switching devices. Transient response has been

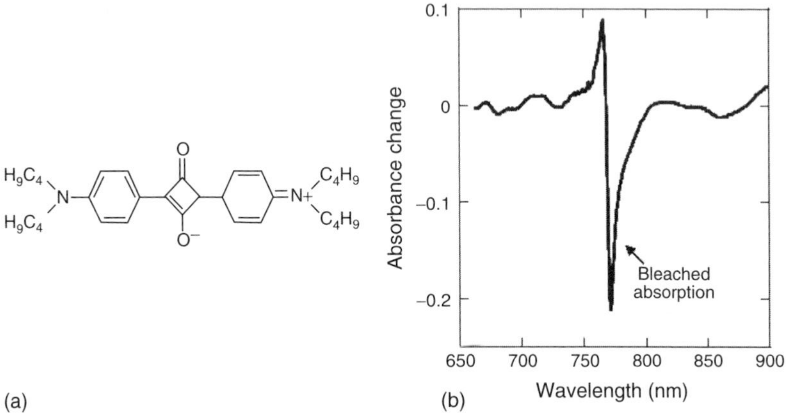


**FIGURE 2.37** (a) An example of chemical formula of SQ dye-substituted with four butyl groups. (b) Absorption change spectrum of an SQ dye thin film under femtosecond pulse excitation. (From Furuki M. et al., *Appl. Phys. Lett.*, 78, 2634, 2001.)

measured using a two-colour pump-and-probe system based on the combination of a Ti:S mode-locked laser and a pulse chirping prism pair. The fastest recovery time obtained by using an 820 nm excitation and 780 nm probe is 98 fs (half decay time), and more importantly the signal recovers to 90% in 320 fs and the recovery completes within 500 fs, as shown in Figure 2.38,

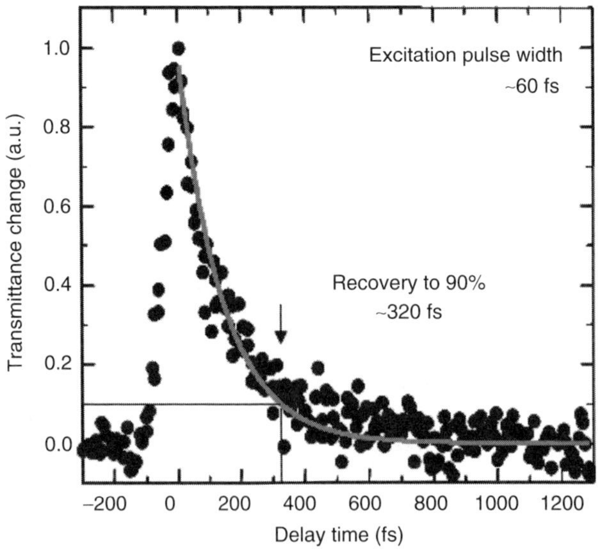


**FIGURE 2.38** Temporal transmission change of an SQ dye thin film measured under off-resonant excitation. (From Furuki M. et al., *Appl. Phys. Lett.*, 78, 2634, 2001.)

which confirms THz repetition rate operation [83]. Typical response time and the absorption modulation ($\Delta T/T$) measured near resonance are 200–300 fs and 30%–40%, respectively. The energy required for such a switching is as low as several tens of fJ $\mu m^{-2}$, which is as efficient as that in semiconductor QW materials as described in the previous sections. More precise evaluation of the nonlinearity has been performed by the Z-scan technique to show that the real and imaginary parts of $\chi^{(3)}$ are $-4.2 \times 10^{-8}$ and $-0.8 \times 10^{-7}$ esu, respectively [84].

Ultrafast demultiplexing operation has been demonstrated by using a system incorporating an SQ-film-based 2D all-optical switch as shown in Figure 2.39a [84,85] . The signal beam consisting of a series of femtosecond pulses with 1 ps intervals (equivalent to a 1 THz repetition rate pulse series) is fed into the 2D switch. A single shot of the control (pump or readout) pulse is obliquely fed into the film, and the film becomes transparent as the control pulse scans over the film. An additional delay plate is used for generating time delay in the perpendicular direction on the switch surface. Owing to such a time-of-fright effect, each of the signal pulses passes through different positions in the film and thus the serial pulses in the time domain are converted into

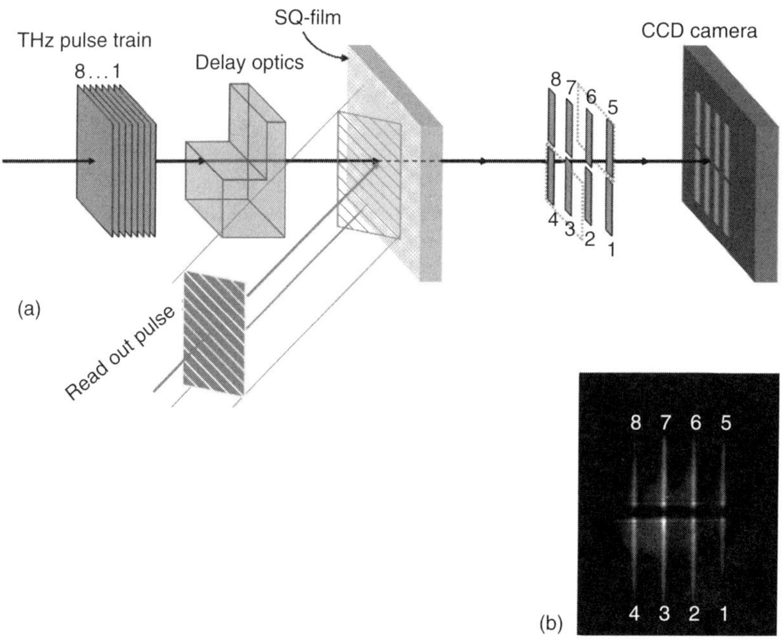


**FIGURE 2.39** (a) Setup for demultiplexing experiment based on the time-to-space conversion using an SQ dye thin-film switch. (b) Image pattern of the transmitted signal pulses spread on a 2D space through SQ switch. Spacing between serial pulse images is 1 mm in this case. (From Iwasa I. et al., *IEICE Trans. Electron.*, E85-C, 167, 2002.)

discrete images in the space domain. The oblique angle used is 17.5° and the spacing produced on the switch plane is 1 mm $ps^{-1}$. Figure 2.39b shows an image of the transmitted 8 bits of signal pulses detected by a CCD camera.

This result indicates the basic operation of demultiplexing operation at 1 Tb $s^{-1}$, although the signal extinction ratio should be still increased by improving the recovery time and nonlinearity. Such a DEMUX operation by a single shot of the control pulse can be extremely useful in ultrafast signal processing such as the header signal recognition as required in photonic routers for packet switching. In such an application, serial to parallel conversion is performed in real time by using this ultrafast 2D switch and then the comparator operation can be performed in a space-parallel manner at a slower speed. For the applications in practical systems, materials must be guaranteed for temperature stability. Preliminary aging test has been carried out on the SQ J-aggregate materials and an extrapolated lifetime at room temperature of 100 000 h [84]. Very recently, the possibility of shifting the operation wavelength to the long-wavelength region has been shown by synthesizing a new dye molecule, di(benzofranonyl) methanolate (BM), which exhibits an absorption peak at 1100 nm [86].

### 2.4.6 Current Status of Ultrafast All-Optical Switches

To provide an overview of the current status of development, Figure 2.40 summarizes various all-optical switch data displayed as a function of the power per bit versus the switching time as has been first used by Smith [87] . Some of the all-optical switch data plotted have been projected from the reported

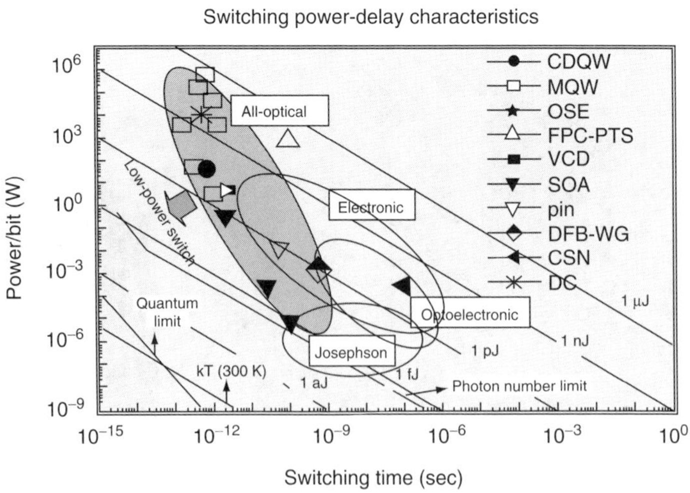


**FIGURE 2.40** A plot of the switching power per bit versus the switching time for different types of switching devices. (From Wada O., *Opt. Quantum Electron.*, 32, 453, 2000.)

response time and switching energy (e.g., absorption saturation energy). Other switching devices including electronic transistors, Josephson junction devices, and optoelectronic switches based on integrated device structures are included for comparison [88]. Recent studies on all-optical switches have shown a response time of 1 ps for the switching energy well below 1 pJ, as already described above, and this is very promising for the application to ultrafast systems beyond 100 Gb $s^{-1}$. This switching energy corresponds to the power consumption of 1 W in a chip when operated at the bit rate of 1 Tb $s^{-1}$, which would cause too much heat generation under continuous operation. Also, considering the plausible use of an integrated array of multichannel all-optical switches operating at Tb $s^{-1}$-class bit rate in the future systems, technical challenges must be directed toward the reduction of switching energy by at least an order of magnitude. Enhancement of nonlinearity without a sacrifice of response speed is mandatory. The introduction of novel concepts including PCs [89], QDs, etc., is expected to play an important role in this direction.

On the basis of compact, low-power ultrafast devices, it will become very important to develop system architecture for further improving the system throughput and flexibility, utilizing the wide bandwidth of ultrashort pulses. Figure 2.41 shows a conceptual drawing of an extremely high throughput OTDM–WDM mixed optical interconnection system based on femtosecond lasers, all-optical switches, and waveguide-based optical integration technology [88]. Femtosecond pulses generated in a laser are multiplexed in time by a waveguide multiplexer and then separated into

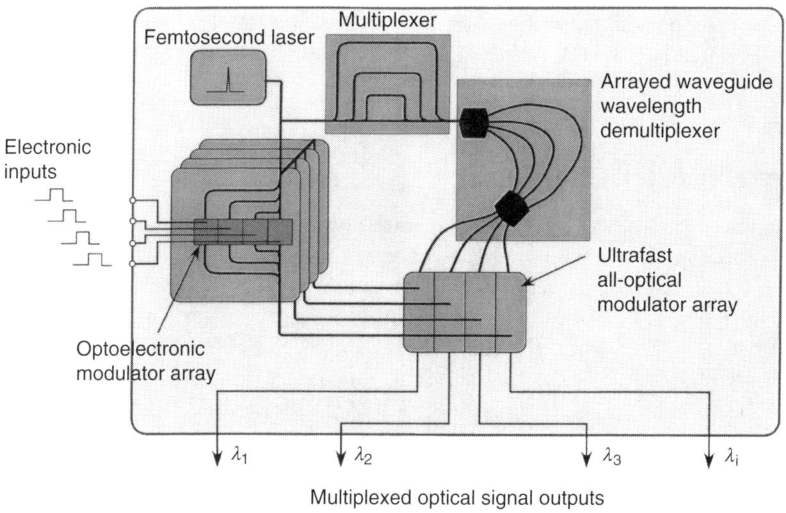


**FIGURE 2.41** Conceptual schematic of an OTDM–WDM mixed architecture of ultrahigh throughput optical interconnection system.

different wavelength channels by an arrayed grating wavelength demultiplexer. On the other hand, coding of ultrashort pulses can be carried out by external electronic signals at a slow rate (equal to the laser repetition rate) using arrays of optoelectronic modulators combined with waveguide delay lines. These coded signals are used as the control signals for arrayed all-optical modulators, so that all electronic signal channels are coded in time and wavelength divisions. When optimized, such mixed use of OTDM and WDM is expected to provide an extremely high throughput as well as a high flexibility of signal coding. Also, the combined use of optoelectronic and all-optical switching schemes is expected to merit the realization of reconfigurable optical interconnection and signal-processing systems.

## 2.5 CONCLUSIONS

The development of compact, stable, ultrafast all-optical devices is a key tool for constructing Tb $s^{-1}$ optical communication systems to meet the increasing needs of information throughput. As has been reviewed in this paper, a variety of all-optical devices for ultrafast pulse generation, dispersion and delay control, switching, reamplification, and wavelength conversion have been demonstrated at a bit rate of 100 Gb $s^{-1}$–1 Tb $s^{-1}$. Some of the devices have reached a technical level where the fabrication cost is the most significant issue for real application at moderate bit rates. Toward Tb $s^{-1}$-class applications, however, ultrafast devices require a still higher repetition rate and lower power consumption. Future technical challenges should continue to exploit novel ultrafast phenomena, materials, device concepts, and also system architecture compatible with all-optical devices, and this will lead to practical ultrafast devices and systems to support the overall throughput in the range of 1–10 Tb $s^{-1}$.

## ACKNOWLEDGMENTS

A major part of this study was supported by the New Energy and Industrial Technology Development Organization within the framework of the Femtosecond Technology (FST) Project. The author is pleased to acknowledge the contributions of his collaborators in the FST project, particularly F. Saito, T. Sakurai, A. Suzuki, K. Asakawa, H. Ishikawa, T. Mozume, N. Georgiev, H. Yoshida, A. Neogi, S. Lan, A.V. Gopal, T. Akiyama, T. Nishimura, A. Tackeuchi, and S. Nishikawa at FESTA Laboratories at Tsukuba. Collaboration with researchers at other FESTA research units, particularly Y. Ogawa (Oki), Y. Lee (Hitachi), S. Wakabayashi (Matsushita), K. Tajima (NEC, currently Tokushima University), and S. Nakamura (NEC), N. Suzuki (Toshiba), T. Isu (Mitsubishi, currently Tokushima University), M. Sugawara (Fujitsu), and L.S. Pu and M. Furuki (Fuji Xerox), is also greatly appreciated.

## REFERENCES

1. Wada, O. et al., 1999. *Femtosecond Technology—From Basic Research to Application Prospects* (eds.) T. Kamiya, F. Saito, O. Wada, and H. Yajima (Berlin: Springer) ch III-1 pp. 59–78.
2. OITDA (www.oitda.or.jp/), 2004. Optoelectronics Industry Trends *OITDA Activity Report* OITDA (www.oitda.or.jp/)
3. Saruwatari, M., 2000. *IEEE J. Sel. Top. Quantum Electron.*, 6, 1363.
4. Nakazawa, M. et al., 2000. *Electron. Lett.*, 36, 2027.
5. Morioka, T. et al., 2004. *Opt. Rev.*, 11, 153.
6. Miller, A, Reid, D.T. and Finlayson, D.M. (ed.), 2004. *Ultrafast Photonics* (Bristol: Institute of Physics).
7. Saito, F., 2002. *IEICE Trans. Electron.*, E85-C, 106.
8. Yajima, H. et al., 1998. *IEICE Trans. Electron.*, E81-C, 95.
9. Diels, J-C., and Rudolph, W., 1996. *Ultrashort Laser Pulse Phenomena—Fundamentals, Techniques, and Applications on a Femtosecond Time Scale* (New York: Academic).
10. Kamiya, T., 1999. *Femtosecond Technology—From Basic Research to Application Prospects* (eds.) T. Kamiya, F. Saito, O. Wada, and H. Yajima (Berlin: Springer) ch I-1 pp. 1–11.
11. Nuss, M.C. et al., 1996. *Electron. Lett.*, 32, 1311.
12. Agrawal, G.P., 2002. *Fiber-optic Communication Systems* (New York: Wiley).
13. Chen, Y.K. et al., 1991. *Appl. Phys. Lett.*, 58, 1253.
14. Arahira, S. et al., 1994. *Opt. Lett.*, 19, 834.
15. Arahira, S. et al., 1994. *Tech. Dig. SSDM 2000* C-1–2.
16. Kunimatsu, D. et al., 1999. *IEEE Photon. Technol. Lett.*, 11, 1363.
17. Arahira, S. et al., 2000. *Electron. Lett.*, 36, 454.
18. Arahira, S. et al., 2002. *IEEE Photon. Technol. Lett.*, 14, 537.
19. Arahira, S. et al., 1996. *IEEE J. Quantum Electron.* 32, 1211.
20. Arahira, S. et al., 2000. *IEICE Trans. Electron.*, E83-C, 966.
21. Matsui, Y. et al., 1999. *IEEE Photon. Technol. Lett.*, 11, 1217.
22. Tsuchiya, M. et al., 2001. *Opt. Quantum Electron.* 33, 751.
23. Arahira, S. et al., 1998. *J. Lightwave Technol.*, 16, 405.
24. Arahira, S. et al., 2001. *Quantum Electron.* 33, 691.
25. Pachel, V. et al., 1995. *Appl. Phys. Lett.*, 67, 2111.
26. Lee, Y., 1998. *Appl. Phys. Lett.*, 73, 2715.
27. Lee, Y., 2002. *IEICE Trans. Electron.*, E85-C, 190.
28. Joannopoulos, J.D., Meade, R.D., and Winn, J.N., 1995. *Photonic Crystals—Modeling the Flow of Light* (Princeton, NJ: Princeton University Press).
29. Nishikawa, Y. et al., 2002. *Opt. Lett.*, 27, 2079.
30. Wakabayashi, S. et al., 2002. *IEICE Trans. Electron.*, E85-C, 135.
31. Tajima, K. et al., 1993. *Jpn. J. Appl. Phys.*, 32, L1746.
32. Nakamura, S. et al., 2000. *IEEE Photon. Technol. Lett.*, 12, 425.
33. Nakamura, S. et al., 2002. *Abstr. Int. Workshop Femtosecond Technol. (June 2002, Tsukuba)*, p. 61.
34. Nakamura, S. et al., 1998. *IEEE Photon. Technol. Lett.*, 11, 1575.
35. Ueno, Y. et al., 2001. *IEEE Photon. Technol. Lett.*, 13, 469.
36. Tackeuchi, A. et al., 1999. *Jpn. J. Appl. Phys.*, 38, 4680.

37. Nishikawa, Y. et al., 1996. *IEEE J. Sel. Top. Quantum Electron.* 2, 661.
38. Nishimura, T. et al., 1999. *Jpn. J. Appl. Phys.*, 38, L941.
39. Tackeuchi, A. et al., 1997. *Appl. Phys. Lett.*, 70, 1131.
40. Wada, O. et al., 1998. *Proc. SPIE*, 3283, 602.
41. Takahashi, R., 2001. *Opt. Quantum Electron.* 33, 999.
42. Iwamura, H. et al., 1999. *Femtosecond Technology—From Basic Research to Application Prospects* (eds) T. Kamiya, F. Saito, O. Wada, and H. Yajima (Berlin: Springer) ch III-5 pp 122–33.
43. Takahashi, R. et al., 2003. *Electron. Lett.*, 39, 1135.
44. Faist, J. et al., 2000. *Intersubband Transitions in Quantum Wells—Physics and Device Applications II* (New York: Academic) ch 1 pp. 1–84.
45. Liu, H.C., 2000. *Intersubband Transitions in Quantum Wells—Physics and Device Applications I* (New York: Academic) ch 3 pp. 129–96.
46. Asano, T. et al., 1998. *Jpn. J. Appl. Phys.*, 37, 6020.
47. Georgiev, N. et al., 2001. *J. Appl. Phys.*, 89, 1064.
48. Yoshida, H. et al., 1998. *Electron. Lett.*, 34, 1008.
49. Yoshida, H. et al., 1999. *Electron. Lett.*, 35, 1103.
50. Gopal, A.V. et al., 2002. *Tech. Dig. OFC2002* paper TuF1.
51. Neogi, A. et al., 2000. *Physica* A, 7, 183.
52. Gopal, A.V. et al., 2002. *IEEE J. Quantum Electron.* 38, 1515.
53. Mozume, T. et al., 2001. *J. Cryst. Growth*, 227–228, 577.
54. Gopal, A.V. et al., 2003. *IEEE J. Quantum Electron.* 39, 299.
55. Akiyama, T. et al., 2002. *IEEE Photon. Technol. Lett.* 14, 495.
56. Iizuka, N. et al., 2000. *Appl. Phys. Lett.*, 77, 648.
57. Gmachl, C. et al., 2000. *Appl. Phys. Lett.*, 77, 3722.
58. Iizuka, N. et al., 2005. *Optics Express*. 13, 3835.
59. Iizuka, N. et al., 2003. *Abstr. Int. Workshop Femtosecond Technol. (July 2003, Makuhari)* p. 22.
60. Akimoto, R. et al., 2002. *Appl. Phys. Lett.*, 81, 2998; Akimoto, R. et al., 2002 *Abstr. Int. Workshop Femtosecond Technol. (June 2002, Tsukuba)*, pp. 83–86.
61. Stegeman, G.I. et al., 1996. *Opt. Quantum Electron.* 28, 1691.
62. Kanbara, H. et al., 1999. *IEEE Photon. Technol. Lett.*, 11, 321.
63. Kikuchi, K. et al., 1992. *IEEE J. Quantum Electron.* 28, 151.
64. Kuwatsuka, H. et al., 1997. *IEEE J. Quantum Electron.* 33, 2002.
65. Shimoyama, T. et al., 2000. *IEEE Photon. Technol. Lett.*, 12, 31.
66. Kuwatsuka, H. et al., 2000. *Proc. ECOC2000 (Sept. 2000, Munich)* paper 7.4.3, p. 64.
67. Akiyama, T. et al., 2002. *IEEE Photon. Technol. Lett.*, 14, 1139.
68. Akiyama, T. et al., 2000. *Appl. Phys. Lett.*, 77, 1753.
69. Akiyama, T. et al., 2002. *Electron. Lett.*, 38, 1139.
70. Sugawara, M. et al., 2002. *Proc. SPIE*, 4905, 259.
71. Akiyama, T. et al., 2005. *IEEE Photonics Tech. Lett.*, 17, 1614.
72. Kita, T. et al., 2002. *Jpn. J. Appl. Phys.*, 41, L1143.
73. Jayavel, P. et al., 2004. *Appl. Phys. Lett.*, 84, 1820.
74. Ishihara, H. et al., 1996. *Phys. Rev.* B, 53, 15823.
75. Ishihara, H. et al., 2002. *Phys. Rev. Lett.*, 89, 017402.
76. Akiyama, K. et al., 2002. *Physica* B, 314, 293.
77. Isu, T. et al., 2005. Inst. Phys. Conf. Series No. 184, Section I (IOP Publishing Ltd.) p. 119 (2005).

78. Isu, T. et al., *Phys. Stat. Sol. (c)*, 3, 671, 2006.
79. Chen, Y-C. et al., 2002. *Appl. Phys. Lett.*, 81, 975.
80. Set, S.Y. et al., 2003. *Proc. OFC2003 (March 2003, USA)* paper FL2.
81. Set, S.Y. et al., 2003. *Proc. OFC2003 (March 2003, USA)* postdeadline paper PD44.
82. Tatsuura, S. et al., 2001. *Appl. Phys. Lett.*, 79, 2517.
83. Furuki, M. et al., 2001. *Appl. Phys. Lett.*, 78, 2634.
84. Iwasa, I. et al., 2002. *IEICE Trans. Electron.* E85-C, 167.
85. Furuki, M. et al., 2000. *Appl. Phys. Lett.*, 77, 472.
86. Furuki, M. et al., 2004. *IEICE Trans. Electron.* E-87-C, 1161.
87. Smith, P.W., 1972. *Bell System Tech. J.*, 61, 1975.
88. Wada, O., 2000. *Opt. Quantum Electron.* 32, 453.
89. Kanamoto, K. et al., 2004. *IEICE Trans. Electron.* E87-C, 1142.

# 3 Direct Modulation of Injection-Locked Semiconductor Lasers

*Erwin K. Lau, Hyuk-Kee Sung, and Ming C. Wu*

## CONTENTS


3.1 Introduction ........ 77
3.2 History ........ 81
3.3 Fundamentals of Optical Injection Locking ........ 84
  3.3.1 Theory ........ 84
    3.3.1.1 Locking Range ........ 85
    3.3.1.2 Resonance Frequency Enhancement ........ 86
    3.3.1.3 Phasor Model ........ 89
  3.3.2 Modulation Frequency Response ........ 91
    3.3.2.1 Strong Injection Locking ........ 92
    3.3.2.2 Injection Locking of VCSELs ........ 95
3.4 Applications ........ 96
  3.4.1 Analog Link Performance Improvement ........ 96
    3.4.1.1 Link Linearity Improvement ........ 96
    3.4.1.2 Link Gain Improvement by Injection-Locked Gain-Lever DBR Laser ........ 97
  3.4.2 Millimeter-Wave Generation ........ 98
  3.4.3 Optical Injection Phase-Locked Loop ........ 100
  3.4.4 Injection Locking of Mode-Locked Lasers ........ 101
  3.4.5 All-Optical Signal Processing ........ 103
  3.4.6 Other Applications ........ 103
3.5 Future Directions ........ 104
3.6 Conclusion ........ 104
References ........ 105


## 3.1 INTRODUCTION

Directly modulated (DM) semiconductor lasers are compact, low-cost transmitters for both digital and analog photonic communication systems. However, their use in high-performance analog photonic systems is limited by several performance issues listed in Table 3.1. As shown in this table, optical

**TABLE 3.1**
**Limitations of Directly Modulated Lasers and Improvements by Optical Injection Locking**

| | Fundamental Limits | Benefit from OIL |
|---|---|---|
| Laser | Mode partition noise (Fabry–Perot laser) | Single mode with side-mode suppression [1] |
| | Relaxation oscillation frequency | Enhanced relaxation oscillation frequency [2–4] |
| | Nonlinear electron–photon coupling | Reduced nonlinearities [5,6] |
| | Amplified spontaneous emission noise | Reduced RIN [6–10] |
| | Wavelength chirp (nonzero $\alpha$ parameter) | Reduced chirp [11–13] |
| Link | Differential quantum efficiency $< 1$ | Increased link gain [14] |
| | Double-sideband modulation | Near-single-sideband modulation [15] |

injection– locking (OIL) systems can improve a host of fundamental limitations of directly modulated lasers and links: single-mode performance and side-mode suppression [1], enhanced bandwidth and relaxation oscillation frequency [2–4], suppressed nonlinear distortion [5,6], reduced relative intensity noise [6–10], reduced chirp [11–13], increased link gain [14], and near-single-sideband modulation [15]. In addition to improving the performance of optical communication links, injection-locked laser systems have many other unique properties. These properties make OIL attractive for applications such as optical frequency reference generation [16], phased-array radars [17], phase modulation [18], and optical signal processing [19], among others.

Figure 3.1a shows an experimental schematic of an injection-locked laser system. The light of a master laser is injected into a slave laser. The light from the slave is the useful output of the system. An isolator is placed between the master and the slave to eliminate light coupling back to the master. There are two possible configurations of injection locking, depending on the choice of outputs of the slave. In transmission-style injection locking (Figure 3.1a), the injected master light enters one slave laser facet and the output is taken from the other facet. This necessitates two coupling systems on the slave laser alone. To simplify the system, a reflection-style setup is used (Figure 3.1b). The output is taken at the same facet as the input of the injected light. An optical circulator is used to ensure only the output beam goes to the photodetector. The reflection-style system output is susceptible to noninjected master light coupling to the output when the incident master light reflects off the slave laser facet. This is important only in the strong injection regime (see Section 3.5) and not an issue for transmission-style implementations. It can be minimized by applying antireflection coating to the slave facet. Additionally, in a fiber system, a polarization controller is necessary to ensure

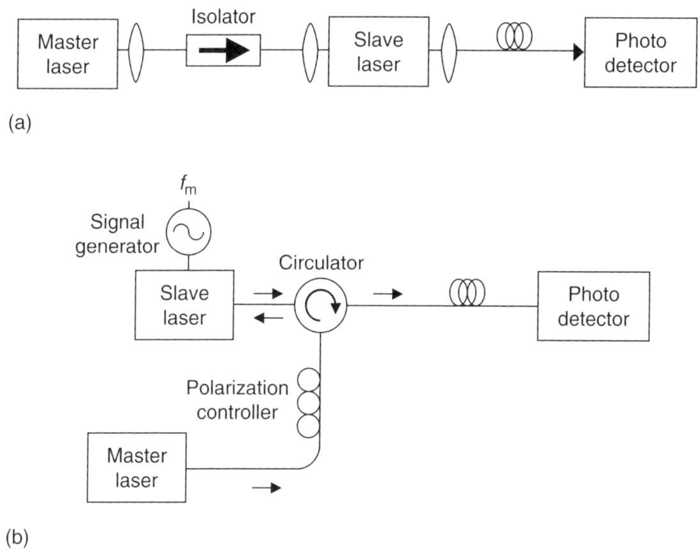


**FIGURE 3.1** Schematic of optical injection–locked laser system: (a) transmission-style (b) reflection-style.

that the master and slave polarizations are matched. Figure 3.1b also shows that direct modulation is typically applied to the slave laser.

When injection-locked, the slave's lasing wavelength is locked to the master's. Figure 3.2a shows the spectrum of a free-running single-mode (SM) laser. Light from the master is then injected into the slave, not necessarily at the same wavelength. Figure 3.2b shows both original slave mode and injected master light before locking. Finally, when the dynamics of the laser settle, the slave wavelength is pulled toward the master wavelength until it equals that of the master, locking both its frequency and phase. Now, if the master laser frequency is changed, the slave will track this frequency until the difference between master and free-running frequencies (detuning frequency, $\Delta f$) becomes too large. At this point, the slave unlocks from the master and lases at its natural wavelength. The span of frequencies that result in a locked state is the locking range. The locking range typically becomes larger as the ratio of master and slave optical powers defined as the injection ratio, $R$, increases. The relative phase between the slave and the master ($\phi$) is fixed though its value depends on the detuning frequency and the injection ratio (see Section 3.3).

A few theoretical models have been developed that explain the basic physical mechanism of injection locking. A graphical phasor model will be described in Section 3.3. A gain competition model is shown in Figure 3.3. The bottom diagram shows the locking range in gray. Outside the locking

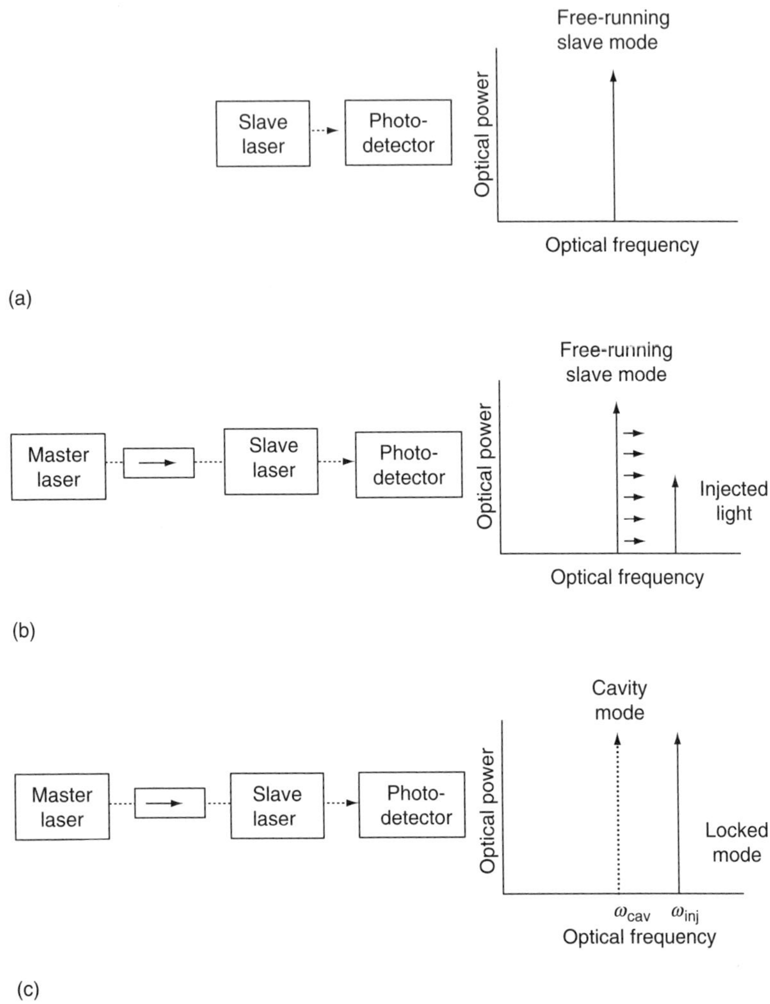


**FIGURE 3.2** Conceptual diagram of injection locking: (a) free-running laser (b) slave with injected light, before locking (c) locked slave laser.

range, the laser power is dominated by amplified spontaneous emission (ASE) in the slave cavity. As in a free-running laser, the ASE of the cavity mode captures the gain and dominates the slave laser power. With external injection, the injected light competes with the spontaneous emission of the slave laser to determine the dominant lasing mode. Within the locking range, the external injection dominates over the spontaneous emission of the slave's natural lasing mode. The injection mode then captures the gain of the laser and the ASE from the other modes is suppressed.

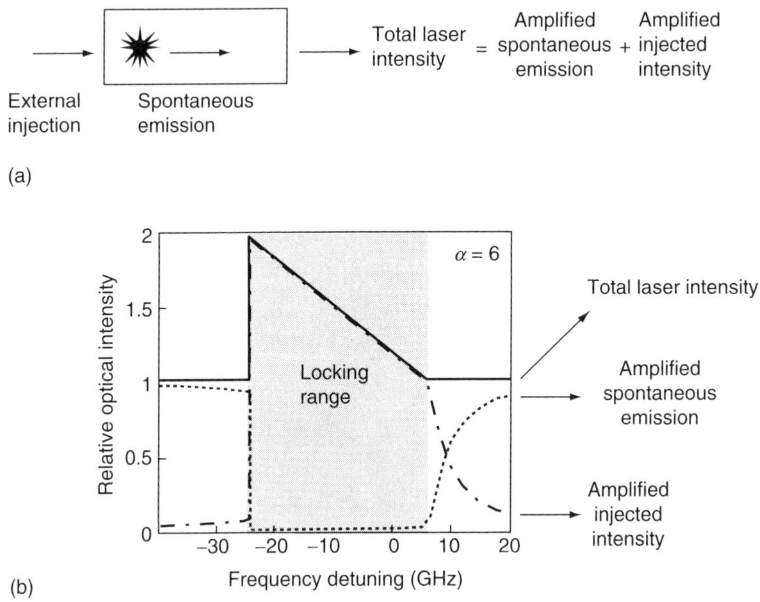


**FIGURE 3.3** Physical gain competition model of injection locking. (a) Illustration of gain model. (b) Detuning dependence on optical intensity, showing competition of ASE and amplified injected intensity, after Henry. (From Henry, C.H., Olsson, N.A., and Dutta, N.K., *IEEE Journal of Quantum Electronics* QE-21 (8), 1152, 1985.)

## 3.2 HISTORY

Perhaps the first observation of injection locking emerged as a thought experiment of the scientist Christiaan Huygens (1629–1695) [20]. Huygens, inventor of the pendulum clock, observed that the pendulums on two clocks mounted on the same wall would eventually lock frequencies, and swing with opposing phase (Figure 3.4). He reasoned that the pendulums must somehow affect each other. He eventually concluded that they were coupled by emitting vibrations passed through the wall that supported them. One pendulum sent vibrations that traveled through the wall and "injected" small perturbations to the other pendulum, eventually "locking" the frequency and phase of the two pendulums together.

Huygens's thought experiment involved injection locking of mechanical systems. The first published work on injection locking was by R. Adler [21] in 1946. Adler injection-locked an electrical oscillator with an external frequency source. The "free-running" oscillator (without injection of external source) will oscillate at its natural frequency, $\omega_0$. Adler showed that when an

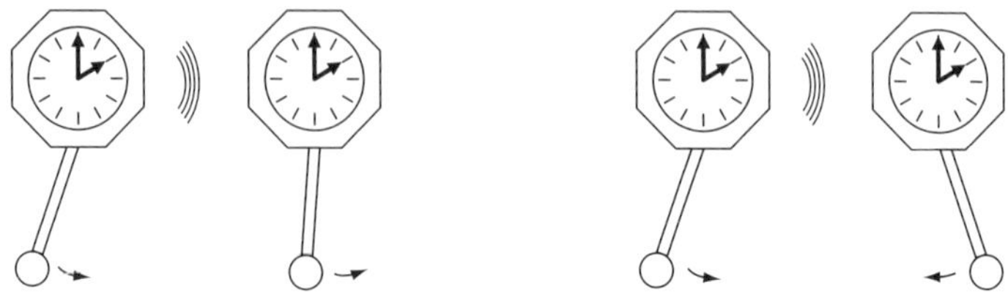

(a) (b)

**FIGURE 3.4** Huygens thought experiment showing injection locking of wall-coupled pendulum clocks. (a) Pendulums are out of phase and frequency, but coupled by wall vibrations. (b) Over time, pendulums eventually lock in frequency with opposite phase.

external signal at frequency $\omega_{inj}$ is injected into the oscillator, the circuit will now oscillate at the injected frequency, provided $\omega_{inj}$ is sufficiently close to the natural oscillator frequency, $\omega_0$.

Finally, 20 years later, injection locking was applied to light, when a source for coherent light was invented in the form of the laser. In 1965, Pantell expanded Adler's injection-locking theory to include lasers [22]. A year later, Stover and Steier demonstrated the first injection-locked laser using two red HeNe lasers [23]. Here, the laser cavity acts as the oscillator and $\omega_0$ is the free-running laser frequency.

Injection-locking work slowed for the next decade. Lasers were themselves incipient and new applications and materials were just being developed. The 1970s saw OIL applied to different laser systems, such as $CO_2$ lasers in Buczek and Freiberg's work in 1972 [24]. The 1970s also saw the development of low-loss optical fiber and the maturation of semiconductor lasers, as well as optical communication schemes such as direct and coherent detection. Injection locking again came into the spotlight in 1980 when the first demonstration of injection locking in semiconductor lasers was reported by Kobayashi and Kimura using GaAs lasers [25]. Since injection locking can pull two lasers to the same wavelength, they made attractive local oscillators in coherent detection systems [26], which was one of the leading methods of long-distance optical communications (before the popularization of optical amplifiers in the early 1990s).

The 1980s saw rapid development of new phenomena and applications for OIL systems. In 1982, Kobayashi and Kimura demonstrated optical phase modulation by direct modulation of the slave laser current [18]. By injection from a master laser, the frequency of the slave is fixed. Changing the slave bias current will cause its locked conditions to change, thus causing the phase difference ($\phi$) between the master and the slave to shift. This phase shifting by current modulation provided an attractive and simple method for achieving phase-shift key (PSK) modulation for coherent detection systems. In the same year, Kobayashi and Kimura demonstrated the effects of injecting modulated light into the slave. The master was frequency-modulated (FM) and injected at weak injection ratios, much lower than the DC power of the slave. The FM was preserved on the slave output while observing up to 30 dB power gain due to the much higher slave output power [27]. Later, Kasapi et al. would use this to develop a subshot-noise FM spectroscopy technique [28]. In 1989, Esman et al. used a similar method of phase modulation, but applied to injection locking of electrical oscillators [29]. They directly modulated a laser to create sidebands and detected the heterodyne beat with a microwave oscillator circuit. The heterodyne beat locked the oscillator and the microwave phase was controlled by changing the laser diode bias. Similar work was done in a pure electrical domain, but with the widespread popularization of optical fiber, this technique allowed for an easy distribution method for frequency references in phased-array radar [17].

Coherent optical communications would eventually be eclipsed by the advent of the erbium-doped fiber amplifier (EDFA) in the late 1980s, making extremely long-haul direct-detection fiber links possible. In the simultaneously developing field of direct detection, OIL made its impact as well. Several groups [12,13,30] in the mid-1980s demonstrated record bit rate-distance (B-L) products, pushing the limits of long-haul optical communications. In 1984, Lin and Mengel found that OIL reduces the frequency chirp in direct amplitude-modulated lasers by holding the slave frequency constant to the master frequency [11]. Olsson et al. demonstrated this reduced chirp by reporting a record 165 Gbit/s-km B-L product [13]. This reduced chirp lessens the linewidth broadening, thus reducing the effects of pulse broadening due to fiber dispersion. This allows for longer maximum transmission lengths and higher B-L product.

Other applications for OIL emerged in the 1980s as well. In 1982, Goldberg et al. developed a method of optical microwave signal generation [31]. The master laser was modulated at a single frequency, $f_m$. The slave was then locked to the weak FM sideband rather than the carrier. When the master and slave light were optically combined and detected, the heterodyned frequencies produced a microwave beat note, stronger than the direct detection of the modulated master alone. Goldberg also developed variant methods that locked two slaves to different modulation sidebands (third order), resulting in frequency multiplication of six to give 35 GHz signals [16]. Applications include distribution of microwave references, frequency multiplexing, and locking of microwave oscillators (see Section 3.4).

The seminal theoretical works on injection locking were done mostly in the 1980s, as applications were simultaneously developed. In 1982, Lang [32] published the first definitive theoretical analysis of OIL lasers, including the three OIL laser master rate equations (see Section 3.3). He was the first to note the effect on the refractive index of the slave laser. This resulted in the discovery

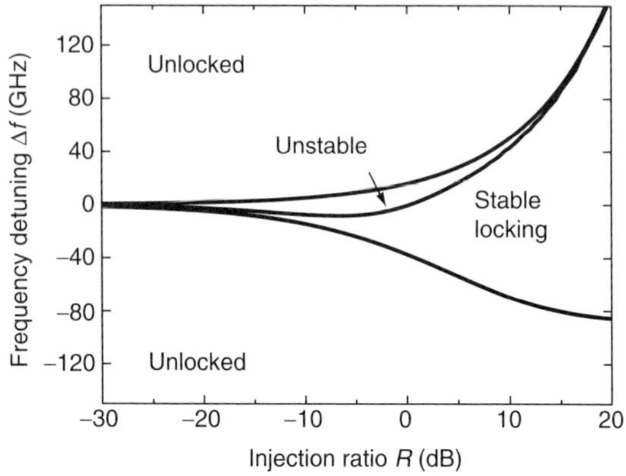


**FIGURE 3.5** Injection-locking stability as a function of injection ratio and frequency detuning. Strong optical injection broadens stable locking range.

of an asymmetric locking range and the unstable locking regime (see Figure 3.5 and Equation 3.2). Henry [2] also published rate equations based on Lang's work, formalizing Lang's theory with Henry's linewidth enhancement factor, $\alpha$. As shown in Section 3.3, Henry also derived an approximate formula for the resonance frequency of OIL lasers. He first discovered this important phenomenon but perhaps did not appreciate its significance until Simpson [4] and Meng [33] in the mid- to late-1990s showed the enhancement of resonance frequency and modulation bandwidth. In 1985, Mogensen et al. published several works, developing a set of master rate equations with the Langevin noise treatment [34–36]. They developed the theory of maximum phase tuning of $\phi$ to less than $\pm\pi$ [36]. In addition, they used the Langevin formulation to derive the FM noise for OIL lasers, finding that the FM noise of the slave evolves to resemble that of the master as the injection ratio increases; thus the slave laser linewidth can be suppressed to that of the master [35].

As briefly stated in the introduction, injection locking has been shown to bestow many attractive improvements to the free-running slave laser. The 1990s brought about the discovery of three of the most significant benefits of OIL systems: noise suppression, reduced nonlinear distortions, and bandwidth enhancement. These will be treated in Section 3.3 and Section 3.4, after a brief formalization of injection-locking theory is introduced.

## 3.3 FUNDAMENTALS OF OPTICAL INJECTION LOCKING

### 3.3.1 THEORY

The dynamics of injection locking can be described by the rate equations that couple the field amplitude, field phase, and carrier number of the slave laser.

Many authors have developed variations of similar rate equations to model noise [7,37]. Here, we use a simple description of OIL that captures the important physical effects. This model is based on single-mode rate equations with a linear gain model, neglecting noise [3,36]:

$$\frac{dA(t)}{dt} = \frac{1}{2} g[N(t) - N_{th}]\, A(t) - \kappa A_{inj} \cos \phi(t) \tag{3.1a}$$

$$\frac{d\phi(t)}{dt} = \frac{\alpha}{2} g[N(t) - N_{th}] - \kappa \frac{A_{inj}}{A(t)} \sin \phi(t) - 2\pi\Delta f \tag{3.1b}$$

$$\frac{dN(t)}{dt} = J - \gamma_N N(t) - \{\gamma_p + g[N(t) - N_{th}]\} A^2(t) \tag{3.1c}$$

where $A(t)$ is the field amplitude, normalized to $A^2(t) = S(t)$, where $S(t)$ is the photon number. $\phi(t)$ is the phase difference between the temporal laser field of the slave laser and the master laser. $N(t)$ is the carrier number and $J$ is the injection current. Other parameters represent typical laser parameters; $g$ is the linear gain coefficient, $N_{th}$ is threshold carrier number, $\gamma_p$ is photon decay rate, $\kappa(=1/\tau_{in})$ is coupling coefficient, $\tau_{in}$ is cavity round-trip time of the slave laser, $\alpha$ is linewidth enhancement factor, and $\gamma_N$ is the carrier decay rate. $N_{th}$ defines the carrier number at the onset of lasing, and contains both transparency carrier number and photon loss rate: $N_{th} \equiv N_{tr} + \gamma_P/g$. Regarding the injection condition, $A_{inj}$ is the field amplitude injected into the slave laser and $\Delta f$ is the lasing frequency difference between the master and the slave laser in the free-running state. Dynamics of the slave laser are governed by the injection-locking parameters, frequency detuning $\Delta f$, and injection power ratio $R$ $(=A_{inj}^2/A_{free}^2)$, where $A_{free}$ is the field amplitude of the free-running slave laser. The field amplitude and carrier equations are similar to the classic laser rate equations found in many textbooks [38]. In injection-locking systems, the injected master and slave light coherently sum inside the slave laser cavity, thus an equation (Equation 3.1b) describing the slave's phase is added.

#### 3.3.1.1 Locking Range

By applying small-signal linear approximation and stability analysis to the above rate equations [2,3,36], the injection-locking range can be derived:

$$-\sqrt{1+\alpha^2}\,\kappa\left(\frac{A_{inj}}{A_0}\right) < \Delta\omega_L < \kappa\left(\frac{A_{inj}}{A_0}\right) \tag{3.2}$$

where $A_0$ is the steady-state amplitude of the slave laser under optical injection and $\Delta\omega_L$ is the range of detuning frequencies ($\Delta\omega$) that result in a locked state.

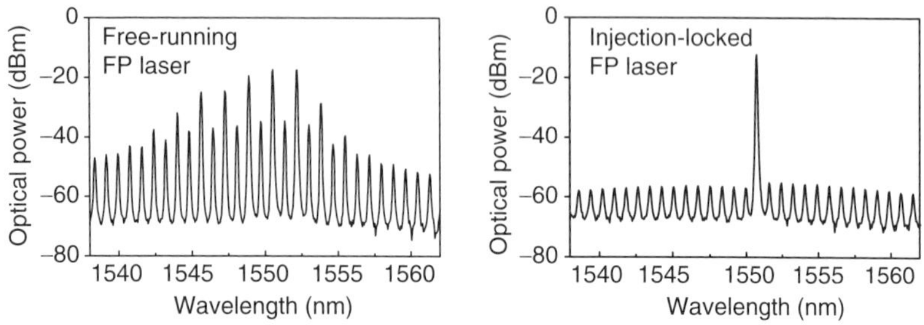


**FIGURE 3.6** Experimentally measured optical spectra of a Fabry–Perot laser under free-running and injection-locking conditions. When injection-locked, the laser oscillates in single mode (the locked mode) and other side modes are suppressed by more than 40 dB.

The stable locking range is shown as a function of injection-locking parameters in Figure 3.5. Equation 3.2 and Figure 3.5 illustrate the trend that stronger optical injection broadens the stable locking range. The locking range itself can be classified into stable and unstable locking regimes. In the unstable locking regime, various nonlinear phenomena have been observed such as chaos, four-wave mixing, and pulsation [39–42]. In the stable locking regime, the free-running slave laser is frequency-locked to the master laser. Therefore, most of the slave lasing power is locked to the master laser and single-mode lasing can be achieved. This characteristic can be utilized to reduce mode-partition noise of multimode lasers [1]. Figure 3.6 shows the optical spectra of a free-running and injection-locked Fabry–Perot (FP) laser. Stable single-mode operation with large side-mode suppression ratio (SMSR) has been achieved by OIL. At large detuning, the master laser frequency may become closer to one of the adjacent longitudinal modes. In that case, the FP slave will prefer to lock to that mode, thus reducing the locking range of the individual mode.

By preemptively suppressing other modes using a distributed feedback (DFB) laser, we can utilize the entire locking range. Figure 3.7 shows mode-evolution across the locking range of an injection-locked DFB laser. Figure 3.7a shows the experimentally measured optical spectrum over the full locking range. The diagonal line marks the frequency position of the master. The broken horizontal line shows the frequency of the slave. Within the locking range, the slave becomes the same frequency as the master. Figure 3.7b shows sample spectra at different points along the locking range.

### 3.3.1.2 Resonance Frequency Enhancement

The relaxation oscillation of a free-running laser is created by the resonant coupling of photons and carriers [38]. In OIL, because of the coherent summation of externally injected and internally generated slave fields, the

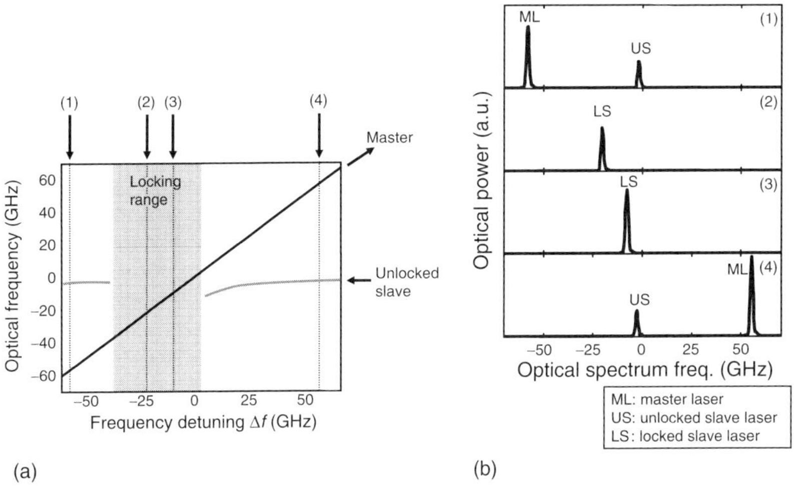


**FIGURE 3.7** (a) Experimentally measured optical spectrum for injection-locked DFB laser across the locking range. Diagonal line is the master. Broken horizontal line is the unlocked slave. Note the locking range asymmetry. (b) Example spectra are shown at different representative points in the locking range. (1) and (4) show unlocked states while (2) and (3) show stable locking points.

phase adds an additional dynamic variable. Consequently, a new resonant coupling between the field amplitude and the phase appears and can dominate the laser resonance. By performing a small-signal linear analysis on Equations 3.1a–c, we can derive an approximate formula for the resonance frequency, $\omega_R$, of the injection-locked laser [2,3]:

$$\omega_R^2 \approx \omega_{R0}^2 + \kappa^2 \left(\frac{A_{inj}}{A_0}\right)^2 \sin^2 \phi_0 \tag{3.3}$$

where $\omega_{R0}$ is the relaxation oscillation frequency of the free-running slave laser and $\phi_0$ is the steady-state phase difference between the injection-locked slave laser and the master laser. Here, we see that increasing the injection ratio will enhance the resonance frequency. Figure 3.8 shows the calculated frequency response curves, where the resonance frequency increases with the increasing injection ratio.

We are able to gain more insight into the origin of resonance frequency enhancement, through theory developed by Murakami et al. [3]. We define the resonance frequency enhancement term, $\Delta\omega_R$, as the last term in Equation 3.3:

$$\Delta\omega_R \equiv -\kappa \frac{A_{inj}}{A_0} \sin \phi_0 \tag{3.4}$$

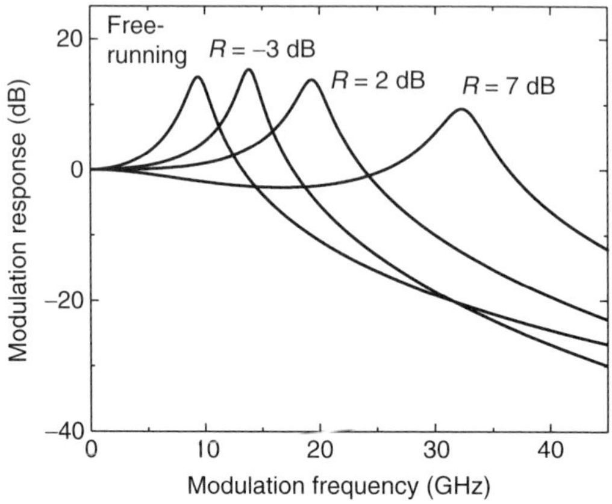


**FIGURE 3.8** Calculated modulation frequency response curves for various injection ratio $R$. Frequency detuning $\Delta f$ is tuned at the edge of positive detuning for prominent resonance peak.

Solving Equation 3.1b in the steady state and using Equation 3.4 gives

$$\Delta\omega_{\mathrm{R}} = -\frac{\alpha}{2} g[N_0 - N_{\mathrm{th}}] + \Delta\omega \tag{3.5}$$

It is clear from Equation 3.5 that the two effects allow injection-locked lasers to significantly enhance the bandwidth. The first term on the right-hand side accounts for the shifting of the slave cavity mode from its free-running frequency by Henry's $\alpha$ parameter [43]. As the steady-state carrier number ($N_0$) can only be less than the threshold, the first term is positive. The primary enhancement for higher injection ratios comes from the second term, $\Delta\omega$. Here, we define the detuning frequency explicitly as $\Delta\omega \equiv \omega_{\mathrm{inj}} - \omega_{\mathrm{fr}}$, and is positive when the master laser frequency is higher than the slave's frequency. From Equation 3.2 and Figure 3.5, we see that stronger injection ratio expands the locking range. This allows the laser to remain locked for extremely large detuning values and $\Delta\omega_{\mathrm{R}}$ can be much larger than the free-running relaxation oscillation frequency. These two effects are graphically shown in Figure 3.9. The enhancement of the resonance, $\Delta\omega_{\mathrm{R}}$, can be seen as the difference between cavity and locked modes. Physically, when directly modulated near or at the frequency $\Delta\omega_{\mathrm{R}}$, the lower FM sideband will lie near the cavity mode. The cavity then resonantly enhances this modulation sideband, causing a resonance peak to appear on the laser's frequency response. An example of this is illustrated in Figure 3.10. Figure 3.10a shows an experimental representation of the concept illustrated in Figure 3.9. Here, we locked the laser to $\Delta f = +34$ GHz, which is well into the positive detuning range. The cavity mode, revealed here by the ASE peak, is shifted by −23 GHz from

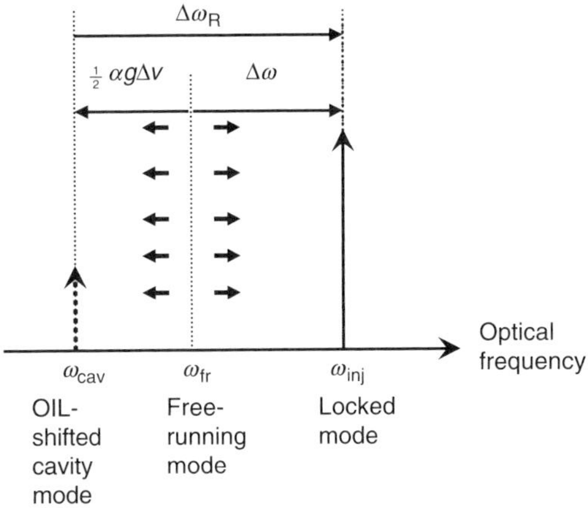


**FIGURE 3.9** Optical frequency domain view of injection locking with enhanced resonance explained.

the free-running frequency (shown in gray). The separation of $\Delta\omega_R = 57$ GHz is responsible for the 57 GHz resonance peak in the modulation frequency response, shown in Figure 3.10b.

### 3.3.1.3 Phasor Model

The model presented in Figure 3.3 argues that the injected light dominates over the slave spontaneous emission generation, thus winning the competition

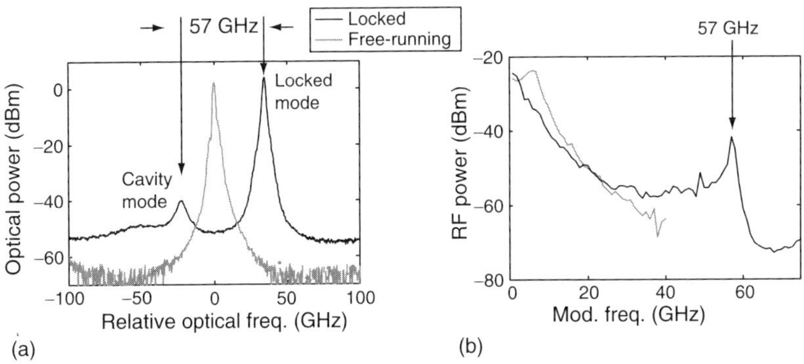


**FIGURE 3.10** Experimental representation of origin of resonance frequency enhancement. (a) Optical spectrum showing shifting of cavity mode ($f_{cav}-f_{fr} = -23$ GHz) and positively detuned locked mode ($\Delta f = +34$ GHz). (b) Modulation frequency response showing resonance peak enhanced to 57 GHz.

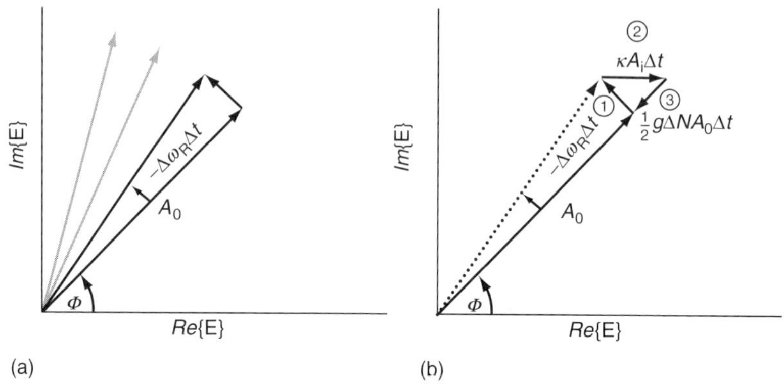


**FIGURE 3.11** Phasor model of injection locking shows how steady state is reached. (a) Shows the evolution of the slave laser if lasing at its cavity mode. Since it is not locked, it rotates at a frequency of $\Delta\omega_R$. (b) When locked to the master laser, the phasor is static by the addition of three vectors: (1) Phasor rotates by difference between the master and the slave frequencies. (2) Injected master light adds a real component. (3) Amplitude decreases due to reduced gain.

of gain and determining the slave's lasing wavelength. This model breaks down when we consider detuning the master laser away from the slave's natural cavity resonance. For instance, we would expect frequency-filtering effects due to the cavity of a DFB laser to reduce the lasing magnitude when we detune away from the free-running frequency. However, this is not the case. We can gain better insight into injection-locking physics by using a phasor model, expanded on the one developed by Henry [2]. Figure 3.11 shows the phasor of the slave laser field, in the reference point of the master laser. Figure 3.11a shows the slave when lasing at its natural cavity mode. Since its frequency differs from the master's by $\Delta\omega_R$, the phasor rotates by $\Delta\omega_R\Delta t$ during each time interval, $\Delta t$. Figure 3.11b shows the slave when locked. In a time interval, $\Delta t$, the sum of three vectors balances to return the phasor to its original position, thus preserving the steady-state requirement. Vector 1 rotates the phasor by the frequency difference of master and slave cavity mode, $\Delta\omega_R$ (where $\Delta\omega_R \equiv \Delta\omega - (\alpha/2)g\Delta N$ includes the effect of cavity-mode shift, as described in Equation 3.4.). Vector 2 adds the contribution of the injected light. This differs in phase to the slave phasor by $\phi$. Since the injected light increases the slave field, the slave's gain must reduce to maintain a steady-state amplitude. This gain reduction is achieved by reducing the carrier number to below threshold. Its effects are shown by vector 3, where the slave's amplitude is reduced by the negative overall gain. Here, $\Delta N$ is the carrier number difference from the threshold ($\Delta N \equiv N_0 - N_{th}$). The three vectors sum to keep the phasor in steady state. Vector 1 iterates the fact that the slave still lases at its natural cavity mode. The constant addition of injected light "kicks" the slave phase to match that of the master, thus locking its wavelength. This kicking effect by the injected light is similar to the Doppler effect for sound waves. A relative velocity change between the sound wave and the observer causes a frequency shift in the perceived sound. This explains why the longitudinal cavity resonance properties do not affect the locked mode, despite being detuned from the free-running value. The slave still lases at the cavity mode but the injected light coherently sums with this light to shift the perceived output frequency. Hence, the laser can be an Fabry perot (FP), Distributed feedback (DFB), Distributed Bragg reflector (DBR), or vertical-cavity surface-emitting laser (VCSEL), and the frequency-filtering effects of the laser structure will not affect the injection-locking dynamics.

### 3.3.2 Modulation Frequency Response

As described in the previous sections, OIL enables the single-mode operation of FP lasers, phase modulation, and laser linewidth reduction. The advantages above are mainly beneficial to coherent optical communication systems. In the 1990s, due to the development of high-power semiconductor lasers with Continuous wave (CW) operation, strong OIL properties were intensively investigated [9,44–46]. In the strong OIL regime, injection-locked lasers exhibit enhanced modulation bandwidth, chirp reduction, and nonlinear distortion reduction as well as maintaining the benefits of weak injection locking. The advantages achieved by strong OIL substantially improve the performance of directly modulated fiber-optic links both in transmitting analog and digital signals. The 3 dB bandwidth is one of the important figure-of-merits that characterize the performance of the lasers. Modulation bandwidth of a directly-modulated semiconductor laser is limited by the relaxation oscillation of the lasers. Typically, the relaxation oscillation frequency is a function of the laser bias current and is in the range of a few gigahertz to a few tens of gigahertz. When the lasers are modulated by the signals containing frequency components close to the relaxation oscillation frequency, the lasers experience nonlinear coupling between carriers and photons. This nonlinear coupling produces signal distortions, resulting in the degradation of system performance, such as spurious free dynamic range (SFDR), intersymbol interference (ISI), and bit-error-rate (BER). Therefore, the high-speed modulation of semiconductor lasers is limited by the relaxation oscillation frequency. Modulation bandwidth enhancement by OIL has been reported by many authors, theoretically and experimentally [2–4,45–48]. The improvement is mainly achieved by enhancement of the relaxation oscillation of the free-running laser. Figure 3.12 shows the measured frequency response of an injection-locked DFB laser. The figure shows the resonance frequency increase and modulation bandwidth enhancement of injection-locked laser by increasing the injection ratio [46]. The resonant frequency has been increased by four times, from 4 GHz to 16 GHz.

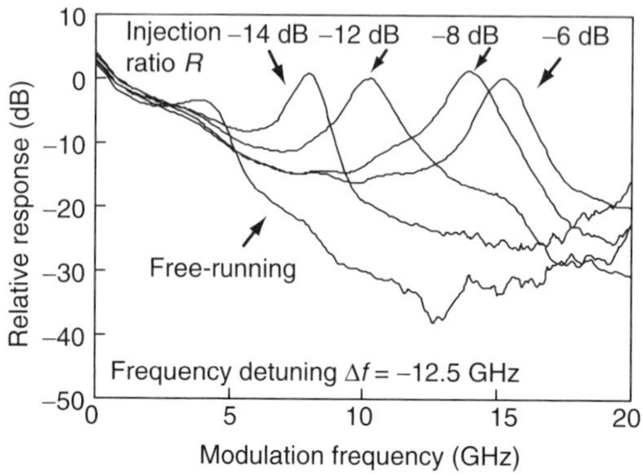


**FIGURE 3.12** Measured frequency response of an injection-locked DFB laser. (after Reference 46.) (From Chrostowski, L., Zhao, X., and Chang-Hasnain, C.J., Microwave Performance of optically injection-locked VCS Els, *Microwave Theory and Techniques, IEEE Transactions on*, 54 (2), 788, 2006. With permission.)

#### 3.3.2.1 Strong Injection Locking

Perhaps the most significant modern advancement for OIL lies in the area of strong OIL for bandwidth enhancement. As shown in Equation 3.3, the laser's resonant frequency is enhanced by injection locking. This can lead to an expanded 3 dB bandwidth of over five times the free-running bandwidth [49]. Work in the 1990s and the early 2000s was limited experimentally to weak ($R < -10$ dB) or strong ($-10$ dB $< R < 0$ dB) injection-locking regimes due to the limited injection power from the master laser and coupling techniques. Consequently, theoretical works failed to extrapolate the consequences of operating at even higher injection ratios. Two significant parallel advancements allowed the community to explore the ultrastrong ($R > 0$ dB) locking regime: development of VCSELs and proliferation of optical fiber amplifiers. We describe these techniques in the following section.

Recently, Liu et al. and Murakami et al. demonstrated the resonant frequency increase of directly modulated semiconductor lasers, both experimentally and theoretically [3,4,50]. However, the analysis and experiment mainly concentrated on the frequency response and the resonance enhancement in the positive frequency-detuning regime, obtaining sharp resonances up to 34 GHz. These results showed a resonance frequency increase but did not show a significant bandwidth increase due to low-frequency damping. Our group has explored the modulation response over the full locking range. We discover three regimes of operation, shown in Figure 3.13. These regimes are only clearly visible with strong injection. In the negative detuning regime ($\Delta f = -60$ GHz), the amplitude response is enhanced (by ~10 dB) while the

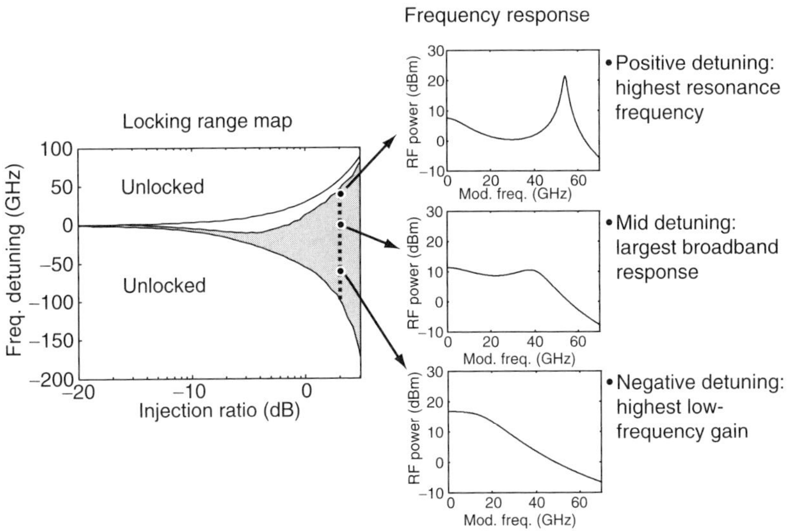


**FIGURE 3.13** The three regimes of modulation response under strong injection locking.

bandwidth does not increase significantly. This can be utilized to increase the RF link gain in analog fiber-optic links. In the positive detuning regime ($\Delta f = +40$ GHz), the response is characterized by a sharp, high-frequency resonance peak. This can be used in microwave photonics applications such as millimeter-wave generation by double-locking or sideband injection locking. This high-frequency oscillation also has its potential in optoelectronic oscillators (OEOs) when combined with fiber feedback loops [51]. The mid-detuning regime ($\Delta f = 0$ GHz) is characterized by a flat frequency response with an extremely high 3 dB bandwidth. This is useful for broadband digital communications.

Plotting modulation frequency responses across the entire detuning range gives us the waterfall plot in Figure 3.14. The highlighted slices are representative response curves for the three regimes shown in Figure 3.13. Additionally, the figure shows the evolutionary increase of the resonance peak frequency as detuning becomes more positive. The positive detuning regime has been lightly explored by the previously mentioned groups. We capitalize on the previously unresearched broadband (middle) regime to optimize bandwidth in our OIL system. Additionally, we use a high-power EDFA to increase the injected power to obtain injection ratios of 18 dB. These improvements allow us to obtain the highest 3 dB bandwidth (44 GHz) for any wavelength laser [49], shown in Figure 3.15a. This increase represents a greater than five times improvement in the free-running bandwidth. Again,

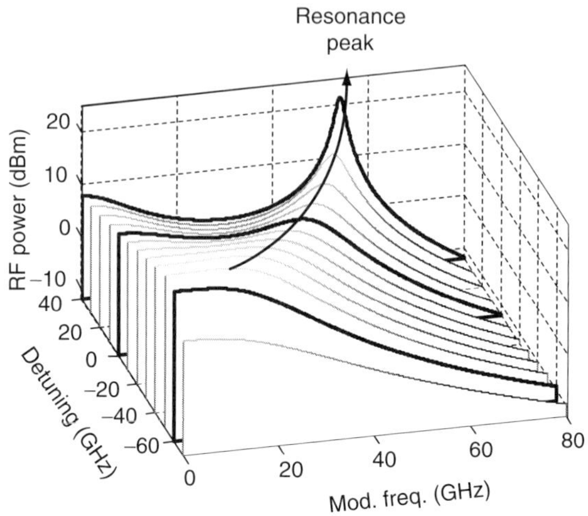


**FIGURE 3.14** Waterfall plot showing modulation frequency responses for different detuning values. The highlighted slices are representative responses for the three modulation regimes.

by use of an EDFA, the locking range increases, allowing us to detune to extremely high values ($\Delta f = +64$ GHz). Optimizing for the positive detuning regime, we also reported the highest resonance frequency (72 GHz) in a semiconductor laser [49] shown in Figure 3.15b.

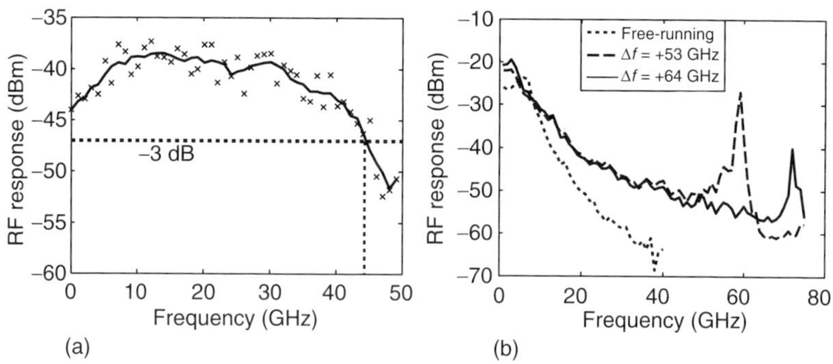


**FIGURE 3.15** Experimental results showing (a) highest 3 dB bandwidth (b) highest resonance frequency of any wavelength semiconductor laser. (From Lau, E.K., Sung, H.K., and Wu, M.C., Optical Fiber Communication Conference, 2006, Technical Digest, 2006.)

#### 3.3.2.2 Injection Locking of VCSELs

VCSELs provide a unique medium for pushing the limits of ultrastrong injection locking. The locking range determines the extent of the resonance frequency enhancement. As shown in Equation 3.2, the locking range increases with the term: $\kappa A_{inj}/A_0$. κ represents the coupling rate (inverse time). One method to define it is to assume the injected field distributes itself evenly throughout the length of the slave cavity:

$$\kappa \equiv \frac{\nu_g}{2L} \tag{3.6}$$

As a typical VCSEL cavity is much shorter than an in-plane laser, the coupling rate can potentially be stronger, leading to higher locking ranges and potentially larger resonance frequencies. Chrostowski et al. have exploited this fact to produce strong injection results without the use of EDFA, achieving 50 GHz resonance frequencies [52,53]. As shown in Figure 3.16, resonance frequency of 50 GHz can be achieved by large positive frequency detuning (large negative wavelength detuning) and high injection

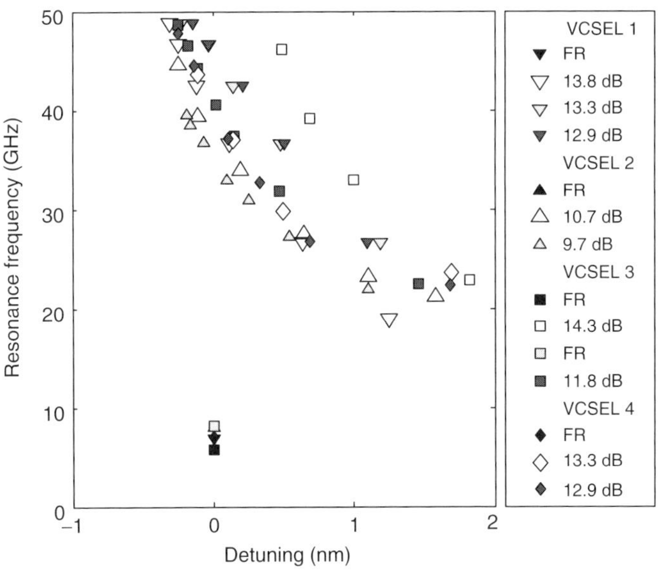


**FIGURE 3.16** Experimental data showing resonance frequency with various detuning and injection ratio for four VCSELs. Legend: FR = Free-running, numbers indicate injection ratio. (From Chrostowski, L., Zhao, X., and Chang-Hasnain, C.J., *IEEE Transactions on Microwave Theory and Techniques* 54 (2), 788, 2006.)

ratio. Further improvement can be achieved as high-speed VCSEL packaging and design advanced.

## 3.4 APPLICATIONS

### 3.4.1 Analog Link Performance Improvement

#### 3.4.1.1 Link Linearity Improvement

In analog links such as cable television (CATV) video distribution and antenna remoting, link performance is characterized by modulation bandwidth, RF link gain, noise figure, and SFDR [54]. Directly modulated semiconductor lasers show significant nonlinear distortion such as harmonic and intermodulation distortion [55,56] due to the nonlinear coupling between carriers and photons. The distortions become severe when the laser is modulated at frequencies close to the relaxation oscillation frequency, where the nonlinear coupling is the strongest. Meng et al. [46] and Chrostowski et al. [57] demonstrated the reduction of nonlinear distortion and SFDR improvement as well as modulation bandwidth increase by injection-locked DFB lasers and VCSELs, respectively. Figure 3.17 shows the SFDR improvement in an injection-locked DFB laser [58]. The improved performance is mainly due to the resonance frequency increase in addition to the suppressed relaxation oscillation of the original free-running laser. Figure 3.18 shows the

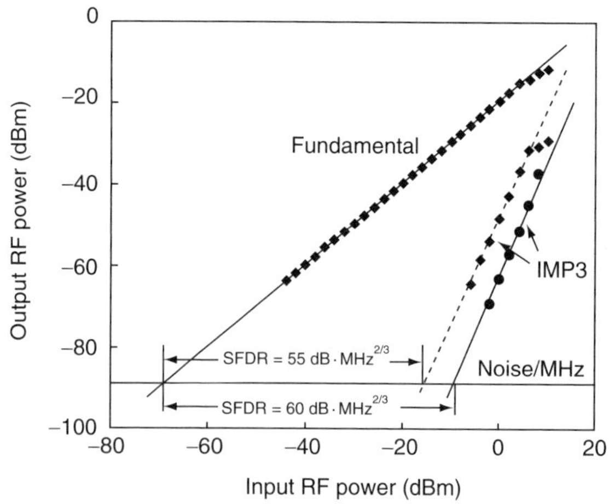


**FIGURE 3.17** Measured RF fundamental power and third-order intermodulation products for free-running lasers and injection-locked DFB lasers. (From Meng, X.J., Chau, T., and Wu, M.C., *IEEE Transactions on Microwave Theory and Techniques* 47 (7), 1172, 1999. With permission.)

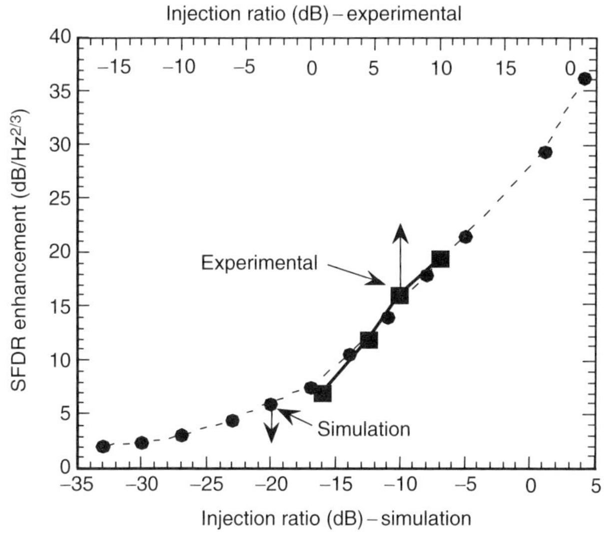


**FIGURE 3.18** Measured SFDR improvement versus injection ratio for injection-locked VCSELs. (From Chrostowski, L., Zhao, X., and Chang-Hasnain, C.J., *IEEE Transactions on Microwave Theory and Techniques* 54 (2), 788, 2006. With permission.)

SFDR improvement as a function of injection ratio. The SFDR improvement by injection ratio is achieved by the increased RF gain under stronger injection.

#### 3.4.1.2 Link Gain Improvement by Injection-Locked Gain-Lever DBR Laser

Aside from the coupling loss between optical components, the link loss of directly modulated links is mainly due to electrical-to-optical (E/O) and optical-to-electrical (O/E) conversion processes in lasers and photodetectors. To reduce the E/O conversion loss and improve amplitude modulation (AM) efficiency of lasers, gain-lever lasers have been proposed by Moore and Lau [59] and Vahala and Newkirk [60]. The increased modulation efficiency is explained in Figure 3.19. The left section is biased below threshold and therefore higher differential gain (high slope on gain curve). To compensate for low gain, the right section is biased above threshold, thereby maintaining an overall threshold gain. The left section is modulated, taking advantage of the higher differential gain to improve the modulation efficiency. However, the gain-lever modulation exhibits significant nonlinear distortion because the improved AM efficiency is achieved at the expense of linearity in the gain versus injection current plot. By combining the gain-lever modulation and injection locking as shown in Figure 3.20, both link gain and linearity can be improved, in addition to increasing the modulation bandwidth [14]. Figure 3.21

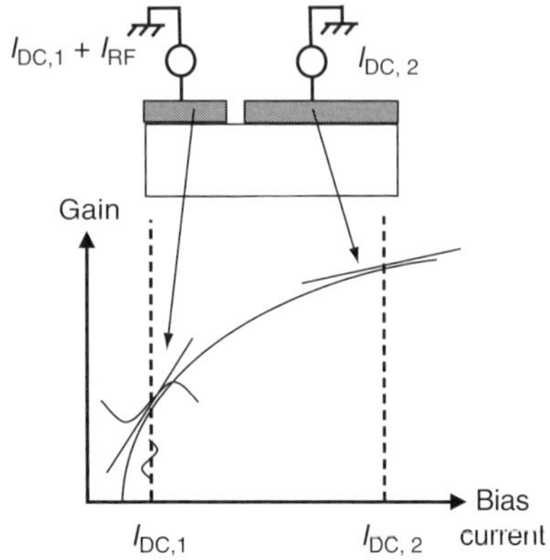


**FIGURE 3.19** Gain-lever effect in semiconductor lasers. Bottom graph is the gain-versus-current curve.

shows the frequency response of gain-lever injection-locked DBR lasers. Injection-locked lasers combined with the gain-lever modulation improve laser modulation bandwidth and efficiency simultaneously.

### 3.4.2 Millimeter-Wave Generation

Distribution of RF signals with high-frequency carriers (~60 GHz) through optical fiber has attracted much interest because optical fiber supports long-distance transmission without significant loss. In radio-over-fiber (RoF) systems, a millimeter-wave signal is generated by optical sources in the central office (CO) and transmitted through fibers to base stations (BS). Several research groups have proposed optical generation of millimeter waves with low-phase noise by injection locking. Goldberg et al. [61] first demonstrated millimeter-wave generation by injection locking between modulation sidebands of a master laser and two slave lasers as shown in Figure 3.22. Two incoherent optical beams from slave laser 1 and slave laser 2 are injection-locked to the sidebands of the RF-modulated master laser so that high-power,

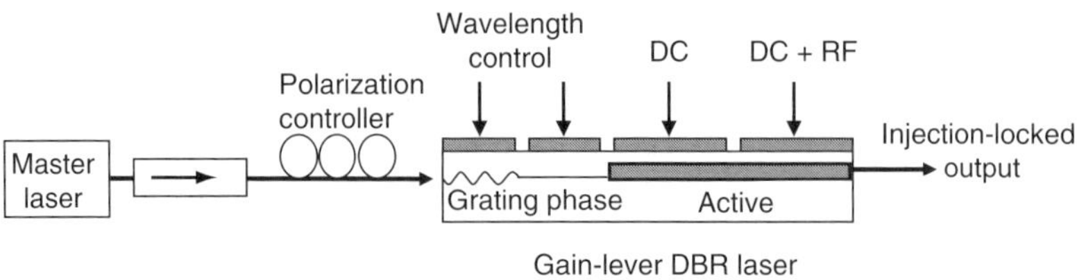


**FIGURE 3.20** Injection locking of gain-lever DBR laser for link gain improvement.

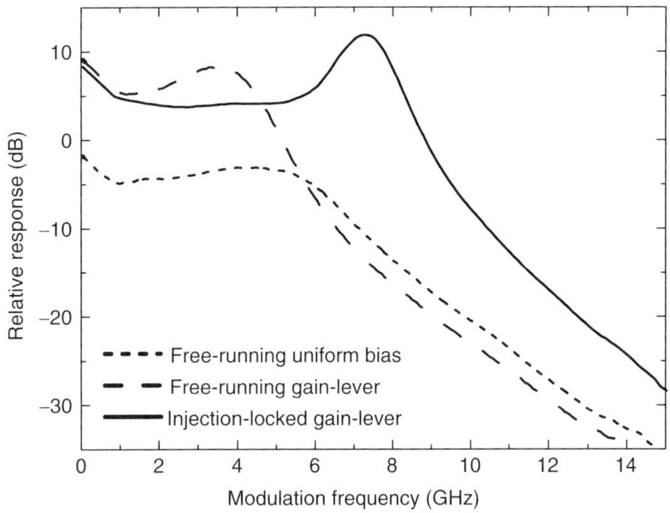


**FIGURE 3.21** Frequency response of the free-running laser under uniform and gain-lever bias conditions and injection-locked gain-lever laser. (From Sung, H.-K., Jung, T., Wu, M.C., Tishinin, D., Liou, K.Y., and Tsang, W.T., Proceedings of 2004 International Topical Meeting on Microwave Photonics, 2004. With permission.)

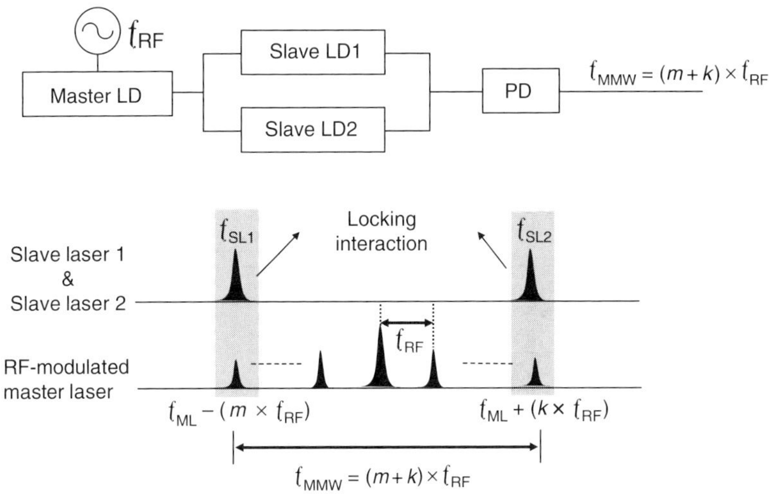


**FIGURE 3.22** Schematic illustrating the use of OIL for millimeter-wave generation.

low-phase noise millimeter-wave signals can be generated optically by beating the optical carriers in the photodetector. In the area of data transmission over the millimeter-wave carrier generated by sideband injection locking, Braun et al. [62] have reported a 155-Mb/s data transmission on a RF carrier of 64 GHz over 12.8 km standard single-mode fiber (SMF). By employing a similar technique but rather using one master laser and one RF-modulated slave laser, Noel et al. [63] generated high-purity tunable millimeter-waves. Furthermore, the sub-harmonic locking for millimeter-wave generation has been extended to injection locking of passively mode-locked DBR lasers [64] and two-section DFB lasers [65,66].

### 3.4.3 Optical Injection Phase-Locked Loop

Synchronization of incoherent lasers in homodyne and heterodyne systems has been an important issue in coherent optical communication system [67] and microwave photonics. The synchronization can be achieved either in the electrical domain by the help of phase-locked loop (PLL) or optical domain by injection locking [25]. Recently, optical injection phase-locked loop (OIPLL) has been demonstrated by A.J. Seeds' group [68]. OIPLL combines phase-locking technique and injection-locking technique as shown in Figure 3.23. Figure 3.24 shows the measured power spectral densities (PSD) for homodyne OPLL and OIPLL systems as well as theoretical PSD for OIPLL system. Clearly, the addition of injection locking to a conventional OPLL improves phase noise performance because improved loop stability by injection locking removes phase noise peaking of OPLL. This combination overcomes the limitation of optical PLL and OIL by presenting better noise suppression and wider stable locking range.

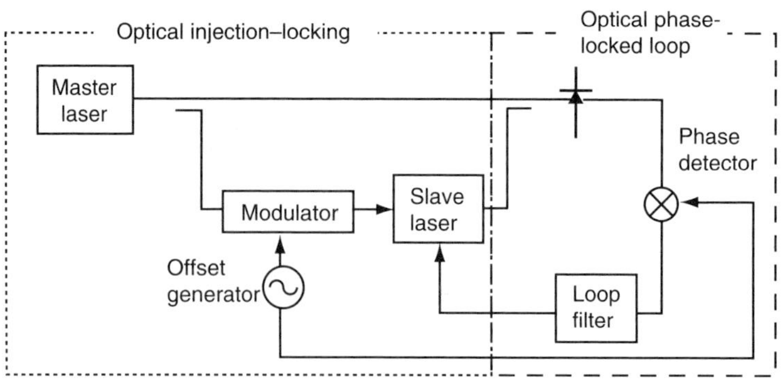


**FIGURE 3.23** Schematic on optical injection phase-locked loop (OIPLL) for heterodyne system. Synchronization is achieved by both OIL and phase-locked loop.

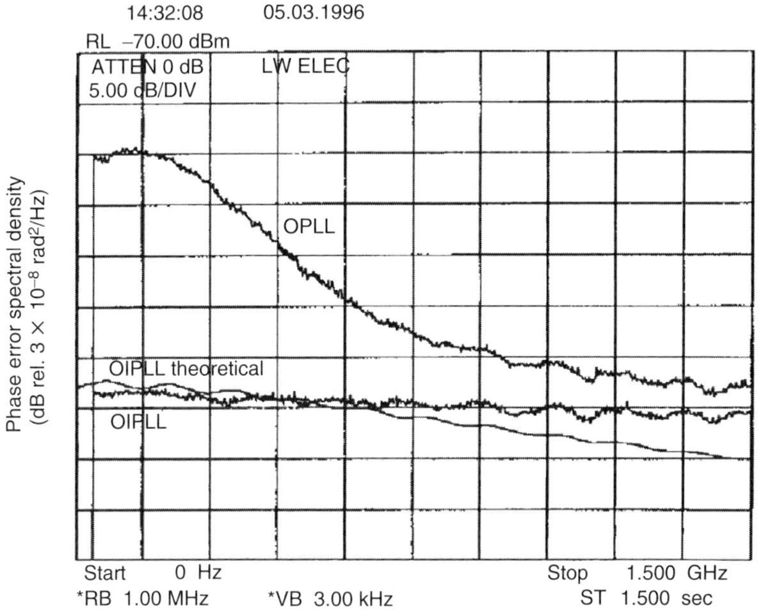


**FIGURE 3.24** Power spectral densities for homodyne OPLL and OIPLL systems. (From Bordonalli, A.C., Walton, C., and Seeds, A.J., *Journal of Lightwave Technology* 17 (2), 328, 1999. With permission.)

### 3.4.4 Injection Locking of Mode-Locked Lasers

Channelization of broadband (~100 GHz) RF signals can be conveniently performed in the optical domain using coherent optical receivers if one can synchronize the RF-modulated optical carrier with an optical frequency comb (OFC) local oscillator. Channelization is a broadband RF signal processing technique that "chops" the RF signals into $N$ consecutive frequency bands and converts them into an array of $N$ intermediate frequency (IF) signals with smaller ($1/N$) bandwidth. The OFC is typically generated by an external cavity mode-locked laser or passively mode-locked laser. To implement such coherent optical channelizers, the OFC needs to be synchronized with the optical carrier of the RF-modulated signals. Injection locking of a mode-locked laser has been demonstrated to generate spatially separated coherent multifrequency sources [69–72]. By injecting the light from a CW laser [69,71] or mode-locked laser [70,72], phase coherence is established between the master and the injected mode, thereby phase locking all modes to the master. Figure 3.25 shows the schematic of the injection-locked OFC for a coherent channelizer. By combining the injection-locked OFC with a dispersive medium such as a diffraction grating, the spatial channelization of

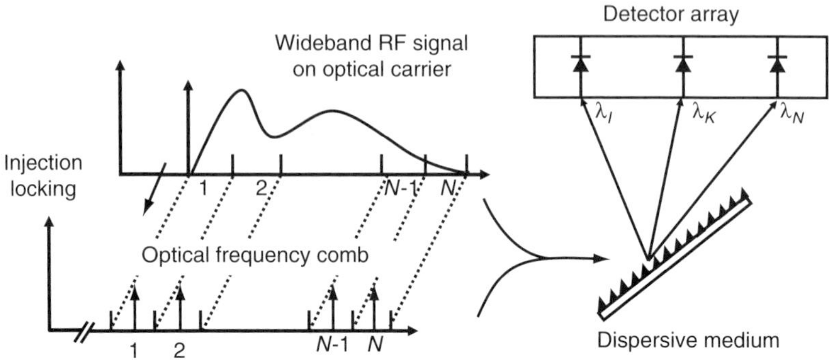


**FIGURE 3.25** Conceptual architecture of coherent channelizer. (From Jung, T., Ji-Lin, S., Tong., D.T.K., Murthy, S., Wu, M.C., Tanbun-EK, T., Wenshen, W., Lodenkamper, R., Davis, R., Lembo, L.J., and Brock, J.C., *IEEE Transactions on Microwave Theory and Techniques* 47 (7), 1225, 1999. With permission.)

broadband optical signals can be attained. Figure 3.26 shows the experimental setup for optical heterodyne detection for sub-GHz filtering using an injection-locked passively mode-locked laser. In Figure 3.26, a passively mode-locked laser is injection-locked to a CW master laser. A single-axial mode of the injection-locked passively mode-locked laser is used as a local oscillator to coherently recover the RF sidebands on the master carrier.

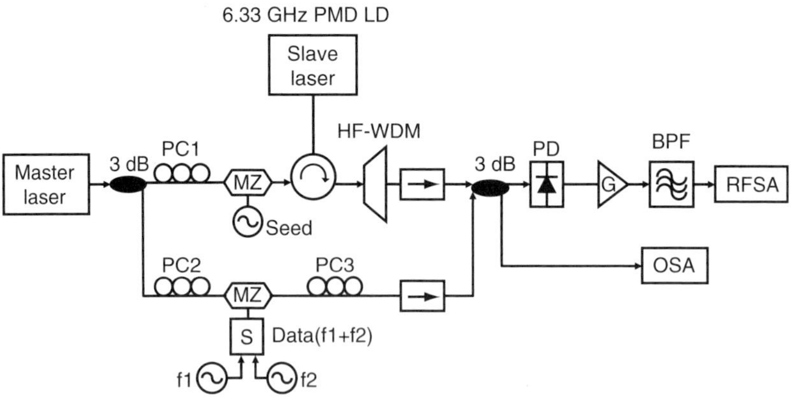


**FIGURE 3.26** Experimental setup for optical heterodyne detection for sub-GHz using an injection-locked passively mode-locked laser as local oscillator. (From Wangkuen, L., Mielke, M., Etemad, S., and Delfyett, P.J., *IEEE Photonics Technology Letters* 16 (8), 1945, 2004. With permission.)

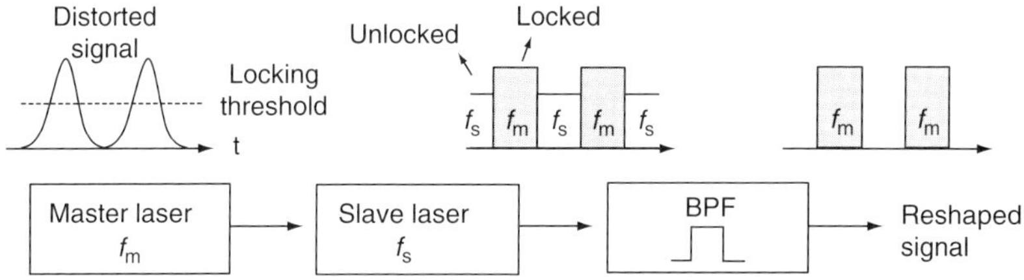


**FIGURE 3.27** Schematic of optical waveform reshaping by injection locking. (From Kuramoto, A., Yamashita, S., *IEEE Journal of Selected Topics in Quantum Electronics* 9 (5), 1283, 2003. With permission.)

### 3.4.5 All-Optical Signal Processing

Clock recovery, an essential component of digital communication, can be achieved by injection-locked passively mode-locked lasers [73]. All-optical regeneration has drawn attention and has been widely investigated by using an injection-locked DFB laser [19], a side-mode injection-locked DFB laser [74], and a two-mode injection-locked FP laser [75]. Figure 3.27 is a schematic of waveform reshaping by injection locking. The reshaping process is based on the switching off between locked and unlocked states, controlled by the injection ratio. As discussed in Figure 3.5 and Equation 3.2, when the frequency detuning is fixed, the locking range depends on the injection power. When the signal with power above locking threshold is injected to a slave laser, the slave laser is locked to the master laser frequency $f_m$. On the other hand, when the injection signal power is small, the slave laser operates at the original frequency $f_s$. Because of the threshold nature of the locking–unlocking process, a distorted signal can be reshaped to an FM signal with reduced noise. The signal from the slave laser is then optically filtered by band-pass filter so that the reshaped signal can be obtained. Injection locking for all-optical processing also provides various functionalities such as optical inverters [76], all-optical format conversion between Return-to-zero (RZ) and Non-return-to-zero (NRZ) [77], and polarization controllers [78].

### 3.4.6 Other Applications

OIL has been widely investigated by either utilizing the locking phenomena itself in the optical domain or the combined locking phenomena in the electrical and optical domain. These flexible approaches enable injection locking to open up wide application areas. For example, clock recovery can also be achieved by optoelectronic injection locking using heterojunction phototransistors [79]. High-performance OEO has been demonstrated by combining the conventional OEO configuration with a master OEO [80]. Injection locking of uncooled VCSELs has potential for low-cost and stable transmitters for wavelength division multiplexed passive optical networks

(WDM-PON) [81]. Optical single-sideband modulation, which substantially reduces fiber chromatic dispersion effects, can be achieved by utilizing the resonantly enhanced cavity mode through strong OIL [15].

## 3.5 FUTURE DIRECTIONS

In the realm of bandwidth improvement for faster and more demanding communications systems, injection locking has proven to add great value to the direct modulated laser. The bandwidth increases by direct modulation have gradually slowed as the laser community reaches the fundamental limits set by laser physics [82]. Injection locking allows us to exceed those fundamental limits, and is a technique applicable to any laser, regardless of design. This ensures universal applicability and performance improvements for any laser system. Improvements to bandwidth and resonance frequency so far have no inherent fundamental limits. Improvements are linked to injection ratio and therefore limited only to the injected power. We foresee continued improvements to both bandwidth and resonance frequency as higher power sources are implemented, or by engineering and optimizing the slave laser for superior injection-locking performance.

Injection locking inherently complicates the basic laser system. System integration becomes an issue when mass production and cost are important. Since the birth of integrated circuits, there has existed a universal trend to single-chip integration; this is not different for semiconductor lasers and photonic systems. We can achieve this integration by building the master and slave lasers monolithically on a single chip [12,14,53,83]. Analog performance benefits were observed despite the lack of optical isolation causing mutual locking. As new technologies of chip-level optical isolation become more prevalent, monolithic integration of injection-locked lasers should become an increasingly attractive method of meeting record-breaking system specifications.

## 3.6 CONCLUSION

We have provided a brief overview of OIL history, from its inception to current record-breaking performance. We also provide several intuitive physical models and describe the most prevalent theoretical model of injection locking. This model contains the most important physical effects, including resonance frequency enhancement, noise reduction, and nonlinearity suppression. We use this theory as a basis for understanding the relatively new regime of strong injection locking. Utilizing our understanding of the strong injection regime, we are able to achieve a bandwidth improvement to 44 GHz, currently the highest bandwidth semiconductor laser. Additionally, we report a record 72 GHz resonance frequency, which is also the highest for any semiconductor laser. Finally, we complete our summary of optical injection–locking by surveying the wide array of applications that are currently being developed. Many of these applications provide benefits or additional functionality to microwave photonic systems. Injection locking shows great promise for improving the performance of analog and digital communication systems. As laser technology grows, we will continue to see further improvements, as the fundamental limits governing injection-locking performance are yet to be discovered.

## REFERENCES

1. Iwashita, K. and Nakagawa, K., Suppression of mode partition noise by laser diode light injection, *IEEE Transactions on Microwave Theory and Techniques* 30 (10), 1657, 1982.
2. Henry, C.H., Olsson, N.A., and Dutta, N.K., Locking range and stability of injection locked 1.54 mm InGaAsP semiconductor lasers, *IEEE Journal of Quantum Electronics* QE-21 (8), 1152, 1985.
3. Murakami, A., Kawashima, K., and Atsuki, K., Cavity resonance shift and bandwidth enhancement in semiconductor lasers with strong light injection, *IEEE Journal of Quantum Electronics* 39 (10), 1196, 2003.
4. Simpson, T.B., Liu, J.M., and Gavrielides, A., Bandwidth enhancement and broadband noise reduction in injection-locked semiconductor lasers, *IEEE Photonics Technology Letters* 7 (7), 709, 1995.
5. Meng, X.J., Chau, T., Tong, D.T.K., and Wu, M.C., Suppression of second harmonic distortion in directly modulated distributed feedback lasers by external light injection, *Electronics Letters* 34 (21), 2040, 1998.
6. Chrostowski, L., Chang, C.H., and Chang-Hasnain, C., Reduction of relative intensity noise and improvement of spur-free dynamic range of an injection locked VCSEL, in 2003 IEEE LEOS Annual Meeting Conference Proceedings (IEEE Cat. No. 03CH37460), IEEE, Part vol. 2, 2003, pp. 706–7 vol. 2. Piscataway, NJ, USA, 2003.
7. Schunk, N. and Petermann, K., Noise analysis of injection-locked semiconductor injection lasers, *IEEE Journal of Quantum Electronics* QE-22 (5), 642, 1986.
8. Espana-Boquera, M.C. and Puerta-Notario, A., Noise effects in injection locked laser simulation: Phase jumps and associated spectral components, *Electronics Letters* 32 (9), 818, 1996.
9. Liu, J.M., Chen, H.F., Meng, X.J., and Simpson, T.B., Modulation bandwidth, noise, and stability of a semiconductor laser subject to strong injection locking, *IEEE Photonics Technology Letters* 9 (10), 1325, 1997.
10. Yabre, G., De Waardt, H., van den Boom, H.P.A., and Khoe, G.D., Noise characteristics of single-mode semiconductor lasers under external light injection, *IEEE Journal of Quantum Electronics* 36 (3), 385, 2000.
11. Lin, C. and Mengel, F., Reduction of frequency chirping and dynamic linewidth in high-speed directly modulated semiconductor lasers by injection locking, *Electronics Letters* 20 (25–26), 1073, 1984.
12. Toba, H., Kobayashi, Y., Yanagimoto, K., Nagai, H., and Nakahara, M., Injection-locking technique applied to a 170 km transmission experiment at 445.8 Mbit/s, *Electronics Letters* 20 (9), 370, 1984.

13. Olsson, N.A., Temkin, H., Logan, R.A., Johnson, L.F., Dolan, G.J., van der Ziel, J.P., and Campbell, J.C., Chirp-free transmission over 82.5 km of single mode fibers at 2 Gbit/s with injection locked DFB semiconductor lasers, *Journal of Lightwave Technology* LT-3 (1), 63, 1985.
14. Sung, H.-K., Jung, T., Wu, M.C., Tishinin, D., Liou, K.Y., and Tsang, W.T., Optical Injection-Locked Gain-Lever Distributed Bragg Reflector Lasers with Enhanced RF Performance, Proceedings of 2004 International Topical Meeting on Microwave Photonics, Ogunquit, ME, 2004.
15. Sung, H.-K., Lau, E.K.L., and Wu, M.C., Near-Single Sideband Modulation in Strong Optical Injection-Locked Semiconductor Lasers, Optical Fiber Communication Conference, Anaheim, CA, 2006.
16. Goldberg, L., Yurek, A.M., Taylor, H.F., and Weller, J.F., 35 GHz microwave signal generation with an injection-locked laser diode, *Electronics Letters* 21 (18), 814, 1985.
17. Seeds, A.J., Blanchflower, I.D., Gomes, N.J., King, G., and Flynn, S.J., New developments in optical control techniques for phased array radar, in 1988 IEEE MTT International Microwave Symposium Digest (Cat. No. 88CH2489-3), IEEE, 1988, pp. 905–8 vol. 2. New York, NY, USA, 1988.
18. Kobayashi, S. and Kimura, T., Optical phase modulation in an injection locked AlGaAs semiconductor laser, *IEEE Transactions on Microwave Theory and Techniques* MTT-30 (10), 1650, 1982.
19. Yamashita, S. and Matsumoto, D., Waveform reshaping based on injection locking of a distributed-feedback semiconductor laser, *IEEE Photonics Technology Letters* 12 (10), 1388, 2000.
20. Pikovsky, A., Rosenblum, M., Kurths, J., and Chirikov, B., Synchronization: A Universal Concept in Nonlinear Sciences, in *Cambridge Nonlinear Science Series* Cambridge University Press, Cambridge, U.K., 2003.
21. Adler, R., A study of locking phenomena in oscillators, *Proceedings of IRE* 34, 351, 1946.
22. Pantell, R.H., The laser oscillator with an external signal, *Proceedings of the IEEE* 53 (5), 474, 1965.
23. Stover, H.L. and Steier, W.H., Locking of laser oscillators by light injection, *Applied Physics Letters* 8 (4), 91, 1966.
24. Buczek, C.J. and Freiberg, R.J., Hybrid injection locking of higher power $CO_2$ lasers, *IEEE Journal of Quantum Electronics* QE-8 (7), 641, 1972.
25. Kobayashi, S. and Kimura, T., Coherence on injection phase-locked AlGaAs semiconductor laser, *Electronics Letters* 16 (7), 668, 1980.
26. Yamamoto, Y., Receiver performance evaluation of various digital optical modulation–demodulation systems in the 0.5–10 mm wavelength region, *IEEE Journal of Quantum Electronics* QE-16 (11), 1251, 1980.
27. Kobayashi, S. and Kimura, T., Optical FM signal amplification by injection locked and resonant type semiconductor laser amplifiers, *IEEE Journal of Quantum Electronics* 18 (4), 575, 1982.
28. Kasapi, S., Lathi, S., and Yamamoto, Y., Sub-shot-noise frequency-modulation spectroscopy by use of amplitude-squeezed light from semiconductor lasers, *Journal of the Optical Society of America B-Optical Physics* 17 (2), 275, 2000.

29. Esman, R.D., Goldberg, L., and Weller, J.F., Optical phase control of an optically injection-locked FET microwave oscillator, *IEEE Transactions on Microwave Theory and Techniques* 37 (10), 1512, 1989.
30. Lin, C., Andersen, J.K., and Mengel, F., Frequency chirp reduction in a 2.2 Gbit/s directly modulated InGaAsP semiconductor laser by CW injection, *Electronics Letters* 21 (2), 80, 1985.
31. Goldberg, L., Taylor, H.F., and Weller, J.F., FM sideband injection locking of diode lasers, *Electronics Letters* 18 (23), 1019, 1982.
32. Lang, R., Injection locking properties of a semiconductor laser, *IEEE Journal of Quantum Electronics* QE-18 (6), 976, 1982.
33. Meng, X.J., Jung, T., Tai, C., and Wu, M.C., Gain and bandwidth enhancement of directly modulated analog fiber optic links using injection-locked gain-coupled DFB lasers, in International Topical Meeting on Microwave Photonics. MWP'99. Technical Digest (Cat. No. 99EX301), IEEE, Part vol. 1, 1999, pp. 141–4 vol. 1. Piscataway, NJ, 1999.
34. Saito, S., Mogensen, F., and Olesen, H., Effective bandwidth for FM noise suppression in an injection-locked semiconductor laser, *Electronics Letters* 21 (24), 1173, 1985.
35. Mogensen, F., Olesen, H., and Jacobsen, G., FM noise suppression and linewidth reduction in an injection-locked semiconductor laser, *Electronics Letters* 21 (16), 696, 1985.
36. Mogensen, F., Olesen, H., and Jacobsen, G., Locking conditions and stability properties for a semiconductor laser with external light injection, *IEEE Journal of Quantum Electronics* QE-21 (7), 784, 1985.
37. Surette, M.R., Hjelme, D.R., Ellingsen, R., and Mickelson, A.R., Effects of noise on transients of injection locked semiconductor lasers, *IEEE Journal of Quantum Electronics* 29 (4), 1046, 1993.
38. Coldren, L.A. and Corzine, S.W., *Diode Lasers and Photonic Integrated Circuits* John Wiley & Sons, Inc., New York, 1995.
39. Annovazzi-Lodi, V., Donati, S., and Manna, M., Chaos and locking in a semiconductor laser due to external injection, *IEEE Journal of Quantum Electronics* 30 (7), 1537, 1994.
40. Kovanis, V., Gavrielides, A., Simpson, T.B., Liu, J.M., and Alsing, P.H., Chaos, period-doubling and reverse bifurcations in an optically injected semiconductor laser, in IEEE Nonlinear Optics: Materials, Fundamentals, and Applications (Cat. No. 94CH3370-4), IEEE, 1994, pp. 30–2, New York, 1994.
41. Simpson, T.B., Liu, J.M., Gavrielides, A., Kovanis, V., and Alsing, P.M., Period-doubling route to chaos in a semiconductor laser subject to optical injection, *Applied Physics Letters* 64 (26), 3539, 1994.
42. Simpson, T.B., Liu, J.M., Huang, K.F., and Tai, K., Nonlinear dynamics induced by external optical injection in semiconductor lasers, *Quantum and Semiclassical Optics* 9 (5), 765, 1997.
43. Henry, C., Theory of the linewidth of semiconductor lasers, *IEEE Journal of Quantum Electronics* 18 (2), 259, 1982.
44. Simpson, T.B., Liu, J.M., and Gavrielides, A., Small-signal analysis of modulation characteristics in a semiconductor laser subject to strong optical injection, *IEEE Journal of Quantum Electronics* 32 (8), 1456, 1996.

45. Yabre, G., Effect of relatively strong light injection on the chirp-to-power ratio and the 3 dB bandwidth of directly modulated semiconductor lasers, *Journal of Lightwave Technology* 14 (10), 2367, 1996.
46. Meng, X.J., Tai, C., and Wu, M.C., Experimental demonstration of modulation bandwidth enhancement in distributed feedback lasers with external light injection, *Electronics Letters* 34 (21), 2031, 1998.
47. Chang, C.-H., Chrostowski, L., and Chang-Hasnain, C.J., Injection locking of VCSELs, *IEEE Journal of Selected Topics in Quantum Electronics* 9 (5), 1386, 2003.
48. Lee, H.L.T., Ram, R.J., Kjebon, O., and Schatz, R., Bandwidth enhancement and chirp reduction in DBR lasers by strong optical injection, in Conference on Lasers and Electro-Optics (CLEO 2000), Technical Digest. Postconference Edition, TOPS Vol. 39 (IEEE Cat. No. 00CH37088), Optical Society of America 2000, pp. 99–100. Salem, MA, 2000.
49. Lau, E.K., Sung, H.K., and Wu, M.C., Ultra-High, 72 GHz Resonance Frequency and 44 GHz Bandwidth of Injection-Locked 1.55-μm DFB Lasers, in *Optical Fiber Communication Conference, 2006, Technical Digest*, 2006.
50. Hwang, S.K., Liu, J.M., and White, J.K., 35-GHz intrinsic bandwidth for direct modulation in 1.3-mm semiconductor lasers subject to strong injection locking, *IEEE Photonics Technology Letters* 16 (4), 972, 2004.
51. Yao, X.S. and Maleki, L., Optoelectronic microwave oscillator, *Journal of the Optical Society of America B* 13, 725, 1996.
52. Chrostowski, L., Zhao, X., Chang-Hasnain, C.J., Shau, R., Ortsiefer, M., and Amann, M.C., 50-GHz Optically Injection-Locked 1.55-μm VCSELs, *IEEE Photonics Technology Letters* 18 (2), 367, 2006.
53. Chrostowski, L., Zhao, X., and Chang-Hasnain, C.J., Microwave Performance of Optically Injection-Locked VCSELs, *IEEE Transactions on Microwave Theory and Techniques* 54 (2), 788, 2006.
54. Cox, C.H., III, Ackerman, E.I., Betts, G.E., and Prince, J.L., Limits on the performance of RF-over-fiber links and their impact on device design, *IEEE Transactions on Microwave Theory and Techniques* 54 (2), 906, 2006.
55. Lau, K.Y. and Yariv, A., Intermodulation distortion in a directly modulated semiconductor injection laser, *Applied Physics Letters* 45 (10), 1034, 1984.
56. Darcie, T.E., Tucker, R.S., and Sullivan, G.J., Intermodulation and harmonic distortion in InGaAsP lasers, *Electronics Letters* 21 (16), 665, 1985.
57. Chrostowski, L., Chih-Hao, C., and Chang-Hasnain, C.J., Enhancement of dynamic range in 1.55-mm VCSELs using injection locking, *IEEE Photonics Technology Letters* 15 (4), 498, 2003.
58. Meng, X.J., Chau, T., and Wu, M.C., Improved intrinsic dynamic distortions in directly modulated semiconductor lasers by optical injection locking, *IEEE Transactions on Microwave Theory and Techniques* 47 (7), 1172, 1999.
59. Moore, N. and Lau, K.Y., Ultrahigh efficiency microwave signal transmission using tandem-contact single quantum well GaAlAs lasers, *Applied Physics Letters* 55, 936, 1989.
60. Vahala, K.J. and Newkirk, M.A., The optical gain lever: A novel gain mechanism in the direct modulation of quantum well semiconductor lasers, *Applied Physics Letters* 54, 2506–2508, 1989.
61. Goldberg, L., Taylor, H.F., Weller, J.F., and Bloom, D.M., Microwave signal generation with injection-locked laser diodes, *Electronics Letters* 19 (13), 491, 1983.

62. Braun, R.P., Grosskopf, G., Rohde, D., and Schmidt, F., Low-phase-noise millimeter-wave generation at 64 GHz and data transmission using optical sideband injection locking, *IEEE Photonics Technology Letters* 10 (5), 728, 1998.
63. Noel, L., Marcenac, D., and Wake, D., Optical millimetre-wave generation technique with high efficiency, purity and stability, *Electronics Letters* 32 (21), 1997, 1996.
64. Ahmed, Z., Liu, H.F., Novak, D., Ogawa, Y., Pelusi, M.D., and Kim, D.Y., Locking characteristics of a passively mode-locked monolithic DBR laser stabilized by optical injection, *IEEE Photonics Technology Letters* 8 (1), 37, 1996.
65. Jin, H. and Rongqing, H., Tunable millimeter-wave generation with subharmonic injection locking in two-section strongly gain-coupled DFB lasers, *IEEE Photonics Technology Letters* 12 (5), 543, 2000.
66. Sung, H.-K., Jung, T., Wu, M.C., Tishinin, D., Liou, K.Y., and Tsang, W.T., Optical generation of millimeter-waves using monolithic sideband injection locking of a two-section DFB laser, in Lasers and Electro-Optics Society, 2003, LEOS Tucsou, Az, 2003, The 16th Annual Meeting of the IEEE, 2003, pp. 1005.
67. Kahn, J.M., 1 Gbit/s PSK homodyne transmission system using phase-locked semiconductor lasers, *IEEE Photonics Technology Letters* 1 (10), 340, 1989.
68. Bordonalli, A.C., Walton, C., and Seeds, A.J., High-performance phase locking of wide linewidth semiconductor lasers by combined use of optical injection locking and optical phase-lock loop, *Journal of Lightwave Technology* 17 (2), 328, 1999.
69. Jung, T., Ji-Lin, S., Tong, D.T.K., Murthy, S., Wu, M.C., Tanbun-Ek, T., Wenshen, W., Lodenkamper, R., Davis, R., Lembo, L.J., and Brock, J.C., CW injection locking of a mode-locked semiconductor laser as a local oscillator comb for channelizing broad-band RF signals, *IEEE Transactions on Microwave Theory and Techniques* 47 (7), 1225, 1999.
70. Wangkuen, L. and Delfyett, P.J., Jr., Optical heterodyne detection based on dual-mode injection locking of modelocked semiconductor lasers, 2004, pp. 1005.
71. Wangkuen, L., Mielke, M., Etemad, S., and Delfyett, P.J., Subgigahertz channel filtering by optical heterodyne detection using a single axial mode from an injection-locked passively mode-locked semiconductor laser, *IEEE Photonics Technology Letters* 16 (8), 1945, 2004.
72. Lee, W. and Delfyett, P.J., Dual-mode injection locking of two independent modelocked semiconductor lasers, *Electronics Letters* 40 (19), 1182, 2004.
73. Mathason, B.K. and Delfyett, P.J., Pulsed injection locking dynamics of passively mode-locked external-cavity semiconductor laser systems for all-optical clock recovery, *Journal of Lightwave Technology* 18 (8), 1111, 2000.
74. Kuramoto, A. and Yamashita, S., All-optical regeneration using a side-mode injection-locked semiconductor laser, *IEEE Journal of Selected Topics in Quantum Electronics* 9 (5), 1283, 2003.
75. Yamashita, S. and Suzuki, J., All-optical 2R regeneration using a two-mode injection-locked Fabry–Perot laser diode, *IEEE Photonics Technology Letters* 16 (4), 1176, 2004.
76. Onishi, Y., Nishiyama, N., Caneau, C., Koyama, F., and Chung-en, Z., Dynamic behavior of an all-optical inverter using transverse-mode switching in 1.55-mm vertical-cavity surface-emitting lasers, *IEEE Photonics Technology Letters* 16 (5), 1236, 2004.

77. Kawaguchi, H., Yamayoshi, Y., and Tamura, K., All-optical format conversion using an ultrafast polarization bistable vertical-cavity surface-emitting laser, in *CLEO 2000*, 2000, pp. 379–380.
78. Hasebe, K., Onishi, Y., and Koyama, F., Novel polarization controller based on injection-locked vertical-cavity surface-emitting laser, in *CLEO 2005*, 2005, pp. 164–165.
79. Kamitsuna, H., Sano, K., Kurishima, K., and Ida, M., 80 GHz electrical clock extraction from 80 Gbit/s RZ optical data stream using direct optical injection locking of InP/InGaAs HPT oscillator, *Electronics Letters* 40 (2), 139, 2004.
80. Weimin, Z. and Blasche, G., Injection-locked dual opto-electronic oscillator with ultra-low phase noise and ultra-low spurious level, *IEEE Transactions on Microwave Theory and Techniques* 53 (3), 929, 2005.
81. Wong, E., Zhao, X., and Chang-Hasnain, C., Uncooled, optical injection-locked 1.55 μm VCSELs for upstream transmitter in WDM-PONs, *OFC 2006*, 2006.
82. Olshansky, R., Hill, P., Lanzisera, V., and Powazinik, W., Frequency response of 1.3 mm InGaAsP high speed semiconductor lasers, *IEEE Journal of Quantum Electronics* 23 (9), 1410, 1987.
83. Nizette, M., Erneux, T., Gavrielides, A., Kovanis, V., and Simpson, T.B., Bistability of pulsating intensities for double-locked laser diodes, *Physical Review E. Statistical Physics, Plasmas, Fluids, and Related Interdisciplinary Topics* 65 (5), 056610/1, 2002.

# 4 Optical Measurement Technologies for High-Frequency Electronics

*Tadao Nagatsuma and Masahiro Tsuchiya*

## CONTENTS

4.1 Advantages of Optical Measurement ... 112
4.1.1 Basic Concept ... 112
4.1.2 Features of Optical Measurement ... 112
4.1.2.1 Ultrafast Operation ... 113
4.1.2.2 Low Invasiveness ... 113
4.1.2.3 Spatial Expansiveness ... 114
4.1.2.4 High Spatial Resolution ... 114
4.2 Key Technologies for Optical Measurement ... 114
4.2.1 Physical Phenomena Used for Optical Measurements ... 114
4.2.1.1 EO Effect ... 115
4.2.1.2 MO Effect ... 115
4.2.2 Measurement Based on EO Effect ... 116
4.2.2.1 Principle of EO Probing ... 116
4.2.2.2 Probing Configurations ... 117
4.2.2.3 EO Materials ... 118
4.2.2.4 EO Probes ... 120
4.2.3 Measurement Based on MO Effect ... 120
4.2.3.1 Principle of MO Probing ... 120
4.2.3.2 Probing Configurations and MO Materials ... 122
4.2.3.3 MO Probes ... 124
4.2.4 Optical Signal Sources for Probing ... 124
4.2.4.1 Pulse Source ... 124
4.2.4.2 CW Light and Sinusoidally Modulated Light with EO Modulators ... 126
4.2.5 Photonic Generation of Electrical Signals and Radio Waves ... 127
4.2.5.1 Requirements for Photonically Assisted Signal Generation ... 127

4.2.5.2 O–E Conversion Devices ........ 128
4.2.5.3 Optical CW Sources ........ 128
4.2.5.4 Integrated Signal Generators ........ 129
4.2.6 Photonic Amplification of EO–MO Signals ........ 130
4.3 Measurement Systems and Their Applications ........ 130
4.3.1 Integrated Circuit Diagnosis Systems ........ 131
4.3.1.1 Noncontact IC-Probing System ........ 132
4.3.1.2 Contact Probing Systems ........ 134
4.3.2 Vector Network Analyzer ........ 137
4.3.3 Measurement Systems for Free-Space Radio Waves ........ 139
4.3.4 Systems for Two-Dimensional Mapping of Evanescent Fields ........ 143
4.3.5 Imaging of Objects ........ 146
4.4 Conclusion and Future Issues ........ 150
References ........ 151

## 4.1 ADVANTAGES OF OPTICAL MEASUREMENT [1]

### 4.1.1 Basic Concept

Although both radio waves and light waves are electromagnetic waves, their frequencies are very different. This is why in many substances the mutual interaction between radio waves and light waves is essentially weak. At the same time, there are (1) materials in which electrons and holes are generated by optical transitions and (2) materials that exhibit electro-optic (EO), magneto-optic (MO), or electroabsorption (EA) effects. In the former, the direction of interaction is from light waves to radio waves, whereas in the latter the interaction is in the opposite direction. The latter group of effects can be applied to a probe tip or sensor used in a measurement so that the radio wave or electrical signal information is projected onto the light wave.

### 4.1.2 Features of Optical Measurement

Conventional methods for measuring electromagnetic signals by electronic techniques are highly sensitive in nature because the electromagnetic interaction between the probe and target of the measurement is sufficiently strong. The strong nonlinear characteristics of electronic devices have been used to provide many useful functions such as frequency conversion and gate operations. These mature technologies stand in contrast to the newly emerging technology of optical measurement that is often known for its drawbacks such as the weak interaction between light and radio waves and the faint nonlinear characteristics of light waves. In fact, an optical measurement cannot be expected to achieve the same level of performance and functionality as an electrical measurement in the conventional sense. It does, however, possess the features described below, which can more than offset the above disadvantages.

#### 4.1.2.1 Ultrafast Operation

Advances in ultrashort optical pulse generation technology have made it possible to achieve optical pulse widths as short as several femtoseconds. This figure is symbolic of the ultrafast characteristics of photonic technologies and is also the ultimate index for the temporal resolution of measurement techniques that use optical pulses. In contrast, the minimum value for electrical pulse widths is of the order of picoseconds. The optoelectronic conversion of femtosecond optical pulses can be used to generate electrical pulses at the subpicosecond level, and while a simple comparison is not really meaningful for the reasons given below, photonic technologies can be said to be about three orders of magnitude faster. Actual systems have limiting factors other than optical pulse width, but subpicosecond operations can still be achieved to a sufficient degree. It is noted here that the high-speed characteristics of photonics are employed in a standard way to detect electromagnetic waves in untapped frequency bands such as the terahertz (THz) region.

#### 4.1.2.2 Low Invasiveness

"Invasiveness" is the degree to which a measurement system influences the electromagnetic field of the system being measured. In this regard, the weak interaction between light waves and radio waves, although a drawback in terms of sensitivity, turns out to be a great advantage in terms of low invasiveness. This is especially true because the interactive area within the material used for the interaction at the probe or sensor tip is limited and may be even microscopic in size. Of course, invasiveness cannot be made infinitely small because the permittivity and permeability of the sensor material differ from that in a vacuum. On the other hand, the format adopted for the optical path up to the sensor material is either free-space propagation or optical fiber propagation, which makes for a high level of noninvasiveness that is considered far superior to that of a metallic probe. With conventional electronic measurement techniques that require metal wiring, it is difficult to reduce the factors that disturb the measurement. The ability to conduct measurements without the use of metal is an outstanding feature of optical technologies.

Electronic measurement techniques may also increase the invasiveness beyond that associated with the probe tip, which again is a problem that may be solved by using optical technologies. In particular, when measuring radio wave signals in the microwave to submillimeter wave bands, it is impossible to avoid the three-dimensional spread of such waves in addition to planar propagation. We cannot dismiss the possibility of coupling between the electromagnetic field components being measured and the measurement system composed of a metallic probe and holder. One effective way of solving this problem is to use optical signal propagation with no electromagnetic coupling, which is another beneficial feature of the optical probe used in the optical fiber–based system described later.

#### 4.1.2.3 Spatial Expansiveness

There is little fear that light propagating through free space or through fiber will be disturbed by the electromagnetic field of the measurement target or an electromagnetic field radiating from the signal source. The reasons for this are essentially the same as those given above, but of particular interest here is that this characteristic can be used to configure a system that significantly extends the distance between the probe tip and the main unit of the measurement system. For example, it is difficult to transmit an electrical picosecond pulse signal over a distance of a meter or more but this becomes easy if it is converted to an optical signal. Optical fiber technology with dispersion control is particularly advantageous and the transmission of subpicosecond pulses over distances of the order of a kilometer will ultimately be possible. Remote and wide-area electromagnetic field distribution measurements using optical fiber networks are no longer a dream.

#### 4.1.2.4 High Spatial Resolution

In general, the wavelength of electromagnetic phenomena applied to measurements is a factor that limits spatial resolution. For light waves, this wavelength is of micrometer order which, when compared with microwaves with centimeter order wavelengths, represents an improvement in spatial resolution of at least four orders of magnitude. This is an extremely beneficial feature when performing near-field measurements on electromagnetic fields emitted from fine devices and circuits. Note that while the beam spot diameter is generally taken to represent spatial resolution, there are other factors that must be considered as described later.

The use of nanotechnology is also expected to contribute to even higher spatial resolution. To begin with, it is expected to help achieve spatial resolutions in the nanometer region, namely shorter than the wavelengths used by near-field optical techniques. In addition, a technique that combines nanometallic probes with optical probes, although resulting in some deterioration in terms of noninvasiveness, should be able to produce higher levels of resolution.

## 4.2 KEY TECHNOLOGIES FOR OPTICAL MEASUREMENT

### 4.2.1 Physical Phenomena Used for Optical Measurements

The optical measurement of high-frequency electrical signals or radio wave signals is illustrated in Figure 4.1. Various physical phenomena that arise from the interaction between light waves and radio waves can serve as the basis for a transducer that converts radio wave signals to optical signals. Typical phenomena include the EO effect, the photoconductive (PC) effect, the EA effect, and the MO effect. In all cases, special materials are needed to

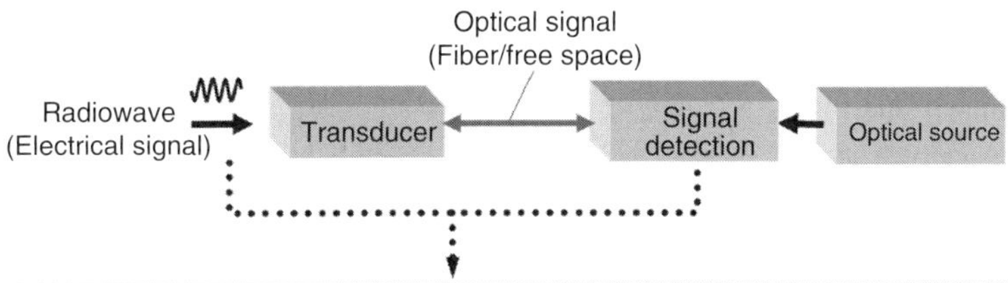


| Methods | Physical Phenomena | Example Materials |
|---|---|---|
| Electro-optic (EO) probing | Change in birefringence in electric field | $LiTaO_3$, CdTe, ZnTe |
| Photoconductive (PC) probing | Change in conductivity by light | LT-GaAs, LT-InGaAs/InAlAs |
| Electroabsorption (EA) probing | Change in light absorption in electric field | GaAs, InGaAs |
| Magneto-optic (MO) probing | Rotation of light polarization by magnetic field (Faraday effect, etc.) | YIG, GaP, TGG, ZnTe |

**FIGURE 4.1** Optical probing techniques and their physical basis.

make use of these various phenomena, and the sensitivity and frequency range of the electric field or magnetic field sensor varies greatly with the type and form of the material. Of these, the EO and MO effects have been most extensively employed in recent studies on optical measurements.

#### 4.2.1.1 EO Effect

Applying an electric field to a transparent solid or liquid changes the refractive index of that medium. Here, a change in the refractive index proportional to the electric field intensity is called the "Pockels effect," whereas that proportional to the square of the electric field intensity is called the "Kerr effect." The response of these effects is extremely fast (with a response time of the order of 0.01 to 0.1 ps), making them applicable to electric field measurements in the terahertz region.

#### 4.2.1.2 MO Effect

Passing linearly polarized light through a transparent medium in a magnetic field in a direction parallel to that field will cause the plane wave to rotate. This phenomenon was discovered by Faraday with lead glass and is sometimes called "Faraday rotation." It originates in an index of refraction that differs between left and right circularly polarized light. The phenomenon whereby the polarized plane of a reflected light rotates due to a substance in a magnetic field or to a magnetic body is called the "magnetic Kerr effect."

The application of the MO effect to optical measurement requires a magnetic body with a large Faraday rotation. However, the phenomenon of domain inversion that provides a large Faraday effect does not respond to magnetic field signals in the microwave frequency region. Obtaining a sufficient response to high-frequency signals therefore requires the use of the Faraday effect in magnetization phenomena based on different principles such as rotational magnetization. When iron garnet crystals are used, the ferromagnetic resonance frequencies of their rotation magnetization (RM) phenomena are a few tens of gigahertz or higher. An appropriate device design makes it possible to apply those crystals to MO probes with a bandwidth up to the resonance frequency [2]. It must be noted, however, that obtaining a response to high-frequency signals is often accompanied by low sensitivity due to insufficient magnetization intensity. Attention must be paid to the trade-off between speed and sensitivity.

## 4.2.2 Measurement Based on EO Effect

### 4.2.2.1 Principle of EO Probing

The optical measurement of electrical signals based on the EO effect is known as EO probing. The principle of EO probing is the modulation of the intensity of light by an electric field or an electrical potential, as illustrated in Figure 4.2 [3–6]. In other words, an unknown electrical signal is measured from changes in optical intensity. Here we explain the mechanism for converting the electrical signal to the optical signal. During EO probing, a special material called an EO material serves as a transducer. The refractive index of the EO material changes instantaneously according to the strength of an applied electric field. Light that passes through the material whose refractive index is varied by an electric field undergoes a change in polarization. That change in polarization can then be converted to a change in light intensity by passing the light through a polarizer. By using pulsed light, it is possible to measure the temporal changes in the waveform of the electrical signal with a time resolution almost equal to the pulse width. This technique is called EO sampling (EOS). Lasers with pulse widths of 0.1–0.3 ps are now readily available, and so it is easy to obtain a subpicosecond temporal resolution, which converts to a bandwidth of about 1 THz.

Using continuous light rather than pulsed light, it is also possible to measure the waveform or frequency with an electrical sampling oscilloscope or a spectrum analyzer after conversion to an electrical signal by means of a high-speed photodiode [7–9]. In such cases, the bandwidth is determined by the performance of the photodiode and the electrical measurement technology behind it. Furthermore, if the measurement band is broadened, optical shot noise increases and so sensitivity decreases. Accordingly, measurement at several gigahertz is appropriate.

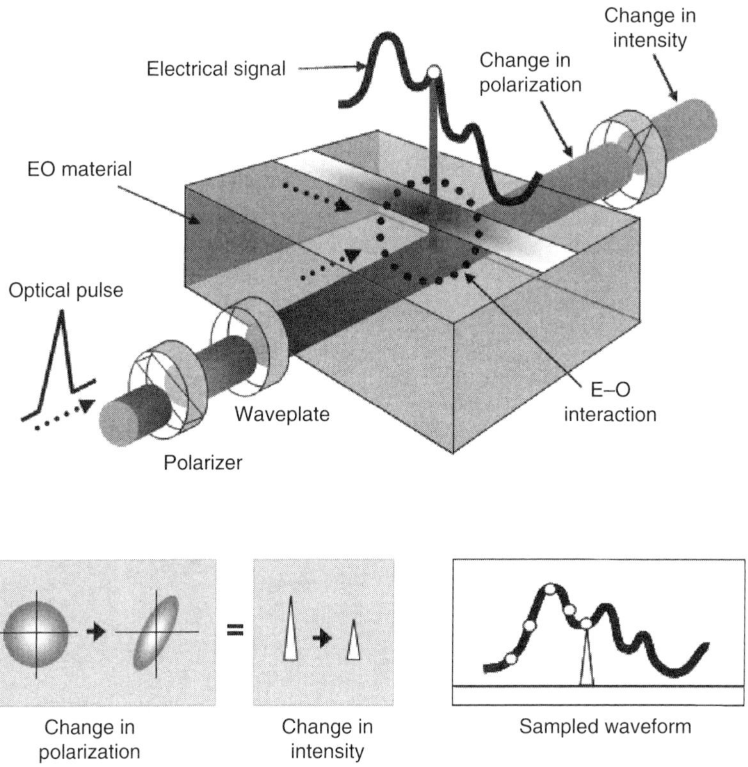


**FIGURE 4.2** Principle of electrical signal (electric field) measurement based on EO effect.

#### 4.2.2.2 Probing Configurations

Figure 4.3 shows the principle of EO probing applied to the measurement of an electrical signal propagating across an IC or circuit board wiring or the measurement of an electric field (radio waves) from an antenna or other such radiating bodies. In Figure 4.3a, an EO material is brought near the wire and the refractive index of the material is changed by the electric field (fringe field) generated by the wire. A dielectric mirror is formed on the bottom surface of the electro-optical crystal to reflect the laser light, and the reflected light is polarized according to the electric field strength. In Figure 4.3b, a small gold needle is attached to the EO material. When that needle comes in contact with the wire, the bottom surface of the material takes on the same electrical potential as the wire and thus becomes more sensitive. In this form, the potential of the wire can be known precisely. For both the contact and noncontact methods, the input impedance of the measurement system is an order of magnitude higher than for conventional electrical methods, so there is extremely little disturbance of the object being measured. When the object is a radiating body, such as an antenna,

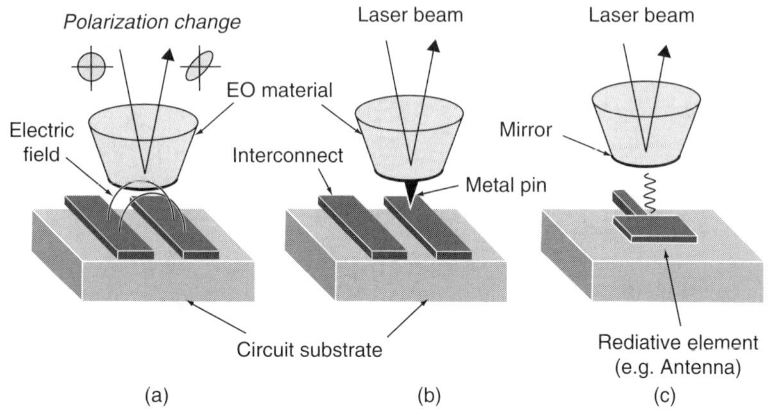


**FIGURE 4.3** Typical configurations for device measurement based on EO probing (a) Non-contact electric-field probing, (b) Contact voltage probing, and (c) Probing of radio waves propagating in free space.

the waveforms and strength of radio waves ranging from microwaves to submillimeter waves can be measured by placing the EO material in the propagation path of the radio waves as shown in Figure 4.3c.

### 4.2.2.3 EO Materials

Typical EO materials and their properties are listed in Table 4.1. The main properties required for EO materials are listed below:

**TABLE 4.1**
**Representative Materials for EO Probing**

| EO Material | $n$ | $r$ (pm/V) | $\varepsilon$ | $n^3r$ | $\lambda$ (μm) | Direction of Electric Field |
|---|---|---|---|---|---|---|
| $LiTaO_3$ | 2.18 | 30.3 | 43 | 313 | >0.4 | L |
| $LiNbO_3$ | 2.20 | 30.8 | 28 | 327 | >0.4 | L/T |
| GaAs | 3.5 | 1.2 | 13 | 51 | >0.9 | L/T |
| KD*P | 1.51 | 24 | 48 | 81 | >0.2 | L |
| KTP | 1.86 | 35 | 15.4 | 224 | >0.35 | L/T |
| BSO | 2.60 | 5.0 | 56 | 78 | >0.4 | L/T |
| BTO | 2.70 | 5.7 | 47 | 112 | >0.5 | L/T |
| ZnTe | 3.1 | 4.3 | 10 | 128 | >0.6 | L/T |
| CdTe | 2.8 | 6.8 | 9.4 | 148 | >0.9 | L/T |
| Poled polymer | 1.63 | >30 | 3.2 | 130 | >0.6 | L |
| DAST | 2.46 | 50 | 7 | 744 | >0.7 | T |

L: longitudinal-field probing, T: transverse-field probing, KD*P: $KD_2PO_4$, KTP: $KTiOPO_4$, BSO: $Bi_{12}SiO_{20}$, BTO: $Bi_{12}TiO_{20}$, DAST: 4-dimethylamino-*N*-methyl-4-stilbazolium tosylate.

1. Large $n^3r$ ($n$: refractive index, $r$: EO constant)
2. High resistivity, $\rho$
3. Low dielectric constant, $\varepsilon$
4. Sensitivity to electric fields of one particular direction
5. No strong absorption at the wavelength of the laser light that is used
6. Easy machine processing (cutting and polishing) and the formation of a film mirror

The first property is related to sensitivity, the second to responsiveness at low frequencies, the third to capacitive load and disturbance, and the fourth to spatial resolution and the cross talk characteristics. Generally, the change in the refractive index of the EO material in response to a change in the electric field is proportional to $n^3r$; hence this value is used as a measure of sensitivity. Also, with noncontact measurement, where there is a gap between the wiring line and the EO material (Figure 4.3a), the sensitivity is lower if the dielectric constant is high, because the electric field is prevented from penetrating the EO material. Therefore, $n^3r/\varepsilon$ is used as the figure of merit for the sensitivity of the material.

Furthermore, with regard to the fourth of the above properties, it is important to decide whether to use a longitudinal type of material, in which the direction of the laser light beam and the direction of the electric field to be detected are parallel, or to use a transverse type of material, in which the laser beam and electric field lines are perpendicular. For EO probing, the longitudinal type is generally preferred to the transverse type, except for special cases where we measure the electric field that is parallel to the IC substrate because the spatial resolution is better.

Of the vertical materials, semiconductors such as CdTe, ZnTe, and GaAs have the merits of relatively high $n^3r$ values and small $\varepsilon$ values. However, because the resistivity is low (no more than $10^9$–$10^{10}$ $\Omega$ cm), electric fields at frequencies lower than about 1 MHz are screened out by free electrons and the sensitivity decreases. Since this is not a problem when use is restricted to the measurement of high frequencies of above 1 GHz, these materials are often used in high-speed IC probes. On the other hand, whereas the oxide crystals BSO and BTO are also vertical-type materials, they have high resistivity values of $10^{14}$–$10^{15}$ $\Omega$ cm, and so are responsive even at low frequencies [10]. The dielectric constant, however, has a high value of 50, and so these materials are rather disadvantageous in terms of sensitivity and disturbance when used in IC probes. Therefore, the use of a metal conductor for direct contact as shown in Figure 4.3b is the better method. As revealed by the above, there is little room as regards the choice of EO material to be used in EO probing. If we assume the use of long-wavelength laser light sources ($>1$ $\mu$m), the choice is restricted to CdTe, ZnTe, BTO, and KTP. Further development of EO-probing technology requires the development of new EO materials. Polymers [11] and organic crystals hold considerable promise as replacements for semiconductors and oxide crystals.

Among various organic crystals, the organic ionic salt crystal 4-dimethylamino-*N*-methyl-4-stilbazolium tosylate (DAST) is a promising nonlinear optical material that exhibits a larger EO coefficient ($r_{11} = 50$ pm/V at 1.55 μm) and a lower dielectric constant ($\varepsilon = 3$ to 5) than conventional EO materials [12]. A new crystal growth technique, called the slope nucleation method (SNM), enables us to control both the nucleation position and the growth orientation of crystals during spontaneous nucleated growth, which provides the highest quality DAST ever reported. Zheng et al. [13] and Adachi et al. [12] demonstrated the S:N ratio improvement of a DAST EO transducer in a sampling measurement of high-speed devices by comparing it with conventional inorganic crystals $LiTaO_3$ and KTP. The huge static birefringence of DAST is sometimes a problem in practical applications, but it can be compensated for by inserting a quarter-wave plate between the crystal and the mirror in the reflection geometry of the sensor [13]. In addition to the EO transducer, DAST can be applied to the radiation of terahertz signals by using the difference frequency generation technique [14].

#### 4.2.2.4 EO Probes

Electric field sensors fabricated from various EO materials based on the principle described above are generally called EO probes. Examples of actual EO probes that employ CdTe are shown in Figure 4.4. Figure 4.4a shows a probe for the noncontact measurement of an electrical signal waveform within an IC chip. The probe tip is fabricated with a diameter of about 100 μm to enable it to be brought to within micron distances of the IC chip. The CdTe is part of the probe tip and its thickness is also about 100 μm. It is fixed to the quartz support (the white part in the figure). The laser beam diameter is about 2–5 μm at this probe tip. The probe shown in Figure 4.4b is used for contact-type IC measurements as described later in this chapter. It has a micropin with a tip diameter of about 1 μm attached to a CdTe crystal, and the laser light is delivered by an optical fiber. Figure 4.4c shows an electric field probe for measuring radio waves propagating in free space. The CdTe crystal in this probe is several millimeters in size.

### 4.2.3 Measurement Based on MO Effect

#### 4.2.3.1 Principle of MO Probing [2,15]

In this section, we describe an alternative method to EO probing. It is based on the MO effect. In other words, the Faraday rotation effect is used to provide a sensing capability of magnetic fields instead of electrical fields. When a contiguous measurement condition is provided, the scheme can realize a function for profiling distributions of currents in devices under test. It is well

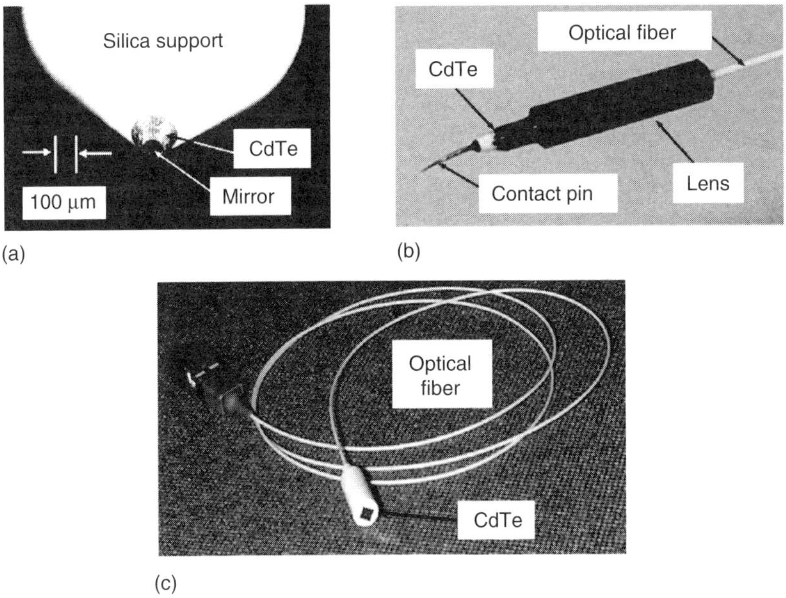


**FIGURE 4.4** Example of EO sensor for (a), (b) IC diagnosis, and (c) free space.

known that in-plane magnetic fields correspond to current distributions in the plane, provided that the field equivalence theorem is effective. This current profiling capability is unique and attractive if an EO measurement is not sufficient. Typical examples include cases where the impedance of the circuit under test is unknown or fairly low.

The complementary combination of MO and EO probing would be useful with regard to the former case, in which circuit patterns are unique or complicated and therefore the local impedances of the circuits are unclear. Although an absolute evaluation of local impedance is not easy because of the natures of the EO and MO probing schemes, its relative value could be obtained via an EO–MO combination scheme. As for the latter case, it is notable that MO probing is more suitable when dealing with high admittance lines where the voltage variation is not well distinguished along the lines. Such cases include networks of power and ground lines in an large scale integrated circuit (LSI) chip. One example of this type of measurement is described later.

Furthermore, it is notable that current information is more important and intuitive than voltage information, as found in many textbooks on electromagnetism, when we deal with electromagnetic waves radiating from antennas or antenna-like circuits. Since MO probing is sensitive to the current distribution of circuits under test, it is expected to be useful when searching for hidden emission sources. This is a required functionality as regards electromagnetic compatibility issues.

In spite of these attractive features, the development of MO probing has stagnated to some extent. This is because sensitive MO effects were assumed to be slow compared with the EO effects. However, the situation has changed since the development of RM-based probes, as described below.

### 4.2.3.2 Probing Configurations and MO Materials

The MO probing configuration is basically the same as that for EO probing although the rotation of the light wave polarization is detected through a polarizer (Figure 4.5a). However, optical pulse sources are seldom used. Instead, a continuous wave (CW) light source or a sinusoidally modulated light source is often employed. This is because the bandwidths of the MO materials are rather narrow, much narrower than those of EO materials, and in most cases the materials do not respond to picosecond pulses. Therefore, an optical pulse source is not necessarily used in the MO probing configuration. At the same time, the main targets of MO probing have been in the microwave frequency range or below.

The properties of MO materials, mostly crystals, required for the photonic sensing of magnetic fields are as follows:

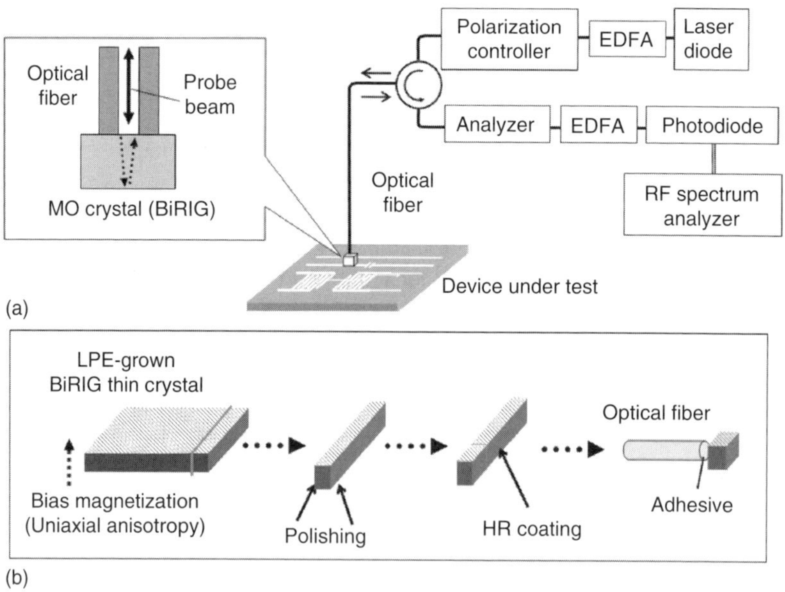


**FIGURE 4.5** (a) Schematic of typical MO probing system with a magnified view of its probe head. (b) Method to fabricate a fiber-facet-type MO probe is indicated.

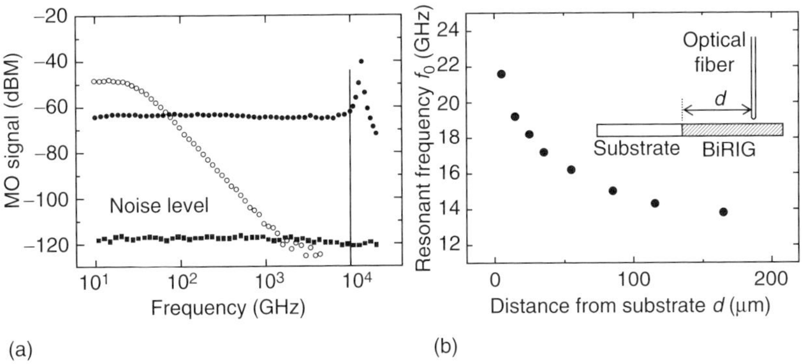


**FIGURE 4.6** (a) Frequency response of RM-based MO probe is depicted with solid dots (signal) and solid squares (noise). 10 GHz bandwidth is indicated. In comparison, that of a domain inversion–based probe is also shown by open dots. (b) Ferromagnetic resonance frequency, the bandwidth of the MO probe in other words, is plotted as a function of distance from MO crystal substrate. The bandwidth can be improved to be beyond 20 GHz.

1. The material is transparent at the probe light wavelength.
2. The material has a large Verde constant.
3. The MO response is fast enough to cover the frequency range of interest.
4. The magnetic axis is uniform within the sensing volume.

Garnet crystals are typically used for this purpose. A commercially available transparent iron garnet crystal with a large Verde constant is often used in optical isolator applications. However, the Faraday rotation in such crystals depends predominantly on the domain inversion magnetization mechanism and its frequency response is usually limited in the megahertz range. Therefore, such crystals cannot be applied to radio frequency field measurements. Elezabbi et al. [16] pointed out that the RM in a Bi-replaced iron garnet crystal is faster and its ferromagnetic resonance frequency is around a few tens of gigahertz or beyond. Furthermore, Yamazaki and his coworkers pointed out that the RM scheme can be applied to high-speed measurement up to the ferromagnetic resonance frequency (Figure 4.6). They invented an elegant method for implementing an RM-based MO probe, in which a tiny single domain MO crystal is sliced off and affixed to the end facet of an optical fiber. Here, the optical fiber facet is used both to launch light waves into the crystal and as a cantilever to suspend and carry the crystal (Figure 4.5b). The detailed fabrication procedure has already been reported and the schematically drawn concept of the fabrication is shown here in the lower part of Figure 4.5b. This fiber facet probe is beneficial if the piece of crystal is thin and tiny, and an RM crystal is often so.

#### 4.2.3.3 MO Probes

The application of the RM scheme and the fiber facet scheme to the MO probe has made it possible to scan magnetic near-fields of RF circuits with a high bandwidth and high spatial resolution. Regarding an example for the latter performance, Figure 4.7 shows a fine measurement result of magnetic field distribution over a 100 μm scale circuit together with a conceptual figure of the sensitive volume. At present, the fiber-facet-type RM MO probe is dominant. However, there is a drawback with that type of probe in that only the device under test (DUT) surface-normal component of the magnetic field is measured. To overcome this, a probe designed to detect the surface-parallel component was proposed and demonstrated.

Unfortunately, the sensitivity of an MO probe is less than that of an EO probe with the same configuration as far as a 50 Ω planar waveguide is concerned. For a low impedance transmission line such as a superconductor line, the situation would be different [17].

### 4.2.4 Optical Signal Sources for Probing

#### 4.2.4.1 Pulse Source

Although there was a boom in research on the optical measurement of high-speed electrical signals in the last half of the 1980s, it went no further than producing tools for optical scientists and physicists. One reason for this is that

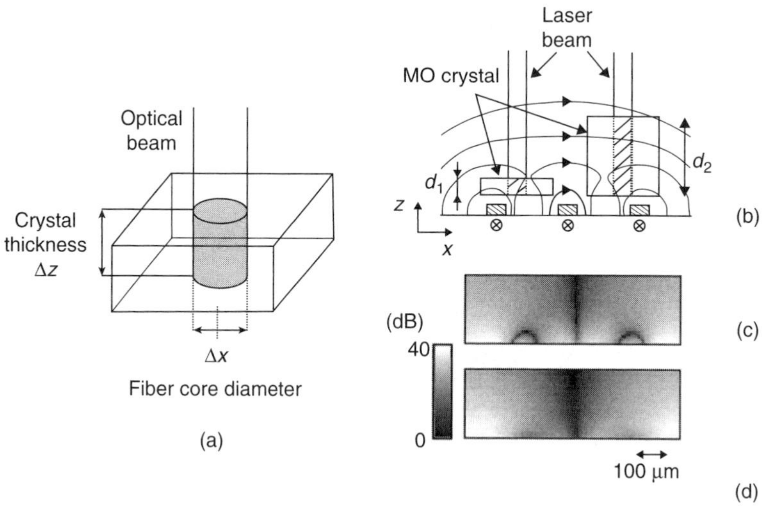


**FIGURE 4.7** (a) Definition of sensitive volume, which is an important parameter of MO probe. (b) Two probes with different crystal thicknesses, different sensitive volumes in other words. (c), (d) Spatial profiles.

dye lasers and Nd:YAG/YLF lasers, which require large-scale facilities and skill in operation, had been used as the optical pulse sources to obtain temporal resolutions of much better than 1 ps. Nevertheless, as the result of progress in the development of compact and stable optical pulse sources based on semiconductor lasers in the 1990s, optical measurement technology reached such a level of practicality that it became the standard for high-speed electrical measurement.

The performance requirements for a pulse source used for optical measurement are listed below:

1. Low repetition frequency: 1 MHz to 1 GHz
2. Narrow pulse width: 100 fs to 10 ps
3. Low jitter: less than the pulse width
4. Easy synchronization with an external signal
5. High output: greater than 10 mW
6. Can be mounted in a module or rack that is resistant to vibration
7. Adjustment and maintenance are not necessary

The first of the above is required for general sampling measurements; the repetition frequency of the sampling pulse must be set lower than the operating frequency of the DUT. To cope with various operating frequencies, an upper limit of 1 GHz for the sampling frequency was considered.

An example of an optical source that satisfies requirements 1 through 7 is shown in Figure 4.8. First an optical pulse of about 10–20 ps is obtained by direct modulation of a semiconductor laser using the gain switching method [18] or by external modulation of the output of a semiconductor laser with an

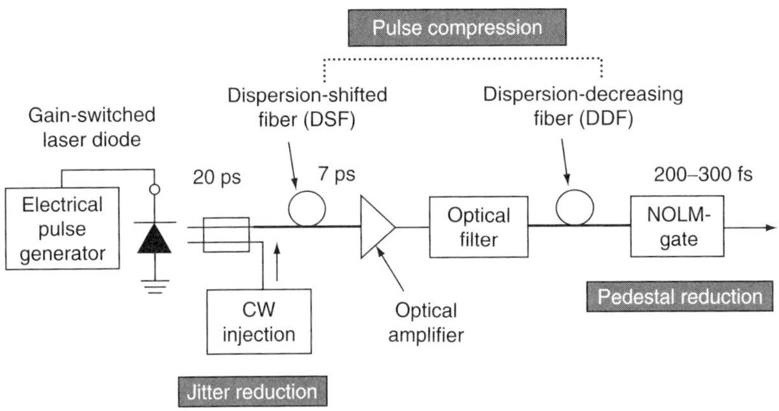


**FIGURE 4.8** Block diagram of repetition-rate variable optical pulse generation with jitter and pedestal reduction scheme.

electric field absorption type modulator [19]. Then the optical fiber is used to compress the pulse to 200–300 fs [20,21]. This optical source allows us to set the repetition frequency of the light pulse at any value because the repetition frequency is determined by the frequency of the driving electrical pulse. Although timing jitter arises in the pulse oscillation process with the gain switching method, this jitter can be reduced to the subpicosecond level by injecting a small amount of continuous light into the laser [18]. Another way to effect pulse compression is to use a nonlinear optical fiber to broaden the light spectrum, and then extract light of the desired spectral width and center frequency with filters to obtain an optical pulse with the ideal transform-limited waveform [22]. Recently, optical pulse sources called fiber lasers, in which the resonator is formed by the optical fiber itself, have been used in optical measurement in addition to semiconductor lasers, thus improving stability of operation and operability [23,24].

#### 4.2.4.2 CW Light and Sinusoidally Modulated Light with EO Modulators

As described above, an ultrashort optical pulse source is effective for measuring electromagnetic signals with ultimately high temporal resolutions. However, on the other hand, we have to deal with the rather sophisticated and costly optical apparatus of the picosecond or subpicosecond pulse source. Alternatives to such a pulse source are a nonmodulated or a sinusoidally modulated light source if the frequency of interest is in the microwave or millimeter-wave (MMW) range. As described below, such schemes offer the simplicity of light sources and other advantageous features.

The former scheme needs a high-speed photodetector and a fast (or high-frequency) electrical analyzer such as an RF spectrum analyzer or a sampling oscilloscope. This rather straightforward method has become more attractive recently as a result of the progress made on photodetectors with high power and high-speed features, which are usually in a trade-off relationship. The unitraveling carrier photodiode is one such photodiode. The high power feature is important since the sensitivity of a probe is proportional to the detectable optical input power of the photodetector, provided that the power of the light source or the optically amplified power is sufficiently high.

One benefit of the latter is that we can utilize the photomixing effect at an EO–MO crystal and, therefore, a slow photodetector can be used as in optical pulse sampling measurements. This situation makes it possible to launch a high optical power into the photodetector, which leads to higher sensitivity. In addition, we can obtain more flexible frequency tunability if we use an RF EO modulator to generate a sinusoidally modulated light, which is difficult to achieve with any optical pulse source. Furthermore, it should be noted that a CW laser is better than a pulsed laser with respect to relative intensity noise. High-frequency EO Mach–Zehnder interferometer modulators are available and the MMW range is already covered by this method. Moreover, their half-wave voltages are becoming lower and, eventually the harmonics of driving electric signals in the optical domain are expected to reach the submillimeter-wave frequency range or beyond.

## 4.2.5 Photonic Generation of Electrical Signals and Radio Waves

### 4.2.5.1 Requirements for Photonically Assisted Signal Generation

For high-frequency measurements, the method of generating the electrical signal that is fed to the DUT is as important as signal detection. With a conventional electrical signal generator, there is an upper limit of about 110 GHz for sinusoidal-wave signals, and of 40 Gb/s for digital pulse signals. Other problems that appear at higher frequencies are the attenuation of the level and distortion of the waveform of the electrical signal as it passes through cables and connectors. The attenuation of the signal level is typically more than 50% in one meter of coaxial cable for use with 50 GHz signals. Expectations are building that photonics technology will exceed the limits of the conventional technology as regards the generation and transmission of high-frequency signals.

Figure 4.9 shows the basic idea of a fully photonics-based measurement system designed to deal with the above issue [25,26]. We believe this to be the ultimate form of future optical measurement. One model of this system has been implemented in a network analyzer (NA) that is described later.

An optical-to-electrical (O–E) converter generates electrical signals that are transmitted as a stimulus through free space or through a transmission line to a DUT. A probe consisting of an electrical-to-optical (E–O) transducer, which is described in the above sections, then optically detects the electrical response of the DUT. Here, lossy coaxial cables and bulky, metallic hollow waveguides for delivering and receiving high-frequency electrical signals are replaced with optical fiber cables.

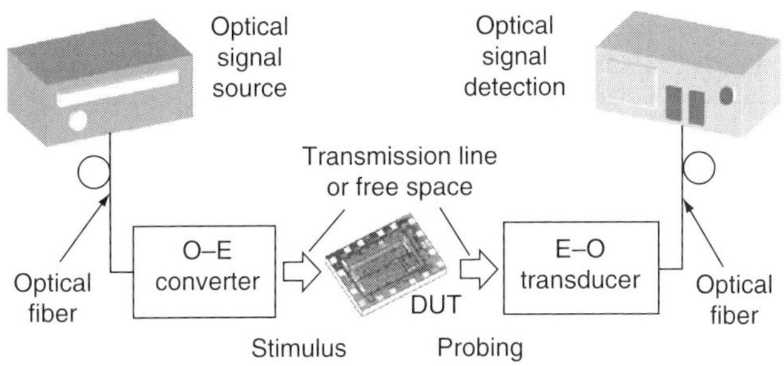


**FIGURE 4.9** Concept of all-photonics-based measurement system based on optical generation, transmission, and detection techniques.

#### 4.2.5.2 O–E Conversion Devices

Photoconductors and photodiodes have been used as O–E converters. Low-temperature-grown (LT-) GaAs is a widely used PC material for pulse and sinusoidal-signal generation at a laser wavelength of <800 nm [27,28]. In the 1.3 to 1.55 μm wavelength region, where the loss and dispersion of fibers are minimal, both the bandwidth and the output power level of PIN photodiodes have recently been greatly improved through the use of waveguide, traveling wave and distributed structures [29,30], and novel carrier dynamics [31,32]. Of these, the unitraveling carrier photodiode (UTC-PD) [5,6] is unique in that it provides both a large bandwidth and a high output current. As regards its operation, only electrons are used as active carriers, and hole transport directly affects neither its PD response nor its output saturation mechanisms (Figure 4.10). The smallest pulse width obtained to date is 0.97 ps, and this corresponds to a 3 dB bandwidth of 310 GHz [33]. Using a UTC-PD and a mode-locked laser diode (MLLD), a 100 GHz CW with an output power of 11 dBm has been obtained [34]. This is comparable to the performance of solid-state electronic devices.

#### 4.2.5.3 Optical CW Sources

With respect to optical CW (sinusoidal or quasi-sinusoidal) sources, we can use an MLLD integrated with saturable absorbers [35–39] and heterodyne laser mixing [40–42]. There are two main types of operation and stabilization scheme for MLLDs: an active MLLD with electrical injection [35,36] and a passive MLLD with optical injection [37–39]. With either scheme, signals of greater than 200 GHz with subharmonic injection frequencies can be

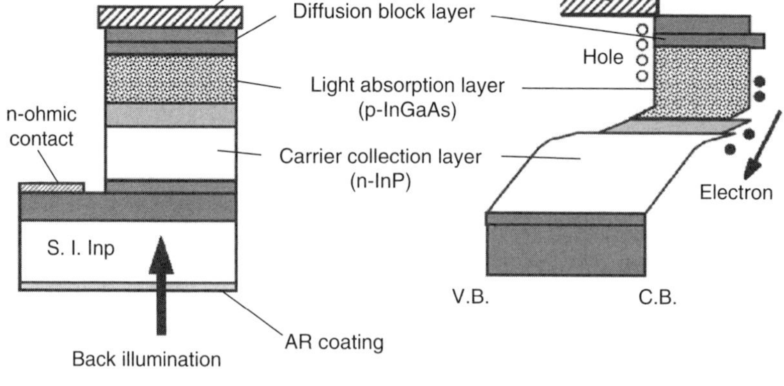


**FIGURE 4.10** Band diagram and epitaxial layer structure of UTC-PD.

generated, as long as the fundamental frequency is within the capability of a microwave synthesizer, that is, from 10 to 60 GHz.

Optical heterodyne mixing employs the beat signal from the interference of two optical frequencies. This scheme allows us to extend the frequency range into the terahertz regime easily with greater tunability of frequency. When two single-frequency (wavelength) lasers are used, the phase-locking of the two lasers is crucial to obtaining sufficiently narrow spectral line widths. The combination of a fiber laser and an external cavity semiconductor laser is one of the best options as regards achieving a high degree of spectral purity over a broad frequency range [40]. The selection of two lines from a single laser that produces lines at multiple wavelengths, such as a Fabry–Perot laser diode, a mode-locked laser, and an optical comb generator, is an easier and simpler way to generate stable beat signals [41,42].

#### 4.2.5.4 Integrated Signal Generators

Figure 4.11 shows two examples of fiber-optic electrical signal generators that employ UTC-PDs. In Figure 4.11a, an O–E conversion probe head for triggering electronic devices and ICs on a wafer is formed by bonding the UTC-PD to a coplanar waveguide (CPW) made on a quartz substrate

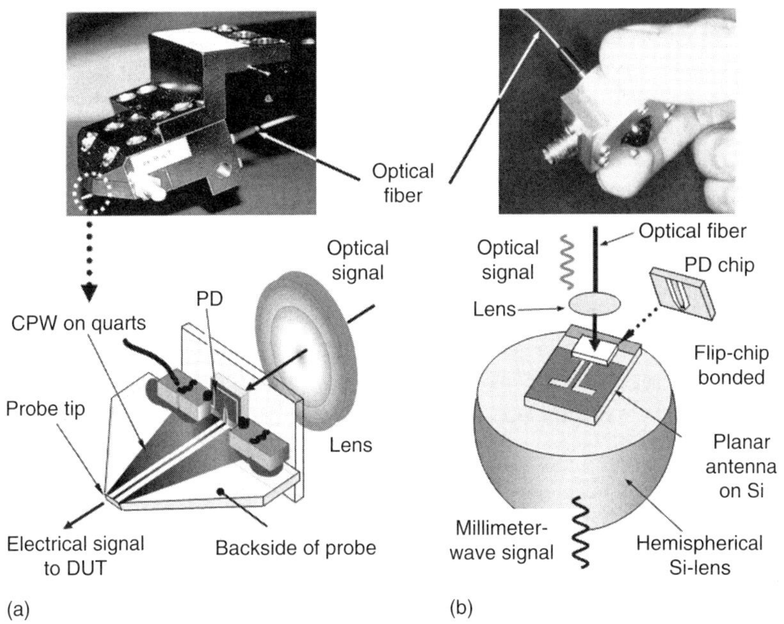


**FIGURE 4.11** Fiber-optic electrical signal generators. (a) O–E conversion probe head for testing electronic devices on wafer. (b) Millimeter-wave signal emitter for free-space applications.

[43]. The electrical signal from the photodiode is delivered to the input pad of the DUT via a low loss, low distortion transmission line. It is possible to generate an electrical signal that has a pulse width of 3 ps (equivalent to a bandwidth of greater than 100 GHz) and an amplitude of 1 V [43], which is sufficiently high to drive digital ICs directly.

A free-space MMW emitter [44,45] for wireless applications is shown in Figure 4.11b. The UTC-PD chip is flip-chip mounted on a planar slot antenna on a silicon substrate. The antenna chip is bonded to a hemispherical silicon lens to collimate MMW signals in the direction opposite to the illumination. At 120 GHz, >10 mW was generated in the PD chip and >8 mW was emitted into free space.

Moreover, the monolithic integration of PDs with antennas on the same substrate has been successfully demonstrated. By eliminating the component bonding process, this approach improves both ease of manufacture and performance [46–48]. Log-periodic Yagi–Uda and patch antennas have been monolithically integrated with the UTC-PD on the InP and Si substrates, respectively [47,48].

### 4.2.6 Photonic Amplification of EO–MO Signals

Since optical fibers are widely used in photonic measurement systems as described above, the application of fiber-optic amplifiers to these systems is both straightforward and attractive. It should be noted that an Er-doped fiber amplifier (EDFA) can generate 30 dB gain for light waves over a fairly high terahertz bandwidth. The gain in the light wave band is doubled after its O–E conversion and corresponds to 60 dB. This level of signal amplification performance is not available with RF domain electronics. Indeed, EDFAs combined with appropriate optical bias settings have proved to be effective in enhancing the sensitivity of photonic probes [49]. Figure 4.12 shows a case for MO probe measurements, in which 46 dB gain by EDFA was achieved via avoiding the gain saturation phenomenon. In addition, Figure 4.13 indicates measured MO signal and noise plotted functions of polarizer angle. Both are well explained by theory and further possible improvements are suggested.

## 4.3 MEASUREMENT SYSTEMS AND THEIR APPLICATIONS

Many measurement instruments and systems based on the EO or MO probing techniques have been developed to evaluate and inspect various types of electronic components, ranging from transistors to circuit boards and modules (Table 4.2). They include a NA for evaluating individual elements such as transistors or filters, and antennas, a noncontact probe, and microprobe for integrated devices and circuits, and a pen-type probe for measuring circuit boards and modules on which the devices are mounted. All these instruments have bandwidths that exceed those of conventional devices. Below, we describe these instruments briefly and present example measurements.

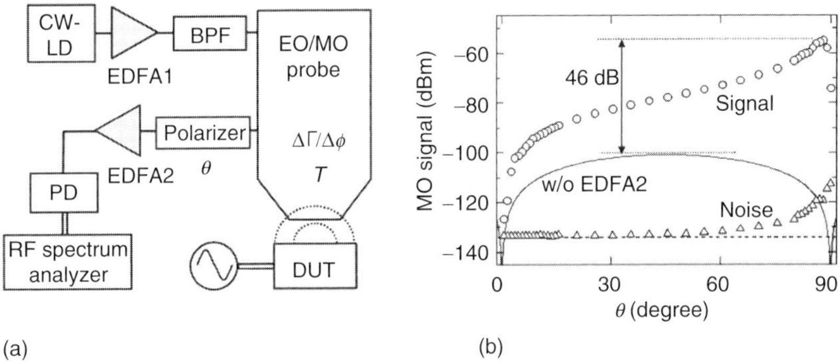


**FIGURE 4.12** (a) Model of photonic probing system. (b) Measured MO signal data plotted as functions of polarizer angle. The use of EDFA2 in an MO system is indicated to lead to 46 dB gain of its MO signal under a condition of fixed (limited) photodetector input.

### 4.3.1 Integrated Circuit Diagnosis Systems

When there is an abnormality in the output signal of the IC and it is judged that the IC is not operating as designed, an effective next step is to investigate the state of the electrical signal inside the IC chip. Many transistors, resistors, capacitors, and the wiring that connects them are integrated inside the chip. By confirming the presence of electrical signals properly in the wires, it is possible to locate the source of the failure.

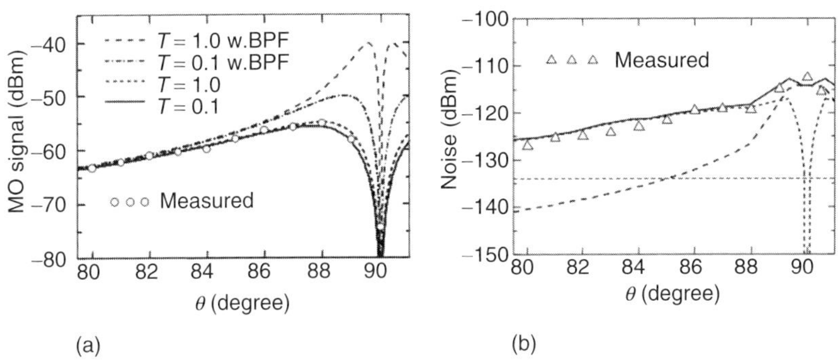


**FIGURE 4.13** (a) Measured MO signal power data, which are indicated by open circles, agree well with a theoretical curve. Also, suppression of optical loss in the probe is suggested to increase the MO signal power further. (b) Measured noise data (triangles) are well explained by a calculation. Main noise components are brought by beat noises.

**TABLE 4.2**
**Comparison of Photonic and Conventional Instrument**

| Device under Test | Transistor | IC Chip | Circuit Board |
|---|---|---|---|
| Instruments | Network analyzer | IC probing system | High-impedance probe |
| Conventional | DC-110 GHz (or band operation) | 5 GHz | 2.5 GHz |
| Photonic | >300 GHz | >1000 GHz | >10 GHz |

#### 4.3.1.1 Noncontact IC-Probing System

##### *4.3.1.1.1 Voltage Signal Monitors*

Figure 4.14a shows a configuration for measuring the electrical signal waveform within an IC chip and Figure 4.14b shows the developed IC-probing system [50,51]. This system can trace the electrical signal within the IC from the input pin to the output pin. Figure 4.15a shows examples of electrical signal waveform measurements made with this system within a digital telecom IC operating at 20 Gbit/s. It can be seen that the 1 and 0 levels within an IC that operates at such high speed are not as clear as we might expect; the waveforms are complex due to such factors as the superimposition of signals leaking from adjacent lines. Using such data obtained from actual measurements as feedback to the design process can contribute to the development of higher-performance ICs and to the shortening of the development and fabrication period.

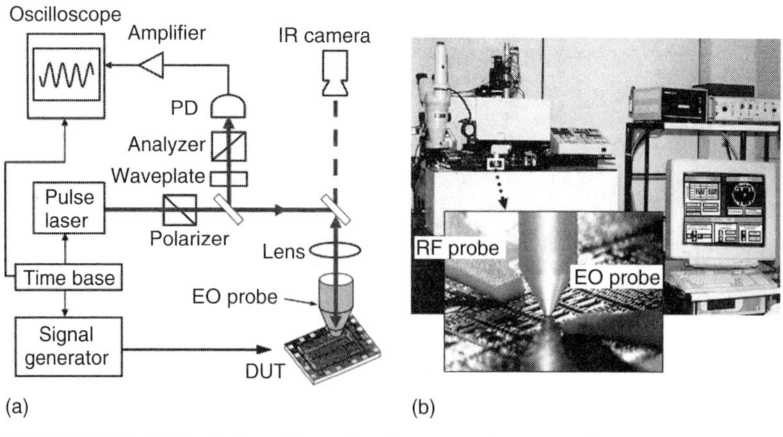


**FIGURE 4.14** (a) Typical configurations for measuring internal-node voltage waveforms in ICs based on EO probing. (b) Fully automated IC testing system.

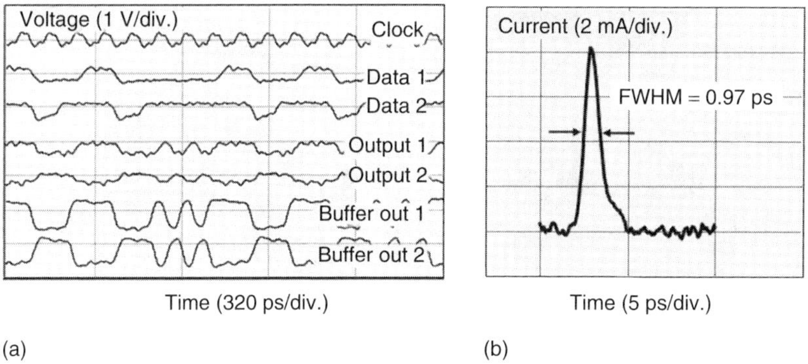


**FIGURE 4.15** Examples of measured temporal waveforms for (a) communication use IC and (b) photoresponse of photodiode.

Figure 4.15b shows the time response waveform of a UTC-PD, which is a high-speed device that has a bandwidth of more than 300 GHz [33]. For this kind of high-speed device, there is no nondisturbing evaluation method other than EOS. Normally, a voltage sensitivity of 0.1–1 mV (unit band conversion) can be obtained.

By moving the EO probe around the DUT, we can map the two-dimensional electric field distribution over the IC. This is a powerful tool especially for diagnosing Monolithic Microwave Integrated Circuits (MMICs) in combination with an electromagnetic field simulator. More details are provided in Section 4.3.4.

#### *4.3.1.1.2 Current Signal Monitors*

It is known that the switching action of a logic gate generates a charge transfer between ground lines and power supply lines in an integrated circuit. This charge transfer phenomenon leads to electrical potential fluctuations in those lines at the harmonics of the clock frequency and therefore has the potential to cause logic circuit malfunctions. This makes it important to reduce such potential fluctuations by employing a decoupling capacitor scheme in an LSI. At the same time, it is necessary to monitor such current behavior and the degree of fluctuation suppression. Here, it should be noted that mainly the top wiring layer of an LSI is visible, and this is usually composed of networks of 10 μm wide or wider ground or power supply lines and consequently is of high admittance nature. An MO probe is a powerful tool for this purpose, namely, for the visualization of current flows in those ground and power supply lines. This is shown in Figure 4.16 [52].

#### *4.3.1.1.3 LSI-Pins Monitors*

The proper installation of an LSI on a printed circuit board is one of the most important issues in the electromagnetic compatibility field. The goal is the sufficient suppression of the unnecessary emission of electromagnetic waves

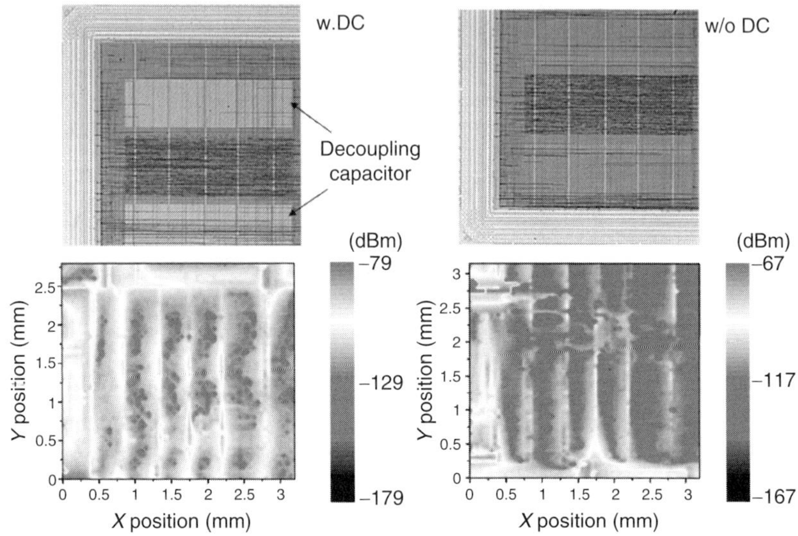


**FIGURE 4.16** Distribution of currents at second harmonic frequency of clock in power and ground lines and areas measured by a fiber-facet edge MO probe with a 10 μm spatial resolution are shown with micrographs of an LSI under test. The top photos show two areas of the LSI with (left) and without (right) decoupling capacitors (DC). The bottom figures are corresponding measured data. Some effects via DC are apparent.

from the printed circuit board. Again, decoupling condensers are effective for this purpose. Here, it is essential to monitor the current at the LSI pins. The fiber facet MO probing scheme is useful because the probe can penetrate into the vicinities of LSI pins without a fatal invasive effect on the circuit of interest. An example of such measurements is shown in Figure 4.17 [53].

### 4.3.1.2 Contact Probing Systems

#### *4.3.1.2.1 Microprobe*

Recently, microprocessor operating speeds of gigahertz order have been achieved, and there is an increased demand for the high temporal resolution observation of the internal waveforms of data processing Ultra-large Scale Integrated Circuits (ULSIs) as well as the waveforms within small-scale telecommunication ICs. Thus, we require a spatial resolution that is at least of submicron order. However, noncontact probes for detecting the electric fields leaking from interconnect lines cannot meet this requirement. Furthermore, high-density LSI chips contain multiple layers of wiring, so it is often desirable to perform measurements with no signal cross talk from lower layers.

To deal with the above problem, we can use a probe consisting of a micropin attached to an EO crystal (Figure 4.3b) [54,55]. To measure

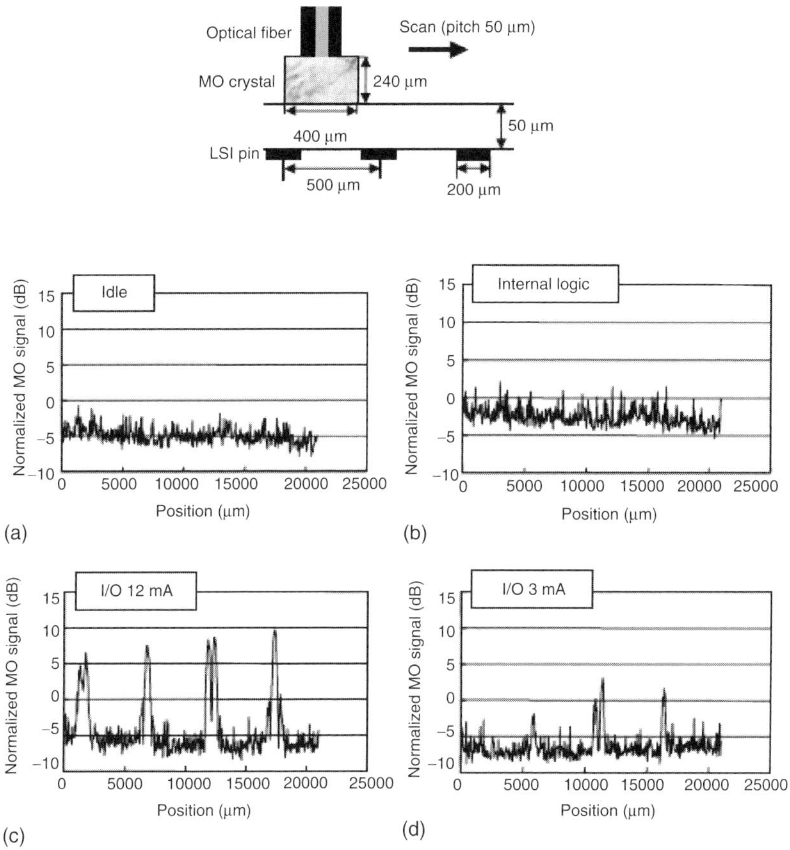


**FIGURE 4.17** Examples of data in LSI pin currents measurements are shown with a variety of operating conditions. It is clearly indicated that an interface circuit in the LSI generates currents at harmonics of clock frequency along power and ground pins.

electrical signals at lower level interconnections, a measurement pad is formed on the lower layer by the focused ion beam (FIB) process. First, a through-hole is opened (Figure 4.18a) and then tungsten via-holes and a measurement pad are formed (Figure 4.18b). The microprobe is placed in contact with the pad (Figure 4.18c). When an 8 mm long tungsten EO microprobe was used, 4.5 GHz was confirmed as the 3 dB bandwidth. Incidentally, as this EO microprobe is mounted on an optical fiber, there is no need for the troublesome alignment of the probe with the laser light source and the optical system for the detection of polarized light.

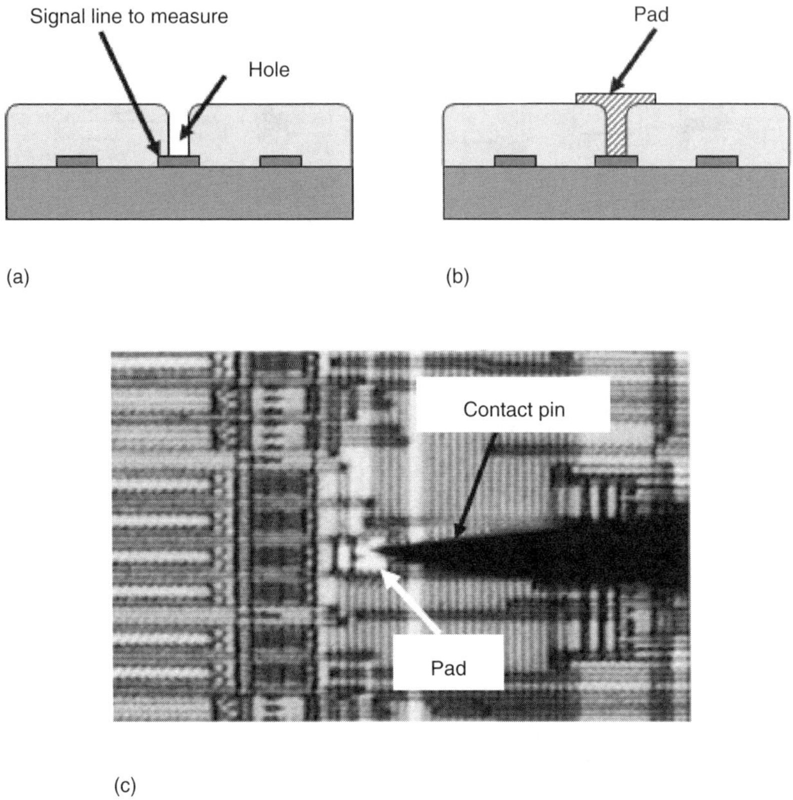


**FIGURE 4.18** ULSI measurement with contact-type EO microprobe. (a) Making through-hole. (b) Making via-holes and measurement pad. (c) Contact pin attached to pad.

Combination with scanning probe microscopy offers <100 nm resolution both in measurement and observability [56]. With the IC-probing system, good observability must be ensured to allow the proper positioning of the probe at the interconnection of interest within a ULSI.

#### *4.3.1.2.2 Hand-Held High-Impedance Probe*

As the operating speeds of ICs and LSIs increase, it is natural that the operating speeds of circuit boards and modules that use multiple ICs and LSIs to implement the desired functions also increase. Undertaking measurements with the methods used to date has become difficult. Therefore, to proceed further with the implementation format in which a metal pin is attached to the EO material (Figure 4.3b), all of the key components from the light source to the polarization-detecting optical system are integrated in a

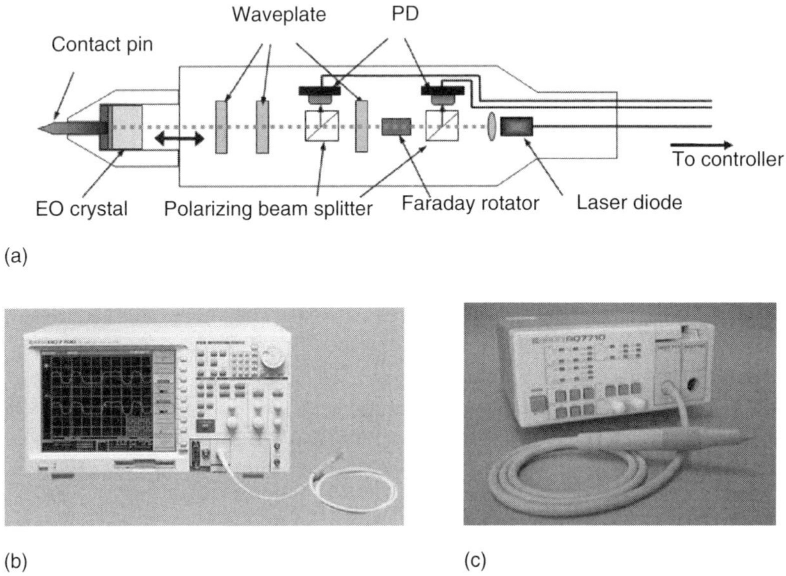


**FIGURE 4.19** Hand-held probes for modules and printed circuit boards. (a) Block diagram of the typical probe head. (b) EO digital sampling oscilloscope. (c) Real-time EO probe unit.

portable module, which is the hand-held probe shown in Figure 4.19a [57,58]. In Figure 4.19b, the probe is connected to a sampling oscilloscope using a pulse laser. In Figure 4.19c, a continuous light source is used and the output from the photodiode in the probe is connected to a conventional electrical sampling oscilloscope to measure waveforms or a spectrum analyzer to measure frequencies.

An important point with respect to the practicality that can be realized by applying photonic technology to the IC-probing system and the hand-held probe described above is one-point contact, namely, there is no need to connect a measurement terminal to a ground line to serve as the voltage reference as required with electrical measurement methods. With previous electrical measurement methods, the input impedance of the measuring probe varies depending on how the grounding is accomplished; hence, these measurement methods suffer from the problem that the waveform is strongly dependent on the grounding method. These sampling probes have already attained bandwidths of up to 10 GHz.

### 4.3.2 Vector Network Analyzer

The NA, which is used to measure the *S* parameters of a device, is a basic measuring instrument on par with the oscilloscope. It is used for a wide range of DUT from active devices such as transistors and amplifiers to passive devices such as transmission lines and filters. With a vector NA, which can measure amplitude and phase, it is currently possible to cover the DC to 110 GHz frequency range with coaxial components in a single setup. In other words, there is a commercially available high-end product. Higher frequencies can be handled by conversion with a mixer or a frequency multiplier and by using waveguides for the input and output of signals to and from the DUT [59,60]. However, this is not a practical solution because of the narrow bandwidth and calibration problems.

Against such a background, an NA based on photonic technology has been proposed with an effective bandwidth of over 300 GHz [61,62]. The concept of the broadband NA is illustrated in Figure 4.20. All of the NA functions are implemented in an integrated probe, as shown in the figure. In other words, the function for (1) the optical generation of the electrical signal that is applied to the DUT (i.e., the stimulus pulse), (2) EOS optical detection of the response, (3) the application of the DC bias to the DUT, and (4) access by

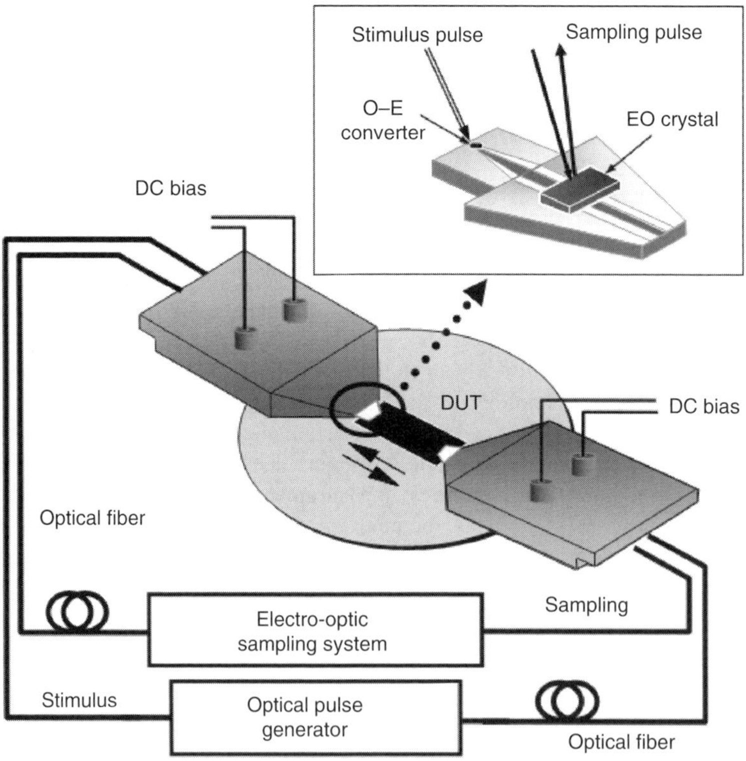


**FIGURE 4.20** System configuration of photonic millimeter-wave network analyzer.

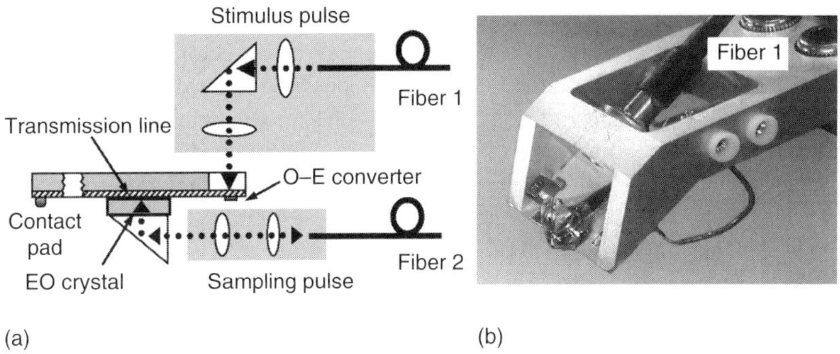


**FIGURE 4.21** (a) Optical arrangements for generating stimulus signal and for detecting electrical response. (b) Photograph of integrated probe head.

CPW that allows the on-wafer measurement of the *S* parameters of two ports, are all integrated in the probe unit.

One key component of this NA probe is an O–E conversion device called a UTC-PD that was described earlier. Making use of the high amplitude of the output allows the evaluation of characteristics even during the high amplitude operation of the DUT, which has been difficult to achieve with previous NAs. The electrical signal generated by the UTC-PD is delivered to the DUT by the 50 Ω transmission lines of the CPW. The signals input to the DUT and reflected from the DUT, and the signal that transmits through the DUT when another probe is used, are detected optically by EOS. This is accomplished by using EO crystals (CdTe) placed at specified positions on the transmission line as electric field sensors.

The details of the optical system of the developed probe and the probe's external appearance are shown in Figure 4.21a and Figure 4.21b, respectively. The results of using this NA system for the on-wafer measurement of a 0.1 μm gate HEMT and an MMW antenna are presented in Figure 4.22. For comparison, a conventional NA was used to measure the *S* parameters for the same device up to 40 GHz. In the low-frequency region, the results from the two instruments were in good agreement. Problems for the future include finding a way to improve the dynamic range of the detection system (currently greater than 30 dB at 100 GHz) and the development of a method for accurate calibration at frequencies above 100 GHz.

### 4.3.3 Measurement Systems for Free-Space Radio Waves

Another important application of the EO or MO probe is in the measurement of radio wave signals propagating in free space. In particular, the EO probe has proven to be a powerful tool for antenna characterization [63]. As

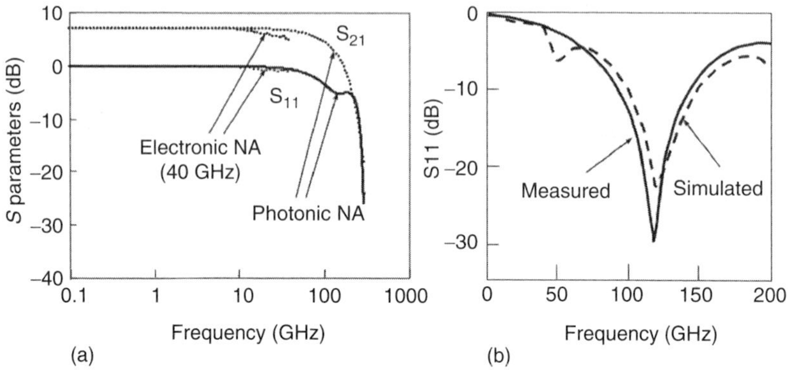


**FIGURE 4.22** Examples of measured frequency characteristics of (a) HEMT and (b) antenna.

frequency increases, the antenna size decreases. With near-field measurements, conventional antenna measurement techniques using an electrical detector (diode) and antenna suffer from invasiveness and poor spatial resolution. Long electrical cables running to the detector also affect the radiation pattern of the antenna, even when a high-impedance line is used. Since the diode rectifies the radio-frequency signal, information about the frequency spectrum and phase is lost. Figure 4.23 shows an EO probe intended for use in MMW antenna characterization [64] with amplitude and phase measurement. The EO crystal tip and the polarization detection optics are integrated into a nonmetallic packaged probe, and the optical input signal and the intensity-modulated optical signals are delivered through optical fiber cables.

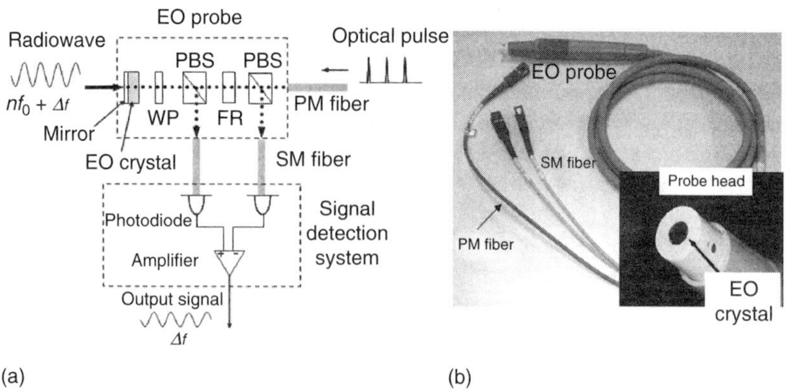


**FIGURE 4.23** EO probe head for free-space radio wave signal measurement. (a) Schematic of probe head. (b) Photograph of probe.

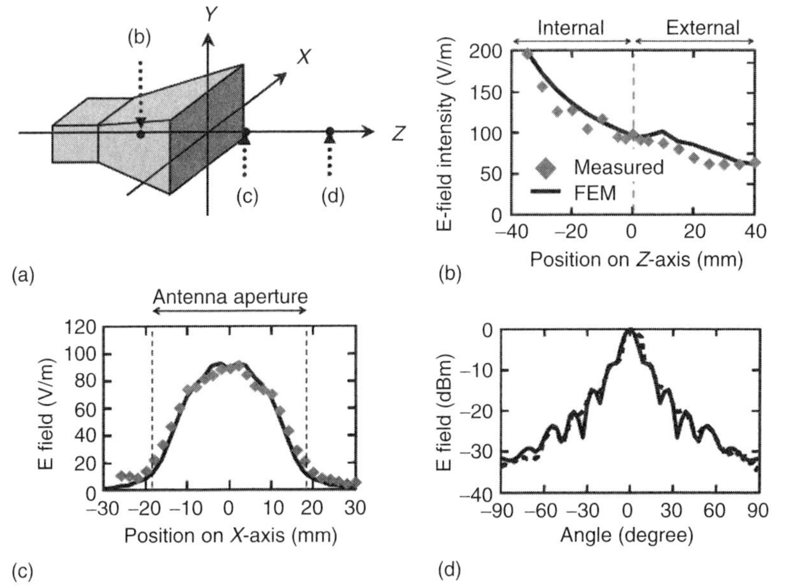


**FIGURE 4.24** Measured electric field distribution of antenna. (a) Measured horn antenna (b) Internal field (c) Near field (d) Far field.

The minimum detectable electric field of the probe was 1 V/m at 60 GHz. Figure 4.24 shows the measured distributions of the electric field that radiated from a horn antenna [64], which were obtained by moving the EO probe around the antenna.

To increase the sensitivity by more than one order, the electric field should be confined in the EO crystal. Electrodes and optical waveguides are fabricated on the EO substrate, which enhances the interaction between the optical signal and the microwave signal. Antennas are connected to this type of EO sensor for the efficient detection of radio wave signals propagating in free space [65,66], and moreover planar antennas are integrated on the EO substrate together with the optical waveguide [67]. Some sensors have also been commercialized and widely applied to antenna characterization, Electromagnetic compatibility (EMC) measurement, and specific absorption rate (SAR) measurement [68].

Figure 4.25 shows an example of an optical waveguide–type electric field sensor fabricated on a $LiNbO_3$ substrate [69]. An optical interferometer was fabricated, thus eliminating the polarization detection optics shown in Figure 4.16. It has a small dipole antenna, whose length is less than one tenth of the wavelength of the radio waves at around 10 GHz. This antenna can be considered a linearly polarized electric infinitesimal antenna, and it was used for the near-field measurement of planar antennas and arrays.

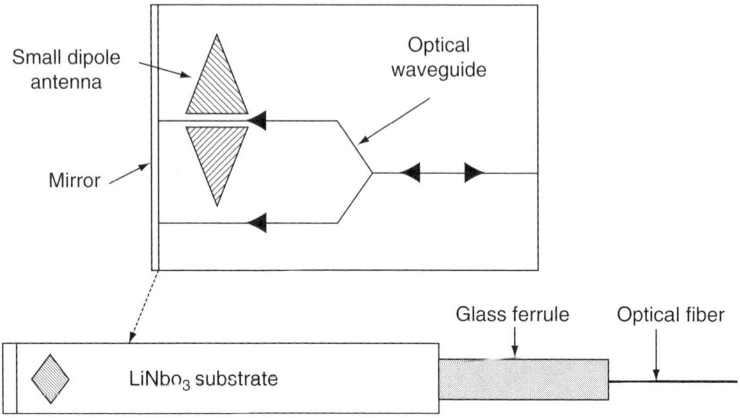


**FIGURE 4.25** Configuration of optical waveguide–type EO sensor integrated with antenna.

As an extension of the above structure, a miniature three-axis probe has been developed to measure the SAR from mobile phones (Figure 4.26) [68]. It can measure fields as low as 10 mV/m. A magnetic field probe has also been demonstrated as shown in Figure 4.27 [70]. It was fabricated by attaching a loop element to the optical waveguide probe in place of the dipole antenna. It

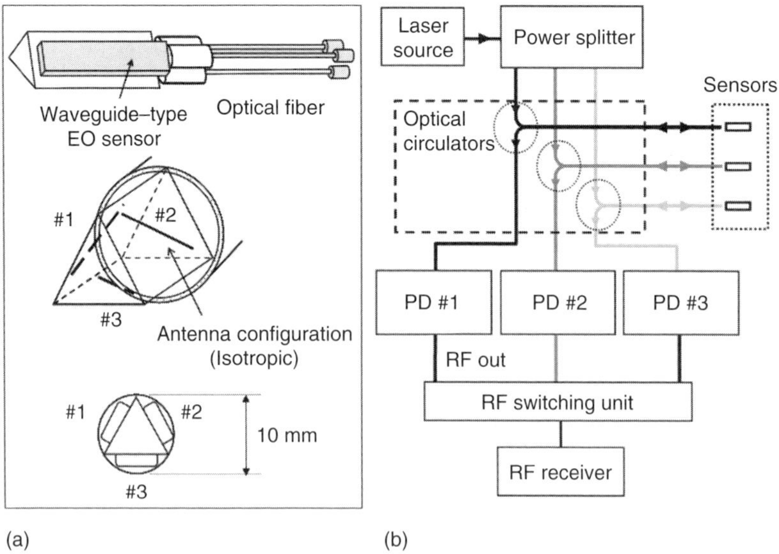


**FIGURE 4.26** Three-axis EO sensor. (a) Head configuration. (b) System block diagram.

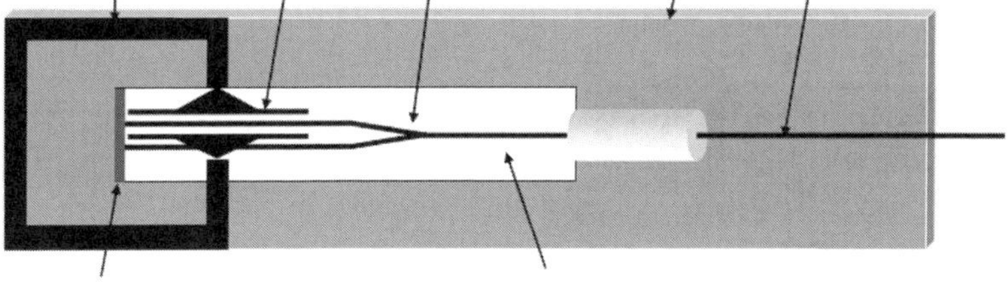


**FIGURE 4.27** Optical waveguide–type EO sensor for magnetic field measurement.

has already been used to measure the magnetic near-field distribution of patch antennas at 2.5 GHz.

Now that the use of various types of electronic information devices is becoming more widespread, the effect of the electromagnetic waves generated by those devices on other equipment and on human health is becoming an issue. The EO probe is expected to serve as a powerful tool in explaining complex EMI phenomena.

### 4.3.4 Systems for Two-Dimensional Mapping of Evanescent Fields [71]

One of the most attractive functionalities of optical probes is their ability to visualize electromagnetic field distributions at high frequencies. In particular, the distribution characterization of near-fields, in other words evanescent fields, emphasizes the advantages of photonic measurement schemes. This is because we can minimize the invasiveness of probes by eliminating metals within the heads of the EO–MO probes. This low invasiveness feature is important with respect to the evanescent field monitoring capability. Otherwise, contiguous probes may disturb the field distribution of interest. Furthermore, it is almost impossible to deduce very near-field distributions with higher special resolutions than the RF wavelength from far-field distributions. One example of such two-dimensional mapping measurements is shown in Figure 4.28. It should be noted here that the MO mapping results are complementary to those of EO mapping as regards the electromagnetic standing wave properties, in other words, the positions of nodes and antinodes [72,73]. Furthermore, a highly spatially resolved MO mapping in another direction of cross section is shown in Figure 4.29, which indicates possible deduction of current distribution in the circuit plane (Figure 4.30).

Near-fields radiated from electronic instruments can be measured in the same way. By bringing the hand-held probe described in Section 4.3.1.2.2

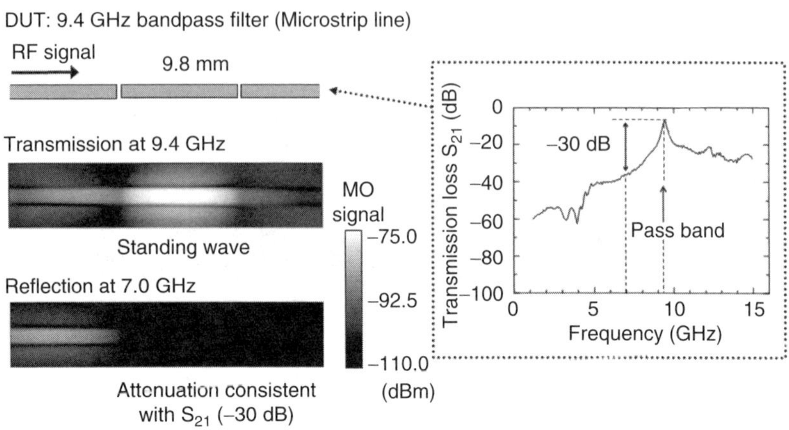


**FIGURE 4.28** Measured two-dimensional distributions of radio frequency evanescent waves are shown with a DUT band-pass filter made of a microstrip line. MO signals are plotted. Natures at pass and stop bands are clearly shown in the visualized standing waves of currents. The frequency response of DUT is also attached.

close to objects, we can measure the electric field intensity distributions for a cellular phone while a call is in progress, and for a printed circuit board resonating at 710 MHz, as shown in Figure 4.31a and Figure 4.31b, respectively.

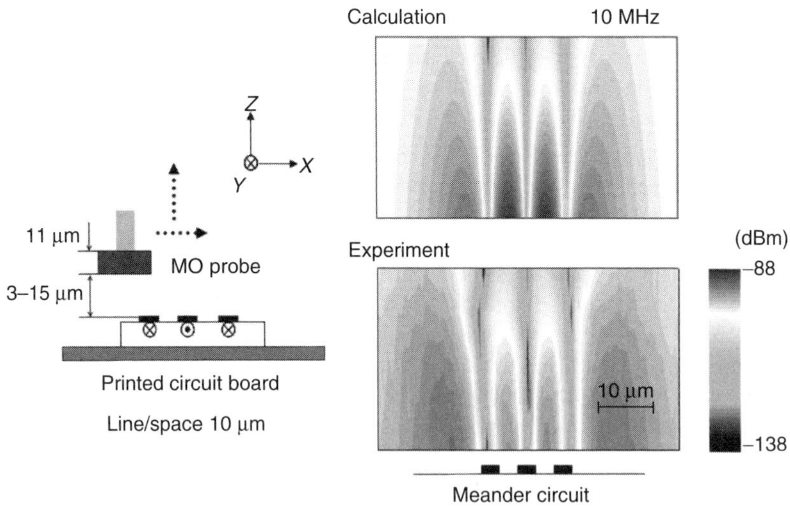


**FIGURE 4.29** Simulated and measured two-dimensional distributions of evanescent waves over a three-folded meander microstrip line. Microscopic distributions of magnetic fields over a 10 μm width circuit are clearly visualized.

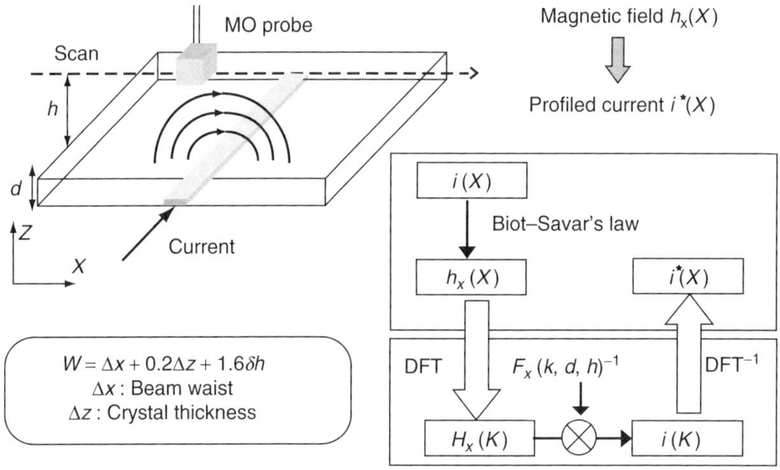


**FIGURE 4.30** A schematic is shown for the current deduction method by the use of measured one-dimensional profile of magnetic field.

One of the biggest technical obstacles to this two-dimensional mapping capability is the length of the data acquisition time. Usually, the data acquisition channel number is unity and a corresponding photonic probe is scanned two-dimensionally employing mechanical translation stages. This is because of the complexities in the highly sensitive detection system of a standard photonic probing system and high cost of their multiplification for parallelism. It typically takes hours and no shorter than a few tens of minutes to obtain a two-dimensionally mapped image. However, we must overcome this time constraint if we wish to realize a really useful system. There have been

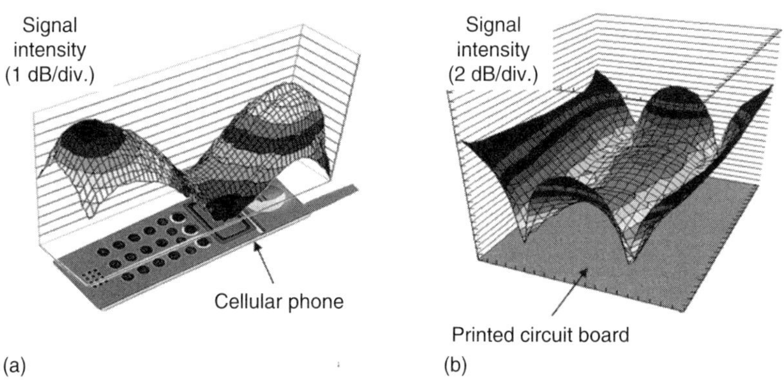


**FIGURE 4.31** Electric field mapping of radiation from (a) cellular phones and (b) printed circuit board measured by hand-held EO probe.

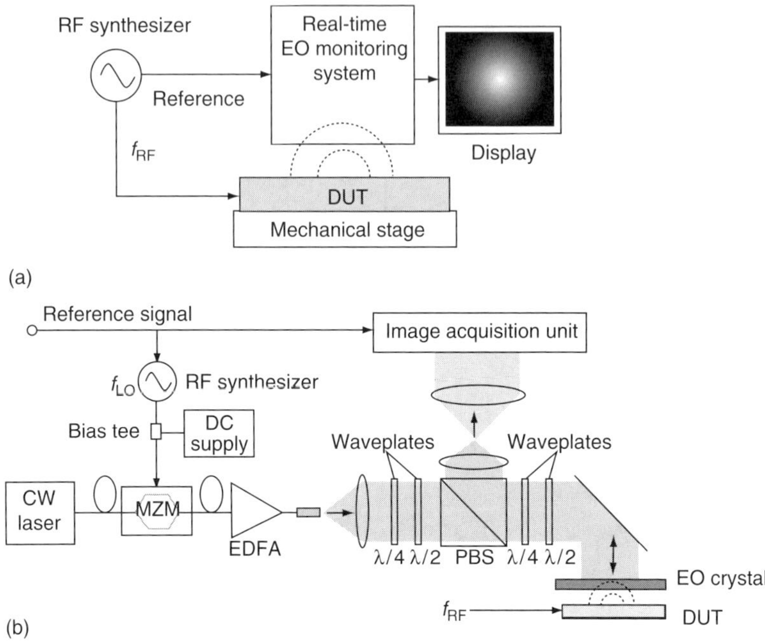


**FIGURE 4.32** (a) Schematic of real-time EO image monitoring system with (b) its internal optical system. The number of EO data acquisition channels is 64 in parallel. A data acquisition rate of 640 pixels per second was achieved.

two trials designed to overcome this obstacle and these are described below. Here, Figure 4.32 shows a new approach that employs a photonic parallelism concept. In this case there are 64 data acquisition channels (pixels). Although the pixel number is small and not satisfactory, it is sufficiently large for certain kinds of two-dimensional visualization of near-fields distributions. Note that it is possible to monitor the distribution image in real time [74]. Examples for those are shown in Figure 4.33.

### 4.3.5 Imaging of Objects

Imaging using electromagnetic waves in the millimeter to submillimeter wavelength range is useful in obtaining information, for example, through clouds, fog, smoke, dust, and other screening conditions, for which visible, IR, and x-ray systems are ineffective. An imaging system using photonically generated pulsed or single-cycle electromagnetic waves with a subpicosecond duration has recently been extensively studied [75,76]. Such imaging is often referred to as T-ray imaging. Terahertz imaging is described in detail in Chapter 10. Possible areas of application include the inspection of the materials and packages, the spectroscopy of gases, solids and liquids, and biological

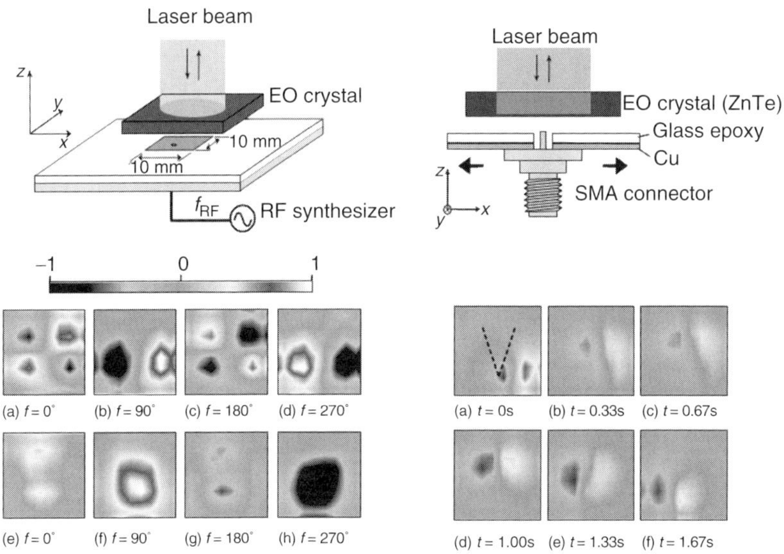


**FIGURE 4.33** Two examples of real-time RF near-field image display are shown. Electric field distributions over a patch antenna are shown with RF signal phase as a parameter (left). A moving RF emitter was also visualized and dynamic evolution of its field distribution change is shown with 0.33 second time intervals (right). A real-time movie is also available.

imaging. This covers a wide range of fields, including science, biomedicine, industry, electronics, and quality control.

Figure 4.34 shows an example setup for imaging with MMWs. By using photonic techniques, widely tunable MMW emitters can be used to illuminate an object. The waves transmitted through or reflected by the object are detected with an EO sensor. Possible areas of application include the inspection of materials and packages. Moreover, because electromagnetic waves in the 100 GHz to 10 THz frequency band interact strongly with molecules and gases, there is an interest in applications to environmental and medical measurement. As described in Section 4.2.5, the photonic approach could lead to the fabrication of frequency-variable MMW sources. In the future, the combination of such frequency-variable sources and EO-based broadband detection may enable us to obtain MMW spectroscopic or MMW color images.

Figure 4.35 shows an example of an MMW imaging system that employs CW millimeter waves. The MMW signal reflected by the object is detected with a waveguide-mounted EO sensor head [77]. The minimum detectable power of the sensor is less than −60 dBm. The object shown in Figure 4.36a is a 5 mm thick slice of pork, and its MMW image at 100 GHz is shown in Figure 4.36b. It is important to note that images with visible light and MMW

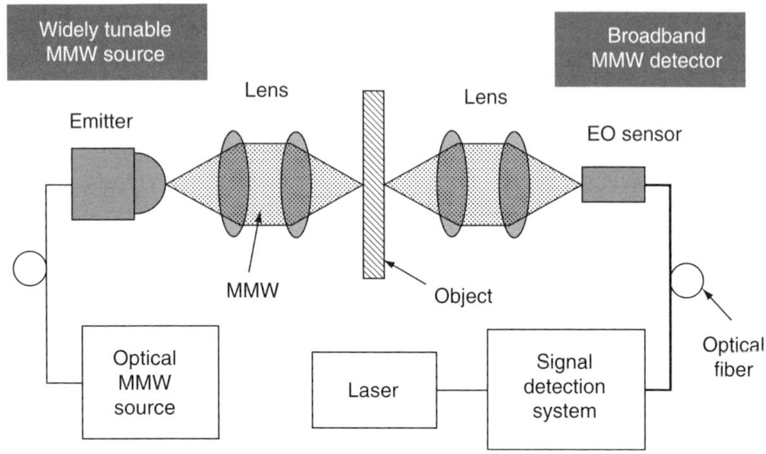


**FIGURE 4.34** Block diagram of MMW imaging based on photonic techniques.

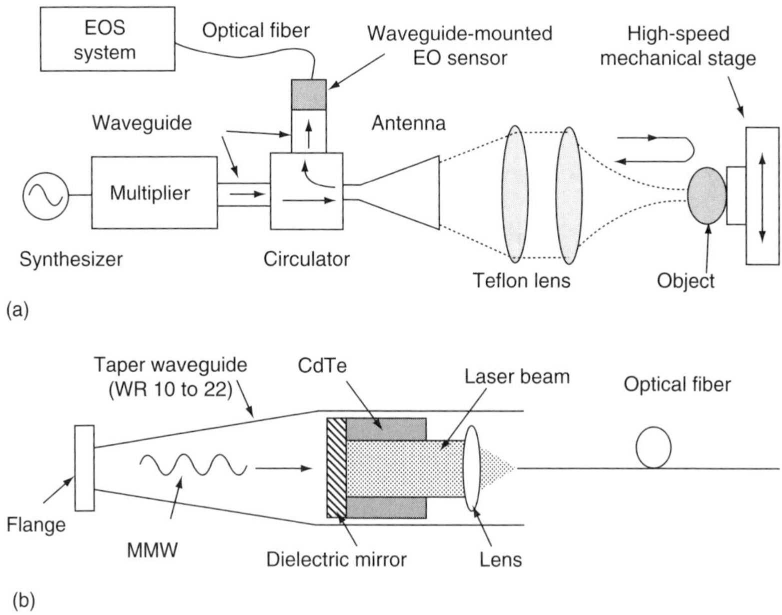


**FIGURE 4.35** Reflection-type MMW imaging system using waveguide-mounted EO sensor. (a) System block diagram. (b) Waveguide-mounted EO sensor.

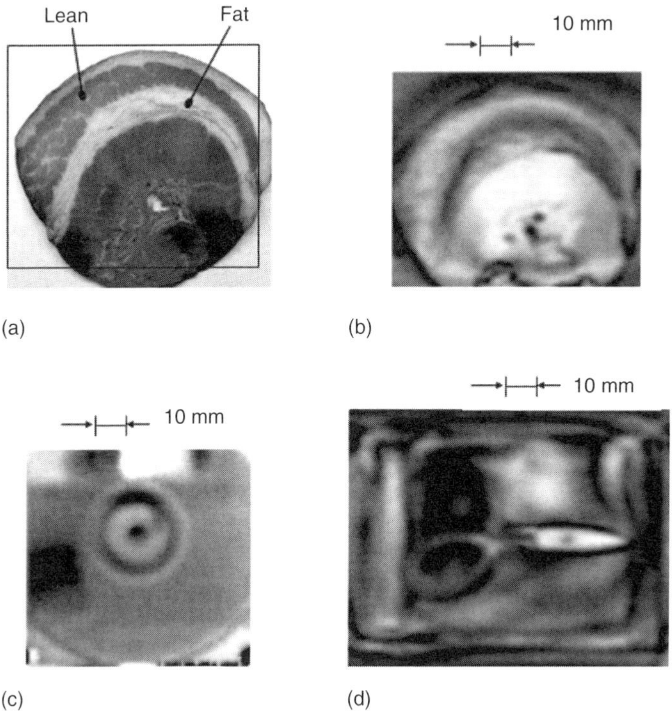


**FIGURE 4.36** (a) Visible image of slice of pork. Obtained MMW images of (b) slice of pork, (c) scissors in purse, and (d) floppy disk.

are rather different in that the MMW signal is absorbed to a much greater degree by the lean meat than by the fatty meat. Figure 4.36c and Figure 4.36d are images of a small scissors in a purse and a floppy disk, respectively. We obtained good images with a diffraction-limited resolution of about 3 mm. The acquisition time for 20,000-pixel amplitude and phase images was less than 10 min, and this is limited by the speed of the motor-driven translation stage. Transmission-type imaging has also been demonstrated [24].

To reduce the measurement or acquisition time of the imaging, large-area EO crystals have been used as a focal plane for the images. To this end, two approaches have been reported: one is the combination of a large EO crystal and an optical scanner [78], and the other is the combination of a large EO crystal and a CCD detector [79]. In the former case, a large aperture EO crystal is scanned with a laser beam as shown in Figure 4.37a. In the latter scheme, the two-dimensional field distribution projected on the EO crystal is read off using an expanded laser beam, which enables real-time imaging as shown in Figure 4.37b. Because of the limited sensitivity of the system, at present the radio wave should be a "pulse" with high intensity.

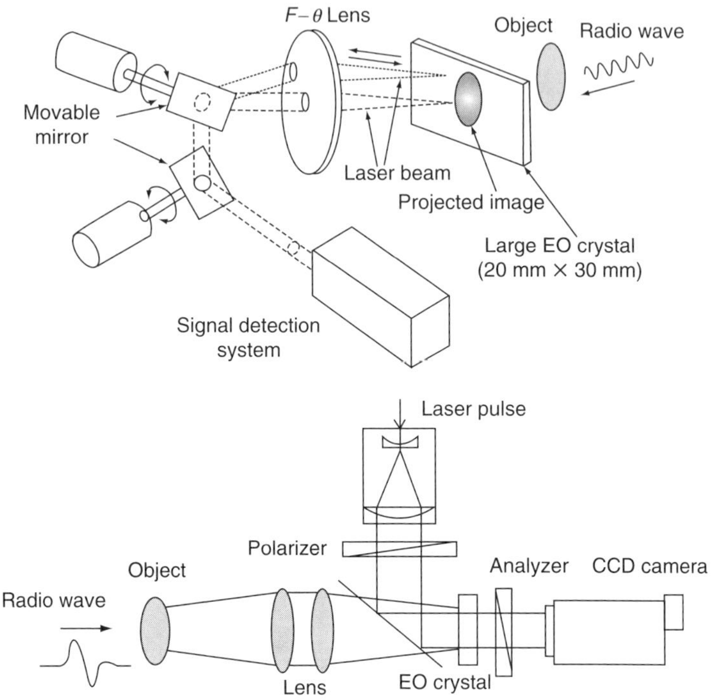


(a)

(b)

**FIGURE 4.37** Examples of 2D imaging system using large aperture EO sensor with (a) optical scanner and (b) CCD camera.

A functional CCD camera system, which has a signal processing capability for each pixel, improves the S/N ratio of the obtained image [80]. A terahertz imaging system based in this scheme is already commercially available [81].

## 4.4 CONCLUSION AND FUTURE ISSUES

We have described photonic technologies for measuring high-speed electrical signals and microwave or MMW signals, focusing in particular on EO and MO probing technology. The basic concept of EO probing was actually first proposed in experiments by Auston et al. when picosecond and femtosecond lasers were in their infancy (1970s). At the beginning of the 1980s, Valdmanis et al. refined the concept as a concrete electrical signal measurement technology that is known as EO sampling, or EOS. However, the pulse lasers of the time were large, special units that required specialized skill to operate, so the EOS technology that used them was very far from practical. In the 1990s, however, advances in optical telecommunication technology made it possible to use pulse laser light sources that employed optical fiber and that operated simply and with high stability at long wavelengths and with a broad bandwidth simply by inserting a switch. Against the background of a strong demand from ultra-high-speed semiconductor electronics as described at the beginning of this chapter, there was also great progress toward practicality in photonic measurement technology based on EOS in the last half of the 1990s.

The IC-probing system, microprobe, and hand-held probe are steadily reaching the level where ordinary users can benefit from their unique functions and performance at their own sites. Nevertheless, there remain a number of points related to cost and ease of use where they do not compare favorably with conventional electronic measuring instruments. As we have shown in this chapter, one reason is that the photonic modules that constitute the measurement system comprise groups of discrete components, and there has been almost no progress on integrating these components. In the conversion to optical transmission in data communication, and in the access system in particular, optical component integration technology is a bottleneck, and in the next generation of photonic measurement technology too, there will be a need to introduce the same kind of technology.

Furthermore, if we are to open up the millimeter and submillimeter wave bands and the terahertz band, we must urgently establish technology for measurement in the unused frequency bands, such as network and spectrum analyzers. In the future, we expect this development to proceed while maintaining a good balance between reducing the cost of photonic measurement (increasing general availability) and increasing performance (pursuing optimum performance).

## REFERENCES


1. Tsuchiya, M. Methods for electromagnetic field measurements with advantageous optical features employed, *Rev. Laser Eng.*, 33, 365, 2005.
2. Yamazaki, E. et al. High-frequency magneto-optic probe based on BiRIG rotation magnetization, *IEICE Trans. Electron.*, E86-C, 1338, 2003.
3. Auston, D.H. Picosecond optoelectronic switching and gating in silicon, *Appl. Phys. Lett.*, 26, 101, 1975.
4. Valdmanis, J.A., Mourou, G., and Gabel, C.W. Subpicosecond electro-optic sampling, *IEEE J Quantum Electron.*, QE-19, 664, 1983.
5. Nagatsuma, T. Measurement of high-speed devices and integrated circuits using electro-optic sampling techniques, *IEICE Trans. Electron.*, E76-C, 55, 1993.
6. Nagatsuma, T. et al. 1.55-μm photonic systems for microwave and millimeter-wave measurement, *IEEE Trans. Microw. Theory Tech.*, 49, 1831, 2001.
7. Takahashi, H., Aoshima, S., and Tsuchiya, Y. Sampling and real-time methods in electro-optic probing system, *IEEE Trans. Instrum. Meas.*, 44, 965, 1995.
8. Wakana, S. et al. Fiber-edge electrooptic/magnetooptic probe for spectral-domain analysis of electromagnetic field, *IEEE Trans. Microw. Theory Tech.*, 48, 2611, 2000.

9. Shinagawa, M. et al. A real-time electro-optic handy probe using a CW laser, *IEEE Trans. Instrum. Meas.*, 50, 1076, 2001.
10. Miyazawa, S. and Tabata, T. Growth of $Bi_{12}TiO_{20}$ single crystals for high-impedance electro-optic probe applications, *NTT Review*, 10, 62, 1998.
11. Yaita, M. and Nagatsuma, T. Optical sampling of electrical signals in poled polymeric media, *IEICE Trans. Electron.*, E76-C, 222, 1993.
12. Adachi, H. et al. High-quality organic 4-dimethylamino-*N*-methyl-4-stilbazolium tosylate (DAST) crystal for electro-optic measurement, *IEICE Trans. Electron.*, E86-C, 1352, 2003.
13. Zheng, X., Wu, S., and Sobolewski, R. Electro-optic sampling system with a single-crystal 4-*N*, *N*-dimethylamino-4-*N*-methyl-4-stilbazolium tosylate sensor, *Appl. Phys. Lett.*, 82, 2383, 2003.
14. Taniuchi, T., Shikata, J., and Ito, H. Tunable terahertz-wave generation in DAST crystal with dual-wavelength KTP optical parametric oscillator, *Electron. Lett.*, 36, 1414, 2000.
15. Wakana, S. et al. Performance evaluation of fiber-edge magnetooptic probe, *IEEE J. Lightwave Technol.*, 21, 3292, 2003.
16. Elezzabi, A.Y. and Freeman, M.R. Ultrafast magneto-optic sampling of picosecond current pulses, *Appl. Phys. Lett.*, 68, 3546, 1996.
17. Iwanami, M. et al. Wideband magneto-optic probe with 10 mm-class spatial resolution, *Jpn. J. Appl. Phys.*, 43, 2288, 2004.
18. Sahri, N. et al. Picosecond synchronous electro-optic sampling with a jitter-reduced gain-switched laser diode, *Jpn. J. Appl. Phys.*, 37, L39, 1998.
19. Guy, M.J. et al. 200fs soliton pulse generation at 10 GHz through nonlinear compression of transform-limited pulses from an electroabsorption modulator, *Electron. Lett.*, 31, 740, 1995.
20. Yaita, M. and Nagatsuma, T. The effect of sampling-pulse pedestals on temporal resolution in electro-optic sampling, *IEICE Trans. Electron.*, E81-C, 254, 1998.
21. Tamura, K.R. and Nakazawa, M. Spectral-smoothing and pedestal reduction of wavelength tunable quasi-adiabatically compressed femtosecond solitons using a dispersion-flatted dispersion-imbalanced loop mirror, *IEEE Photon. Technol. Lett.*, 11, 230, 1999.
22. Morioka, T. et al. Transform-limited, femtosecond WDM pulse generation by spectral filtering of gigahertz supercontinuum, *Electron. Lett.*, 30, 1166, 1994.
23. Takizawa, T. et al. Application of fiber laser to EOS probing system, *Proc., Electron. Soc. Conf. IEICE*, Tokyo, 1997, C-12-7.
24. Sasaki, A. and Nagatsuma, T. Millimeter-wave imaging using an electro-optic detector as a harmonic mixer, *IEEE J. Select. Topics Quantum Electron.*, 6, 735, 2000.
25. Nagatsuma, T. et al. All optoelectronic generation and detection of millimeter-wave signals, *Tech. Dig. Microw. Photonics*, New Jersey, 1998, 5–8.
26. Nagatsuma, T. Photonic measurement technologies for high-speed electronics, *Meas. Sci. Technol.* 13, 1655, 2002.
27. Verghese, S., McIntosh, K.A., and Brown, E.R. Optical and terahertz power limits in the low-temperature-grown GaAs photomixer, *Appl. Phys. Lett.*, 71, 2743, 1997.
28. Matsuura, S. et al. A traveling-wave THz photomixer based on angle-tuned phase matching, *Appl. Phys. Lett.*, 74, 2872, 1999.

29. Stöhr, A. Ultra-broadband Ka to W-band 1.55-μm travelling-wave photomixers, *Electron. Lett.*, 36, 970, 2000.
30. Chau, T. et al. Generation of millimeter waves by photomixing at 1.55 μm using InGaAs-InAlAs-InP velocity-matched distributed photodetectors, *IEEE Photon. Technol. Lett.*, 12, 1055, 2000.
31. Ishibashi, T. et al. Uni-traveling-carrier photodiodes, *Tech. Dig. Ultrafast Electron. Optoelectron.*, 83, 1997.
32. Ishibashi, T. et al. InP/InGaAs uni-traveling-carrier photodiodes, *IEICE Trans. Electron.*, E83-C, 938, 2000.
33. Ito, H. et al. InP/InGaAs uni-traveling-carrier photodiode with 310-GHz bandwidth, *Electron. Lett.*, 36, 1809, 2000.
34. Nagatsuma, T. et al. Characterization of uni-traveling-carrier-photodiode monolithically integrated with matching circuit, *Electron. Lett.*, 37, 1246, 2001.
35. Sato, K. et al. Active mode locking at 50 GHz repetition frequency by half-frequency modulation of monolithic semiconductor lasers integrated with electro-absorption modulators, *Appl. Phys. Lett.*, 69, 2626, 1996.
36. Sato, K. et al. High-repetition frequency pulse generation at 102 GHz using mode-locked lasers integrated with electroabsorption modulators, *Electron. Lett.*, 34, 790, 1998.
37. Arahira, S., Matsui, Y., and Ogawa, Y. Mode-locking at very high repetition rates more than terahertz in passively mode-locked distributed-Bragg reflector laser diodes, *IEEE J. Quantum Electron.*, 32, 1211, 1996.
38. Arahira, S. et al. Stabilization and timing jitter reduction of 160-GHz colliding-pulse mode-locked laser diode by subharmonic-frequency optical pulse injection, *IEICE Trans. Electron.*, E83-C, 966, 2000.
39. Wen, Y.J. et al. Generation of 140-GHz optical pulses with suppressed amplitude modulation by subharmonic synchronous mode-locking of Fabry–Perot semiconductor, *Electron. Lett.*, 37, 581, 2001.
40. Payne, J., Shillue, B., and Vaccari, A. Photonic techniques for use on the Atacama large millimeter array, *Tech. Dig. Microw. Photonics*, Melbourne, 1999, 1.
41. Swanson, E.A. et al. 100-GHz soliton pulse train generation using soliton compression of two phase side bands from a single DFB laser, *IEEE Photon. Technol. Lett.*, 6, 1194, 1994.
42. Gough, O.P., Silvia, C.F.C., and Seeds, A.J. Exact millimeter-wave frequency synthesis by injection locked laser comb line selection, *Tech. Dig. Microw. Photonics*, Melbourne, 1999, 61.
43. Otsuji, T. et al. A 105-GHz bandwidth optical-to-electrical conversion stimulus probe head employing a uni-traveling-carrier photodiode, *IEEE Photon. Technol. Lett.*, 11, 1033, 1999.
44. Nagatsuma, T. et al. A 120-GHz integrated photonic transmitter, *Tech. Dig. Microw. Photonics*, Oxford, 2000, 225.
45. Hirata, A., Ishii, H., and Nagatsuma, T. Design and characterization of a 120-GHz millimeter-wave antenna for integrated photonic transmitters, *IEEE Trans. Microw. Theory Tech.*, 49, 2157, 2001.
46. Royter, Y. et al. Ultrahigh-speed 1.55-μm photodiodes on Si with low-loss millimeter-wave waveguides, *Tech. Dig. IEEE LEOS*, Puerto Rico, 2000, 283.
47. Minotani, T. Three-dimensional millimeter-wave photonic integrated circuits on Si, *Tech. Dig. IEEE MTT-S Int. Microw. Symp.*, Phoenix, 2001, 57.

48. Hirata, A., Furuta, T., and Nagatsuma, T. Monolithically integrated Yagi–Uda antenna for photonic emitter operating at 120 GHz, *Electron. Lett.*, 37, 1107, 2001.
49. Mitani, S. et al. Sensitivity enhancement by optical pre-amplification combined with optical bias optimization in fiber-optical probing systems for electromagnetic field measurements, *The 4th Korea–Japan Joint Workshop on Microwave and Millimeter-Wave Photonics*, P-14, 157, 2003.
50. Shinagawa, M. and Nagatsuma, T. An automated electro-optic probing system for ultrahigh-speed IC's, *IEEE Trans. Instrum. Meas.*, 43, 843, 1994.
51. Nagatsuma, T. Electro-optic testing technology for high-speed LSIs, *IEICE Trans. Electron.*, E79-C, 482, 1996.
52. Iwanami, M. et al. Magneto-optical mapping of near-field distributions over an LSI chip, *Tech. Dig. of Seventh Int. Symp. Contemp. Photon. Technol. (CPT2004)*, Tokyo, J-4, 167, 2004.
53. Iwanami, M. et al. Magnetic near field measurements over LSI package pins by fiber-edge magnetooptic probe, *IEEE J. Lightwave Technol.*, 21, 3273, 2003.
54. Shinagawa, M. and Hashimoto, T. Development of an EO micro-probe for measuring waveforms within a GHz band LSI, *IEICE Tech. Rep.*, MW2001-44, 63, 2001.
55. Hashimoto, C. et al. Observation of the internal waveforms in high-speed high-density LSIs using an EOS prober, *Microelectron. Reliability*, 41, 1203, 2001.
56. Ozaki, K. et al. Novel optical probing system with submicron spatial resolution for internal diagnosis of VLSI circuits, *Proc. Int. Test Conf.*, Washington, 1996, 20.
57. Shinagawa, M. and Nagatsuma, T. A novel high-impedance probe for multi-gigahertz signal measurement, *IEEE Trans. Instrum. Meas.*, 45, 575, 1996.
58. Shinagawa, M. et al. A real-time electro-optic handy probe using a CW laser, *IEEE Trans. Instrum. Meas.*, 50, 1076, 2001.
59. Oleson Microwave Labs Model V05VNA-T/R (140–220GHz), Morgan Hill, CA.
60. AB Millimetre Model MVNA 8–350 with options ESA, Paris, France.
61. Sahri, N. and Nagatsuma, T. Application of 1.55-μm photonic technologies to practical millimeter-wave network analyzer, *IEICE Trans. Electron.*, E82-C 1307, 1999.
62. Sahri, N. and Nagatsuma, T. Packaged photonic probes for an on-wafer broad-band millimeter-wave network analyzer, *Photon. Technol. Lett.*, 12 1225, 2000.
63. Whitaker, J.F. et al. Electro-optic probing for microwave diagnostics, *IEICE Trans. Electron.*, E86-C, 1328, 2003.
64. Togo, H. et al. Characterization of millimeter-wave antenna using photonic measurement techniques, *Int. J. RF Microw. Comput. Aided Eng.*, 14, 290, 2004.
65. Loader, B. et al. An optically modulated field sensor to make TEM cells traceable to the NPL calculable standard dipole antenna, *Dig. Conf. Precision Electromagn. Meas.*, Sydney, 2000, TUP7-5.
66. Tajima, K. Development of optical isotropic E-field sensor operating more than 10 GHz using Mach-Zehnder interferometers, *IEICE Trans. Electron.*, E85-C, 961, 2002.
67. http://www.nec-tokin.com/
68. Loader, B.G. et al. An optical electric field probe for specific absorption rate measurements, *Proc'03, EMC*, Zurich, 2003, 12C3.

69. Hirose, M. et al. Planar near field measurement using photonic sensor for microstrip array antenna at X band, *Tech. Rep. IEICE*, AP2002-147, 77, 2003.
70. Takahashi, M. et al. Magnetic near-field distribution measurements above a patch antenna using an optical waveguide probe, *Proc'04, EMC*, Sendai, 2004, 2C4-3.
71. Wakana, S. et al. Study of the crystal size effect on spatial resolution in three-dimensional measurement of fine electromagnetic field distribution by optical probing, *Jpn. J. Appl. Phys.*, 42, 6637, 2003.
72. Ohara, T. et al. Two-dimensional field mapping of microstrip lines with a band pass filter or a photonic bandgap structure by fiber-optic EO spectrum analysis system, *Tech. Dig. Int. Topical Meeting on Microw. Photonics*, Oxford, UK. WE2.17, 210, 2000.
73. Tsuchiya, M. et al. Microscopic distribution measurements of microwave frequency magnetic fields by fiber-edge magneto-optic (FEMO) probing, *J. Magnetics Soc. Jpn.*, 26, 128, 2002.
74. Sasagawa, K. and Tsuchiya, M. Real-time monitoring system of RF near-field distribution images on the basis of 64-channel parallel electro-optic data acquisition, *IEICE Electronics Express*, 2, 600, 2005.
75. Hunsche, S. et al. New dimensions in T-ray imaging, *IEICE Trans. Electron.*, E81-C, 269, 1988.
76. Chen, Q. et al. Recent development of free-space THz imaging, *Proc. IEEE Int. Conf. Terahertz Electronics*, 73–6, 1999.
77. Sasaki, A. and Nagatsuma, T. Reflection-type millimeter-wave imaging with a high-sensitivity waveguide-mounted electro-optic, *Jpn. J. Appl. Phys.*, 41, L83, 2002.
78. Wu, Q. et al. Two-dimensional electro-optic imaging of THz beams, *Appl. Phys. Lett.*, 69, 1026, 1996.
79. Sasaki, A. and Nagatsuma, T. Electric-field scanning system using electro-optic sensor, *IEICE Trans. Electron.*, E86-C, 1345, 2003.
80. Yonera, T. et al. THz Imaging System Using CMOS Camera Operated in differential Mode, *Tech. Dig. 28th Int. Conf. Infrared and Millimeter Waves*, W8-3, 295, 2003.
81. Usami, M. et al. Development of a THz imaging system, *Phys. Med. Biol.*, 47, 3749, 2002. *http://www.tochigi-nikon.co.jp/*

# 5 Hybrid Fiber Radio—Concepts and Prospects

*Dalma Novak, Ampalavanapillai Nirmalathas, Christina Lim, Rodney Waterhouse, Masuduzzaman Bakaul, and Teddy Kurniawan*

## CONTENTS

5.1 Introduction ..... 157
5.2 Evolution of Wireless and Optical Communication Networks ..... 158
5.3 Integration of Radio and Optical Networks ..... 160
5.3.1 Hybrid Fiber Radio Architectures ..... 161
5.4 Technologies for Distributing Radio Signals over Fiber ..... 164
5.4.1 Signal Transport Schemes ..... 164
5.4.2 Optimizing System Performance ..... 168
5.5 Integration of HFR Systems with WDM Networks ..... 174
5.6 Base Station Technologies in HFR Systems ..... 176
5.7 Concluding Remarks ..... 181
Acknowledgments ..... 181
References ..... 182

## 5.1 INTRODUCTION

Radio networks provide users the attractive feature of untethered connectivity for a range of applications including cellular communications, wireless local area data networks, and broadband fixed wireless access. The application of optical fiber in these networks for the transport and distribution of radio signals brings a number of benefits, including high-performance remoting links and reduced infrastructure costs. In this chapter, we present an overview of hybrid fiber radio systems with a focus on the basic enabling technologies and concepts for such systems, including signal transport schemes and network architectures. We also discuss the technical challenges and potential solutions associated with implementing hybrid fiber radio systems operating at higher frequencies and review the current prospects for their practical realization.

## 5.2 EVOLUTION OF WIRELESS AND OPTICAL COMMUNICATION NETWORKS

The global evolution and development of telecommunication networks is being driven by the continuing increase in customer demand for bandwidth-intensive applications, mobile communications, as well as the desire for untethered connectivity in personal communications. The need for higher data rates is also extending broadband communications from wired to wireless and mobile networks. Radio networks provide users the attractive feature of untethered connectivity for a range of applications including cellular communications, wireless local area networks (WLANs), broadband fixed wireless access, and short-range connectivity to devices. The diversity of wireless networks and radio standards is evolving at a rapid rate, as shown in Figure 5.1.

With the proliferation of a variety of new data communication services, the demand for broadband wireless networks is increasing rapidly. The current generation of mobile communication systems has limited available spectral bandwidth, which prohibits the provision of broadband services for a large customer base. To overcome this problem, the use of smaller radio coverage areas in microcellular and picocellular wireless networks is attracting attention as a means of increasing capacity by enabling more efficient use of the limited available bandwidth. The increasing demand for broadband services and the transmission of higher data rates have also led to the consideration of wireless networks operating at higher frequencies and extending well into the millimeter-wave band (26 GHz and above) where the total capacity of an

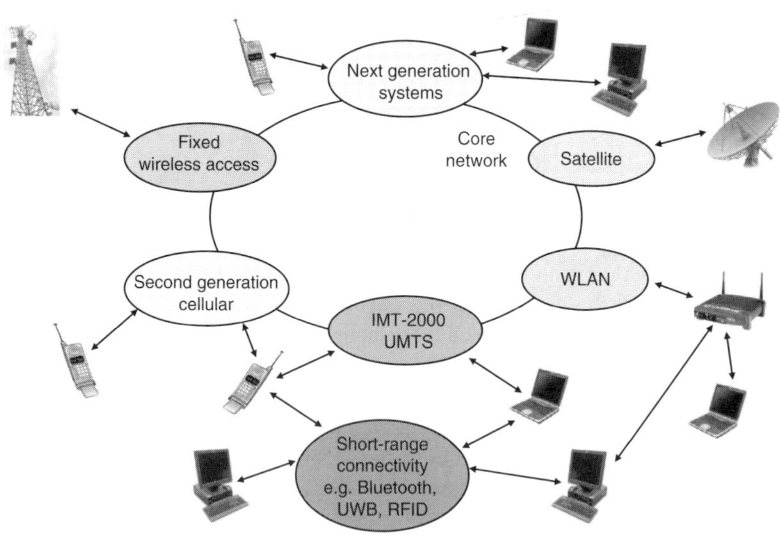


**FIGURE 5.1** The diversity of radio communication networks.

antenna base station (BS) can approach 1–5 Gb/s [1]. Such radio networks offer the ability to provide truly broadband services to users by utilizing the enormous bandwidth available in a number of these higher frequency bands. Because of the large atmospheric absorption that occurs at millimeter-wave frequencies, such radio networks operate with significantly smaller wireless coverage areas, which also enable efficient radio frequency reuse schemes.

Along with the progress in increasing the bandwidth offered to users in wireless networks, the nature of the data has changed dramatically, as has also occurred in wired telecommunication networks. As data traffic now exceeds voice traffic in wired networks, the dominant data transport protocol in such networks is now the Internet protocol (IP). As a result, the application of IP to wireless networks is currently being investigated for the efficient transport of data over noisy wireless channels. Figure 5.2 shows the radio-operating frequencies for a variety of current and next generation wireless networks, including those found in the USA, Europe, and Japan. The current digital cellular radio networks encompass systems operating around 800–900 MHz, 1.8–1.9 GHz, and 2 GHz, with data rates from 9.6 (GSM-900) to 384 kb/s (DCS-1800). Wireless data systems such as WLANs cover the 2.4 and 5.8 GHz bands while higher frequency radio networks comprise fixed wireless access (LMDS at 28 GHz) as well as wireless personal area networks (WPANs around 60 GHz) and next generation mobile broadband systems (MBS at 66 GHz).

The proliferation of new radio services introduces a number of requirements for the implementation of future radio communication systems. Ideally, the deployment of a new wireless network infrastructure would enable the simultaneous support of different radio standards so that a variety of access

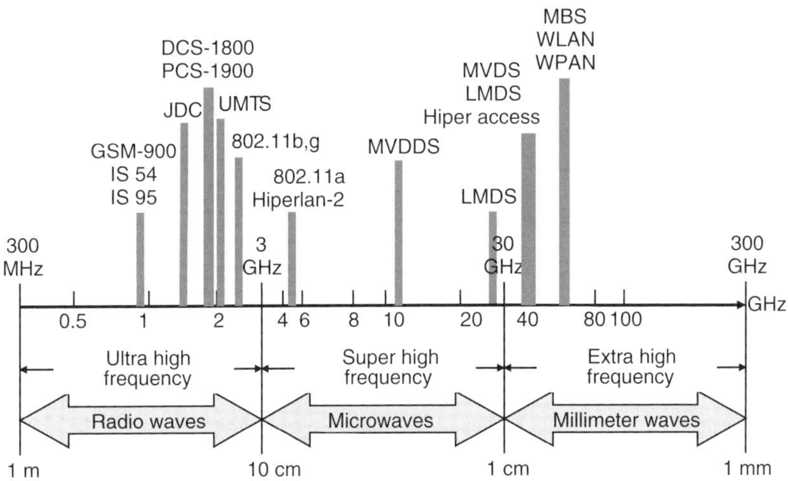


**FIGURE 5.2** Radio spectrum utilized by a variety of current and next generation wireless networks.

networks that serve customers may coexist. The wireless networks should provide for seamless roaming between different networks with unique radio services, ensuring an uninterrupted high quality of service with ubiquitous coverage. In addition, future radio networks should enable the provision of new services in a straightforward manner, with flexible control of the radio interface that can adapt to different wireless modulation and air interface access formats. Finally, next generation wireless networks should also be able to seamlessly integrate with existing wired networks. The application of optical fiber networks for the transport and distribution of radio signals in wireless networks can provide some of the enabling technologies that satisfy these requirements. The integration of wired fiber-optic and wireless communication networks is the focus of the next section.

## 5.3 INTEGRATION OF RADIO AND OPTICAL NETWORKS

The application of fiber-optic networks for the distribution of radio signals in wireless communication systems has been an area of intense research and investigation for the last 15 years. The advantages of the optical fiber as a transmission medium, such as low loss, large bandwidth characteristics, small size, and low cable cost, make it the ideal solution for efficiently transporting radio signals from the central office (CO) to the remote antenna sites. In addition, optical fiber has significant weight advantages over coaxial cable. Hybrid fiber radio (HFR) is an essential technology for the integration of broadband wireless and optical access networks and enables a flexible access network infrastructure capable of offering broadband wireless connectivity to a range of services and applications. Figure 5.3 shows the general concept for the integration of optical and wireless networks.

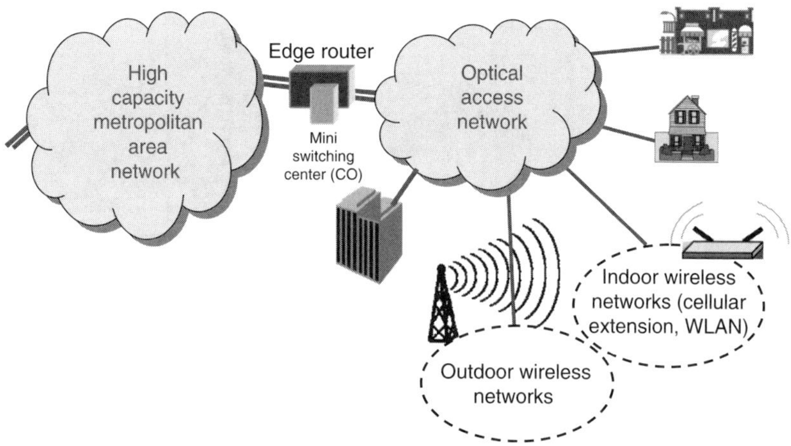


**FIGURE 5.3** The integration of optical fiber and wireless communication networks.

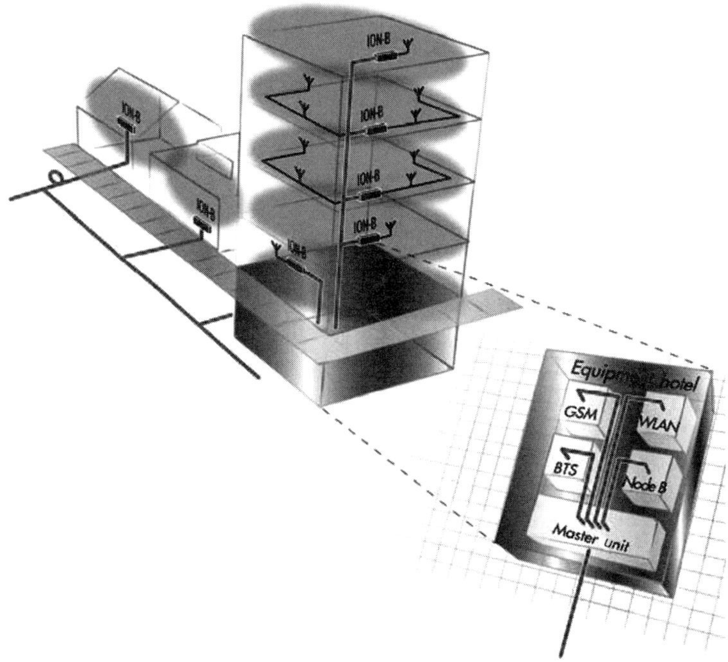


**FIGURE 5.4** Andrew Corporation's ION family of optical distributed antenna systems.

HFR is now well and truly a commercial reality for indoor applications in the form of distributed antenna systems as well as outdoor wireless systems. For example, HFR can provide specialized coverage of wireless communications, extending the range and capacity (the "RF footprint") of current wireless network services. HFR can also be used to establish an appropriate backbone network for the interconnection of multiple BSs with the CO that provides the wireless network switching and routing functionality. An example of a commercial HFR system manufactured by Andrew Corporation for indoor wireless signal distribution is shown in Figure 5.4 [2]. This optical distributed antenna system (the ION) provides uniform radio coverage and is capable of transporting a variety of radio signals at frequencies ranging from 800 to 2500 MHz, covering cellular and WLAN applications.

### 5.3.1 Hybrid Fiber Radio Architectures

Another motivation for the merging of optical and wireless networks lies in the nature of the optical network itself. Much progress has been made in the last few years in the development of optical wavelength division multiplexed

(WDM) networks for the provision of high-capacity communications between major telecommunication nodes in metropolitan areas. In such metropolitan area networks (MANs), nodes are connected by optical fiber links that have the capability to support multiple wavelength channels, each carrying high data rates up to 10 Gb/s. In addition, self-healing WDM ring topologies have been developed to give fully interconnected connectivity with a very high level of built-in protection against unexpected failures in some parts of the network. Given the large amount of fiber infrastructure already installed close to the customer premises for the introduction of new services in the metropolitan area, the possibility of exploiting these networks to support the development of future broadband wireless access networks is another driver for the interest in HFR technology.

Current wireless network architectures are characterized by centralized switching nodes that are interconnected to geographically distributed antenna BSs via microwave links. In future broadband picocellular network architectures, because of the large capacity and large number of BSs, radio networks may be dimensioned in such a way that there will be a number of clusters of BSs serviced by a switching node. For a large metropolitan area, a number of such switching nodes will therefore be required that will be interconnected via an optical MAN ring architecture, as shown in Figure 5.5. For each cluster of BSs, the switching center would provide the role of a CO, performing the switching and routing functions for the remote antenna BS or access point. For wireless networks operating at higher frequencies, in the millimeter-wave frequency range, the smaller radio coverage areas and the subsequent need for more BSs have driven the development of system architectures where the functions relating to the access point in a multiuser

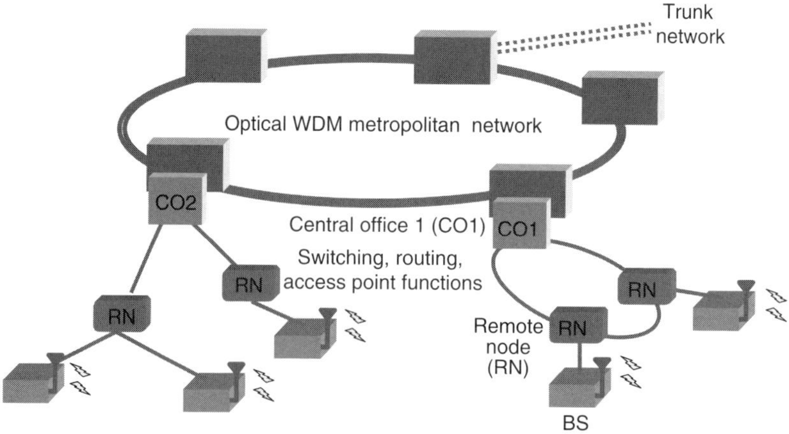


**FIGURE 5.5** WDM HFR and WDM MAN interface.

environment are also carried out at the CO. Such a centralized control scheme enables the deployment of BSs with significantly reduced functionality and complexity.

In the architecture shown in Figure 5.5, WDM channels could be dropped or added into the optical WDM MAN via the switching center. As depicted in Figure 5.5, CO1 is connected to several remote nodes (RNs) via an optical ring while the second central office (CO2) feeds its RNs via a star-tree arrangement [3]. In either arrangement, individual wavelengths are demultiplexed from the WDM signal by being dropped from the optical WDM MAN via the remote nodes, which direct the optical signals to the remote antenna BSs for signal detection and radio distribution. In the upstream direction, radio signals generated at the customer site are up-converted in frequency and radiated to the BS where an electrical-to-optical conversion process takes place before the signal is directed to the CO for further processing.

The application of wavelength division multiplexing concepts and technologies in HFR networks offers a number of advantages. For example, WDM and photonic routing can simplify the network architecture because it is possible to have a shared feeder in multipoint networks. Also, WDM offers flexible traffic routing per cell (different wavelengths directed to a particular BS can carry different data rates), independent handling of broadcast and interactive services, and other network management advantages including protection features. WDM also provides increased flexibility and scalability when upgrading capacity in the HFR network, whether it is by deploying additional BSs, introducing new services, increasing the quality of service, or ensuring transparency to the service that is introduced.

One of the key challenges in implementing HFR systems is developing suitable optical network architectures for efficient distribution of the radio signals while also maintaining a simple and compact BS configuration. This is particularly important for radio carrier frequencies in the millimeter-wave frequency range, where the deployment of HFR systems will require the installation of a larger number of BS units. At the antenna BS, the ability to integrate the optical devices, electronic components, as well as the antennas as the radiating elements, will enable the design and development of simple and lightweight BSs for easy installation on building walls and corners, streetlights, and telephone poles. The particular architecture of the optical network in the HFR system will ultimately determine the required hardware at the CO, the RN, and the BS itself.

As discussed earlier, hybrid fiber radio systems enable the exploitation of existing or planned high-capacity optical networks to efficiently establish high bandwidth interconnections between BSs. They provide a flexible approach for remotely interfacing with multiple BSs and enable the possibility to reduce system complexity with a centralized architecture that incorporates a simplified BS located closer to the customer. HFR technologies also provide a "future-proofing" capability where by proper design, multiple radio services and standards can be accommodated. Ultimately there will be trade-off between the complexity of the electronic/radio frequency (RF) and the optoelectronic interfaces in the CO and antenna BS. In the following section, we describe some of the enabling technologies for the implementation of HFR networks including approaches for distributing the radio signals over fiber as well as the requirements for the various devices and components in the system.

## 5.4 TECHNOLOGIES FOR DISTRIBUTING RADIO SIGNALS OVER FIBER

### 5.4.1 Signal Transport Schemes

There are several possible approaches to transporting radio signals over optical fiber in HFR systems, as shown in Figure 5.6, with each technique having advantages and disadvantages. The three generic techniques can be described as "RF-over-fiber," "IF-over-fiber," and "baseband-over-fiber," where the two former schemes are examples of analog photonic links and the latter is a digital photonic link. HFR analog photonic links are typically multichannel in nature and require high power compared to digital schemes because of the increased carrier-to-noise ratio (CNR) requirements. The performance, including CNR and capacity, of HFR systems employing analog optical links is limited by the noise of the various optical and electrical components in the link as well as by device nonlinearities, which introduce intermodulation and distortion products that create interference with other radio channels.

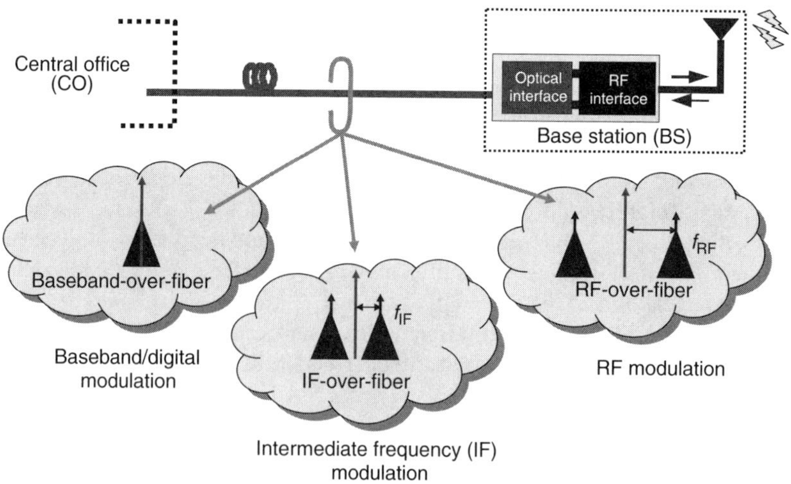


**FIGURE 5.6** Radio signal transport schemes for HFR systems.

RF-over-fiber is probably the most straightforward radio signal distribution scheme because the wireless signals are transported directly over the fiber at the radio carrier transmission frequency without the need for any subsequent frequency up- or down-conversion at the remote antenna BSs. The key optical and RF devices required at the CO and the BS for downstream signal transmission in an HFR system based on RF-over-fiber is shown in Figure 5.7. As can be seen from this diagram, this transport scheme enables centralized control and remote monitoring of the radio signal distribution via the fiber backbone network and reduces the complexity of the BS implementation. In Figure 5.7, the wireless data obtained from the trunk network are modulated onto a number of lower intermediate frequency (IF) carriers, which are then combined to form a subcarrier multiplexed (SCM) signal. This SCM signal is up-converted to the radio transmission frequency using a local oscillator (LO) source located at the CO and is then modulated onto an optical carrier. At the remote BS, the analog optical signal is detected, amplified, filtered, and directed to an antenna for free-space transmission. Note that Figure 5.7 depicts only the required CO and BS hardware for downstream wireless transmission. Upstream radio transmission to the BS and subsequently back to the CO will require a mechanism for modulating an optical source located at the BS at the radio carrier frequency, and photodetection of this signal back at the CO.

Commercial fiber radio systems based on RF-over-fiber signal transport schemes have been developed, however, currently only for wireless networks operating at lower frequencies (<3 GHz). Typically, these systems work on the direct modulation of low-cost lasers, either multimode Fabry–Perot (FP) or single-mode distributed feedback (DFB) devices, depending on the application and the required linearity or dynamic range of the link. The spurious free dynamic range (SFDR) required in HFR systems is influenced by the radio environment such as the wireless standard specifications and coverage range as well as the specific signal transport approach, and can vary from 95 to 115 dB-Hz$^{2/3}$. The use of externally modulated isolated DFB lasers in cooled packages

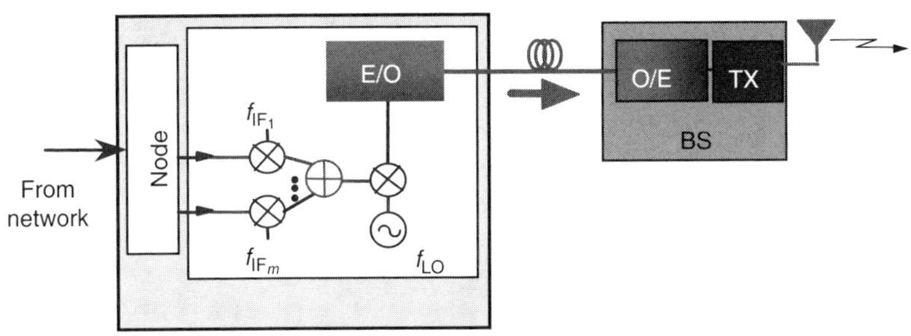


**FIGURE 5.7** CO and BS hardware in an HFR system for downstream signal transmission via RF-over-fiber.

can provide SFDR in excess of 100 dB-Hz$^{2/3}$ at frequencies up to several gigahertz, whereas lower cost, unisolated, uncooled FP lasers are typically limited to less than this value. The SFDR of an analog photonic link can also be improved via the use of a linearization scheme and several linearization techniques for improving HFR performance have been recently reported [4,5].

Figure 5.8 shows a schematic diagram of the basic hardware required at the CO and remote BS for downstream signal transmission in an HFR system based on the distribution of the radio signal at a lower IF, the so-called "IF-over-fiber" signal transport scheme. IF signal transport schemes offer the advantage that the readily available mature microwave hardware can be utilized at the BS, although the requirement for frequency conversion at the BS increases the complexity of the BS architecture particularly as the frequency of the wireless application moves into the millimeter-wave frequency region. The BS hardware now requires LOs and mixers for the frequency conversion processes, which may limit the ability to upgrade or reconfigure the radio network with the provision of additional radio channels or the implementation of required changes in RF frequency. IF-over-fiber radio signal transport schemes do allow transmission over multimode fiber (MMF) and several commercial HFR products are based on the distribution of radio signals over MMF since many buildings have legacy optical fiber infrastructure networks based on multimode fibers. For example, LGC Wireless' product LGCell is based on this concept [6]. Recently HFR research has also considered the use of new vertical cavity surface emitting lasers (VCSELs) operating at 850 nm, which are currently being developed for a range of applications including gigabit ethernet. VCSELs are low-cost devices and their application in analog optical links has been the subject of several papers in the last few years and SFDR values exceeding 90 dB-Hz$^{2/3}$ (for frequencies <5 GHz) have been reported [7]. The transmission of RF signals over MMF using VCSELs has been reported for HFR applications, which include fiber remoting links for WLANs [8].

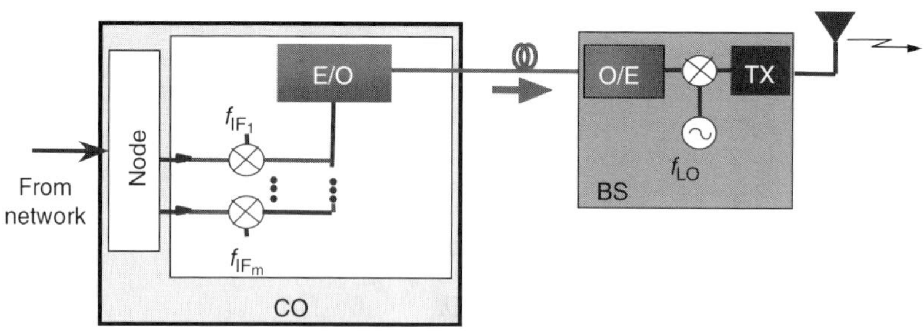


**FIGURE 5.8** CO and BS hardware in an HFR system for downstream signal transmission via IF-over-fiber.

The third technique that can be used to transport data-carrying radio signals between the CO and the remote antenna BS in HFR systems is via a baseband-over-fiber approach depicted in Figure 5.9. As shown here, the radio information for the radio carriers is transported to the BS as a time-division multiplexed digital data stream. The individual data channels are then demultiplexed, up-converted to IFs, before undergoing an additional frequency up-conversion to the required radio frequency band via an LO located at the BS. Upstream signal transport via baseband-over-fiber can also be accomplished by down-converting the received wireless carriers at the BS to the baseband before transmission back to the CO. As with IF-over-fiber, HFR systems based on baseband-over-fiber transport schemes can readily exploit the use of mature and reliable RF and digital hardware for signal processing at the CO and BS as well as low-cost optoelectronic interfaces. However, the need for frequency conversion at the BS complicates the BS architecture design as the air interface frequency increases. The additional LO source and extensive signal processing hardware (frequency conversion and multiplexing and de-multiplexing (MUX–DEMUX) of signals from many users) in the antenna BS may also limit the upgradeability of the overall fiber radio system.

Given the availability of digital, microwave, and millimeter-wave hardware with the required performance, wireless networks ranging in operating frequency from 900 MHz (cellular) to millimeter-wave WLANs at 58 GHz may be able to incorporate baseband-over-fiber signal transport schemes. One currently available commercial HFR system product based on digital-over-fiber transport is the Digivance system developed by ADC, which provides an integrated platform for transporting and distributing both 800 and 1900 MHz radio signals [9]. In this system, the CO digitizes the RF spectrum, then digitally transports it over single-mode fiber to remote antenna sites that can be located up to 20 km away. The ADC digital RF transport scheme is capable of handling multiple radio access standards and modulation schemes

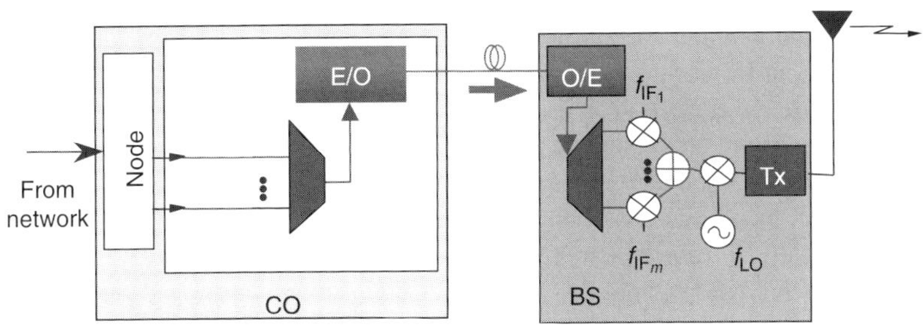


**FIGURE 5.9** CO and BS hardware in an HFR system for downstream signal transmission via baseband-over-fiber.

simultaneously, and the distance between the CO and the BSs is limited only by the time delay requirements of the wireless air interface standards.

### 5.4.2 Optimizing System Performance

As is clear from Figure 5.7, the RF-over-fiber signal transport scheme has the direct advantage of realizing simpler BS architectures in HFR systems. However, these links also require optical components that exhibit good analog performance over the bandwidth required by the wireless network characteristics. Although this is more straightforward to achieve at lower radio frequencies using well-established approaches for linearization of directly modulated and externally modulated optical sources [4,5], the task becomes more difficult for millimeter-wave HFR systems [10]. Another drawback of the RF-over-fiber signal distribution scheme is the increased impact of fiber chromatic dispersion (at a wavelength of 1.55 μm) on the detection of the optical signal and subsequent recovery of the RF carriers, both at the BS and the CO. In contrast, dispersion effects are less of an issue in IF-over-fiber and baseband-over-fiber signal distribution schemes, with chromatic dispersion having a minimal effect on the delivery of sub-Gb/s data streams over links that are less than 100 km in length.

In a conventional intensity-modulated analog optical link, the effect of fiber chromatic dispersion is to impose an RF power penalty on the recovered electrical signal, which also increases with increasing fiber length and RF frequency. The resulting RF power variations will lead to degradations in the link CNR. As the RF frequency increases, the impact on CNR can become dramatic; for example, a 3 dB power penalty in detected RF power occurs for a fiber length of 6 km in an externally modulated link operating at 20 GHz, with the same penalty arising after a distance of only 0.7 km for a radio system operating at 60 GHz. Considerable research effort has been devoted towards developing RF-over-fiber transport schemes that are either dispersion-tolerant [11,12] or based on dispersion compensation techniques commonly used in optical networks such as chirped fiber Bragg grating (FBG) filters [13].

A key advantage of the radio signal transport scheme that is dispersion tolerant is its flexibility, because ideally it can be made independent of the fiber length and wireless carrier frequency, and thus does not require a priori knowledge of the characteristics of either the optical distribution architecture or the wireless network. Figure 5.10 shows one such dispersion-tolerant modulation technique based on the external modulation of an optical carrier using a dual-electrode Mach–Zehnder modulator (DE-MZM) [14]. In Figure 5.10, the DE-MZM is biased at quadrature and driven with signals at the two RF electrodes which are 90° out of phase. This arrangement produces an optical modulation format known as optical single sideband with carrier (OSSB+C), which has also been demonstrated via the use of an optical FBG [15]. In contrast to the filtering scheme, the DE-MZM approach for

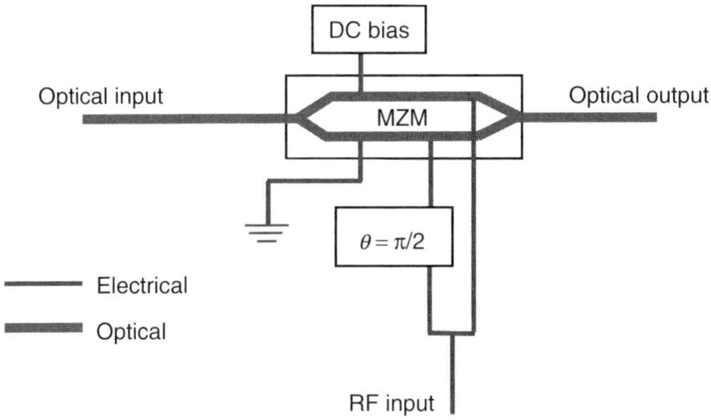


**FIGURE 5.10** Dispersion-tolerant signal transport scheme for HFR systems incorporating OSSB+C modulation via a single DE-MZM.

generating OSSB+C modulation does not rely on the characteristics of the optical filter and the suppression of the sideband is limited only by the precision with which the OSSB+C generator can be practically implemented (the two RF drive signals should be matched in amplitude to achieve the maximum sideband suppression).

The implementation of HFR systems based on RF-over-fiber signal transport for operation at higher wireless transmission frequencies, particularly in the millimeter-wave range, presents a number of challenges. Such HFR architectures require suitable high-speed optical modulation techniques that have the ability to generate milimeter-wave modulated optical signals as well as high-speed photodetection techniques. The simplest and most mature method for generating a high-frequency modulated optical signal is to modulate the optical carrier via an external modulator. However, for the application of high-speed external modulators in millimeter-wave HFR systems, it is imperative that these devices exhibit a number of characteristics in addition to the required broad bandwidth. These features include low-drive voltages, good linearity, high bias stability, high optical power-handling ability, and also low optical insertion loss. Millimeter-wave photodetectors (PDs) with high saturation powers and large electrical output powers are also very useful in the development of the antenna BS with fewer electronic amplifiers. The PD saturation power is limited by the nonlinearity of its response and it has been shown that this nonlinearity is caused by a decrease in the electric field and reduction in the carrier velocity due to the space-charge effect. A PD structure that can overcome the drawback of a low saturated carrier velocity is the unitraveling carrier (UTC) PD in which only electrons traveling at a velocity much higher than the saturation velocity contribute to the space-charge effect. An optical receiver module incorporating a UTC PD with bandwidth exceeding 120 GHz was recently reported [16].

As discussed earlier, HFR system performance limitations in analog optical links will arise from the cumulative effects of noise and distortions from device nonlinearities as well as the crosstalk arising from impairments in the optical network. An improved metric for defining the performance in such systems is the carrier-to-interference-and-noise ratio (CINR), which takes into account the impact of both parameters. Figure 5.11 shows the importance of considering CINR in establishing the optimum performance of an HFR system [17]. This graph shows the predicted CINR as a function of optical modulation index (OMI) for a local multipoint distribution system (LMDS) at 28 GHz, employing an RF-over-fiber transport scheme based on OSSB+C modulation. LMDS operates within the 28–33 GHz frequency band and can deliver data rates to users of up to 40 Mb/s occupying noise equivalent bandwidths of 20 MHz when quadrature phase-shift keying modulation is used. An LMDS wireless network with a 50 m cell radius and line-of-sight (LOS) radio transmission will require an RF signal power of 12 dBm and a CINR ratio of 21 dB to achieve the target bit error rate (BER) of $10^{-6}$.

In Figure 5.11, the number of channels in the HFR LMDS system has been varied from 2 to 8 with the simulation including typical values for the currently available commercial optoelectronic devices. As can be seen in the plots, the maximum CINR in the system reduces with increasing channel number, peaking at a lower value of OMI. This is to be expected because although the CNR in the analog optical link increases with increasing input RF power applied to the external modulation, the CINR decreases with increasing number of channels and RF power, because of the increasing intermodulation distortion levels. The combination of these conflicting effects leads to the result shown in Figure 5.11.

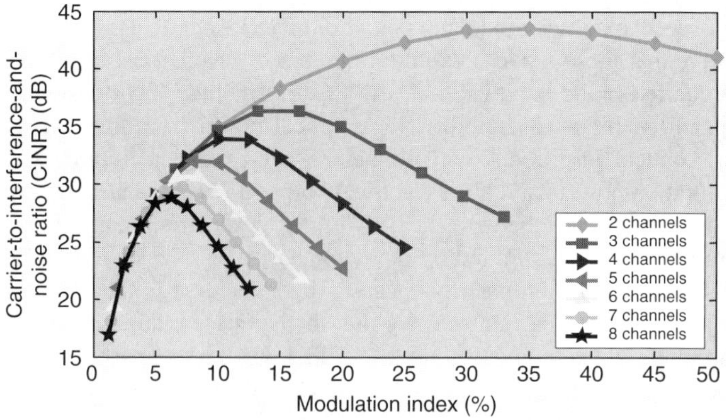


**FIGURE 5.11** Predicted CINR versus OMI for an LMDS system incorporating an RF-over-fiber scheme with the transmission of three RF subcarriers.

In considering the best approach for radio signal transport in HFR systems, a proper assessment of the various tradeoffs with each technique also requires a consideration of the performance differences between the schemes. Figure 5.12 shows four possible HFR signal transport architectures for comparison. Figure 5.12a and Figure 5.12b are RF-over-fiber schemes while Figure 5.12c and Figure 5.12d are IF-over-fiber links. The key difference

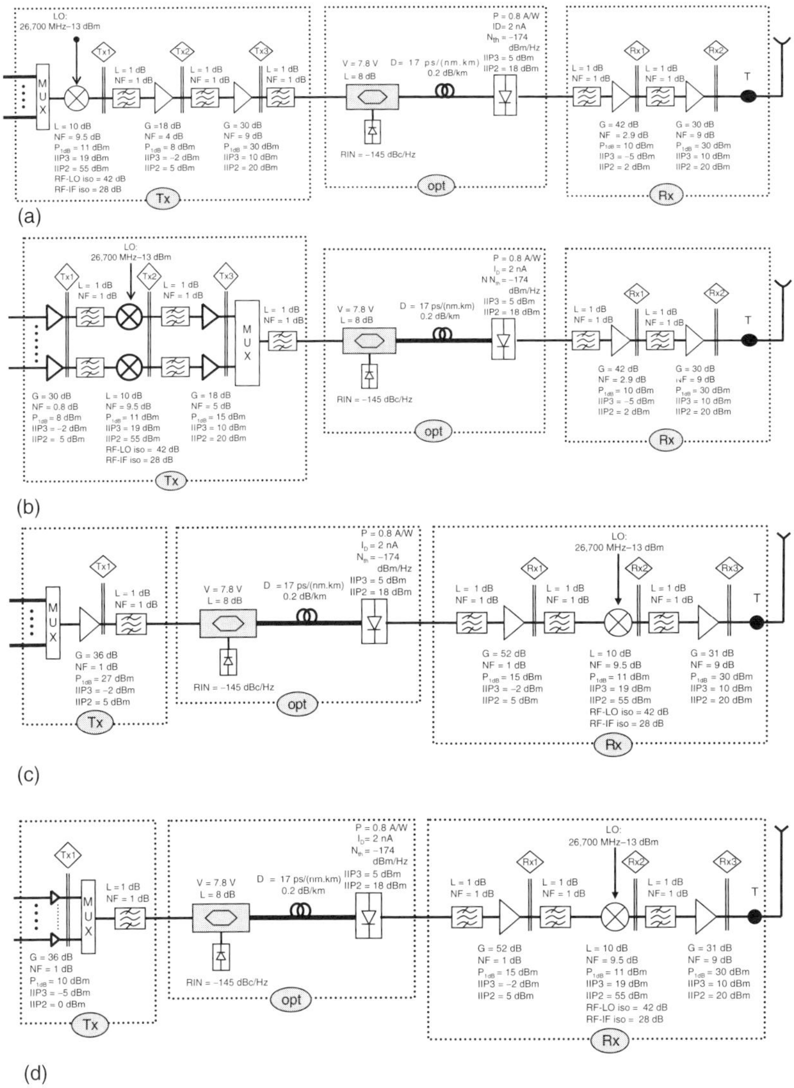


**FIGURE 5.12** Modeled HFR system link configurations: (a) RF-1 (b) RF-2 (c) IF-1 (d) IF-2. (From Kurniawan, T., Nirmalathas, A., Lim, C., Novak, D., and Waterhouse, R., *IEEE Trans. Microw. Thy. & Tech.*, 54, 2006. With permission.)

**TABLE 5.1**
**Predicted CINR for the Different Subsystems in the Four HFR Link Configurations Shown in Figure 5.12**

| | CINR at TX (dB) | CINR at Opt (dB) | CINR at RX (dB) |
|---|---|---|---|
| RF-1 | 49.6 | 35.6 | 31.8 |
| RF-2 | 73.3 | 36.1 | 32.6 |
| IF-1 | 42.4 | 34.0 | 28.3 |
| IF-2 | 73.3 | 36.2 | 29.6 |

From Kurniawan, T., Nirmalathas, A., Lim, C., Novak, D., and Waterhouse, R., *IEEE Trans. Microw. Thy. & Tech.*, 54, 2006. With permission.

between the configurations lies in the arrangement of the components in the transmitter circuits; two of the configurations incorporate multiplexing of the RF channels prior to amplification (and frequency up-conversion for the RF-over-fiber link), thereby reducing the number of required RF amplifiers and potentially the cost of the system.

Table 5.1 shows the predicted performance characteristics for the four different HFR link configurations for three input RF channels of equal power. The CINR has been calculated for the three key subsystem blocks in the link architecture: the transmitter (Tx), optical (Opt), and receiver (Rx). The values shown in Table 5.1 correspond to the power in the middle of the three subcarriers and represent the worst-case scenarios for interference. Comparing the results in Table 5.1 it is evident that the overall link performance is limited by the performance of the optical subsystem, mainly due to the nonlinear transfer function of the MZM incorporated in the link. Comparing the RF-over-fiber link performance with that of IF-over-fiber, it can also be seen that the RF-over-fiber transport scheme gives more than 3 dB improvement in CINR. This is due to the need for additional amplification at the BS in the IF-over-fiber HFR link as well as the required signal processing, which introduces additional nonlinearities. Table 5.2 shows the predicted SFDR as a

**TABLE 5.2**
**Predicted SFDR for the Four HFR Link Configurations Shown in Figure 5.12**

| | SFDR (dB-Hz$^{2/3}$) | | |
|---|---|---|---|
| | N = 3 | N = 5 | N = 8 |
| RF-1 | 83.2 | 78.8 | 76.0 |
| RF-2 | 83.8 | 79.5 | 76.6 |
| IF-1 | 81.4 | 78.0 | 74.6 |
| IF-1 | 82.4 | 78.5 | 74.9 |

From Kurniawan, T., Nirmalathas, A., Lim, C., Novak, D., and Waterhouse, R., *IEEE Trans. Microw. Thy. & Tech.*, 54, 2006. With permission.

function of the number of subcarriers, $N$, for the four different HFR link configurations depicted in Figure 5.12. Slightly higher dynamic range performance is achieved in the RF-over-fiber configurations, with the SFDR degrading with increasing number of RF subcarriers.

Ultimately, the dynamic range performance of the HFR system will be determined by the nonlinearity performance of the various optical and RF components incorporated in the implementation. Although linearization of the external modulator will lead to an increased SFDR, performance improvements can also be achieved by increasing the third-order and second-order intercept points for the RF devices. In the RF-over-fiber architecture most performance improvement will be obtained by increasing the linearity of the particular device, which appears as the last stage in the receiver configuration. In contrast, increasing the linearity of the second stage in the IF-over-fiber receiver, namely the mixer, will provide the maximum benefit in SFDR improvement.

It is also interesting to compare the performance for a baseband-over-fiber signal transport scheme in an LMDS HFR system. Figure 5.13 shows measurements of BER for all three generic transport schemes for LMDS signal transmission at 29 GHz over 0–50 km of single-mode fiber with 10 km length increments. From the results, it can be seen that the power penalties for transmission over 50 km are less than 0.5 dB for the three transport techniques with the sensitivity of the baseband-over-fiber technique approximately 4 dB better than IF-over-fiber and about 7 dB better than RF-over-fiber. These results are not unexpected because, compared to the baseband- and IF-over-fiber transport methods, RF-over-fiber will require the most received optical power on detection to obtain the same BER of the recovered data due to the

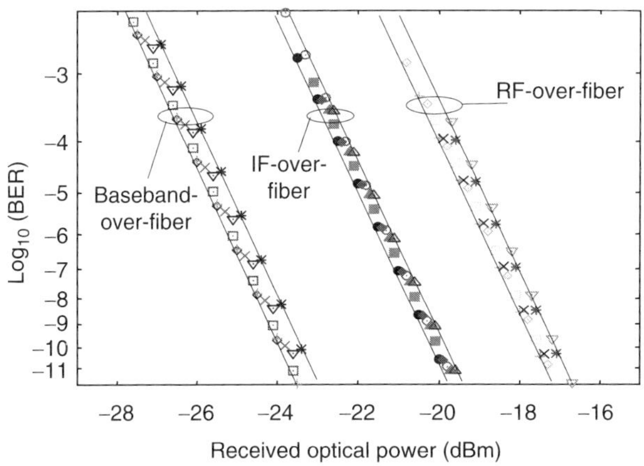


**FIGURE 5.13** Measured BER for an LMDS HFR system incorporating baseband-over-fiber (622 Mb/s, $f_{LO}$ = 29 GHz), IF-over-fiber ($f_{IF}$ = 2 GHz, $f_{LO}$ = 27 GHz), and RF-over-fiber ($f_{RF}$ = 29 GHz with OSSB+C modulation).

decreased optical link efficiency as the radio carrier frequency increases. At the same time, however, the improved sensitivity of the baseband- and IF-over-fiber transport methods for millimeter-wave fiber radio systems comes at the expense of requiring a stable high-frequency LO with low phase noise at the antenna BS.

## 5.5 INTEGRATION OF HFR SYSTEMS WITH WDM NETWORKS

As discussed in Section 5.3, wavelength division multiplexing in hybrid fiber radio systems is being explored as a means to exploit the various benefits of optical networking, as well as to simplify the management of the radio network, the fiber radio network architecture, and the design of the BS itself. WDM can also play a key role in the design of millimeter-wave HFR systems, where LOS constraints imposed by signal propagation characteristics at millimeter-wave frequencies introduce the need for a sectorized antenna interface at the BS to optimize the antenna coverage area [18]. Here the distribution scheme can be further simplified using WDM in conjunction with subcarrier multiplexing, where a different WDM channel feeds each BS and different SCM channels can drive the individual antenna sectors. A key issue that arises in the application of WDM to millimeter-wave HFR systems is the need to merge or integrate with current WDM infrastructures [19], in particular metropolitan area and local optical access networks. In addition, there is a challenge to improve the resulting optical spectral efficiency in the network because WDM channel spacings in excess of 100 GHz will be required if using conventional double sideband optical modulation techniques, whereas the radio signal (spaced at the millimeter-wave frequency from the optical carrier) itself is only modulated with an information bandwidth of less than 2 GHz.

The use of OSSB+C modulation to overcome the fiber dispersion effects in millimeter-wave HFR systems naturally leads to an increase in spectral efficiency because one of the modulation sidebands is removed; however, the spectral efficiency can be further improved via the use of novel wavelength filtering schemes. One such approach is shown in Figure 5.14 where wavelength interleaving is used to increase the optical spectral efficiency and hence the overall capacity of the WDM network [20]. Here the WDM channels that originate at the CO are multiplexed together in a spectral overlapping arrangement via a wavelength-interleaved multiplexer (WI-MUX) and transported to RNs feeding a particular antenna BS. The RN selects the optical carrier at the specified wavelength and the corresponding optical sideband using a wavelength-interleaved optical add-drop multiplexer (WI-OADM). In the architecture shown in Figure 5.14, the upstream radio signals are carried by a WDM channel at the same wavelength and inserted back into the network via the same WI-OADM. The particular filtering

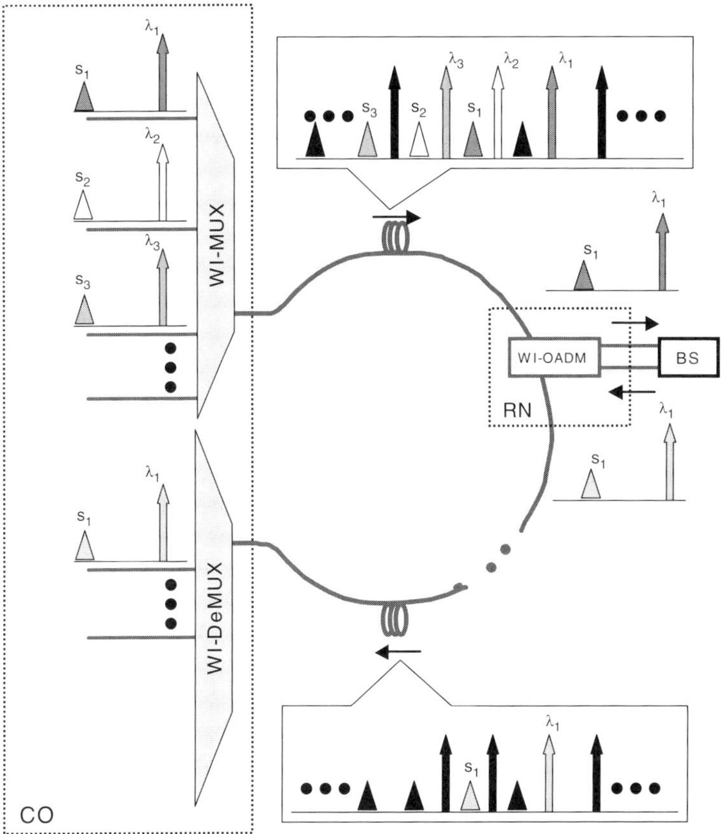


**FIGURE 5.14** Wavelength interleaving technique to increase optical spectral efficiency in HFR systems employing OSSB+C modulation.

characteristics required from such a WI-OADM can be achieved via the implementation of multiple FBGs in cascade to filter out the desired carrier wavelengths and their sidebands, or via novel FBG designs such as one grating incorporating multiple phase-shifts [21].

To leverage off the existing wired WDM infrastructure as much as possible and enable cost-effective realizations of hybrid fiber radio systems, it is essential that such networks should be able to merge and integrate with conventional WDM technologies [22]. In an HFR architecture such as that shown in Figure 5.5, a WDM HFR system incorporating wavelength interleaving in an optical ring network arrangement would interconnect with an optical WDM MAN backbone with 100–200 GHz channel spacing. Multiple blocks of 100 GHz of the optical spectrum containing the interleaved WDM channels would be dropped from the MAN backbone via the CO, distributed to RNs, and a selection of wavelengths then dropped at each RN. At the RN, a wavelength band containing a few wavelength-interleaved channels will then be distributed to the designated BS to feed the sectorized antenna interface in a fiber-to-the-antenna configuration. It has been shown that the arrangement and allocation of interleaved WDM channels within the wavelength block can have a significant impact on the performance of such a WDM HFR radio system [19,22].

WDM HFR systems employing dense wavelength division multiplexing (DWDM) have also been recently investigated. The successful implementation of such wavelength-interleaved DWDM (WI-DWDM) feeder networks in millimeter-wave HFR systems is largely dependent on the realization of suitable and efficient MUX–DEMUX schemes. Simple multiplexing schemes that efficiently interleave DWDM millimeter-wave HFR channels separated by 25 GHz have been recently proposed [23,24], although a suitable demultiplexing scheme required additional wavelength-selective pre- and post-processing hardware. Recently, an alternative approach to realizing a demultiplexing functionality for wavelength-interleaved millimeter-wave HFR system that overcomes some of these limitations has been reported based on the realization of a multifunctional WDM optical interface with simultaneous MUX–DEMUX [25]. The MUX–DEMUX technique has the potential to realize simple, compact, and cost-effective RNs and CO by avoiding the use of wavelength selective pre- and postprocessing hardware. Essentially the wavelength MUX–DEMUX functionalities in the CO and the RN are combined into a single device, which reduces complexity and enables more cost-effective optical architectures to be realized. As shown in Figure 5.15, the scheme incorporates an arrayed waveguide grating (AWG) with multiple loop-backs between the input and the output ports, in addition to multiple optical circulators and optical isolators. The multiplexing functionality of this architecture also incorporates a carrier subtraction technique that reduces the carrier-to-sideband ratios of the multiplexed channels and improves the overall link performance. The interface shown in Figure 5.15 effectively adds or drops the desired channel to and from the WI-DWDM feeder network, and also simplifies the BS by removing the light source from the uplink signal path.

## 5.6 BASE STATION TECHNOLOGIES IN HFR SYSTEMS

A key aspect of the successful deployment of hybrid fiber radio systems is the development of appropriate antenna BS technologies. In addition to the required optical and electrical performance, these optical–electrical–air interfaces should ideally feature at least three interrelated characteristics: high efficiency, low cost, and small size. It is important that the BSs are as efficient as possible in terms of both power consumption and link efficiency. Achieving these

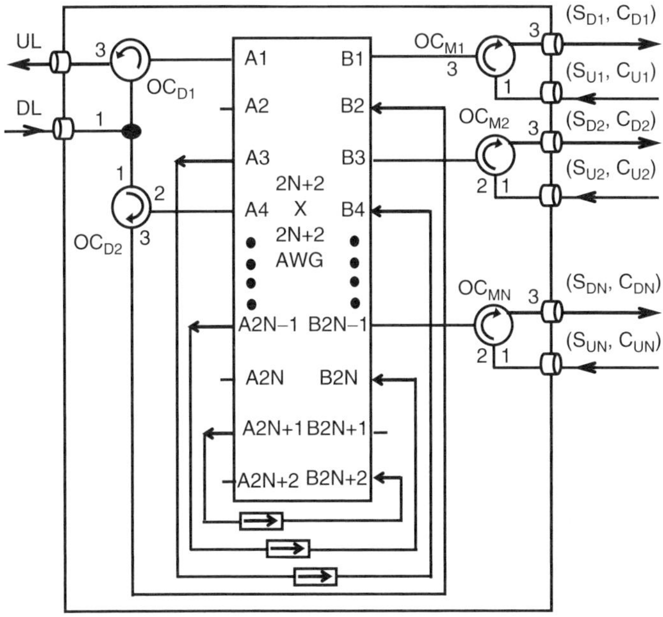


**FIGURE 5.15** Simultaneous multiplexing and demultiplexing of wavelength-interleaved channels in a DWDM millimeter-wave HFR scheme.

goals for millimeter-wave HFR systems presents some major challenges; however, it is imperative to the realization of these networks in the future.

As HFR systems operating in the millimeter-wave frequency region will typically require a larger number of BSs because of the smaller coverage area at these frequencies, the amount of DC power required to operate the remote antenna units has to be minimized. In practice this is typically achieved by reducing the component count within the antenna module itself. From this perspective, one could expect that the RF-over-fiber signal transport scheme would enable the development of antenna BSs with lower power consumption due to the smaller component count required. However, link efficiency must also be taken into account. For example, large bandwidth PDs are typically less efficient than smaller bandwidth devices and therefore higher gain wideband RF amplifiers may be required to boost the detected RF power to the appropriate levels required for free-space wireless transmission. Such high-gain amplifiers will consume more power than lower gain components thereby offsetting some of the perceived power dissipation advantages of millimeter-wave HFR BSs with optoelectronic interfaces based on RF-over-fiber signal transport.

The concept of a completely passive BS for HFR systems operating at frequencies below 2 GHz has been demonstrated based on the use of a single electroabsorption modulator (EAM) [26]. In [26] the one EAM transceiver is operated as both a receiver for the downlink path by detecting only a fraction of the downstream optical signal and a modulator for the upstream radio signal using the remainder of the downstream optical carrier, which was then sent back to the CO. To enable full duplex operation of the HFR system incorporating a single EAM at the BS, a frequency division duplex arrangement is required. The use of a single EAM for dual photodetection and optical modulator functions in an active BS has also been reported for higher frequency HFR systems based on RF-over-fiber in combination with additional dispersion compensation [27].

Reuse of the downstream optical carrier can also be carried out in HFR systems employing OSSB+C modulation for dispersion-tolerant RF-over-fiber signal transport [28]. Figure 5.16 shows an optical interface of an antenna BS that could be incorporated in a WDM hybrid fiber radio system. In this diagram the optical add-drop multiplexer could be in a different location, for example, a RN, or at the antenna BS itself depending on the rollout plan for the HFR network. There are several techniques that could be implemented for the recovery of the downstream optical carrier in Figure 5.16, using either low-cost optical couplers and FBGs or more specialized grating filters in conjunction with an optical circulator [28]. In the example shown in Figure 5.16, OSSB+C modulation is utilized for dispersion-tolerant RF-over-fiber signal transport in both the downstream and upstream

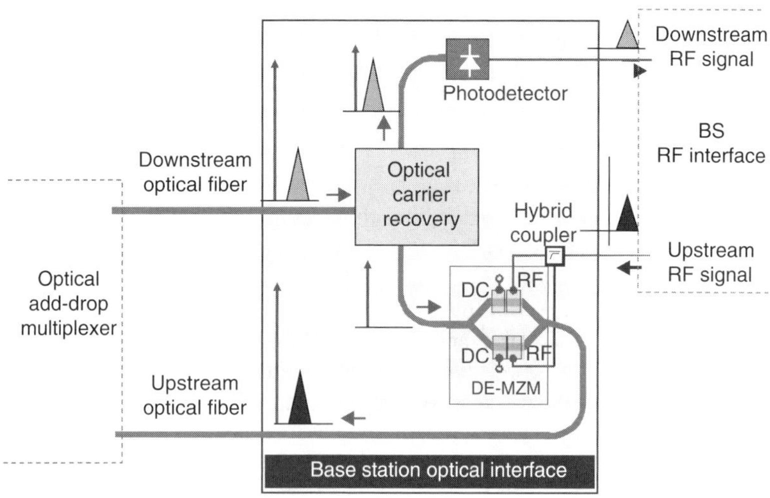


**FIGURE 5.16** Simplified optical interface of an HFR base station incorporating optical wavelength reuse.

directions. An optical interface for HFR antenna BSs that is dispersion tolerant incorporates wavelength reuse, and also supports a wavelength-interleaved architecture, has also been demonstrated [29].

As described earlier, one of the key benefits provided by the RF-over-fiber signal transport scheme for HFR systems is the ability to simplify the BS hardware. This reduction in BS equipment functionality along with the need for low profile BS terminals requires the BSs to be both compact and the various interfaces (optical, RF, and antenna) to be integrated efficiently together. The BS terminal aims to efficiently undertake the process of optical-to-radio conversion for the downlink and vice versa for the uplink. Three potential integration approaches can be implemented to realize a miniaturized BS: hybrid, hybrid integrated circuit (HIC), and monolithic integrated circuit (MIC) technologies, each with its advantages and disadvantages. Because of its maturity, the hybrid approach is currently the most commonly used integration technique and also offers the advantage that the mature component technologies used in other systems can be used. However, the hybrid approach is very labor-intensive and the interconnection of separate packaged modules, each mounted in different transmission technologies, can lead to reliability and parasitic issues.

The simplest integrated circuit approach is based on HIC technology where each subelement function in the BS is implemented on individually optimized islands of materials and these laminates are then interconnected by small sections of planar transmission lines and bond wires. All of these subelements are housed on a common mount to minimize the overall size of the module. The benefits of incorporating HIC technology include its flexibility and cost-effectiveness, although BSs using this component integration approach often require manual "trimming" to achieve optimum performance and therefore can be labor-intensive. In addition, for the development of BSs in millimeter-wave HFR systems, accurate bond wiring procedures must be implemented.

An example of the hybrid integration approach is shown in Figure 5.17, which shows a photograph of a broadband (28–43 GHz) antenna BS module developed for downstream radio signal transmission in a millimeter-wave HFR system offering multiservice capability [30]. The overall size of this unit is $45 \times 20 \times 10$ mm, which also includes a diplexer and a filter for the recovery of a high-frequency LO signal delivered remotely from the CO. It is clear that by incorporating printed circuit technology (also used for fabrication of the wideband millimeter-wave antenna) and developing appropriate alignment methods, a BS module with extremely small size can be achieved. A number of HICs for high-frequency HFR systems have been reported covering a range of millimeter-wave frequencies and incorporating various functionalities. A packaged 60 GHz optical transceiver incorporating an EAM has been reported [31], whereas recently a photoreceiver module based on

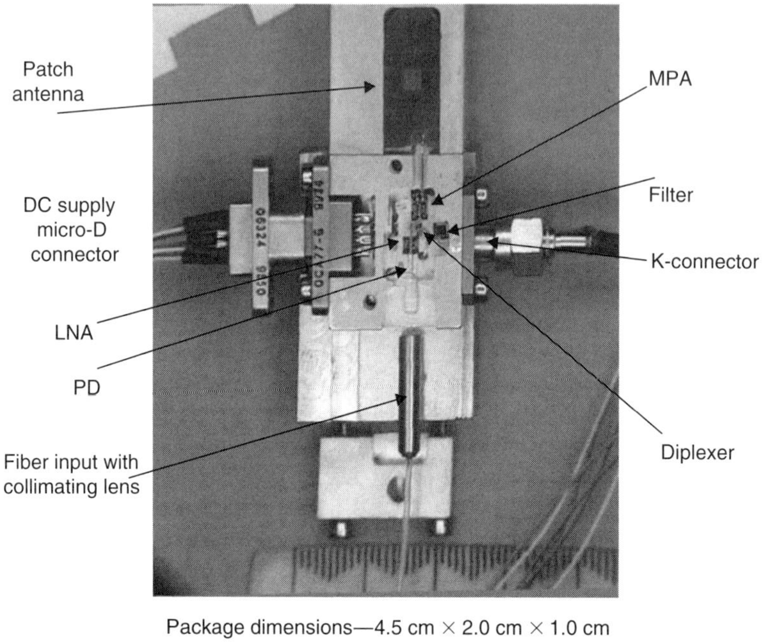


**FIGURE 5.17** Photograph of a broadband antenna BS based on a hybrid integration approach for millimeter-wave HFR systems.

HIC technology comprising a UTC photodiode and an HEMT preamplifier with 22 GHz bandwidth at 120 GHz was demonstrated [16].

The optimum solution to the realization of a low-cost HFR BS terminal is most likely to be MIC technology that enables all the optical and electrical functions to be performed on a single-chip technology. MIC technology offers several advantages including the potential for low-cost manufacture because the amount of manual labor is significantly reduced, as well as the reduction of parasitic effects associated with packaging and interconnects. The MIC approach also ensures device uniformity. Several MIC photoreceivers (incorporating a PD and an amplifier) with good conversion efficiencies have been demonstrated that may be suitable for millimeter-wave HFR systems, although there are still some key technical issues associated with MIC technology that must be addressed. These include low yields, limited heat dissipation for the RF amplifiers, and the ability to directly integrate the electrical and optical components with the antennas.

The poor efficiency of the MIC technique is a fundamental challenge because materials that make efficient antennas unfortunately make poor passive RF components, whereas those that enable efficient RF amplifiers do not make good optoelectronic or electro-optic transducers. Several high-performance efficient antennas that could be integrated with the BS photonic and RF devices have been reported based on the electromagnetic coupling of low dielectric constant material to the monolithic microwave or optoelectronic integrated circuit wafers [32]. A monolithically integrated narrowband 60 GHz photonic emitter has also been demonstrated comprising a UTC PD, a bias circuit, and a patch antenna fabricated on InP [33]. Despite these recent successful implementations of MIC technology for millimeter-wave HFR applications, more work must still be done to ensure that all the required BS functions are efficiently developed using a common material. Until then, the hybrid integration approach appears to be the most cost-effective solution currently available for the development of antenna BSs with optimal performance for HFR systems [34].

## 5.7 CONCLUDING REMARKS

Hybrid fiber radio systems are being considered as a viable technology for providing access to broadband wireless communications, in a range of applications, including last mile solutions, extension of existing radio coverage and capacity, and next generation ultrabroadband data communications. The application of optical fiber networks for the distribution of radio signals in HFR systems has been an intense area of research and investigation for the last 15 years and this technology is now a commercial reality for indoor distributed antenna systems and the remoting of radio signals for current generation mobile networks.

Much progress has been made in the development of HFR technologies for wireless networks operating at a range of frequencies, including next generation systems that extend well into the millimeter-wave frequency region. These new enabling technologies include high-performance optical and RF components, new signal generation and data modulation techniques for down- and upstream signal transport, novel architectures incorporating optical networking concepts, and innovative approaches for the integration of optical and RF components to realize high-performance compact antenna BSs. There are still challenges, however, in improving the performance of these systems as well as the cost of implementation. Achieving these goals will ultimately lead to increased adoption of HFR technology and wider-scale deployment of these systems in the wireless communications market.

## ACKNOWLEDGMENTS

The authors wish to thank their many colleagues and students who, over the years, have contributed to the study of hybrid fiber radio technology.

## REFERENCES

1. Ogawa, H., Polifko, D., and Banba, S., Millimeter-wave fiber optics systems for personal radio communications, *IEEE Trans. Microw. Thy. Tech.*, 40, 2285, 1992.
2. www.andrew.com
3. Smith, G.H., Novak, D., and Lim, C., A millimeter-wave full-duplex fiber-radio star-tree architecture incorporating WDM and SCM, *IEEE Photon. Technol. Lett.*, 10, 1650, 1998.
4. Faccin, P., Notargiacomo, M., and Roselli, L., Laser predistortion for radio over fiber systems, in *Proc. IEEE LEOS Ann. Meet.*, Sydney, 2005, 575.
5. Lee, J.J., Park, S.H., and Choi, Y.W., Enhanced ACPR of W-CDMA signals in optical feedforward transmitter by optimization, in *Proc. IEEE Int. Topical Meet. Microw. Photon.*, Seoul, 2005, 59.
6. www.lgcwireless.com
7. Hartmann, P., Qian, X., Penty, R.V., and White, I.H., High performance uncooled 1.3 μm VCSELs for RF-over-fiber applications, in *Proc. IEEE LEOS Annu. Meet.*, Sydney, 2005, 125.
8. Chia, M.Y.W., Luo, B., Lee, M.L., and Hao, E.J.Z., Radio over multimode fibre transmission for wireless LAN using VCSELs, *Electron. Lett.*, 39, 1143, 2003.
9. www.ADC.com
10. Lim, C., Nirmalathas, A., Novak, D., and Waterhouse, R., Dynamic range of a multi-section laser in a millimetre-wave fibre-radio uplink, *Electron. Lett.*, 36, 975, 2000.
11. Smith, G.H., Novak, D., and Ahmed, Z., Overcoming chromatic dispersion effects in fiber-wireless systems incorporating external modulators, *IEEE Trans. Microw. Thy. Tech.*, 45, 1410, 1997.
12. Lim, C., Novak, D., Nirmalathas, A., and Smith, G.H., Dispersion-induced power penalties in millimeter-wave signal transmission using multi-section DBR semiconductor lasers, *IEEE Trans. Microw. Thy. Tech.*, 49, 288, 2001.
13. Kitayama, K., Fading-free transport of 60 GHz optical DSB signal in non-dispersion shifted fiber using chirped fiber grating, in *Proc. IEEE Int. Top. Meet. Microw. Photon.*, Princeton, 1998, 223.
14. Smith, G.H., Novak, D., and Ahmed, Z., Technique for optical SSB generation to overcome dispersion penalties in fibre-radio systems, *Electron. Lett.*, 33, 74, 1997.
15. Marra, C., Nirmalathas, A., Novak, D., Lim, C., Reekie, L., Besley, J., and Baker, N., The impact of grating dispersion on transmission performance in a millimeter-wave fiber-radio system, *IEEE Photon. Technol. Lett.*, 14, 1345, 2002.
16. Ito, H., Furuta, T., Hirata, A., Kosugi, T., Muramoto, Y., Tokumitsu, M., Nagatsuma, T., and Ishibashi, T., Pre-amplifier integrated uni-travelling-carrier photodiode module for operation in 120 GHz band, *Electron. Lett.*, 41, 360, 2005.
17. Kurniawan, T., Nirmalathas, A., Lim, C., Novak, D., and Waterhouse, R., Performance analysis of optimized millimeter-wave fiber radio system, *IEEE Trans. Microw. Thy. & Tech.*, 54, 2006.
18. Waterhouse, R.B., Novak, D., Nirmalathas, A., and Lim, C., Broadband printed sectorized coverage antennas for millimeter-wave wireless applications, *IEEE Trans. Ant. Propn.*, 50, 12, 2002.
19. Lim, C., Nirmalathas, A., Attygalle, M., Novak, D., and Waterhouse, R., On the merging of millimeter-wave fiber-radio backbone with 25 GHz WDM ring networks, *J. Lightwave Technol.*, 21, 2203, 2003.

20. Lim, C., Nirmalathas, A., Novak, D., Tucker, R., and Waterhouse, R., Novel technique for increasing spectral efficiency in millimetre-wave fibre-radio systems, *Electron. Lett.*, 37, 1043, 2001.
21. Marra, C., Nirmalathas, A., Novak, D., Lim, C., Reekie, L., Besley, J.A., Weeks, C., and Baker, N., Wavelength-interleaved OADMs incorporating optimized multiple phase-shifted FBGs for fiber-radio systems, *J. Lightwave Technol.*, 21, 32, 2003.
22. Lim, C., Nirmalathas, A., Novak, D., and Waterhouse, R., Capacity analysis and the merging of a WDM ring fiber-optic backbone incorporating wavelength interleaving with a sectorized antenna interface, *IEICE Trans.*, E86-C, 1184, 2003.
23. H. Toda et al., 25-GHz channel spacing DWDM multiplexing using an arrayed waveguide grating for 60-GHz band radio-on-fiber systems, in *Proc. IEEE Int. Top. Meet. Microw. Photon.*, Budapest, 2003, 287.
24. Bakaul, M., Nirmalathas, A., Lim, C., Novak, D., and Waterhouse, R., Simplified multiplexing scheme for wavelength-interleaved DWDM millimeter-wave fiber-radio systems, in *Proc. Eur. Conf. Optical Commun.*, Glasgow, 2005.
25. Bakaul, M., Nirmalathas, A., Lim, C., Novak, D., and Waterhouse, R., Simplified multiplexing and demultiplexing scheme for wavelength-interleaved DWDM millimeter-wave fiber-radio systems, in *Proc. Int. Top. Meet. Microw. Photon.*, Seoul, 2005.
26. Wake, D., Johansson, D., and Moodie, D.G., Passive pico-cell—A new concept in wireless network infrastructure, *Electron. Lett.*, 33, 404, 1997.
27. Stöhr, A., Kitayama, K., and Jäger, D., Full duplex fiber optic RF subcarrier transmission using a dual-function modulator/photodetector, *IEEE Trans. Microw. Thy. & Tech.*, 47, 1338, 1999.
28. Nirmalathas, A., Novak, D., Lim, C., and Waterhouse, R.B., Wavelength reuse in the WDM optical interface of a millimeter-wave fiber-wireless antenna base station, *IEEE Trans. Microw. Thy. & Tech.*, 49, 2006, 2001.
29. Bakaul, M., Nirmalathas, A., and Lim, C., Dispersion tolerant novel base-station optical interface for WDM future fibre radio systems, in *Proc. Conf. Opt. Internet & Austn. Conf. Opt. Fibre Technol.*, Sydney, 2003, Paper We-B3.3.
30. Smith, G.H., Waterhouse, R.B., Nirmalathas, A., Novak, D., Lim, C., and Sevimli, O., A broadband integrated photonic-antenna interface for multiservice millimeter-wave fiber-wireless applications, in *Proc. IEEE Int. Top. Meet. Microw. Photon.*, Long Beach, 2002, 173.
31. Kuri, T., Kitayama, K., and Takahashi, Y., 60-GHz-band full-duplex radio-on-fiber system using two-rf-port electroabsorption transceiver, *IEEE Photon. Technol. Lett.*, 12, 419, 2000.
32. Rowe, W.S.T., Waterhouse, R.B., Nirmalathas, A., and Novak, D., Integrated antenna base station design for hybrid fiber radio networks, in *Proc. IEEE Int. Top. Meet. Microw. Photon.*, Melbourne, 1999, 47.
33. Takahata, K., Muramoto, Y., Fukushima, S., Furuta, T., and Ito, H., Monolithically integrated millimetre-wave photonic emitter for 60-GHz fiber-radio applications, in *Proc. IEEE Int. Top. Meet. Microw. Photon.*, Oxford, 2000, 229.
34. Waterhouse, R.B., Rowe, W.S.T., Novak, D., Nirmalathas, A., and Lim, C., Integratable antennas for millimeter-wave fiber-wireless applications, in *Proc. IEEE LEOS Annu. Meet.*, Glasgow, 2002, 803.

# 6 High Dynamic Range, 100 km Digital Radio-over-Fiber Links

*Eric E. Funk, Vincent J. Urick, and Frank Bucholtz*

## CONTENTS

6.1 Introduction ... 185
6.2 Motivation and Applications ... 186
6.3 Link Design ... 187
6.3.1 Noise Figure and Gain ... 187
6.3.2 Spurious Free Dynamic Range ... 191
6.3.3 Chromatic Dispersion ... 193
6.3.4 Stimulated Brillouin Scattering ... 195
6.3.5 Quadrature Amplitude Modulation ... 198
6.3.6 QAM Channel Impairments ... 200
6.3.7 Optical Amplifier Placement ... 201
6.4 Link Demonstrations ... 202
6.4.1 EDFA-Amplified 110 km Link Setup ... 202
6.4.2 EDFA-Amplified 110 km QAM Transmission ... 204
6.4.3 Raman-Amplified 105 km Link Setup ... 205
6.4.4 Raman-Amplified 105 km QAM Transmission ... 208
6.5 Future Capabilities ... 210
Acknowledgments ... 210
References ... 211

## 6.1 INTRODUCTION

Radio-over-fiber (ROF) is one of the most practical and largest scale applications of microwave photonics to date. Current applications range from the back-haul of radar signals from remote antennas to radio frequency (RF) signal transport in third generation digital wireless systems.

ROF systems can now be designed with sufficient dynamic range and signal-to-noise ratio (SNR) to transport many of the widely used digital RF modulation formats over considerable distances. In this chapter we focus on the quadrature amplitude modulation (QAM) format, a spectrally efficient digital format used widely for digital video and cable modems.

## 6.2 MOTIVATION AND APPLICATIONS

We begin by defining ROF as a physical layer transport mechanism whereby an RF signal is modulated onto an optical carrier for transport via optical fiber. The source RF signal is modulated in either an analog (AM, FM, etc.) or digital (QAM, QPSK, etc.) manner.

Cable television (CATV) has been the main commercial sector application for ROF. Most modern CATV systems are hybrid fiber coaxial cable (HFC) systems. At a typical HFC head end, well over 100 individual 6 or 8 MHz bandwidth television channels are frequency division multiplexed together within the range of approximately 52–800 MHz. The multiplexed signal is then intensity-modulated onto an optical carrier. The optical carrier is transported over fiber to a local neighborhood. In the neighborhood a photoreceiver converts the optical signal back to an RF signal where it is sent to the home over a conventional coaxial cable drop.

Fibers provide an economic alternative to coaxial trunking in HFC systems. Coaxial cables suffer from high attenuation and, therefore, RF amplifiers must be placed at regular intervals. The amplifiers require an external power source, degrade the SNR, require regular maintenance, and may cause spurious radiation. Fiber trunking eliminates many of these problems.

Other commercial ROF products are being designed for coverage extension of RF systems such as cellular telephone, local multipoint distribution service (LMDS), and wireless local area networks (WLAN). Likewise, as the reach of high dynamic range ROF links is extended to the 100 km range, ROF becomes a very promising alternative to conventional baseband digital fiber in the long haul.

In radio astronomy and radar systems alike, ROF may be used to link spatially diverse antennas into a phased-array system. The benefit of fiber is its low loss at microwave frequencies and ease of deployment when compared to microwave waveguide.

Furthermore, ROF can provide low loss true time delay of signals. Time delay is an essential operation required for beam steering in wideband systems and is also used for autocorrelation and cross correlation. Over small bandwidths, phase shifters and digital signal processing can be used to create time delay. However, ROF opens up the possibility of obtaining frequency-independent time delays over tens of gigahertz of bandwidth. With ROF links reaching over 100 km, it becomes possible to generate low loss, high dynamic range, ultrawideband RF delays of hundreds of microseconds. This will enable new ultrawideband and high dynamic range signal analysis capabilities.

## 6.3 LINK DESIGN

### 6.3.1 Noise Figure and Gain

As with any radio link, noise figure is critical to the design of a practical ROF link. It is therefore useful to understand the relationship between RF noise figure and optical system parameters including optical noise figure.

We now proceed to derive this relationship for an intensity-modulated ROF link. We assume that the optical carrier is intensity-modulated by an RF tone with a Mach–Zehnder modulator (MZM). The MZM is biased at quadrature. The optical signal propagates through a fiber link with optical amplification and is then directly detected in a p-i-n photodiode.

We further assume that the fiber is optically linear. Such a link is described mathematically by first expressing the optical field entering the MZM as $E(t) = \kappa\sqrt{2P_{\mathrm{o}}}\mathrm{e}^{\mathrm{i}\omega_{\mathrm{o}}t}$ where $\omega_{\mathrm{o}}/2\pi$ is the center frequency of the laser, $P_{\mathrm{o}}$ is the laser output power, and $\kappa$ is a constant relating electric field to power. The MZM drive voltage is expressed as follows: $V_{\mathrm{drive}} = V_{\mathrm{b}} + V_{\mathrm{rf}} \sin \Omega t$, where $V_{\mathrm{b}}$ is the DC bias voltage, $V_{\mathrm{rf}}$ is the RF signal amplitude, and $f = \Omega/2\pi$ is the RF signal frequency. Assuming a push–pull (balanced) MZM, the RF signal produces an equal and opposite phase shift in each arm of the MZM, where the magnitude of the phase shift in each arm is given by $\phi(t)/2 = \phi_{\mathrm{b}}/2 + (\phi_0/2) \sin \Omega t$. The phase shift is related to the applied voltage $V_{\mathrm{drive}}$ by $\phi = \pi V_{\mathrm{drive}}/V_{\pi}(f)$ where $V_{\pi}(f)$ is the peak voltage required to produce a peak phase shift of $\pi$ radians. $V_{\pi}(f)$ is a frequency-dependent characteristic of the MZM.

Assuming no loss, the electric fields at the two outputs of the MZM are calculated using

$$\begin{bmatrix} E_{\mathrm{out}\ 1}(t) \\ E_{\mathrm{out}\ 2}(t) \end{bmatrix} = \frac{1}{\sqrt{2}} \begin{bmatrix} 1 & i \\ i & 1 \end{bmatrix} \begin{bmatrix} \mathrm{e}^{\mathrm{i}\phi(t)/2} & 0 \\ 0 & \mathrm{e}^{-\mathrm{i}\phi(t)/2} \end{bmatrix} \frac{1}{\sqrt{2}} \begin{bmatrix} 1 & i \\ i & 1 \end{bmatrix} \begin{bmatrix} E(t) \\ 0 \end{bmatrix}. \tag{6.1}$$

We then have for the two MZM outputs

$$\begin{bmatrix} E_{\mathrm{out}\ 1}(t) \\ E_{\mathrm{out}\ 2}(t) \end{bmatrix} = \frac{1}{2}\kappa\sqrt{2P_{\mathrm{o}}}\,\mathrm{e}^{\mathrm{i}\omega_{\mathrm{o}}t} \begin{bmatrix} \mathrm{e}^{\mathrm{i}\phi(t)/2} - \mathrm{e}^{-\mathrm{i}\phi(t)/2} \\ \mathrm{i}\mathrm{e}^{\mathrm{i}\phi(t)/2} + \mathrm{i}\mathrm{e}^{-\mathrm{i}\phi(t)/2} \end{bmatrix}. \tag{6.2}$$

Taking $\mathrm{E}_{\mathrm{out}\ 1}$ as the field that is propagated, we calculate the optical power incident on the photodetector using the convention $P_{\mathrm{opt}} = g_{\mathrm{opt}} \mid E_{\mathrm{out}\ 1}(t) \mid^2/2\kappa^2$ as

$$P_{\mathrm{opt}} = \frac{g_{\mathrm{opt}}P_{\mathrm{o}}}{4}\left(2 - \mathrm{e}^{\mathrm{i}\phi(t)} - \mathrm{e}^{-\mathrm{i}\phi(t)}\right) = \frac{g_{\mathrm{opt}}P_{\mathrm{o}}}{2}(1 - \cos\phi(t)) = \frac{I_t}{\Re}, \tag{6.3}$$

where $g_{\mathrm{opt}}$ is the cascaded optical gain, $I_t$ is the total photocurrent, and $\Re$ is the detector responsivity (A/W). Assuming quadrature bias, $\phi_{\mathrm{b}} = \pi/2$, and using the identity, $\sin(x \sin\theta) = 2\sum_{n=0}^{\infty} J_{2n+1}(x)\sin[(2n+1)\theta]$, we arrive at

$$I_t = \Re g_{\text{opt}} P_{\text{o}} \left( \frac{1}{2} + \sum_{n=0}^{\infty} J_{2n+1}(\phi_0) \sin[(2n+1)\Omega t] \right), \tag{6.4}$$

where $J_m$ represents $m$th order Bessel functions of the first kind. Note that Equation 6.4 contains a DC term, $I_{\text{dc}} = \Re g_{\text{opt}} P_{\text{o}}/2$, and an RF term with components at the fundamental RF frequency, $\Omega/2\pi$, and its odd harmonics, $I_{\text{rf}} = \Re g_{\text{opt}} P_{\text{o}} \sum_{n=0}^{\infty} J_{2n+1}(\phi_0) \sin[(2n+1)\Omega t] \equiv \sum_{n=0}^{\infty} I((2n+1)\Omega) \sin[(2n+1)\Omega t]$. It is important to point out that Equation 6.4 is exact only for an ideal MZM with no insertion loss and biased at quadrature. Knowing the photocurrent from Equation 6.4, we can calculate the RF power delivered into the load, Z. The RF power delivered at the fundamental frequency is given by

$$P_{\text{rf-out}} = \frac{1}{2}(I(\Omega))^2 Z = \frac{1}{2}\left[\Re g_{\text{opt}} P_{\text{o}} J_1(\phi_0)\right]^2 Z = 2I_{\text{dc}}^2 J_1^2(\phi_0) Z. \tag{6.5}$$

If we now assume that $V_{\text{rf}} \ll V_\pi$ ($\phi_0 \ll 1$), we can use the approximation $J_1(\phi_0) \approx \phi_0/2$ and rewrite Equation 6.5 as

$$P_{\text{rf-out}} = \frac{1}{2} I_{\text{dc}}^2 Z \left( \frac{\pi V_{\text{rf}}}{V_\pi} \right)^2. \tag{6.6}$$

Recalling that $V_{\text{rf}}$ is the input RF amplitude and assuming that the MZM input impedance is also $Z$, we can use $V_{\text{rf}}^2 = 2\text{P}_{\text{rf-in}}\text{Z}$ in Equation 6.6, which gives us the following expression for the RF gain:

$$g_{\text{rf}} = \frac{P_{\text{rf-out}}}{P_{\text{rf-in}}} = \left( \frac{I_{\text{dc}}}{V_\pi} \right)^2 \pi^2 Z^2, \tag{6.7}$$

where $\text{P}_{\text{rf-in}}$ is the RF input power.

Taking $Z = 50\ \Omega$, the link gain can be expressed in decibel form as follows:

$$G_{\text{rf}}[\text{dB}] = -16 - 20\log(V_\pi[\text{V}]) + 20\log(I_{\text{dc}}[\text{mA}]). \tag{6.8}$$

Figure 6.1 shows $G_{\text{rf}}$ as a function of $V_\pi$ and $I_{\text{dc}}$ as calculated using Equation 6.8.

To derive the RF noise figure, we consider the following sources of noise in the long-haul ROF link: input thermal noise, output thermal noise, detector shot noise, laser relative intensity noise (RIN), and noise due to the optical amplification. Thermal or Johnson noise [1] is a spectrally white noise associated with a resistor in thermal equilibrium. For matched resistors in thermal equilibrium at a temperature of $T$, the thermal noise power spectral density is given by $P_{\text{th}}$ (W/Hz) $= k_{\text{B}}T$, where $k_{\text{B}} = 1.381 \times 10^{-23}$ J/K

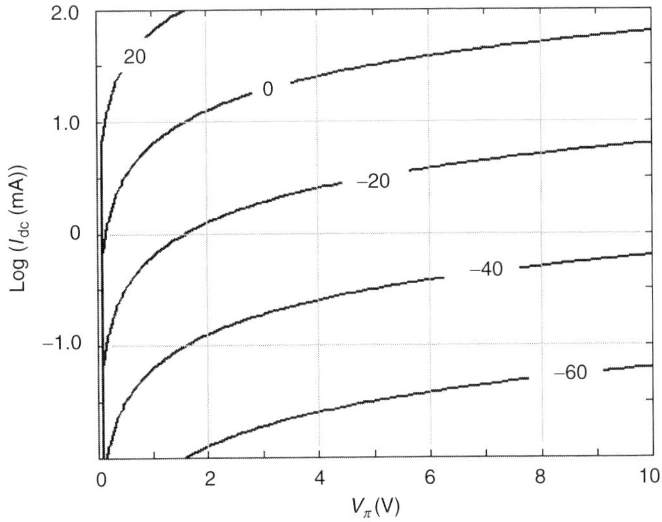


**FIGURE 6.1** RF link gain in decibels as a function of MZM, $V_\pi$, and DC photocurrent for an MZM biased at quadrature.

is Boltzmann's constant. For $T = 300$ K, $P_{th} = -174$ dBm/Hz. Using Equation 6.8, the noise power measured at the photodiode output due to input thermal noise at the MZM is given by

$$N_{ith}[\text{dBm/Hz}] = -174 - 16 - 20\log(V_\pi[\text{V}]) + 20\log(I_{dc}[\text{mA}]), \quad (6.9)$$

while the thermal noise at the output load is simply,

$$N_{th}[\text{dBm/Hz}] = -174. \quad (6.10)$$

Often, $G_{rf}$ (dB) $\ll 0$, and input thermal noise can be neglected. There are also several noise sources inherent in the optical system including detector shot noise, spontaneous emission, and RIN.

Like thermal noise, detector shot noise [2] is spectrally white. The root-mean-squared current spectral density for shot noise is given by $i_{sh}^2(\text{A}^2/\text{Hz}) = 2eI_{dc}$, where $e = 1.602 \times 10^{-19}$ C and $I_{dc}$ is the DC photocurrent. For the system discussed here and with $Z = 50\ \Omega$, the shot noise power spectral density is given by

$$N_{shot}[\text{dBm/Hz}] = 10\log(2eZI_{dc}[\text{mA}]) = -168 + 10\log(I_{dc}[\text{mA}]). \quad (6.11)$$

Next we consider sources of optical noise. Not surprisingly, optical amplification adds additional noise. The primary source of optical noise is spontaneous emission. In the photodiode, spontaneous emission beats (heterodynes) with the optical carrier, generating RF noise.

Using the equations in Ref. [3] we derive the following expressions for the output RF noise power spectral density due to heterodyne mixing between the optical carrier and spontaneous emission

$$n_{\text{sig-sp}}(\text{W/Hz}) = \Re^2 h\nu P_o g_{\text{opt}}^2 nf_{\text{opt}} Z, \tag{6.12}$$

where $g_{\text{opt}}$ is the optical gain, $h = 6.626 \times 10^{-34}$ J·s is Planck's constant, $\nu = \omega_o/2\pi$ is the optical carrier frequency, and $nf_{\text{opt}}$ is the optical noise factor. Note that Equation 6.12 can be applied not only to a single amplifier but also to an entire system of cascaded optical gain and loss.

Using Equation 6.12 and taking a 1550 nm optical wavelength ($\nu = 193$ THz), $Z = 50\ \Omega$, and $I_{\text{dc}} = \Re g_{\text{opt}} P_o/2$, the signal-spontaneous noise power spectral density can be expressed in decibel form as

$$N_{\text{sig-sp}}[\text{dBm/Hz}] = -169 + 10\log(\Re I_{\text{dc}}[\text{mA}]) + G_{\text{opt}}[\text{dB}] + NF_{\text{opt}}[\text{dB}]. \tag{6.13}$$

Similarly, spontaneous emission will beat with spontaneous emission in the photodiode, producing spontaneous-spontaneous RF noise, $n_{\text{sp-sp}}$. Again, using Ref. [3],

$$n_{\text{sp-sp}}(\text{W/Hz}) = \left(\Re h\nu g_{\text{opt}} nf_{\text{opt}}\right)^2 B_o \left(1 - \frac{f}{B_o} - \frac{B_e}{2B_o}\right) Z, \tag{6.14}$$

where $B_o$ is the optical bandwidth and $B_e$ is the electrical bandwidth.

The spontaneous–spontaneous noise power spectral density is obtained in decibel form by using Equation 6.14 with the assumptions that $f \ll B_o$ and $B_e \ll B_o$:

$$\begin{aligned} N_{\text{sp-sp}}[\text{dBm/Hz}] = {} & -241 + 20\log(\Re) + 2G_{\text{opt}}[\text{dB}] + 2NF_{\text{opt}}[\text{dB}] \\ & + 10\log(B_o[\text{GHz}]). \end{aligned} \tag{6.15}$$

For an optical amplifier that is compressed (large $P_o$), the signal-spontaneous noise (Equation 6.13) will dominate over the spontaneous-spontaneous noise (Equation 6.15). If, however, the amplifier is uncompressed, an optical filter is typically employed to filter the amplified spontaneous emission (to limit $B_o$) such that the dominant noise term is still signal-spontaneous noise.

The final source of noise that we will consider is laser RIN. Laser RIN is due to many statistical processes associated with a laser cavity such as amplified spontaneous emission and side modes. Here RIN includes any intensity noise that the laser adds to the system, whatever the cause.

RIN is not white, but rather typically has a peak at the relaxation frequency of the laser. Laser RIN is defined as follows: $RIN_{\text{laser}}(\text{Hz}^{-1}) \equiv \langle \Delta P^2 \rangle / \langle P \rangle^2$, where $\langle \Delta P^2 \rangle$ is the mean square optical power fluctuation and $\langle P \rangle$ is the average optical power [4]. $RIN_{\text{laser}}$ is typically specified relative to the power in the optical carrier in units of dBc/Hz. Knowing the detected DC electrical signal power, the RF noise power from RIN into a 50 Ω load is calculated in decibel form as follows:

$$N_{\text{laser}}[\text{dBm/Hz}] = -13 + 20\log(I_{\text{dc}}[\text{mA}]) + RIN_{\text{laser}}[\text{dBc/Hz}]. \quad (6.16)$$

Assuming that the input RF SNR is limited by thermal noise, the RF noise figure is calculated as

$$\begin{aligned} NF_{\text{rf}}[\text{dB}] &\equiv SNR_{\text{in}}[\text{dB}] - SNR_{\text{out}}[\text{dB}] \\ &= 174 - G_{\text{rf}}[\text{dB}] + N_{\text{t}}[\text{dBm/Hz}], \end{aligned} \quad (6.17)$$

where $N_{\text{t}}$ is the total RF noise at the output. Equation 6.9 through Equation 6.11, Equation 6.13, Equation 6.15, and Equation 6.16 describe the electrical noise power spectral density at the output of the ROF link from each of the primary noise sources. Typically signal-spontaneous noise is at least an order of magnitude larger than other noise sources. In this limit, $N_{\text{t}} \approx N_{\text{sig-sp}}$. Using Equation 6.13 and Equation 6.17, the RF noise figure is expressed as

$$\begin{aligned} NF_{\text{rf}}[\text{dB}] = 21 &+ 20\log(V_{\pi}[\text{V}]) - 10\log(I_{\text{dc}}[\text{mA}]) + G_{\text{opt}}[\text{dB}] \\ &+ NF_{\text{opt}}[\text{dB}] + 10\log(\Re). \end{aligned} \quad (6.18)$$

Equation 6.18 describes the relationship between the RF noise figure and optical parameters in the signal-spontaneous noise limit.

### 6.3.2 Spurious Free Dynamic Range

Two-tone spurious free dynamic range (SFDR) is a commonly used metric for RF link linearity. When driven by two closely spaced RF tones, a nonlinear RF link will produce third-order intermodulation distortion products. If the small signal distortion is extrapolated to higher powers, the fundamental and third-order distortion products will meet at the third-order intercept point as shown in Figure 6.2.

The output SFDR is defined as the difference between the signal output power and the noise power at the point where third-order distortion products

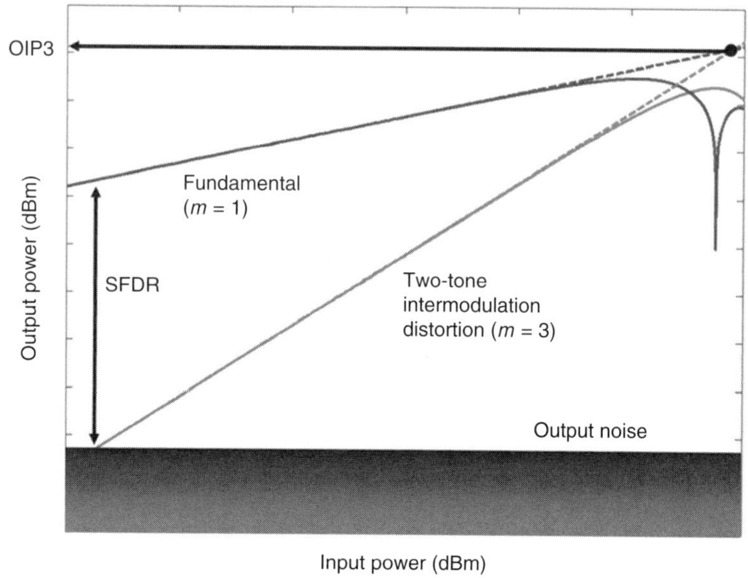


**FIGURE 6.2** Graphical representation of the SFDR for an ROF link limited by the third-order intermodulation product of an MZM.

are equal to the output noise power. In the ROF link, signal-spontaneous noise (Equation 6.13) is dominant. Hence, the SFDR can be written as follows:

$$\text{SFDR}\left[\text{dB}\cdot\text{Hz}^{2/3}\right] = \frac{2}{3}\left(\text{OIP3}[\text{dBm}] - N_{\text{sig-sp}}[\text{dBm/Hz}]\right), \tag{6.19}$$

where OIP3 is the output-referenced third-order intercept point. The relationship expressed by Equation 6.19 can be seen in Figure 6.2 by observing the geometric relationship between the third-order distortion product with a slope of $m = 3$ and the fundamental with a slope of $m = 1$.

For an ROF link employing an ideal MZM biased at quadrature, the OIP3 is determined by distortion in the MZM [5–7]. To derive an analytical expression for OIP3, we first rewrite the total photocurrent (Equation 6.3) with quadrature bias ($\phi_b = \pi/2$) as $I_t = I_{dc}\,(1 + \sin\phi'(t))$. To evaluate the link linearity, we assume sinusoidal drive voltages to the MZM such that $\phi'(t) = \phi_1 \sin \Omega_1 t + \phi_2 \sin \Omega_2 t$ with $\phi_{1,2} = \pi V_{1,2}/V_\pi$ and $f_{1,2} = \Omega_{1,2}/2\pi$. $f_1$ and $f_2$ are closely spaced tones, such that $V_\pi$ is the same at both frequencies. This results in a total photocurrent given by $I_t = I_{dc}\,[\,1 + \sin(\phi_1 \sin \Omega_1 t + \phi_2 \sin \Omega_2 t)]$. Inserting these expressions into Equation 6.3, we arrive at an expression for the RF photocurrent:

$$I_{\text{rf,2-tone}} = 2I_{\text{dc}} \begin{cases} J_0(\phi_2) \sum\limits_{k=0}^{\infty} J_{2k+1}(\phi_1) \sin[(2k+1)\Omega_1 t] \pm \sum\limits_{k=0}^{\infty} \sum\limits_{m=1}^{\infty} J_{2k+1} \\ \{(\phi_1) J_{2m}(\phi_2) \sin[(2m\Omega_2 \pm (2k+1)\Omega_1)t]\} \\ +J_0(\phi_1) \sum\limits_{l=0}^{\infty} J_{2l+1}(\phi_2) \sin[(2l+1)\Omega_2 t] \pm \sum\limits_{l=0}^{\infty} \sum\limits_{n=1}^{\infty} J_{2l+1} \\ \{(\phi_2) J_{2n}(\phi_1) \sin[(2n\Omega_1 \pm (2l+1)\Omega_2)t]\}. \end{cases} \quad (6.20)$$

If $V_1 = V_2 \ll V_\pi$, then $\phi_1 = \phi_2 = \phi_A \ll 1$ and we can use the Bessel approximation $J_n(\phi) \approx \phi^n/(2^n n!)$. These approximations give the following expression for the fundamental and third-order distortion terms of Equation 6.20:

$$\begin{aligned} I_{\text{rf,2-tone}} &= I_{\text{dc}} \phi_A \sin(\Omega_{1,2} t) && \text{(fundamental)} \\ &\pm \frac{I_{\text{dc}} \phi_A^3}{8} \sin[(2\Omega_{1,2} \pm \Omega_{2,1})t] && \text{(third-order distortion).} \end{aligned} \quad (6.21)$$

The third-order intercept point occurs at the intersection of these linear approximations, that is, where the amplitude of the fundamental is equal to the amplitude of the third-order distortion. Taking $I_{\text{dc}}\phi_A = I_{\text{dc}}\phi_A^3/8$ results in a third-order intercept amplitude of $V_{\text{OIP3}} = 2\sqrt{2}V_\pi/\pi$. Using the linear approximation for power in the fundamental Equation 6.6 gives an output intercept power of $\text{OIP3} = 4I_{\text{dc}}^2 Z$. Assuming $Z = 50\ \Omega$, this can be written as

$$\text{OIP3[dBm]} = -7 + 20\log(I_{\text{dc}}[\text{mA}]). \quad (6.22)$$

Note that Equation 6.22 is independent of $V_\pi$ and depends only on the received photocurrent. Inserting Equation 6.22 and Equation 6.13 into Equation 6.19 gives the desired expression for the SFDR of an ROF link in the signal-spontaneous beat noise limit as

$$\text{SFDR}\left[\text{dB} \cdot \text{Hz}^{2/3}\right] = \frac{2}{3}\{162 + 10\log(I_{\text{dc}}[\text{mA}]) - G_{\text{opt}}[\text{dB}] - NF_{\text{opt}}[\text{dB}] - 10\log(\Re)\}. \quad (6.23)$$

### 6.3.3 Chromatic Dispersion

Chromatic dispersion is caused by photon–electron interactions in the fiber and is usually described by the dielectric function. This gives rise to a frequency-dependent index of refraction and, hence, a frequency-dependent phase shift. For double-sideband modulation, the carrier and each of the sidebands are at different optical frequencies, implying that the RF photocurrents generated by the upper and lower sidebands have a frequency-dependent phase relationship. Mathematically, this is described by including the mode propagation constant explicitly in the electric field expression.

The electric field at the output of an MZM as given by Equation 6.2 treats only the temporal dependence of the field and this is completely adequate for the previous discussion. However, to describe chromatic dispersion in an ROF link, we will include the spatial dependence of the field as well. We write the field, $E_{\text{out }1}$, from Equation 6.2, in an expanded form, including the spatial dependence as

$$\begin{aligned} E_{\text{out }1}(z,t) = {} & \kappa\sqrt{P_\text{o}}J_0(\phi_0/2)\sin(\omega_\text{o}t - \beta(\omega_\text{o})z) \\ & - \kappa\sqrt{P_\text{o}}\sum_{\substack{m=-\infty \\ m\neq 0 \\ m\text{even}}}^{\infty} J_m(\phi_0/2)\sin[(\omega_\text{o}+m\Omega)t - \beta(\omega_\text{o}+m\Omega)z] \\ & + \kappa\sqrt{P_\text{o}}\sum_{\substack{k=-\infty \\ k\text{odd}}}^{\infty} J_k(\phi_0/2)\cos[(\omega_\text{o}+k\Omega)t - \beta(\omega_\text{o}+k\Omega)z], \end{aligned} \tag{6.24}$$

where $\beta(\nu)$ is the propagation constant at optical frequency, $\nu$. Equation 6.24 describes a comb of optical frequencies spread about the optical carrier and spaced at the RF modulation frequency. The mode propagation constant $\beta(\nu)$ may be expanded in a Taylor series about the optical carrier frequency $\omega_\text{o}$ as in Ref. [8]. The lowest order terms of the expansion are

$$\beta(\nu) = \beta_\text{o} + \beta_1(\nu-\omega_\text{o}) + \frac{1}{2}\beta_2(\nu-\omega_\text{o})^2. \tag{6.25}$$

The terms in the expansion are given by [9] as $\beta_m = (\text{d}^m\beta/\text{d}\nu^m)_{\nu=\omega_\text{o}}$. Furthermore, the group velocity is given by $\nu_\text{g} = 1/\beta_1$ and the chromatic dispersion parameter of the fiber is $D = -2\pi c\beta_2/\lambda^2$, where $c$ is velocity of light in vacuum and $\lambda$ is the optical carrier wavelength. Using Equation 6.25 and the relations in Equation 6.24, the following expression can be obtained for the RF power at the fundamental frequency

$$P_{\text{rf,out}}(f) = \frac{1}{2}I_{\text{dc}}^2 Z\left(\frac{\pi V_{\text{rf}}}{V_\pi}\right)^2\cos^2\left(\frac{\pi D\lambda^2 Lf^2}{c}\right), \tag{6.26}$$

where $L$ is the propagation length and a small RF drive ($\phi_0 \ll 1$) is assumed. Note that Equation 6.26 is equivalent to Equation 6.6 when $D = 0$. However, for $D \neq 0$, Equation 6.26 details a frequency-dependent output power with complete extinction at certain RF frequencies.

Figure 6.3 shows the calculated and measured dispersion response for 50 km length of standard single-mode fiber with $D = 17$ ps/nm/km at $\lambda = 1551$ nm. The additional measured power penalty is due to a roll-off in the frequency responses of both the MZM and the detector. It is important to note that for small modulation, the fundamental exhibits the $\cos^2(\pi D\lambda^2 Ln^2 f^2/c)$ frequency-dependence shown in Equation 6.26 while the second harmonic follows a $\sin^2(\pi D\lambda^2 L^2 f^2/c)$ dependence. Consequently,

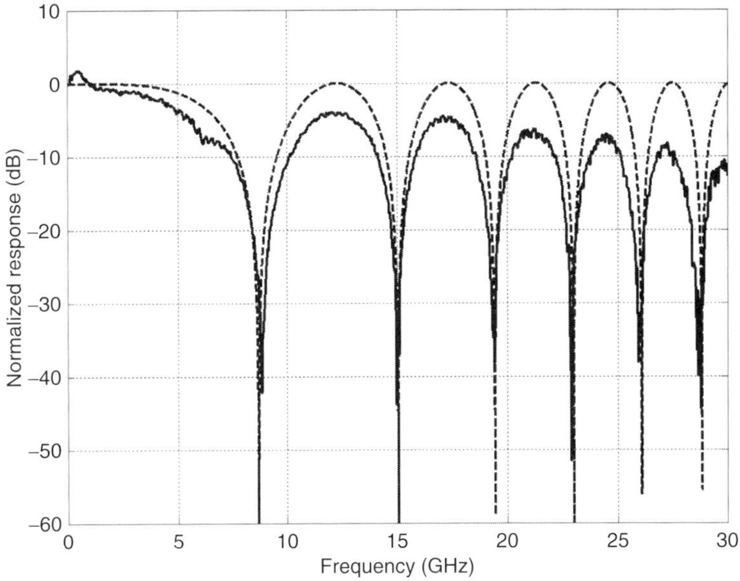


**FIGURE 6.3** Measured (solid) and calculated (dashed) dispersion response for an ROF channel through 50 km of SMF-28 fiber.

for frequencies at which the fundamental is relatively unaffected by dispersion, there may be a large increase in the level of the second harmonic and intermodulation products.

Considering all of this, it is desirable to have as low a net dispersion ($D \times L$) as possible when building a wideband ROF link. As will be demonstrated in Section 6.4, this may be accomplished by building the link with sections of fiber having nearly equal but opposite net dispersion.

### 6.3.4 Stimulated Brillouin Scattering

Unlike the photon–electron interactions that cause chromatic dispersion in fiber, stimulated Brillouin scattering (SBS) is due to photon–phonon scattering. Brillouin scattering is nonelastic scattering whereby photons lose energy to the crystal lattice. Scattered photons are shifted down in frequency relative to incident photons and propagate in the opposite direction of the incident light. The scattered photons are referred to as a Stokes wave. This nonlinear process exhibits a distinct incident optical power threshold.

To understand the adverse effects that SBS can have on a long-haul ROF link, Rayleigh scattering must also be introduced. Rayleigh scattering is elastic, occurring in all spatial directions in the fiber, and is the dominant source of loss in the low-power regime. As will be described, Rayleigh scattering works in concert with SBS to generate noise in ROF links in the high-power limit.

In long-haul ROF links, SBS is troublesome for two reasons. First, signal photons that are meant to propagate forward through the fiber to the receiver are scattered in the reverse direction. In this sense, SBS limits the amount of power that can be transmitted through the fiber. Second, the SBS process adds RF noise to an ROF system.

At low input power, the input and output powers are linearly related by a loss coefficient, which is typically dominated by Rayleigh scattering in the fiber. However, as the power is increased, this relationship becomes nonlinear due to SBS. In fact, SBS can completely saturate the output power of the fiber. SBS is a stimulated process with an associated threshold power, which can be written as in Ref. [9]

$$P_{\text{th}} \approx \frac{21 A_{\text{eff}}}{g_{\text{B}} L_{\text{eff}}}, \tag{6.27}$$

where $A_{\text{eff}}$ is the effective area of the fiber, $L_{\text{eff}}$ is the effective length of the fiber, and $g_{\text{B}}$ is the peak value of the Brillouin gain. The effective fiber length is given by $L_{\text{eff}} = (1 - \text{e}^{-\alpha L})/\alpha$, where $L$ is the physical length and $\alpha$ is the loss coefficient for the fiber. The SBS threshold therefore depends on the fiber characteristics and decreases with physical length. It should also be noted that if the line width of the signal laser is large, $P_{\text{th}}$ will increase further.

It is perhaps more instructive to analyze some measured data pertaining to SBS. Figure 6.4 shows the measured transmitted and scattered power as a function of optical input power in a 20 km link. The link was comprised of standard single-mode fiber (Corning® SMF-28™). A 1551 nm DFB semiconductor laser was used as the source. For input powers below 6 dBm, the output and backscattered power are linearly related to the input power. The SBS threshold is reached at approximately 7 dBm. As the input power is increased further, the backscattered power increases while the output power saturates.

Just as the forward-propagating wave experiences Rayleigh scattering in the reverse direction, the reverse-propagating SBS wave is Rayleigh scattered and copropagates with the signal. The forward-scattered SBS wave beats with the signal wave in the photodiode, generating an RF noise spectrum centered at the SBS frequency. In the electrical domain, the spectrum is a convolution of the laser line width with the net phonon spectrum of the fiber. The spectrum for 20 km of SMF-28™ fiber is depicted in Figure 6.5, showing that the SBS frequency shift is approximately 10.82 GHz. Note that the data in Figure 6.4 and Figure 6.5 pertain to the same optical setup. We observe that SBS causes an increase in RF noise across a wide range of frequencies [10].

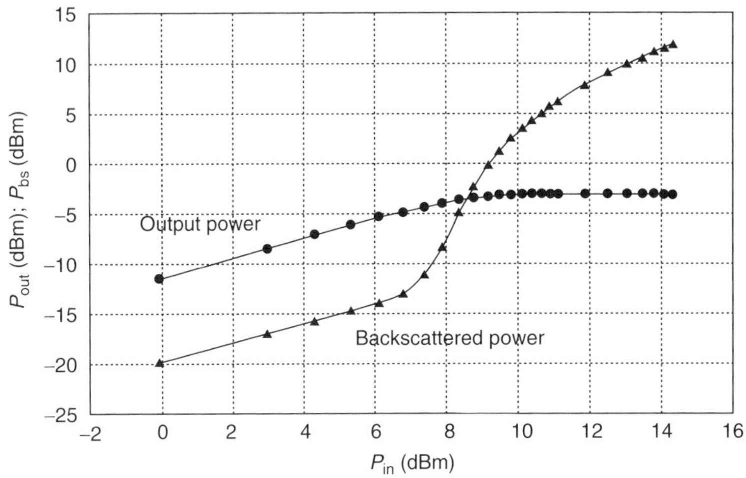


**FIGURE 6.4** Optical transmission curve for 1551 nm DFB semiconductor laser through 20 km of SMF-28™, having an SBS threshold of approximately 7 dBm. The output optical power $P_{out}$ as a function of input optical power $P_{in}$ is shown with circles and the backscattered optical power $P_{bs}$ with triangles.

Equation 6.27 holds for a uniform length of fiber. However, by splicing together different types of fiber with different SBS frequencies, we can decouple the stimulated resonance and therefore increase the SBS threshold. This technique will be revisited and demonstrated in Section 6.4.

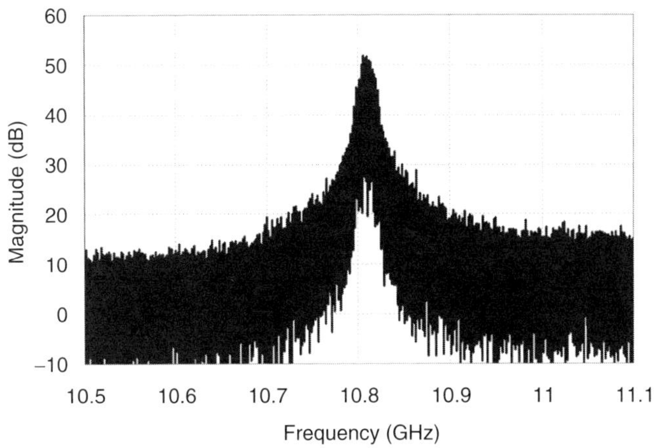


**FIGURE 6.5** Measured SBS RF noise spectrum for 20 km of SMF-28 using a 1551 nm DFB semiconductor laser as the probe.

### 6.3.5 Quadrature Amplitude Modulation

Long-haul ROF links with adequate analog bandwidth, SFDR, and RF noise figure will support many RF digital modulation formats. Such hybrid analog and digital ROF links have numerous military and commercial applications. Consequently, novel digital modulation formats are of interest for currently deployed and future long-haul ROF links. Of particular importance for such links is a given format's spectral efficiency. Recently, very high spectral efficiencies have been achieved in multichannel digital links employing phase-modulated binary formats, namely differential phase-shift keying (DPSK) and variations of DPSK [11–14]. However, multilevel modulation formats are more desirable for ROF links because of their high spectral efficiency. One of the most attractive multilevel formats is $m$-ary quadrature amplitude modulation ($m$-QAM). For example, 256-QAM can exhibit a spectral efficiency of up to 8 b/s/Hz. Spectral efficiency is particularly important in wireless applications where regulatory requirements limit bandwidth occupancy. In addition, $m$-QAM generally has less stringent SNR requirements as compared to $m$-ary phase-shift keying ($m$-PSK) formats. For example, 64-PSK requires 10 dB more SNR than 64-QAM for the same error probability [15]. Here, we provide the analysis of $m$-QAM signals in terms of error-vector magnitude (EVM), modulation-error ratio (MER), and the more common bit-error ratio (BER).

$m$-QAM is a multilevel modulation format in which the amplitude and phase of the carrier is modulated, resulting in $m$ unambiguous symbol points. The aggregate bit rate for $m$-QAM is $\log_2(m) \times S$, where $S$ is the symbol rate. For example, 16-QAM at a 1 Mbaud (1 MSymbol/s) symbol rate gives a data rate of 4 Mb/s. Consider a Gray-Coded [15] 16-QAM constellation diagram, which is arranged in the I-Q plane as shown in Figure 6.6. To quantify the errors in transmission and reception, an error vector ($\vec{r}_e$) is defined. The error vector represents the vector difference between the actual received signal vector and the ideal signal vector in the constellation of Figure 6.6.

The EVM is the time averaged root-mean-square (rms) value of the error vector. We have adopted the convention of defining EVM as the magnitude of the error vector relative to the voltage at highest-voltage symbol points ($V_{max}$). For example, the normalized $V_{max}$ for the constellation in Figure 6.6 is $(V_{max})_{normalized,16\text{-QAM}} = \sqrt{3^2 + 3^2} = \sqrt{18}$. From the measured EVM, the MER and BER can be calculated. The MER is defined as the average SNR per symbol and is related to the EVM as

$$\mathrm{MER[dB]} = -20\log(\mathrm{EVM}) + 10\log\left(\langle V^2\rangle / V_{max}^2\right), \qquad (6.28)$$

where $\langle V^2\rangle$ is the average squared voltage for all points in the constellation. Note that the second term in Equation 6.28 is constant for a particular $m$. For example, the normalized voltage coordinates in Figure 6.6 can be used to

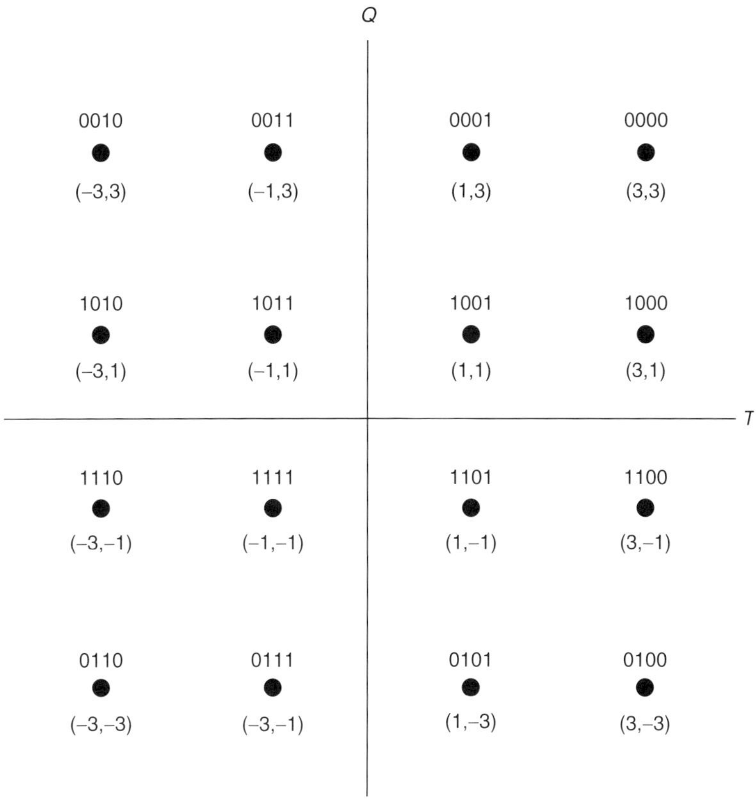


**FIGURE 6.6** Constellation diagram for 16-QAM, showing the normalized voltage vector coordinates for each of the symbol points as well as Gray-coded words for each symbol point. (From J.G. Proakis, *Digital Communications*, 4th ed., Boston: McGraw Hill, 2001.)

calculate the normalized $\langle V^2 \rangle$ for 16-QAM as $\langle V^2 \rangle_{\text{normalized},16\text{-QAM}} = (10 + 10 + 18 + 2)/4 = 10$, giving the second term in Equation 6.28 for 16-QAM as $10\log\left(\langle V^2 \rangle / V^2_{\max}\right)_{16\text{-QAM}} = 10\log(10/18) = -2.5527$. When gray coding is employed (as in Figure 6.6), BER can be estimated from the MER as in Ref. [16],

$$\text{BER} = \frac{2\left(1 - \frac{1}{\sqrt{m}}\right) \cdot \text{erfc}\left(\sqrt{\frac{3\text{MER}}{2(m-1)}}\right)}{\log_2(m)}. \tag{6.29}$$

The BER given by Equation 6.29 is an upper bound and exact only for square $m$-QAM constellations ($\log_2(m)$ even), but serves as a good estimate for

$\log_2(m)$ odd. Equation 6.28 and Equation 6.29 will be used to characterize the long-haul transmission of m-QAM at various symbol rates in Section 6.4.

### 6.3.6 QAM Channel Impairments

MER, the SNR per symbol, is an excellent figure of merit for the evaluation of channel quality. In the ROF link the measured MER will be influenced by both the RF noise figure and intermodulation distortion (IMD). IMD manifests itself as an increase in both adjacent-channel and in-channel noise. In our single channel ROF system, it is the in-channel noise that will degrade the MER.

Whereas Equation 6.21 relates intermodulation distortion to DC photocurrent in a link with two RF tones, we do not have an equivalent analytic expression for the distortion of an $m$-QAM signal in our channel. In electronically amplified QAM systems adjacent-channel noise measurements can be used to predict the mean in-channel noise level [17]. The relationship between in-channel and adjacent-channel distortion has not yet been clearly established for an ROF system with an MZM. However, we can still use adjacent-channel power measurements as a figure of merit in a QAM ROF system.

Figure 6.7 shows the distorted spectrum of a 1 GHz, 5 Mbaud 16-QAM signal after passing through an ROF link with an MZM intensity modulator.

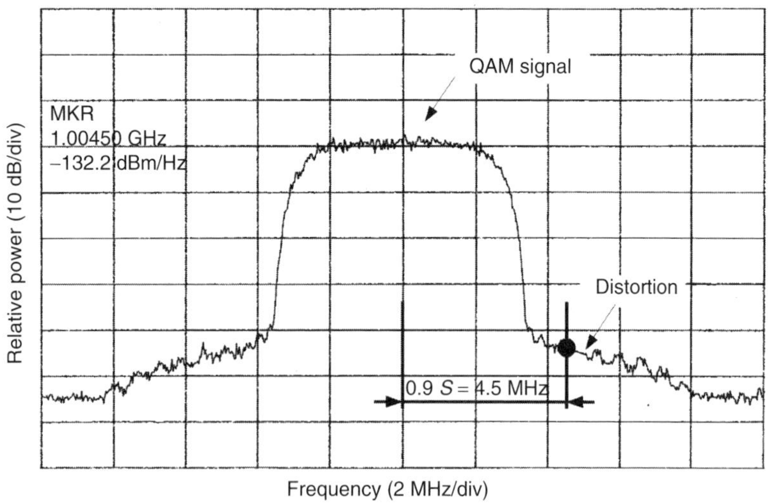


**FIGURE 6.7** Spectral regrowth of 1.0 GHz center frequency $S = 5$ MBaud 16-QAM signal in link ROF with MZM modulator. Distortion power spectral density was measured at marker (circle) in adjacent channel at $0.9\ S = 4.5$ MHz offset from the carrier frequency.

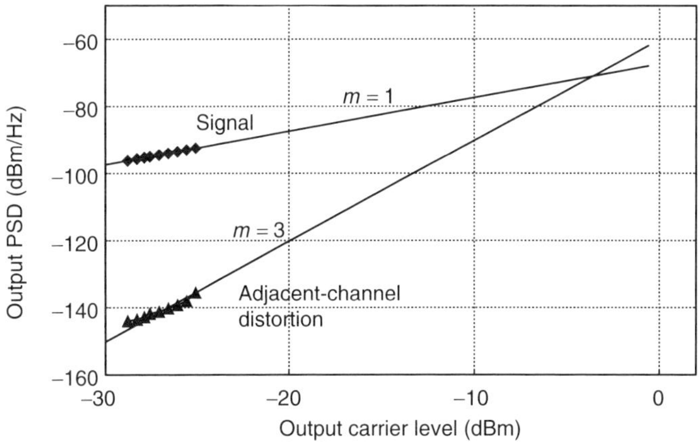


**FIGURE 6.8** Distortion of 16-QAM signal in link with MZM. Output power spectral density (PSD) of desired in-band signal and spurious adjacent-channel distortion are plotted vs. output carrier level. Linear trend with slope of $m = 3$ is evident in adjacent-channel distortion.

Note the presence of out-of-band spectrum regrowth in the adjacent channels. In-band distortion is masked by the useable signal.

Following convention [17], we compared the signal level at the center of the channel to the noise level at an offset of $f_o = 0.9S$, where $S$ is the symbol rate. As shown in Figure 6.8, the slope of distortion vs. carrier power was fit to a line with slope of $m = 3$. This indicates that third-order distortion dominates. The linear fit was extended until the noise power and signal power intercept met at 1.4 dBm. This adjacent-channel output distortion intercept point was approximately 4 dB lower than the measured two-tone OIP3. Furthermore, we verified that the adjacent-channel distortion and two-tone third-order distortion intercept points tracked each other as the photocurrent was increased from 1 to 5 mA. Hence, in this case, higher OIP3 translates into lower distortion of the QAM signal.

### 6.3.7 Optical Amplifier Placement

The placement of optical amplifiers strongly influences the dynamic range of an ROF link. When photodiode saturation and SBS are also considered, amplifier placement becomes critical. For example, we may consider placing a single high-gain optical amplifier at the transmit side of a 100 km link to offset fiber attenuation and reduce the link noise figure. However, the power output from the amplifier would likely exceed the fiber's SBS threshold. Similarly, if we were to place a high-gain amplifier at the link terminus, we would likely obtain sufficient photocurrent but poor optical noise figure.

Incremental or distributed amplification is an excellent alternative. By placing optical amplifiers at discrete intervals along the fiber span, the fiber losses can be offset incrementally without exceeding the SBS threshold. Raman amplification takes distributed amplification to the next level, distributing gain throughout the complete long-haul span and eliminating the need for mid-span amplification hardware.

## 6.4 LINK DEMONSTRATIONS

### 6.4.1 EDFA-Amplified 110 km Link Setup

The principles discussed here were put into practice in the design of a long-haul, 20 GHz, 110 km optically amplified link. The link was tested by the transmission of a 256-QAM-modulated RF signal. 256-QAM is particularly demanding, requiring a high SNR and high linearity when compared to more common digital formats such as bipolar PSK. Therefore, the design required careful management of chromatic dispersion, noise figure, gain, and SBS.

High SFDR was achieved using the amplifier placement strategy discussed in Section 6.3.7. The experimental setup is shown in Figure 6.9. The optical source was a distributed feedback laser (DFB). The laser was modulated by a 20 GHz RF signal in a quadrature-biased MZM. The 110 km fiber link was divided into four discrete spans. Spans 1–3 were assembled with a

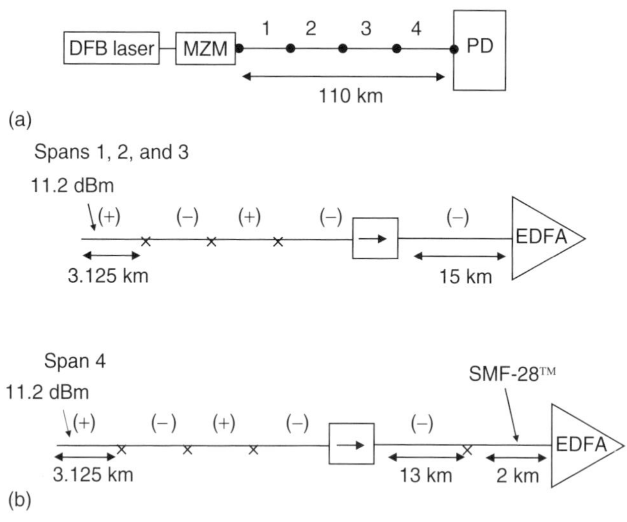


FIGURE 6.9 (a) Link architecture. MZM: Mach–Zehnder modulator, EDFA: erbium-doped fiber amplifier, PD: photodiode. (b) Detail of hybrid spans. (+): Lucent True-wave® Positive fiber, (−): Lucent Truewave® Negative Fiber, boxed arrows: isolators.

combination of fiber types, Lucent Technologies' Truewave® Positive fiber and Truewave® Negative fiber.

An erbium-doped fiber amplifier (EDFA) was placed at the end of each span. The EDFA gain was adjusted to just offset the approximately 7 dB of loss in a single span. Hence, the power launched into each span was maintained at a constant 11.2 dBm, just below each span's SBS threshold. Since the fiber types had different SBS resonance frequencies, the hybrid combination led to an observed increase in SBS threshold of up to 6 dB when compared to a single fiber type of the same span length.

Span 4 included a small additional section of SMF-28™ fiber. This provided a final adjustment of the total link dispersion. The total dispersion was only 8 ps/nm at a wavelength of 1551 nm. The final EDFA amplifier in Span 4 was adjusted to provide 10 mA of photocurrent in the photodiode. This gave a sufficient dynamic range, but was less than the photodiode's saturation current. The photodiode had a responsivity of $\Re = 0.8$ A/W and an RF insertion loss of 6 dB.

The SFDR of the link was measured using the conventional two-tone method. Two equal power tones near 18 GHz were transmitted through the link. The fundamental and third-order intermodulation product output powers were measured over a range of input powers. As shown in Figure 6.10, a linear extrapolation of the data points yields an OIP3 of 7 dBm at a DC photocurrent of 10 mA. After correcting for the 6 dB insertion loss of the photodiode, this is as predicted by Equation 6.22. The measured noise from

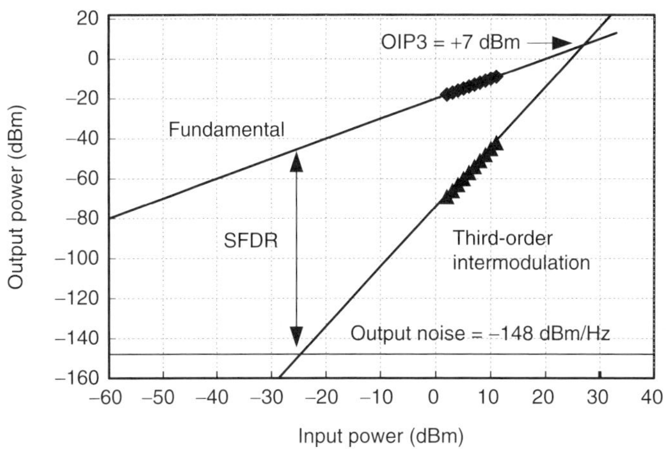


**FIGURE 6.10** Measured fundamental (diamonds) and third-order intermodulation (triangles) response for the 110 km EDFA-amplified ROF link. The measured link noise floor is −148 dBm/Hz and the OIP3 is +7 dBm, resulting in an SFDR of 103 dBHz$^{2/3}$.

the photodiode, $N_{sig\text{-}sp}$, was-148 dBm/Hz. Plugging the measured OIP3 and $N_{sig\text{-}sp}$ into Equation 6.19, the calculated SFDR is 103 dB/Hz$^{2/3}$.

At $I = 10$ mA, we measured $G_{opt} = 1$ dB and $N_{sig\text{-}sp} = -148$ dBm/Hz + 6 dB = −142 dBm/Hz. From Equation 6.13 the link's optical noise figure is $NF_{opt} = 17$. Based on these measurements, the analytic expression (Equation 6.23) predicts an SFDR of 103 dBHz$^{2/3}$. This agrees with the value determined from the two-tone test.

### 6.4.2 EDFA-Amplified 110 km QAM Transmission

The quality of the 110 km EDFA-amplified link was tested with an $m$-QAM modulated 20 GHz RF signal [18]. The $m$-QAM signal was generated by a vector signal generator (VSG) at an output frequency of 2 GHz. The VSG signal was converted up to 20 GHz, amplified, and sent to the MZM. We used a symbol rate of 5 Mbaud (bit rate of 40 Mb/s for $m = 256$). Applying a raised root cosine (rrc) filter with $\alpha = 0.2$ resulted in an occupied bandwidth of 6 MHz. This configuration is typical of digital CATV transmission.

The $m$-QAM signal was sent across the ROF link and received at the photodiode. The received signal was down-converted to a 500 MHz IF and analyzed on a vector signal analyzer (VSA). The VSA was configured with a matched rrc filter. The measured MER was 30.5, 29.3, and 28.9 dB for $m = 16$, 64, and 256, respectively. The measured constellations for $m = 16$ and $m = 256$ are shown in Figure 6.11.

In order to determine the contribution of the ROF link to any observed impairments, the fiber link was entirely removed. In this case, the measured back-to-back MER for $m = 256$ was 29.5 dB. The MER of the back-to-back link was limited by residual noise and distortion in the up- and down-converters, the VSG, and the VSA. The MER of the 110 km ROF link was only 0.6 dB less than the back-to-back link, indicating that the ROF link is almost transparent.

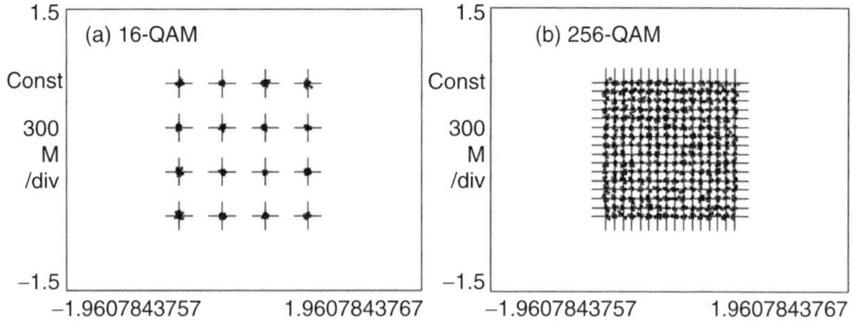


**FIGURE 6.11** Received constellation diagrams for 110 km EDFA-amplified QAM transmission. (a) 16-QAM at 5 Mbaud with MER 30.5 dB (b) 256-QAM at 5 Mbaud with MER = 28.9 dB.

### 6.4.3 Raman-Amplified 105 km Link Setup

In some cases the EDFA-amplified system discussed above is not practical. There are applications where placement of mid-span hardware cannot be realized. Furthermore, the gain spectrum of an EDFA is fixed. Therefore, if optical gain is required at wavelengths outside of the C-band, as in a 1.3 μm ROF link, EDFAs cannot be used. In these cases Raman amplification may be necessary [19]. Assuming the availability of a pump at the appropriate wavelength, a Raman amplifier can provide gain at almost any signal wavelength. More importantly, the fiber can be pumped remotely, at either the transmission or the reception end, such that the transmission fiber is continuous and itself serves as the gain medium.

We have developed and tested a Raman long-haul ROF link as well. The architecture of our 105 km Raman-amplified ROF link is shown in Figure 6.12.

The transmitter for the link was a DFB semiconductor laser operating at 1551 nm with an MZM ($V_\pi = 6$ V at 5 GHz) biased at quadrature. The transmission span was 105 km in length and was comprised of positive and negative dispersion fiber as shown in Figure 6.12. This hybrid fiber combination serves the dual purpose of chromatic dispersion compensation and SBS mitigation, much like in the 110 km EFDA link. The detector for the link was a p-i-n photodiode ($\Re = 0.8$ A/W with 6 dB RF insertion loss).

The net dispersion for this link was measured as 105 ps/nm and the SBS threshold was measured as 8.8 dBm. The normalized RF response for the link is shown in Figure 6.13. The calculated response (Equation 6.26) of a 105 km long link comprising SMF-28™ fiber is also shown for comparison. The 3 dB bandwidth for the link is 17.5 GHz.

Note that the fiber was symmetrically arranged such that it could be pumped from either the transmission or the reception side of the link. We determined that reverse pumping at the reception side resulted in lower RF

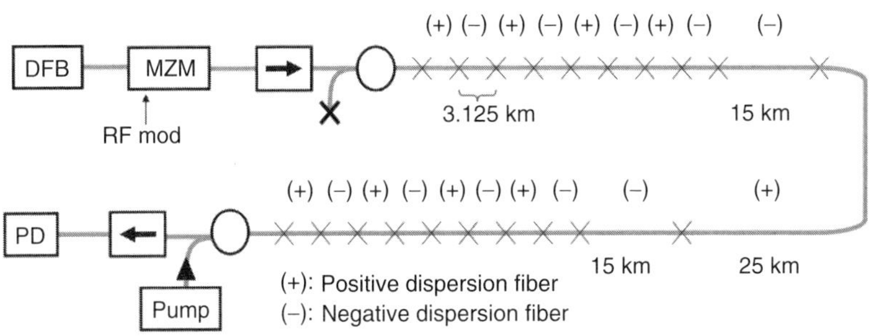


**FIGURE 6.12** Architecture for the 105 km Raman-amplified ROF link. DFB: distributed feedback semiconductor laser at 1551 nm, MZM: Mach–Zehnder modulator biased at quadrature, Pump: Raman laser at 1455 nm, PD: p-i-n photodiode, boxed arrows: isolators, and circles: 1455 nm/1551 nm wavelength division multiplexers.

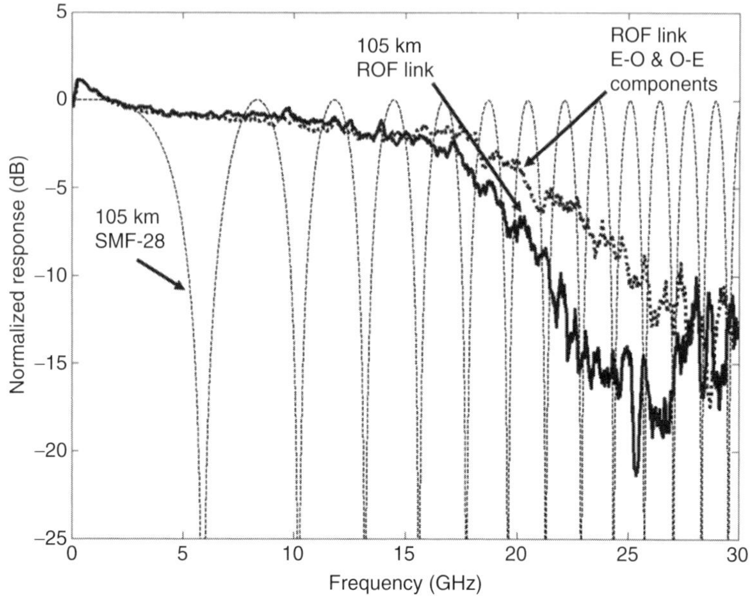


**FIGURE 6.13** The normalized response for the 105 km Raman-amplified ROF link from measured transmission loss data (solid line). The normalized back-to-back response (dotted line, measured) and the normalized response for 105 km of SMF-28™ fiber (dashed line, calculated) are shown for comparison.

noise at our 5.25 GHz link frequency when compared to forward pumping at the transmission side. This is an intuitive result. Forward pumping concentrates gain near the launch side where SBS is problematic. In addition, forward pumping is susceptible to noise induced by intensity and polarization fluctuations of the pump laser. Pump location can be of great practical relevance in antenna remoting operations.

For pumping at the reception end, the optimal pump power for 8.5 dBm of launch power (just below the SBS threshold) was determined to be 25.2 dBm. As will be detailed in the following section, this configuration yielded a 95 dB·Hz$^{2/3}$ SFDR and supported 5 Gb/s QAM transmission in a single channel.

The SFDR for the Raman-amplified 105 km link was measured by applying two tones of equal magnitude to the MZM and measuring the fundamental and third-order intermodulation responses on an RF spectrum analyzer. For the SFDR measurement, the optical launch power was 8.5 dBm and the pump power was 25.2 dBm, resulting in a DC photocurrent of $I = 2$ mA. The tone frequencies were set at 5.00 and 5.02 GHz with equal amplitudes and varied in 1dB increments to obtain the responses shown in Figure 6.14.

The fundamental small-signal RF gain at 5 GHz was measured as −32 dB in agreement with Equation 6.8 for $I_{dc} = 2$ mA and $V_{\pi} = 6$ V after including the

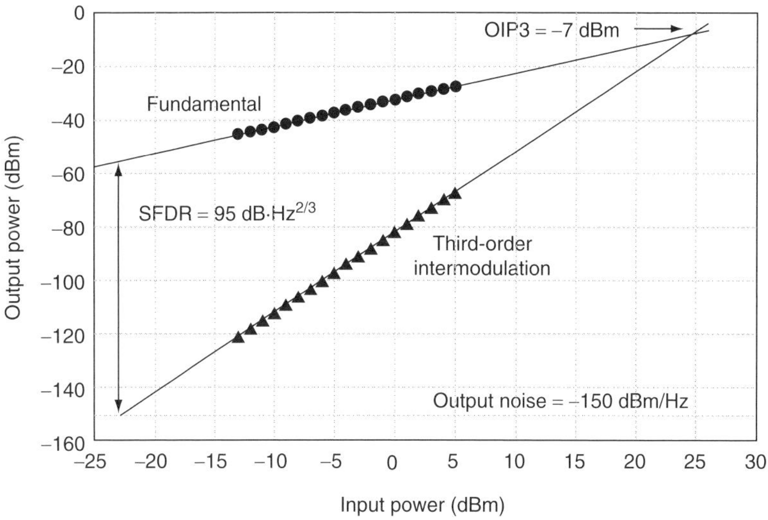


**FIGURE 6.14** Measured fundamental (circles) and third-order intermodulation (triangles) response for the 105 km Raman-amplified ROF link. The measured link noise floor is −150 dBm/Hz and the OIP3 is −7 dBm, resulting in an SFDR of 95 dB·Hz$^{2/3}$. The optical launch power was 8.5 dBm and the pump power was 25.2 dBm, resulting in a DC photocurrent of $I_{dc} = 2$ mA. The tone frequencies were set at 5.00 and 5.02 GHz with equal amplitudes and varied in 1 dB increments to obtain the responses shown in the plot. At these frequencies, $V_\pi \cong 6$ V.

6 dB of RF insertion loss in the photodetector. The fundamental and distortion data were extrapolated to obtain the OIP3 = −7 dBm. After including the 6 dB of photodiode RF insertion loss, the result is in agreement with Equation 6.22. The RF noise floor at 5 GHz was −150 dBm/Hz. Hence, the SFDR was 95 dB·Hz$^{2/3}$.

At $I = 2$ mA, the optical gain and noise figure of the link were measured as $G_{opt} = -4.5$ dB and $NF_{opt} = 27.5$ dB. From Equation 6.23, the predicted SFDR is 95 dB·Hz$^{2/3}$, which is in agreement with the value determined from two-tone measurements. Note that we have followed the convention used in our EDFA link where $G_{opt}$ and $NF_{opt}$ are the cascaded gain and optical noise figure of the *system* including optical loss in the fiber. This is not to be confused with the effective gain and effective noise figure of the Raman amplifier where the contribution of fiber loss is treated differently. The SFDR of the Raman-amplified link, although smaller than the SFDR of the EDFA ROF link above, is still adequate for QAM signal transport. The reduced SFDR is due primarily to the small photocurrent of only 2 mA at the photodiode. Adding additional Raman gain would increase the photocurrent, but at the risk of causing SBS. Pumping of the link mid-span or at both ends may yield a higher SFDR, but this was not yet explored experimentally at the time of publication.

### 6.4.4 Raman-Amplified 105 km QAM Transmission

After the Raman link was constructed we obtained access to a novel custom vector signal generation and analysis system [20]. The custom system included software signal predistortion and postdistortion allowing for the reliable transmission of a remarkably high, 1.5 GHz bandwidth QAM signal [20].

The architecture of the system including the ROF link is shown in Figure 6.15. The results of transmission through 105 km are shown in Figure 6.16 for 16-QAM and 32-QAM modulation, both at 1 Gbaud. The measured EVM for the transmission was 4.99% and 5.65% rms, for 16- and 32-QAM, respectively. For the 16-QAM, the measured EVM yields an MER of 24 dB and a predicted BER bound of $9 \times 10^{-12}$, where Equation 6.28 and Equation 6.29 have been used. The calculation for 32-QAM with MER $=$ 23 dB yields a BER bound of $7 \times 10^{-6}$.

Note that both of the calculated BERs do not include error correction. Error correction coding is normally used in radio links and a BER as high as $10^{-4}$ may be acceptable in radio links with moderate error correction coding. We therefore conclude that this system can achieve error-free transmission of 4 Gb/s (16-QAM at 1 Gbaud) and 5 Gb/s (32-QAM at 1 Gbaud) in a single ROF channel. In addition, given the instantaneous signal bandwidth of 1.5

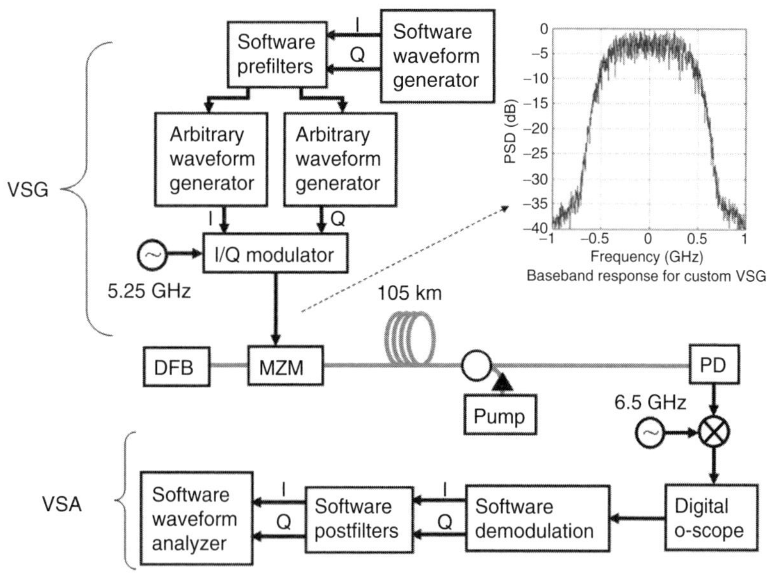


**FIGURE 6.15** The wideband QAM ROF architecture, detailing the custom vector signal generator (VSG) and vector signal analyzer (VSA). (From J. Qiu, D. Abe, T.M. Antonsen Jr., B.G. Danly, and B. Levush, "Calibration of a gigahertz-bandwidth vector signal generation and analysis system," in *IEEE MTT-S Int. Microw. Symp.*, Philadelphia, pp. 1659–1662, June 2003.)

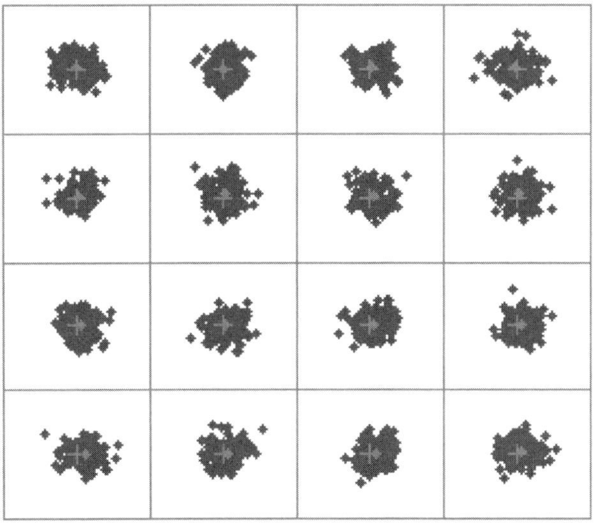

(a)

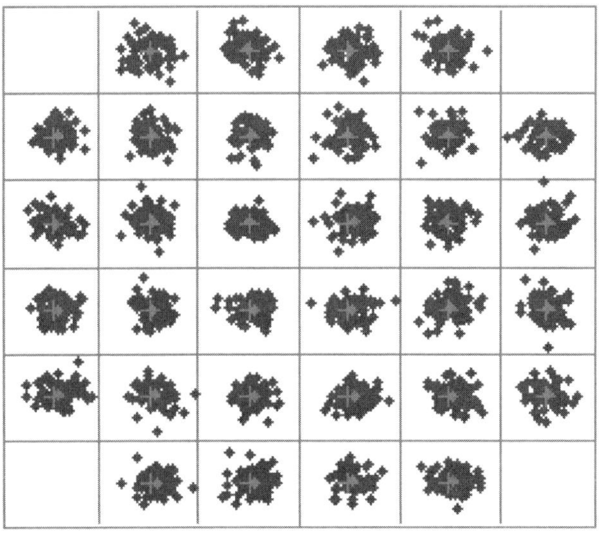

(b)

**FIGURE 6.16** Constellation diagrams for 105 km Raman-amplified QAM transmission. (a) 16-QAM at 1 Gbaud with BER $= 9 \times 10^{-12}$, (b) 32-QAM at 1 Gbaud with BER $= 7 \times 10^{-6}$.

GHz, the RF spectral efficiencies are 2.7 and 3.3 b/s/Hz for 16- and 32-QAM, respectively. At the time of publication, to the best of the authors' knowledge, these are the highest single-channel QAM-over-fiber data rates and RF spectral efficiencies.

## 6.5 FUTURE CAPABILITIES

Historically ROF has been used to transport RF signals between microwave transmitters, antennas, and receivers. However, spectrally efficient microwave digital modulation schemes such as QAM can make ROF a viable alternative to baseband digital data transport. We have demonstrated that Gbaud symbol rates can be supported over fiber spans exceeding 100 km. However, vector modulation and reception at Gbaud rates require a custom vector modulation system [20]. We have been investigating optical vector modulation techniques that may also prove useful [21] for achieving high RF baud rates.

In this chapter we have focused on single channel links. However, throughput can be further improved by frequency multiplexing. In this case, interchannel cross talk is an additional consideration. Second harmonic distortion (2HD) is one of the major channel impairments in both HFC systems [3] and multichannel microwave ROF links [22] where RF bandwidths exceed one octave. Wavelength division multiplexing is another useful means of increasing capacity. However, cross talk between wavelength division multiplexed (WDM) channels must be mitigated [3,22,23].

Also, we note that we have only discussed optical amplitude modulation with an MZM modulator in this chapter. We have also implemented phase-modulated analog systems [24] and have demonstrated QAM channel transport over phase-modulated links [25] with improved SFDR, but at the expense of a somewhat more complex optical receiver.

Many of the ROF techniques discussed in this chapter will be relevant to future digital fiber link design. Traditional wireless modulation formats such as differential phase keying are now being adopted [12–14] in conventional digital fiber links. Furthermore, channel spacing in hyperfine WDM systems is now narrow enough to be in the microwave range. Coding and digital signal processing for microwave radio channels is advancing rapidly, spurred by demand in the wireless sector. It is our hope that advances in all of these fields will lead to enhanced fiber link capacity as well as new hybrid fiber and microwave applications.

## ACKNOWLEDGMENTS

The authors wish to acknowledge Joe X. Qiu of the Electronics Technology Division at NRL for providing the custom vector signal generation and analysis system and also for his assistance in performing the wideband QAM measurements. The authors also wish to acknowledge the contributions to the work presented here by Keith J. Williams, Shane J. Strutz, James L. Dexter, Christopher McDermitt, and Jennifer Witkowsky.

## REFERENCES


1. P. Horowitz and W. Hill, *The Art of Electronics*, 2nd ed., New York: Cambridge, pp. 430–431, 1989.
2. W.V. Sorin, Noise sources in optical measurements, in *Fiber Optic Test and Measurement*, D. Derickson (ed.), Upper Saddle River: Prentice Hall, pp. 597–613, 1998.
3. M.R. Phillips and T.E. Darcie, Lightwave analog video transmission, in *Optical Fiber Telecommunications IIIA*, San Diego: Academic, 1997.
4. K.Y. Lau and A. Yariv, Ultra-high speed semiconductor lasers, *IEEE J. Quantum Electron.*, vol. QE-21, no. 2, pp. 121–138, February 1985.
5. B.H. Kolner and D.W. Dolfi, Intermodulation distortion and compression in an integrated electrooptic modulator, *App. Opt.*, 26(17), 3676–3680, 1987.
6. W.E. Stephens and T.R. Jospeh, System characteristics of direct modulated and externally modulated RF fiber-optic links, *J. Lightwave Technol.*, LT-5(3), 380–387, 1987.
7. C.H. Cox III, G.E. Betts, and L.M. Johnson, An analytical and experimental comparison of direct and external modulation in analog fiber-optic links, *IEEE Trans. Microw. Theory Tech.*, 38(5), 501–509, 1990.
8. B. Christensen, J. Mark, G. Jacobsen, and E. Bodtker, Simple dispersion measurement technique with high resolution, *Electron. Lett.*, 29(1), 132–133, 1993.
9. G.P. Agrawal, *Nonlinear Fiber Optics*, 3rd ed., San Diego: Academic Press, 2001.
10. J. Zhang and M.R. Phillips, Modeling intensity noise caused by stimulated Brillouin scattering in optical fibers, in *CLEO Technical Digest*, Baltimore, vol. 1, pp. 140–142, May 2005.
11. A.H. Gnauck, G. Raybon, S. Chandrasekhar, J. Leuthold, C. Doerr, L. Stulz, and E. Burrows, 25 $\times$ 40-Gb/s copolarized DPSK transmission over 12 $\times$ 100-km NZDF with 50-GHz channel spacing, *IEEE Photonics Technol. Lett.*, 15(3), 467–469, 2003.
12. C. Wree, N. Hecker-Denschlag, E. Gottwald, P. Krummrich, J. Leibrich, E. Schmidt, B. Lankl, and W. Rosenkranz, High spectral efficiency 1.6-b/s/Hz transmission (8 $\times$ 40 Gb/s with a 25-GHz grid) over 200-km SSMF using RZ-DQPSK and polarization multiplexing, *IEEE Photonics Technol. Lett.*, 15(9), 1303–1305, 2003.
13. B. Zhu, L.E. Nelson, S. Stulz, A.H. Gnauck, C. Doerr, J. Leuthold, L. Gruner-Nielsen, M.O. Pedersen, J. Kim, and R.L. Lingle Jr., High spectral density long-haul 40-Gb/s transmission using CSRZ-DPSK format, *J. Lightwave Technol.*, 22(1), 208–214, 2004.
14. C. Xu, X. Liu, and X. Wei, Differential phase-shift keying for high spectral efficiency optical transmissions, *IEEE J. Selected Topics Quantum Electron.*, 10(2), 281–293, 2004.
15. J.G. Proakis, *Digital Communications*, 4th ed., Boston: McGraw Hill, 2001.
16. F. Xiong, *Digital Modulation Techniques*, Boston: Artech House, 2000.
17. O. Gorbachov, Y. Cheng, and J.S.W. Chen, Noise and ACPR correlation in CDMA power amplifiers, *RF Design*, 38–44, 2001.

18. E.E. Funk, V.J. Urick, S.J. Strutz, J.L. Dexter, and K.J. Williams, 110 km 256-QAM digital microwave over fiber link, in *IEEE MTT-S Int. Microw. Symp.*, Philadelphia, vol. 1, pp. 269–272, June 2003.
19. V.J. Urick, J.X. Qiu, F. Bucholtz, C. McDermitt, and K.J. Williams, 4 Gbit/s transmission over single-channel Raman-amplified 105 km link, *Electron. Lett.*, 40(8), 495–496, 2004.
20. J. Qiu, D. Abe, T.M. Antonsen Jr., B.G. Danly, and B. Levush, Calibration of a gigahertz-bandwidth vector signal generation and analysis system, in *IEEE MTT-S Int. Microw. Symp.*, Philadelphia, pp. 1659–1662, June 2003.
21. W.D. Jemison, A.J. Kreuzberger, and E. Funk, Microwave photonic vector modulator for high-speed wireless digital communications, *IEEE Microw. Wireless Components Lett.*, 12(4), 125–127, 2002.
22. E.E. Funk, A.L. Campillo, and D.A. Tulchinsky, Nonlinear distortion and crosstalk in microwave fiber-radio links, *IEEE MTT-S Digest*, 3, 1691–1693, 2002.
23. M.S. Rogge, V.J. Urick, F. Bucholtz, K.J. Williams, and P. Knapp, Comparison of Crosstalk in amplitude- and phase- modulated hyperfine WDM fiber optic links, in *CLEO Tech. Dig.*, Baltimore, vol. 1, pp. 128–130, May 2005.
24. V.J. Urick, F. Bucholtz, J.L. Dexter, K.J. Williams, and C. McDermitt, Increased spurious-free dynamic range for an all-Raman 105 km link using phase modulation and balanced detection, in *CLEO Tech. Dig.*, San Francisco, vol. 2, May 2004.
25. V.J. Urick, J.X. Qiu, and F. Bucholtz, Wide-band QAM-over-fiber using phase modulation and interferometric demodulation, *IEEE Photonics Technol. Lett.*, 16(10), 2374–2376, 2004.

# 7 Photonic Synthesis of Ultrabroadband Arbitrary Electromagnetic Waveforms

*Jason D. McKinney and Andrew M. Weiner*

## CONTENTS

7.1 Introduction ... 214
7.1.1 General Relations Governing Optical Pulse Shaper–Based Arbitrary Electromagnetic Waveform Generators ... 216
7.2 Arbitrary Millimeter Waveform Synthesis ... 218
7.2.1 Experimental Apparatus ... 218
7.2.2 Broadband Burst and Continuous-Periodic Millimeter Waveforms ... 221
7.3 Ultrabroadband Microwave Waveform Synthesis ... 227
7.3.1 Ultrawideband Systems, Applications, and Challenges ... 227
7.3.2 Experimental Apparatus ... 229
7.3.3 Ultrabroadband Radio-Frequency and Microwave Signal Synthesis ... 233
7.3.4 Spectral Engineering of Ultrabroadband Radio-Frequency and Microwave Waveforms ... 236
7.3.5 Amplification of Ultrawideband Waveforms ... 243
7.4 Application of Photonically Synthesized Waveforms to Compensate for Broadband Antennae Phase Distortions ... 247
7.5 Conclusion ... 255
Acknowledgments ... 255
References ... 255

## 7.1 INTRODUCTION

Femtosecond pulse shaping for generating nearly arbitrarily shaped ultrafast optical pulses is now a well-established technology [1] and has been widely adopted for applications ranging from high-speed communications to coherent laser control of chemical reactions. Arbitrary waveform generation (AWG) capabilities for millimeter-wave and microwave electromagnetic signals, however, are quite limited. Commercial radio-frequency AWG instrumentation is currently limited to ~2 GHz bandwidth. Here, we review work at Purdue in which shaped optical pulses are used to drive an optical-to-electrical (O-E) converter. This leverages our femtosecond optical AWG technology to achieve cycle-by-cycle synthesis of arbitrary voltage waveforms in the range of 1–50 GHz. Such capabilities could open new possibilitics for applications in areas such as wireless communications, electronic countermeasures, sensing, and pulsed radar.

The fundamental concept of our technique is the use of shaped broadband time domain optical waveforms to drive an O-E converter; thus, through user specification of the input optical waveform, we craft arbitrarily shaped electromagnetic output waveforms. Our technique is shown schematically in Figure 7.1. With proper choice of the input laser source, optical pulse shaper, and O-E converter our technique is readily applied to synthesis of electromagnetic waveforms spanning the ~1 GHz to several THz range. Several example waveforms illustrating the capabilities of our technique—an ultra-wideband (UWB) RF impulse, an abruptly frequency-modulated millimeter waveform (50/25 GHz), and a ~1 THz phase-modulated waveform—are also shown in Figure 7.1.

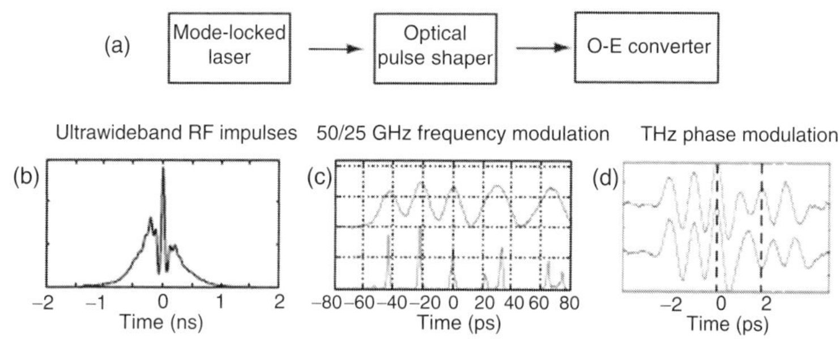


**FIGURE 7.1** (a) Schematic representation of our O-E waveform generation technique. Ultrashort optical pulses from a mode-locked laser are converted to user-defined optical waveforms in one of several optical pulse–shaping geometries. Subsequently, these tailored optical waveforms are converted into arbitrary electromagnetic waveforms in the ~1–1.5 THz range via high-speed O-E conversion. Several example waveforms: (b) ~135 ps ultrawideband RF impulse, (c) abruptly frequency-modulated millimeter sinusoid (top) and the driving optical waveform (bottom) (From McKinney, J.D., Seo, D.S., Leaird, D.E., and Weiner, A.M., *J. Lightwave Technol.*, 21, 3020, 2003. With permission.), and (d) ~1 THz waveforms exhibiting a π phase shift.

The work we discuss here focuses primarily on synthesis of ultrabroadband electromagnetic signals spanning the 1–50 GHz frequency range [2–5]. We term our waveforms ultrabroadband to emphasize that the fractional bandwidth (%BW, defined as the ratio of the signal bandwidth to the center frequency) of our signals may be tuned from ~15% to over 100% in the above-mentioned frequency band, and to avoid the limitations imposed by the term "UWB" (which implies certain defined limits on signal frequency content and fractional bandwidth as discussed in subsequent sections). Our emphasis on large fractional bandwidth signals is in contrast with the vast majority of photonic-synthesis techniques for arbitrary electromagnetic waveforms in the gigahertz to tens of gigahertz range that focus on generation of spectrally pure narrowband electrical signals. To achieve this goal, most systems employ narrowband optical sources and interference techniques. For example, an optoelectronic oscillator [6] utilizing a tunable diode laser source, electro-optic modulator, and dispersive fiber delay lines has been demonstrated to produce tunable RF signals in the range from 0.5 to 9 GHz. Optical heterodyning of different longitudinal modes in a mode-locked laser diode [7] has also been utilized to produce narrowband signals in the 12.4–37.2 GHz range. This technique also allows individual modes to be phase- or amplitude-modulated allowing creation of arbitrarily shaped beat signals. These techniques, while producing extremely pure radiation, are inherently narrowband. Modulation of the underlying periodic microwave or millimeter-wave signals occurs over many cycles of the carrier. Wideband techniques for generation of signals in the low gigahertz range have also been demonstrated; for example, a wavelength division multiplexing scheme combined with a series of optical delay lines to create an optical pulse sequence which, subsequently, drives a photodiode has been used to create an ~6 GHz phase-modulated waveform [8]. This technique, while demonstrating cycle-by-cycle waveform synthesis is limited by the need to carefully control the length of the fiber delay lines—a serious limitation in terms of reprogrammability. Broadband sinusoidal RF and microwave waveforms in the 1–12 GHz range have also been demonstrated through a Fourier transform (FT) optical pulse shaping and dispersive stretching technique [5,9,10].

Here, we give an overview of our research in arbitrary millimeter (tens of GHz) and RF and microwave ($\leq$10 GHz) waveform synthesis; this chapter is, therefore, organized in the following manner. We conclude our introduction with a general discussion of the capabilities of our technique and the primary factors that determine the achievable electrical frequency content in optical pulse shaper–based arbitrary electromagnetic waveform generators. In Section 7.2, we discuss our work in arbitrary millimeter waveform synthesis with an overview of our millimeter waveform generator (Section 7.2.1)—which includes a review of the direct space-to-time (DST) optical pulse shaper that forms the core of our apparatus—and several intriguing examples of broadband burst and continuously periodic millimeter waveforms synthesized in our system (Section 7.2.2). Section 7.3 details our experiments in ultrabroadband RF and microwave signal synthesis. This section includes a detailed discussion of the FT and optical frequency-to-time technique that forms the basis of our photonic time domain electromagnetic signal generator (Section 7.3.2, a parallel discussion is given in Chapter 9) and presents several representative waveforms aimed at UWB systems applications, such as wireless communications and impulse radar (Section 7.3.3). A discussion of how our technique enables user-defined RF power spectra to be synthesized via spectral engineering—for example, to address spectral occupancy constraints in RF wireless applications—is also included in this section (Section 7.3.4). Finally, we present application of our RF waveform synthesis technique as an enabling technology to combat impulsive waveform distortions arising from broadband antennae in Section 7.4 and conclude with discussions of areas of future work and potential applications of our technique in Section 7.5.

### 7.1.1 General Relations Governing Optical Pulse Shaper–Based Arbitrary Electromagnetic Waveform Generators

The frequency content of the arbitrary waveforms synthesized using our technique is, fundamentally, determined by the O-E converter utilized (highest attainable frequency) and our optical pulse shapers (lowest attainable frequency). The particular pulse-shaping geometries utilized in our work will be discussed in some detail in subsequent sections; however, a review of several key aspects of the general "optical pulse shaper" will help elucidate the capabilities (and certain challenges) of our technique. Optical pulse shapers are, by definition, arbitrary optical filters designed to manipulate the complex optical spectrum of a short input pulse to produce a user-defined output temporal waveform. The following relations then hold for these optical filters: the finest temporal feature in the output waveform ($\delta t$) is proportional to the inverse of the filter bandwidth ($\Delta F$) and the time aperture of the output waveform ($\Delta T$) is proportional to the inverse of the minimum spectral feature ($\delta f$). Assuming Gaussian spectral and temporal features for simplicity, the uncertainty relations below summarize these statements in terms of the intensity full-width at half-maximum (FWHM) temporal and spectral features

$$\delta t \cdot \Delta F \sim 0.44 \tag{7.1}$$

$$\Delta T \cdot \delta f \sim 0.44 \tag{7.2}$$

Pictorially, these relations are shown in Figure 7.2a.

Generally, optical pulse shapers are constructed to accommodate the majority of the optical bandwidth of the input pulse; in this case, the finest temporal feature in the output waveform is then limited to the duration of the

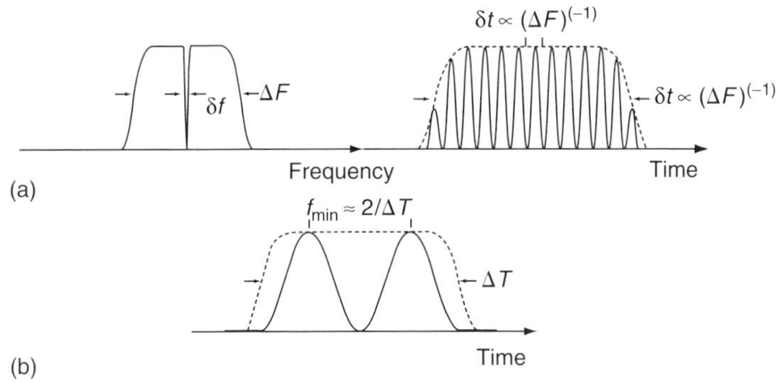


**FIGURE 7.2** (a) Relations governing the minimum temporal feature ($\delta t \propto (\Delta F)^{(-1)}$) and time aperture ($\Delta T \propto (\delta f)^{(-1)}$) of the output temporal waveform from an optical pulse shaper. (b) The minimum electrical frequency from an arbitrary electrical waveform generator employing an optical pulse shaper is $f_{\min} \approx 2/\Delta T$.

input pulse ($\delta t$ ~100 fs–1 ps for typical short pulse lasers). When the temporal output of the shaper is used to drive an O-E converter, the minimum temporal feature in the electrical waveform is then limited by the impulse response (bandwidth) of the O-E converter. Typical commercially available photodiodes have a bandwidth on the order of 60 GHz and research devices, such as unitraveling-carrier photodiodes [11], have demonstrated bandwidths over 100 GHz. For submillimeter signals, photoconductive antennae exhibit bandwidths exceeding 2 THz. The high-frequency limits in optical pulse shaper–based arbitrary waveform generators are clearly determined by the particular choice of O-E converter.

Limits on the lowest attainable frequency are, however, determined by the optical pulse shaper employed in these systems. As illustrated in Figure 7.2a, the resolution of the optical pulse shaper determines the time aperture ($\Delta T$) of the pulse shaper optical output and, hence, the electrical waveform duration. To obtain reasonable user-control over low-frequency signal content, that is, controlling the electrical power spectral density at a particular frequency, the practical limit on the minimum achievable frequency is approximately $f_{\min} \approx 2/\Delta T$ (this allows roughly two electrical cycles within the time aperture $\Delta T$). The time aperture for several classes of optical pulse shapers and the corresponding minimum electrical frequency for arbitrary waveform generators employing these shapers are given in Table 7.1.

As is evident from Table 7.1, the resolution of the optical pulse shaper utilized in these systems must increase (minimum spectral feature $\delta f$ must decrease) to achieve larger time apertures and lower-frequency electrical signals. This is the primary challenge in achieving RF and microwave signals

**TABLE 7.1**
**Time Aperture and Minimum Electrical Frequency for Optical Pulse Shaper–Based Electrical Arbitrary Waveform Generation Systems**

| Pulse Shaper | $\Delta T$ | $f_{min}$ |
|---|---|---|
| Typical (grating-based) | ~25 ps | ~80 GHz |
| High-resolution (grating-[14,15], AWG*-based [46]) | ~100 ps | ~20 GHz |
| Other techniques [5,9,16] | ~2 ns | ~1 GHz |

*"AWG" refers to the arrayed-waveguide gratings typically used as wavelength demultiplexers in optical wavelength division multiplexed communication systems.

($f \leq$ 10 GHz) from pulse shaper–based systems. While typical resolution diffraction grating–based pulse shapers are readily applied to synthesis of arbitrary terahertz waveforms [12,13], high-resolution shapers [14,15] are required to achieve millimeter waveforms (tens of GHz) [2–4] and pulse shapers employing alternative spectral dispersers [16] or all together new approaches [5,9] are required to achieve time apertures in the ns regime and electrical signals below 10 GHz. We will detail aspects of our high-resolution pulse shapers and new approaches for reaching RF frequencies in later sections.

## 7.2 ARBITRARY MILLIMETER WAVEFORM SYNTHESIS

To achieve arbitrary electromagnetic waveforms in the 30–50 GHz range, our technique appeals to previous work in the terahertz frequency range [12,13] where ultrafast optical pulse sequences were used to drive a photoconductive antenna. Optical-to-electrical conversion by the photoconductive dipole antenna resulted in freely-propagating terahertz radiation with temporal structure determined by the driving optical pulse sequence. Our experiments in the millimeter-frequency range utilize optical pulse sequences generated via a novel DST pulse shaper [14,17] to drive a fast photodiode (O-E converter). Here, we exploit the limited bandwidth of the O-E converter and electrical measurement apparatus to suppress harmonics in the optical pulse sequence. This effectively allows sequences of isolated optical pulses to be converted to smooth millimeter-wave sinusoidal waveforms.

### 7.2.1 EXPERIMENTAL APPARATUS

A schematic representation of our millimeter arbitrary waveform generator is shown in Figure 7.3. Short optical pulses from one of several mode-locked fiber lasers (~300 fs @ 40 MHz or ~1 ps @ 10 GHz) are first amplified with

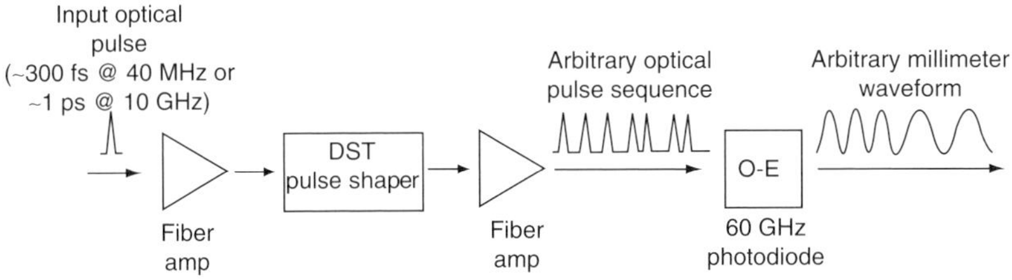


**FIGURE 7.3** Schematic representation of our O-E waveform generator.

an erbium-doped fiber amplifier (EDFA) and then converted to high-rate arbitrary optical pulse sequences in a novel DST optical pulse shaper. These pulse sequences are subsequently amplified by a second EDFA and converted to arbitrary millimeter electrical waveforms via O-E conversion by a 60 GHz photodiode; the resulting waveforms are measured with a 50 GHz sampling oscilloscope. The choice of source laser depends on the desired output electrical waveforms, that is, broadband burst or continuous-periodic millimeter waveforms; the effect of the source laser will be discussed further below.

The user-defined input for either pulse-shaping geometry (FT or DST) is fundamentally a spatial mask (either amplitude, phase, or both) designed to manipulate the optical field of a short pulse inside the pulse shaper. In the FT pulse shaper [1], the spatial mask functions as a user-defined optical spectral filter function and manipulates the dispersed optical frequency components in amplitude, phase, or both. The temporal output of the FT shaper is then essentially given by the FT of the applied spectral filter function. This pulse-shaping geometry may be used to synthesize virtually arbitrary optical waveforms, including square pulses [1], pseudorandom noise bursts [18], as well as arbitrary high-rate optical pulse sequences [12]. Although the FT pulse shaper is widely utilized for a variety of applications (including photonic synthesis of arbitrary electrical waveforms as discussed in Section 7.3)—specifically for its versatility—when the desired temporal output is a high-rate optical pulse sequence, the DST pulse shaper offers significantly simpler operation. In contrast to the FT pulse shaper, in the DST pulse shaper the mapping between the user-defined input spatial mask and the output temporal waveform is direct. The simplicity of this mapping allows (potentially) more rapid reprogramming of the pulse shaper temporal output for applications, such as time domain optical communications, where the time required for the FT calculation could degrade system performance.

Figure 7.4 shows a schematic illustration of the DST shaper. The basic structure of the DST pulse shaper is that of a simple spectrometer [19]. Given a broadband optical source at the input, the grating angularly disperses the optical frequency components; the lens stops this dispersion and focuses

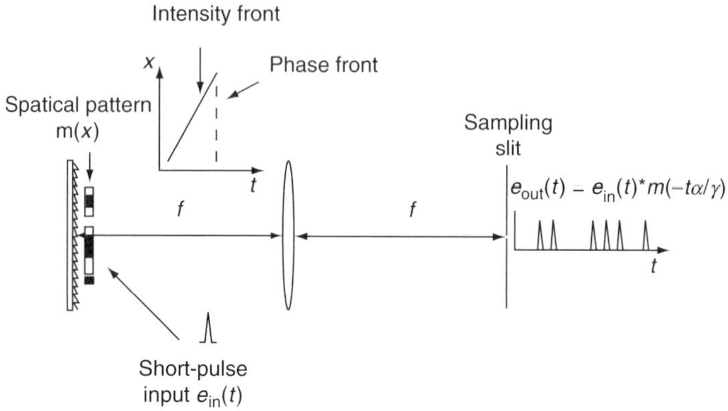


**FIGURE 7.4** Schematic representation of the direct space-to-time optical pulse shaper.

the dispersed optical frequencies at the output slit plane. In the case of an ideal spatial delta-function slit, the spectrometer output is monochromatic with the wavelength of interest selected by the transverse slit position. If the broadband input is a short pulse, the output field after the slit will be a longer pulse, with a duration determined by the delay introduced through diffraction and the input beam size. This diffraction-induced delay is known as pulse tilt [20] and describes the fact that the pulse intensity- and phase-front are no longer coplanar after diffraction from the grating. The result is that the tilted intensity front sweeps across the sampling slit at the pulse shaper output; this effectively maps different spatial locations at the grating to different temporal locations at the pulse shaper output. If the input beam is spatially patterned—for example by a spatial amplitude mask $m(x)$—the field at different temporal locations at the output is effectively turned either on or off. The result is that the temporal waveform at the pulse shaper output is essentially the input spatial pattern evaluated as a scaled function of time. Mathematically, it can be shown [17] that the temporal output of the apparatus is given by the convolution of the short pulse input with the spatial pattern evaluated as a scaled function of time

$$e_{\text{out}}(t) \propto e_{\text{in}}(t)\ m\left(-\frac{\alpha}{\gamma}t\right). \tag{7.3}$$

In this expression,

$$\frac{\gamma}{\alpha} = \frac{\lambda}{cd\cos\theta_i} \tag{7.4}$$

is the space-to-time conversion constant in ps/mm and includes the effects of both the angular dispersion and the change in beam size (astigmatism) introduced by the grating [21]. This constant is determined by the system center wavelength ($\lambda$), the spatial period of the diffraction grating ($d$), the angle of incidence of the input spatial pattern on the grating ($\theta_i$), and the speed of light ($c$). With space-to-time conversion constants ranging from 3–30 ps/mm, pulse trains with repetition-rates ranging from tens of GHz [14] to 0.5 THz [17] are readily achieved.

The work presented here relies on a novel telescopic configuration of the DST pulse shaper [14] operating in the 1.55 μm optical communications-wavelength band. The telescopic design, which increases the spectral resolution of the pulse shaper by the telescope magnification, allows generation of high-rate optical pulse sequences spanning a time aperture in excess of 100 ps. For our system the center wavelength of $\lambda = 1558$ nm, grating period $d = 1/600$ mm, and incident angle of $\theta_i \sim 12°$ yield a space-to-time conversion constant of $\gamma/\alpha \approx 3.1$ ps/mm. The input diffractive optical element (DOE) spatial patterns in this work are generated via a diffractive optical element that functions as a one-dimensional spot generator producing periodic spatial patterns ranging from 8 to 21 spots. An amplitude mask is used subsequent to the DOE to block individual spots and produce arbitrary input spot sequences. The space-to-time functionality then maps these arbitrary input spot patterns to an arbitrary optical pulse sequence at the apparatus' output. The typical spatial extent of our input patterns at the grating is ~32 mm and the minimum spot spacing is approximately 3.2 mm. After space-to-time conversion, these result in optical pulse sequences that span roughly a 100 ps time aperture and exhibit pulse repetition-rates up to ~100 GHz.

### 7.2.2 Broadband Burst and Continuous-Periodic Millimeter Waveforms

We now present several illustrative examples of millimeter waveforms synthesized in our system. Here, we demonstrate the use of our DST pulse shaper, in concert with moderate and high repetition-rate fiber lasers, to achieve cycle-by-cycle synthesis of arbitrary amplitude-, phase-, and frequency-modulated electrical waveforms in the 30–50 GHz range. The first experiments [2] we present focus on the synthesis of arbitrary broadband burst millimeter voltage waveforms in the 30–50 GHz range. To achieve such bursts, a passively mode-locked erbium fiber laser, producing ~300 fs pulses at a repetition-rate of 40 MHz is used as the system input. Each pulse from the source laser is converted to a ~100 ps duration optical pulse sequence in the DST pulse shaper and subsequently converted to an arbitrary millimeter waveform via O-E conversion by a 60 GHz photodiode.

One example of a periodic ~48 GHz sinusoidal burst is shown in Figure 7.5. Here, a six-spot pattern with a spatial period of ~6.7 mm applied to the

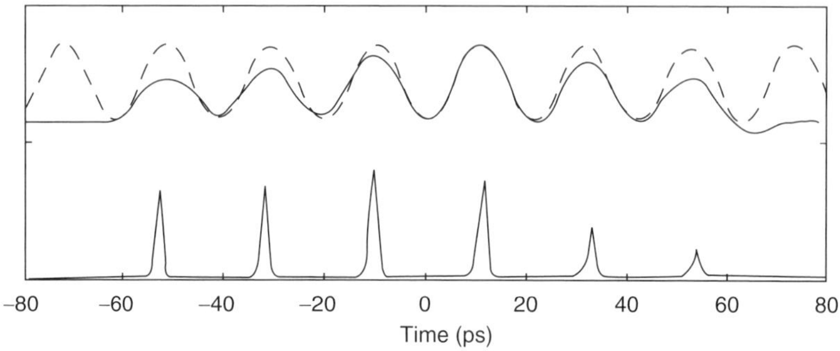


**FIGURE 7.5** Burst ~48 GHz millimeter waveform (solid line), 48 GHz sinusoidal fit (dashed line), and an intensity cross correlation measurement of the driving optical pulse sequence (bottom trace). (From McKinney, J.D., Seo, D.S., Leaird, D.E., and Weiner, A.M., *J. Lightwave Technol.*, 21, 3020, 2003. With permission.)

pulse shaper input results in an output optical pulse sequence consisting of six pulses at a repetition-rate of approximately 48 GHz after space-to-time conversion (optical cross-correlation measurement with a ~100 fs reference pulse directly from the source laser, bottom trace). After O-E conversion by the photodiode, the measured electrical waveform (top trace) is seen to be a smooth sinusoid that agrees extremely well with a 48 GHz sinusoidal fit (dashed trace). Comparison of the electrical and optical waveforms clearly demonstrates that each optical pulse contributes one cycle to the output millimeter waveform. Thus, by changing the periodicity of the driving optical pulse sequence we may tune the frequency of the output electrical waveform.

The mapping of one optical pulse to one electrical cycle enables synthesis not only of periodic millimeter burst waveforms, but phase- and frequency-modulated sinusoids as well. For example, to synthesize a sinusoidal waveform exhibiting a $\pi$ phase-shift at a particular time location, we introduce an extra half-period delay between two optical pulses in the driving optical pulse sequence. After O-E conversion, the resulting electrical waveform exhibits a $\pi$ phase-shift at the location of the extra delay. Figure 7.6 presents several phase-modulated millimeter waveforms synthesized in this manner. Figure 7.6a shows the driving optical pulse sequences used to synthesize the ~48 GHz phase-modulated waveforms in Figure 7.6b. The top trace of Figure 7.6a is a periodic 48 GHz optical pulse sequence that yields a 48 GHz sinusoidal burst waveform after O-E conversion (top trace of Figure 7.6b). In each subsequent trace (Figure 7.6a), an extra ~10 ps delay is introduced after the second, third, and fourth pulses in the driving optical pulse sequence. Correspondingly, in the measured electrical waveform (Figure 7.6b) a $\pi$ phase-shift is introduced after the second, third, and fourth electrical cycle. Thus,

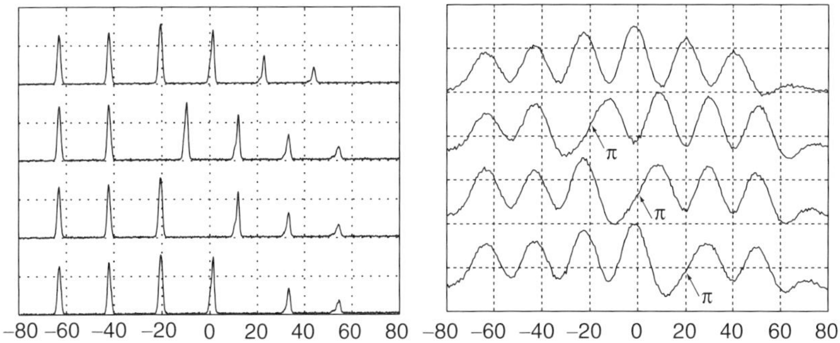


**FIGURE 7.6** ~48 GHz phase-modulated waveforms. The position of the optical pulses in the driving optical pulse sequence (a) controls the phase of the resulting electrical waveform (b) after O-E conversion. Therefore, addition of an extra half-period delay in the driving optical pulse sequence results in a $\pi$ phase-shift of the electrical waveform at the corresponding temporal location. For example, an extra ~10 ps delay after the second, third, or fourth optical pulse in a six-pulse sequence leads to a $\pi$ phase-shift after the second, third, or fourth electrical cycle in the measured millimeter waveform. (From McKinney, J.D., Seo, D.S., Leaird, D.E., and Weiner, A.M., *J. Lightwave Technol.*, 21, 3020, 2003. With permission.)

by controlling the temporal position of pulses in the driving optical pulse sequence, we are able to control the phase of the resulting millimeter waveform. If instead we wish to synthesize a frequency-modulated waveform, we alter the pulse spacing in the driving optical pulse sequence. The idea is to drive the photodiode with several pulses spaced so closely that they cannot be individually resolved in the electrical measurement. Thus, the electrical cycles from each individual pulse blend together to form a single electrical cycle twice the duration of that resulting from single-pulse excitation. An example waveform illustrating this process is shown in Figure 7.7. Here, a series of three optical pulses at a repetition-rate of ~48 GHz is followed by two pulse pairs. The interpulse spacing in the pulse pairs is ~10 ps—too close in time to be individually resolved by the 60 GHz photodiode. After O-E conversion, the measured electrical waveform consists of three ~48 GHz cycles followed by two cycles at ~24 GHz. There is some amplitude distortion evident in the frequency-modulated waveform of Figure 7.7a; specifically, the first 24 GHz cycle is approximately twice the amplitude of the preceding 48 GHz cycles. This is not unexpected because the electrical system response is greater at 24 GHz than at 48 GHz and the photodiode sees twice the optical energy of a single pulse when excited with a pulse pair. To equalize our waveforms, the driving optical pulse sequence may be predistorted by appropriately tailoring the spatial masking function in the DST pulse shaper.

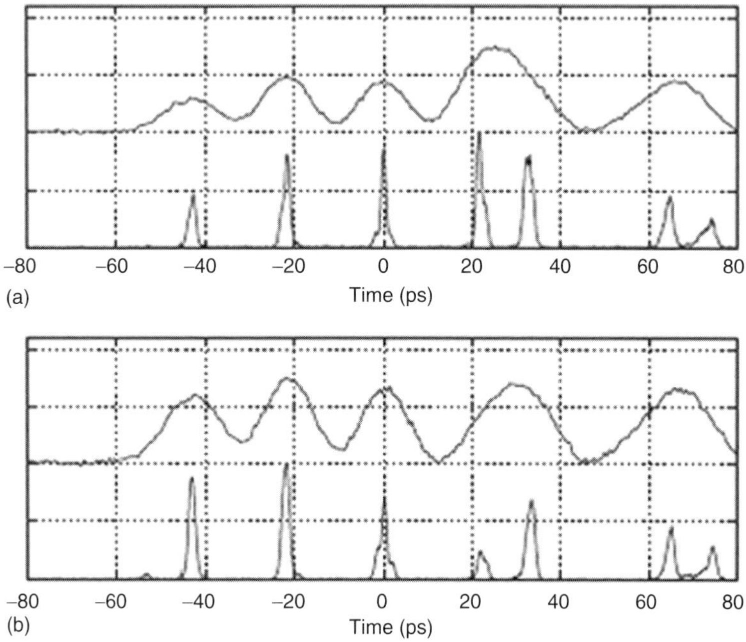


**FIGURE 7.7** By altering the interpulse spacing in the driving optical pulse sequence we achieve abrupt millimeter-frequency modulation. (a) ~48/24 GHz frequency-modulated waveform and driving optical pulse sequence (bottom trace). The pulse pairs driving the 24 GHz cycles are comparable in amplitude to the single optical pulses driving the higher-frequency portion of the waveform, which results in the 24 GHz cycles exhibiting a larger amplitude than the preceding 48 GHz cycles. (b) Equalized ~48/24 GHz frequency-modulated waveform. Here the optical pulse pairs are attenuated to decrease the amplitude of the 24 GHz cycles in the electrical waveform. Predistortion of the driving optical pulse sequences allows the amplitude of each electrical cycle to be tailored independently. (From McKinney, J.D., Seo, D.S., Leaird, D.E., and Weiner, A.M., *J. Lightwave Technol.*, 21, 3020, 2003. With permission.)

As shown in Figure 7.7b, when the first pulse pair is attenuated, the electrical waveform is effectively equalized. Thus, as shown schematically in Figure 7.8, the electrical waveform may be tailored as desired (for example, equalized) through proper predistortion of the driving optical pulse sequence.

The peak voltage amplitudes in these experiments are roughly 3 mV as determined by the very modest optical power incident on the photodiode (~0.5 μW, avg.), the photodiode responsivity (~0.3 A/W), and the electrical measurement system response. To increase the signal levels for applications, either wideband electrical amplification or increased optical excitation power may be used.

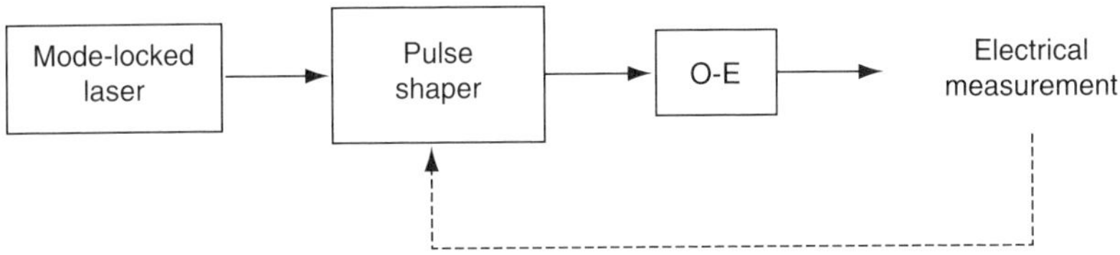


**FIGURE 7.8** The measured electrical waveform may be tailored as desired by predistortion of the driving optical pulse sequence. (From McKinney, J.D., Seo, D.S., Leaird, D.E., and Weiner, A.M., *J. Lightwave Technol.*, 21, 3020, 2003. With permission.)

In the preceding experiments, the source laser was a passively mode-locked erbium fiber laser with a repetition-rate of 40 MHz. Thus, the electrical waveforms synthesized in our apparatus exhibit arbitrary sinusoidal structure on the timescale of tens of picoseconds over a time aperture of ~100 ps determined by the optical pulse shaper. These broadband burst waveforms are essentially isolated in time as they repeat at the 40 MHz (every 25 ns) repetition-rate of the input laser source. For applications requiring higher data rates, arbitrary electrical waveforms that repeat on the timescale of the pulse-shaping window, that is "continuous" millimeter waveforms, are desirable.

To achieve such waveforms, we replace the source laser in our experiments with an actively mode-locked erbium fiber laser producing ~1 ps pulses at a repetition-rate of 10 GHz. By combining the ability to generate arbitrary optical pulse sequences over a 100 ps time aperture with a source of comparable pulse period, it becomes possible to stitch the arbitrary optical pulse sequence synthesized from one source pulse together with that synthesized from the subsequent source pulse. Thus, provided the optical pulse sequence duration and source repetition-rate are ideally matched, continuous optical pulse sequences at rates of ~100 GHz may be obtained from our pulse-shaping apparatus. After O-E conversion of these optical pulse sequences, continuous millimeter sinusoids in the range of 30–50 GHz are obtained.

Figure 7.9 illustrates arbitrary optical pulse sequences and millimeter waveforms synthesized in this manner. In Figure 7.9a we present three arbitrary optical pulse sequences generated in our DST pulse shaper. In the top trace (i), each source pulse is converted to one 100 ps frame consisting of a sequence of 5 optical pulses with a pulse spacing of 20 ps (50 GHz). This frame is then stitched together with that arising from the subsequent source pulse; the frame boundary is denoted by the dashed line. Proper choice of the input spatial pattern in the pulse shaper allows the frame duration to be ideally matched to the pulse period of the source laser (note, this operation is fundamentally limited by the timing jitter of the source laser). Thus, after this stitching operation, we observe a continuous optical pulse train at a

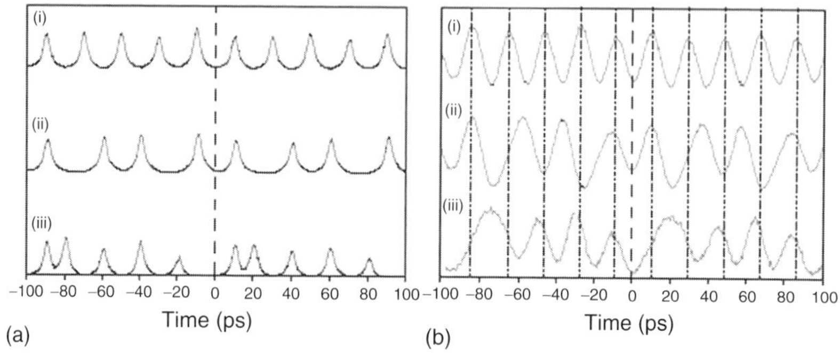


**FIGURE 7.9** (a) Driving optical pulse sequences used for synthesis of continuous millimeter waveforms. The optical pulse sequence is arbitrary within each frame and repeats every 100 ps. (i) Frame consisting of 5 pulses at a repetition-rate of 50 GHz; (ii) frame consisting of a 4-pulse sequence exhibiting an extra 10 ps delay after the first and third pulses; (iii) frame consisting of a 100 GHz pulse pair followed by three pulses at 50 GHz. (b) Millimeter wave forms resulting from O-E conversion of the optical pulse sequences. (i) Continuous 50 GHz sinusoid; (ii) periodically phase-modulated 50 GHz waveform exhibiting a $\pi$ phase-shift after the first and third electrical cycles within each frame; (iii) frequency-modulated sinusoid consisting of a single 25 GHz cycle followed by three 50 GHz cycles within each frame. (From McKinney, J.D., Seo, D.S., Leaird, D.E., and Weiner, A.M., *J. Lightwave Technol.*, 21, 3020, 2003. With permission.)

repetition-rate of 50 GHz. After O-E conversion of this optical pulse sequence, the continuous ~50 GHz electrical waveform of Figure 7.9b (i) is obtained. Here, the stitching error between adjacent frames is minimized, resulting in a smooth millimeter sinusoid.

Phase and frequency modulations of these continuous waveforms may also be achieved in the same manner as used in our burst millimeter waveform experiments. For example, Figure 7.9a (ii) illustrates a continuous optical pulse sequence where an extra 10 ps delay has been introduced after the first and third pulse in each frame. After O-E conversion, the measured millimeter waveform (Figure 7.9b (ii)) is a 50 GHz sinusoid that exhibits a $\pi$ phase-shift after the first and third electrical cycle within each frame. To obtain a 25/50 GHz frequency-modulated electrical waveform, we synthesize an optical frame (Figure 7.9a (iii)) consisting of a 100 GHz pulse pair, followed by three pulses at 50 GHz. The resulting millimeter waveform, shown in Figure 7.9b (iii), consists of a 25 GHz cycle followed by three 50 GHz cycles within each frame.

We note our waveforms are effectively continuously periodic, that is, the electrical (equivalently, optical) modulation is arbitrary within one 100 ps frame and then repeats at the source repetition-rate of 10 GHz. With advances in high-speed optical modulator technology, for example, parallel modulator arrays operating at rates $\geq$10 GHz, frame-by-frame reprogramming of the driving optical pulse sequences (and, hence, arbitrary millimeter electrical waveforms) could be achieved resulting in truly continuous arbitrary optical and electrical waveforms. In these experiments, our peak-to-peak voltage amplitudes are on the order of 20 mV, again determined by the input optical excitation power, photodiode responsivity, and electrical system response. As with our burst millimeter waveforms, optical or wideband electrical amplification could be used to achieve larger signal amplitudes for applications.

In the following section, we discuss our recent work in arbitrary ultrabroadband radio-frequency waveform synthesis. Through a combination of femtosecond optical pulse shaping and optical frequency-to-time conversion in a dispersive medium, we demonstrate synthesis of multigigahertz instantaneous bandwidth signals aimed at applications in UWB systems.

## 7.3 ULTRABROADBAND MICROWAVE WAVEFORM SYNTHESIS

### 7.3.1 Ultrawideband Systems, Applications, and Challenges

In recent years, UWB RF and microwave systems have seen renewed interest. These systems, which employ short bursts of radiation, exhibit several advantages over more common narrowband technologies. In particular, the wide bandwidth (short time duration) and significant low-frequency content offer desirable signal properties such as increased immunity to multipath interference, low probability of intercept (in the frequency domain), and increased material penetration. The benefits make UWB signals attractive for a number of applications including impulse radio [22], ground-penetrating radar (GPR) [23], radio-frequency identification (RFID), precision asset location, and vehicular radar [24]. With the February 2002 Federal Communications Commission (FCC) decision to allow unlicensed operation of UWB devices in the 3.1–10.6 GHz frequency band [25], UWB systems currently garner significant scientific and commercial research interest.

A major concern voiced by the opponents of UWB systems is interference with the existing narrowband RF and microwave systems. Interference with the global positioning system (GPS) (~1.2 and 1.5 GHz) and direct broadcast satellite (DBS, ~11.7–12.5 GHz), both of which neighbor the FCC-specified UWB band of 3.1–10.6 GHz, is of particular concern. To address such concerns, UWB systems are required to adhere to a stringent spectral mask that specifies the allowable in-band and out-of-band signal levels. As an example, the FCC-specified spectral emission levels (spectral mask) for high-frequency imaging systems (for example, GPR, medical imaging systems, wall- and through-wall imaging systems, as well as surveillance systems) are shown in Figure 7.10. This plot shows the allowable average effective isotropically radiated power (EIRP = power radiated $\times$ antenna

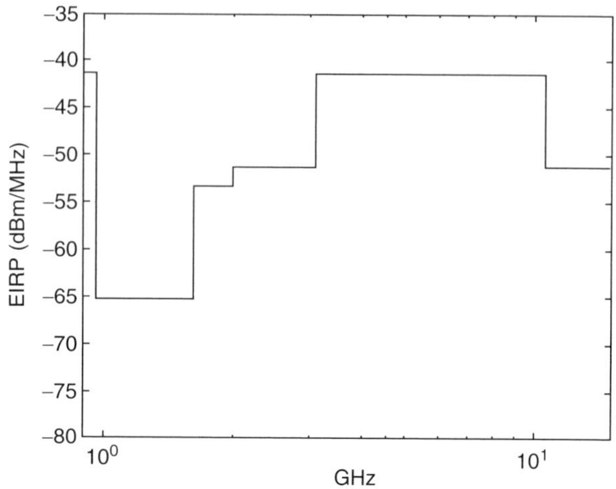


**FIGURE 7.10** FCC spectral mask for high-frequency UWB imaging systems (excludes −75.3 dBm/MHz levels in the 1.164–1.240 GHz and 1.559–1.610 GHz bands). For frequencies above 1.610 GHz, this spectral mask also corresponds to that specified for indoor UWB communication systems.

gain) per megahertz at a distance of 3 m from the transmitter for such imaging systems. In these systems, the intentional signal content must fall in the 3.1–10.6 GHz band, with extraneous out-of-band signal content limited to the levels indicated. Clearly, to adhere to such radiation limits while maximizing the energy in the 3.1–10.6 GHz band requires careful signal and system optimization.

In addition to spectral occupancy and emission constraints, UWB systems face hardware challenges as well. In particular, broadband antennae, though optimized to radiate over a decade in some cases, are rarely optimized with respect to phase response. This nonlinear spectral phase response leads to easily recognizable signal dispersion [26–28], which leads to distortion of the time domain signal on radiation. Such dispersive effects have the potential to significantly degrade signal detectability, multipath immunity, and range resolution in a variety of UWB applications. Frequency-dependent propagation effects, in particular diffraction, also lead to distortion of the intended signal in wireless applications. Given the likelihood of changing regulations regarding spectral content and hardware difficulties associated with the broadband nature of UWB signals, reprogrammable UWB signal generators with control over both the temporal and spectral properties of UWB waveforms are extremely desirable.

Research and commercial electronic techniques for generation of UWB signals exhibiting multigigahertz instantaneous bandwidths typically employ step-recovery [29] or tunnel diodes [27] followed by a pulse-shaping network, or simply an antenna, to achieve UWB impulsive waveforms. Pulse-shaping networks are also frequently utilized to achieve the desired pulse shape and RF spectral properties for these waveforms. Although these techniques have provided intriguing examples for applications such as GPR [23] and short-range medical radar applications [30], once the pulse-shaping network is implemented, the temporal and spectral properties of the UWB signal are necessarily fixed. As a result, current electronic techniques for UWB signal synthesis are not easily adapted to address changing regulations or system concerns such as antenna distortions.

In the following sections, we discuss our work in ultrabroadband RF and microwave signal synthesis via optical pulse shaping. Our technique relies on established optical AWG technology and enables a variety of UWB waveforms outside the current reach of electronic techniques to be readily—and reprogrammatically—synthesized. We note that our waveforms typically far exceed the minimum definitions [25] for UWB signals (%BW>20% or minimum bandwidth of 500 MHz); hence we use the term ultrabroadband to make this distinction. Here, we discuss how our technique is applied to RF spectral shaping to facilitate conformance to RF spectral occupancy constraints as well as optical and electronic methods to achieve increased RF signal amplitudes. In addition, we discuss how our technique has been applied to the first-ever experimental demonstration of multigigahertz bandwidth UWB signals designed to compensate for the dispersive phase response of broadband antennae.

### 7.3.2 Experimental Apparatus

Our method relies on the ability to shape the optical power spectrum in an FT pulse shaper [1] followed by frequency-to-time conversion in a dispersive medium as demonstrated previously [9,10]. A block diagram of our experimental apparatus is illustrated in Figure 7.11. Short pulses from a mode-locked erbium fiber laser (~100 fs, ~30 nm bandwidth, 50 MHz repetition-rate) are spectrally filtered in a reflective FT pulse shaper. This allows us to impress an arbitrary filter function onto the optical spectrum. These shaped pulses are then dispersed in 5.5 km of single-mode fiber. The chromatic dispersion of the fiber gives us a frequency-dependent linear time delay that results in a time domain optical waveform exhibiting the shape of the optical filter function applied in the pulse shaper. Thus, after O-E conversion of the time domain optical waveform, the measured RF waveform exhibits the shape of the filter function applied to the optical power spectrum. By manipulating the optical filter function of the pulse shaper, we directly modulate the optical spectrum and, hence, the time domain RF waveform.

At the core of our apparatus is a reflective-configuration FT optical pulse shaper as illustrated in Figure 7.12. Short pulses from the laser source enter

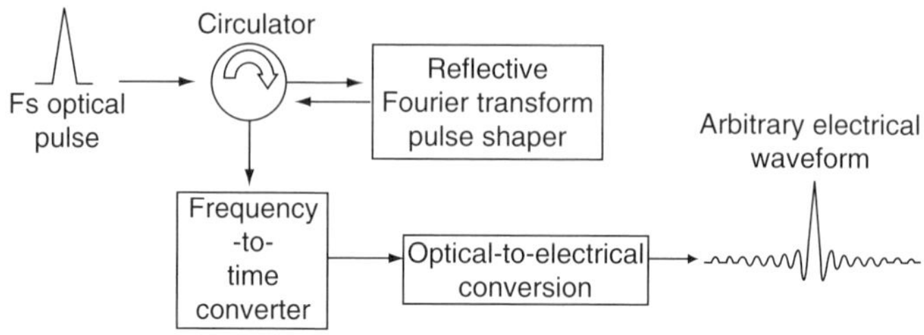


**FIGURE 7.11** Experimental apparatus for ultrabroadband microwave signal synthesis. (From Lin, I., McKinney, J.D., and Weiner, A.M., *IEEE Microw. Wireless Compon. Lett.*, 15, 226, 2005. With permission.)

the pulse shaper from a circulator through a polarizing beam splitter. Individual frequency components of the input pulse are angularly dispersed by an 830 L/mm diffraction grating. A 190 mm focal length lens stops this dispersion and focuses the dispersed frequency components along the retroreflecting mirror at the back focal plane of the lens. A 128-pixel, single-layer liquid crystal modulator (LCM) is placed immediately before the mirror. The dispersed frequency components are amplitude-modulated in parallel under voltage control by the combination of the 128-pixel LCM and the polarizing beam splitter at the output. After modulation, the frequency components are recombined by the lens and grating combination, and exit the pulse shaper through the polarizing beam splitter and circulator. The source laser bandwidth (~30 nm) and the total chromatic dispersion of the fiber stretcher

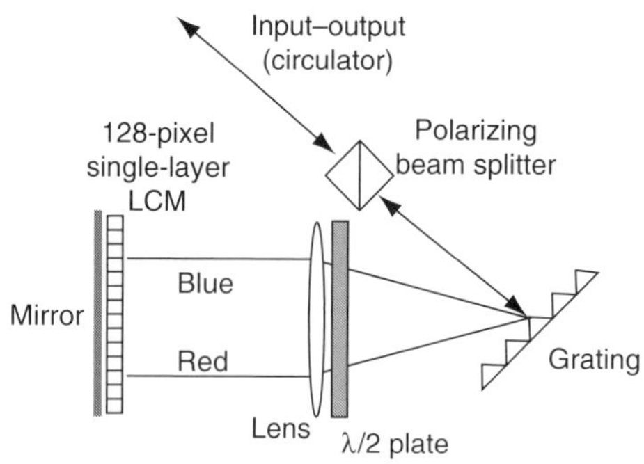


**FIGURE 7.12** Reflective-configuration FT optical pulse shaper. (From Lin, I., McKinney, J.D., and Weiner, A.M., *IEEE Microw. Wireless Compon. Lett.*, 15, 226, 2005. With permission.)

(~96 ps/nm) enable a time aperture of approximately 3 ns for our RF waveforms. After O-E conversion with a high-speed photodiode our temporal RF waveforms are measured with a 50 GHz sampling oscilloscope and the RF spectra of these waveforms are measured with a 50 GHz RF spectrum analyzer. Here, so long as the electrical bandwidth of the photodiode exceeds the frequency of the desired RF waveform, the particular photodiode bandwidth is not a significant concern; the sole function of the photodiode is to provide a faithful electrical replica of the input optical intensity profile. The photodiode may then be chosen based on responsivity or power-handling capability. In our apparatus we typically utilize a 60 GHz photodiode, though lower bandwidth (22 GHz) devices have been used as well. The key difference in our work compared to previous demonstrations of this technique [9] is that our system operates in an open-loop configuration. Thorough characterization of our pulse shaper and dispersive stretcher enables real-time arbitrary RF waveform realization—without the need for iteration.

We note that—as long as the assumption of a linear frequency-dependent time delay in the fiber stretcher is valid—the time aperture and temporal resolution of our system may be configured by the user by adjusting the relation of the optical bandwidth, pulse shaper resolution, and dispersion of the fiber stretcher. A simple way to validate the assumption of a linear frequency-dependent time delay (i.e., the stretcher introduces a heavy chirp [31] to the input optical signal) in the stretcher is to look at the time-bandwidth product of the optical features after the stretcher. As long as the time-bandwidth product is significantly greater than the bandwidth limit, the optical signal may be considered heavily chirped and the assumption of linear time-delay is valid. For a particular root-mean-square (RMS) optical frequency bandwidth $\delta f$ and particular RMS temporal duration $\delta t$, this condition is [32,33]

$$\delta t \delta f \gg \frac{1}{\pi}. \tag{7.5}$$

This condition is readily satisfied for large optical bandwidths even with modest fiber lengths—it is more challenging to satisfy this condition as the optical bandwidth decreases. To illustrate the relation between the above system parameters (optical bandwidth, pulse shaper resolution, and dispersion of the fiber stretcher), we consider the following example. An optical source at a center wavelength of $\lambda_o = 1550$ nm with a bandwidth of $\Delta\lambda = 34.5$ nm is spectrally filtered in an FT pulse shaper employing an $N = 128$ pixel optical modulator array. The optical bandwidth modulated by each pixel in the array is then $\delta\lambda = \Delta\lambda / N \approx 0.27$ nm. In terms of frequency, $\delta f \approx 33.7$ GHz ($\delta f = c\delta\lambda / \lambda_\sigma^2$, where $c$ is the speed of light). If we wish to distinguish one pixel from the next in the time domain waveform, by the condition of Equation 7.5 each pixel must correspond to a temporal feature $\delta t \gg 9.4$ ps.

Reasonable system performance is achieved if the lower limit for "much-much-greater" is taken as a factor of 5; this means to resolve the temporal features arising from adjacent pixels, the temporal duration of each pixel must be approximately $\delta t = 47$ ps. This minimum temporal feature may then be related to the pixel bandwidth $\delta\lambda$ (nm), the fiber dispersion parameter $D$ (ps/nm/km), and the fiber length $L$ (km) as

$$\delta t = \delta\lambda DL \text{ (ps)}. \tag{7.6}$$

Assuming the fiber stretcher uses standard single-mode fiber ($D \sim 17$ ps/nm/km); to achieve single-pixel resolution in the time domain requires a stretcher of length $L \geq 10.24$ km. If we choose the length to be $L = 10.24$ km, we obtain a temporal resolution (Equation 7.6) of $\delta t \sim 47$ ps and a time-aperture $\Delta T = N\delta t$ of approximately 6 ns. By the relations given in Section 7.1.1, the minimum and maximum electrical frequencies achievable from the apparatus are then $f_{\text{min}} \sim 333$ MHz and $f_{\text{max}} \sim 10.6$ GHz. For this same system, choice of a longer fiber stretcher will maintain single-pixel and result in lower operational frequencies. Choice of a shorter stretcher would raise the lowest frequency of the system and reduce the number of independent control elements in the time domain waveform (i.e., single-pixel resolution may not be achieved subject to the time-bandwidth condition of Equation 7.5).

It is interesting to note that here—in contrast to the relations governing bandwidth, temporal resolution, spectral resolution, and bandwidth discussed in Section 7.1.1—the temporal resolution is directly proportional to the minimum spectral feature ($\delta t = \delta\lambda DL$) and the time aperture is directly proportional to the bandwidth ($\Delta T = N\delta\lambda DL = \Delta\lambda DL$). These relations are a direct consequence of the optical frequency-to-time mapping, which essentially relates the temporal waveform at the output of the fiber to the FT of the temporal waveform at the fiber input. Mathematically, this relation may be explained in terms of chirp transforms (convolution with quadratic phase exponentials)—an excellent treatment is given in Ref. [34]. The relation between the RF time aperture and temporal resolution of the system (time-bandwidth product) of the signal generator is determined by the number of independent control elements in the modulator used in the optical pulse shaper. Our current apparatus utilizes a 128-element modulator array that sets the upper bound on the time-bandwidth product to 64. Given optical modulators with 640 independent elements are commercially available, the time-bandwidth product of the system could be easily scaled toward 320.

In this technique, it is the optical power spectral density and O-E converter (photodiode) sensitivity that determine the voltage amplitudes of the output electrical waveforms. Since the peak time domain optical intensity decreases as the inverse of the waveform time aperture, it is the available optical energy and photodiode responsivity that ultimately limit the longest time aperture and, hence, lowest frequency waveforms achievable in this technique. The finest temporal feature and highest electrical frequencies remain limited by the O-E converter bandwidth. Given the typical power levels and optical bandwidths of communications-wavelength fiber lasers, a reasonable estimate of the longest time aperture would be ~60 ns with a resolution determined by the optical modulator array. As a reasonable estimate for the highest frequencies available, the use of high-speed unitraveling-carrier photodiodes [11] (though these are research devices for the moment) could extend this technique to greater than 100 GHz with the time aperture determined by the optical modulator array.

### 7.3.3 Ultrabroadband Radio-Frequency and Microwave Signal Synthesis

We now present several representative waveforms generated in our system [5]. As a first simple example, Figure 7.13 shows a single-tone burst RF waveform. In Figure 7.13a, we show the filtered optical spectrum that exhibits a sinusoidal amplitude modulation with a period of 4.2 nm. After dispersive stretching and O-E conversion the measured RF waveform shows the shape of the optical power spectrum as clearly illustrated in Figure 7.13b. The ~400 ps period of the time domain waveform is determined by the total fiber dispersion (~96 ps/nm) and the applied periodic optical spectral filter function. Good correspondence between optical spectrum and time domain RF waveform is illustrated in Figure 7.13a and Figure 7.13b. The roll-off in the time domain waveform is due to the underlying shape of the optical spectrum and can be eliminated through proper equalization. In the frequency domain, a ~(−50) dBm peak is quite evident at ~2.5 GHz in the RF spectrum of Figure 7.13c (10 kHz resolution bandwidth (RBW)). The peak amplitude of this and subsequent waveforms is ~14.0 mV as determined by the input optical power (~5 μW, average), photodiode responsivity, and 20 dB (power) electrical amplification. The sidelobes in Figure 7.13c are due to the roughly square time aperture of the sinusoidal waveform. A technique to remove these sidelobes through apodization of the time domain waveform is discussed later in this section. We have obtained similar results (not shown) for burst sinusoidal waveforms with center frequencies throughout the range of 1–5 GHz.

We also demonstrate the ability to modulate our waveforms on a timescale not achievable via electronic techniques. Figure 7.14 shows an example of a broadband frequency-modulated waveform composed of ~1.25/2.5/5 GHz cycles. The top trace, Figure 7.14a, again shows the filtered optical spectrum that has been patterned with a sinusoidal shape varying discretely in period from ~8.4 nm (0.5 cycle) to ~4.2 nm (2.5 cycles), to ~2.1 nm (4 cycles). After stretching, this spectrum gives rise to an RF waveform (Figure 7.14b) with abrupt frequency modulation, exhibiting 0.5 cycles at ~1.25 GHz, 2.5 cycles at ~2.5 GHz, and 4 cycles at ~5 GHz. In the RF spectrum (Figure 7.14c, 10 kHz RBW), three main peaks centered at ~1.25/ 2.5/5 GHz are clearly seen.

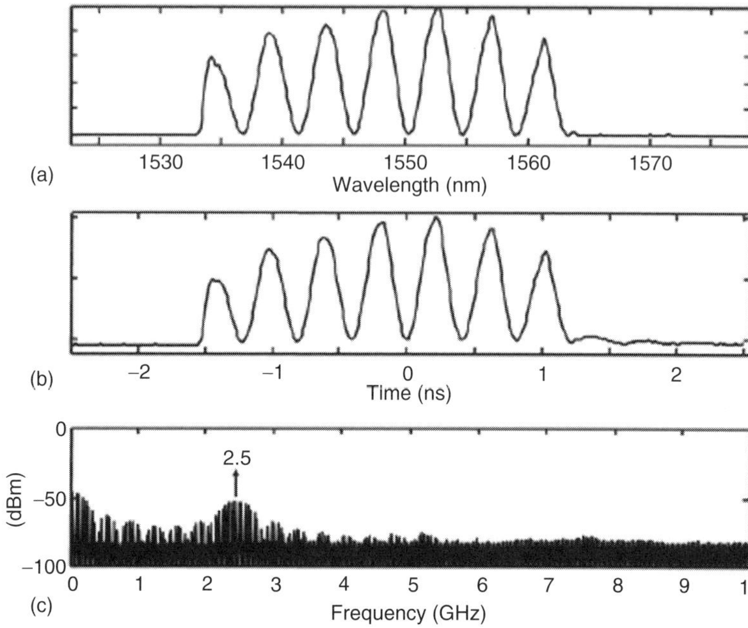


**FIGURE 7.13** Single-tone RF burst waveform. (a) Shaped optical spectrum showing a periodic ~4.2 nm sinusoidal amplitude modulation. (b) Time domain RF waveform measured after optical frequency-to-time conversion and O-E conversion with a photodiode. The temporal period is ~400 ps as determined by the total fiber dispersion of ~96 ps/nm. (c) Measured RF power spectrum showing a broad spectral peak at 2.5 GHz (10 kHz RBW). The sinc-like ripples in the RF spectrum arise from the steep rising and falling edges of the time domain RF waveform. (From Lin, I., McKinney, J.D., and Weiner, A.M., *IEEE Microw. Wireless Compon. Lett.*, 15, 226, 2005. With permission.)

The broad RF spectral bandwidth shown here is achieved by our ability to change frequency on a cycle-by-cycle basis—an ability not achieved with current electronic techniques.

To further illustrate the capability to generate extremely broadband RF waveforms in our system, we turn now to demonstration of impulsive waveforms geared for ultrabroadband RF systems applications, such as UWB communications. An example of an ultrabroadband impulsive waveform is shown in Figure 7.15. Here, we specifically appeal to UWB applications by demonstrating a burst of monocycle waveforms (Figure 7.15a) similar to those frequently used in UWB systems. Several unique capabilities are illustrated in Figure 7.15. First, our monocycles are on the order of ~200 ps in duration, already shorter than those typically generated electronically [35]. Additionally, the time resolution of our system enables the spacing between

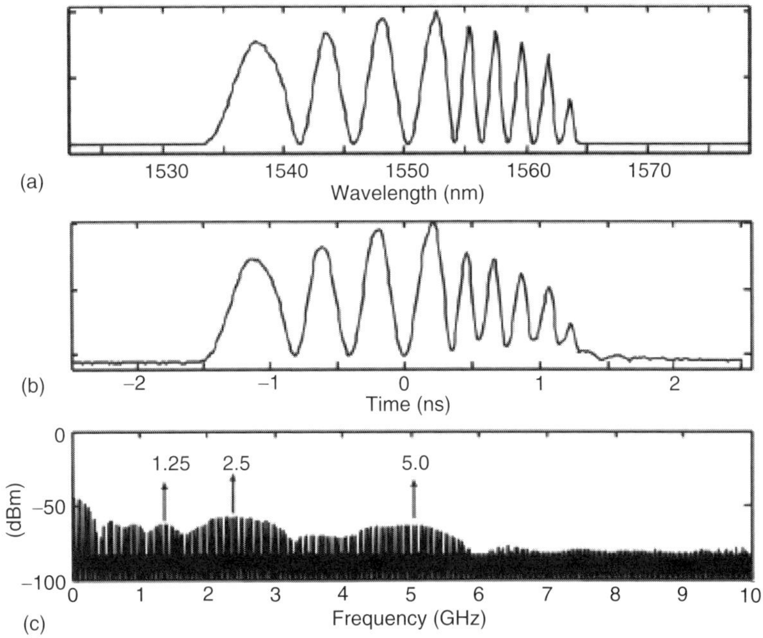


**FIGURE 7.14** Frequency-modulated RF waveform with abrupt frequency changes from 1.25–2.5–5 GHz. (a) Shaped optical spectrum. (b) Time domain RF waveform. (c) Measured RF power spectrum (10 kHz RBW). (From Lin, I., McKinney, J.D., and Weiner, A.M., *IEEE Microw. Wireless Compon. Lett.*, 15, 226, 2005. With permission.)

adjacent pulses to be controlled to within approximately 30 ps; this enables pulse-position modulation encoding with a much finer resolution than is available from current electronic devices. Another capability unique to our system is the potential for arbitrary polarity reversals as illustrated by the second monocycle in the sequence. We expect this could enable the use of phase-shift keying in UWB systems, again with more flexible control than provided by purely electronic techniques. From Figure 7.15b, the ultrabroad bandwidth of the monocycle sequence is clearly evident—the measured RF spectrum (10 kHz RBW) shows nonzero spectral content ranging from DC to 8 GHz.

We note the RF spectra presented in this and subsequent sections reflect the repetition-rate of the input laser source, that is, the ultrabroadband spectral envelope determined by the FT of our waveforms is superposed with a 50 MHz frequency comb. For the typical resolution bandwidths (1–30 kHz) and frequency spans (1–15 GHz) of our spectral measurements, the video display bucket width of our spectrum analyzer (span/601 display points) ranges from ~10 to 24 MHz. As the resolution bandwidth and bucket width

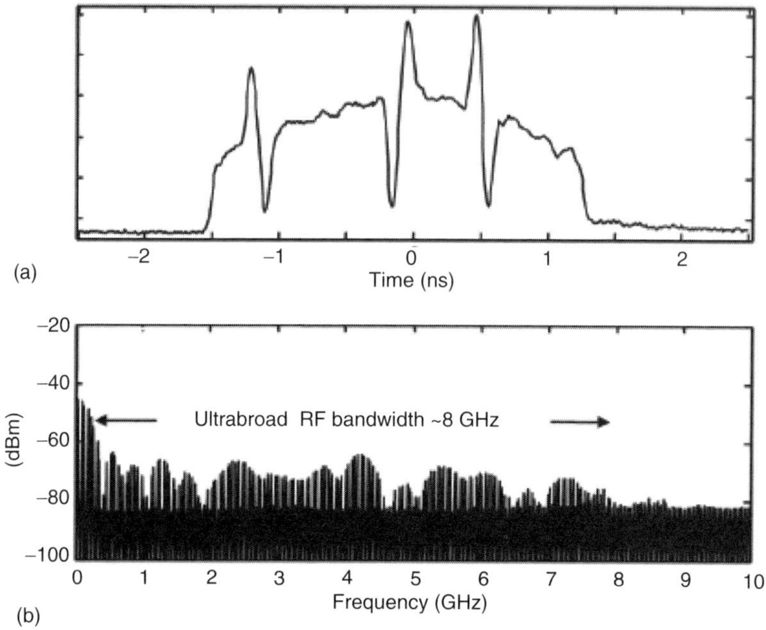


**FIGURE 7.15** Burst of monocycle waveforms exhibiting variable monocycle spacing (resolution of approximately 30 ps) and arbitrary polarity reversals not achievable via electronic techniques. (a) Time domain waveform. (b) Measured ultrabroadband RF power spectrum (10 kHz RBW). (From Lin, I., McKinney, J.D., and Weiner, A.M., *IEEE Microw. Wireless Compon. Lett.*, 15, 226, 2005. With permission.)

are always less than 50 MHz, we are able to resolve individual comb lines. For frequency spans that yield display buckets wider than 50 MHz, or resolution bandwidths in excess of 50 MHz, the displayed spectrum is given by the envelope predicted via FT of the time domain data.

### 7.3.4 Spectral Engineering of Ultrabroadband Radio-Frequency and Microwave Waveforms

To fully utilize the available bandwidth in the 3.1–10.6 GHz UWB band, while adhering to an application-specific spectral mask (such as the example in Figure 7.10) requires precise control over the time domain waveforms (and, hence, spectral content) utilized in a particular system. In the following sections, we demonstrate tailoring of the RF power spectra of ultrabroadband RF waveforms through careful design of the time domain shape of these waveforms. Specifically, we demonstrate that the RF spectral shape, center frequency, and bandwidth may be chosen as desired [36]. In addition, for a particular spectral bandwidth and center frequency, we demonstrate proper choice of the time domain waveform can lead to energy and power spectral density (PSD) increases of over an order of magnitude [37].

We are particularly interested in ultrabroadband signals that conform to the power and spectral content limits specified for UWB communication applications. For these systems, the intentional signal center frequency and −10 dB bandwidth must fall in the 3.1–10.6 GHz frequency band. To achieve tailored waveforms satisfying this condition, we use the following design process. We first define a target RF power spectrum (bandwidth, center frequency, and spectral shape) in a suitable numeric computation package (for example, Matlab). We then perform an inverse fast-FT of the frequency domain data to determine the shape of the corresponding temporal waveform. As our apparatus maps a tailored optical intensity profile to an electrical waveform, our waveforms are required to be positive-definite signals. To achieve oscillatory waveforms subject to this constraint, a minimal DC component is added to the time domain waveform. The resulting signal is sampled and applied as the optical filter function in our apparatus. In the work presented here, our target RF spectrum is typically a super-Gaussian function aimed to achieve optimal spectral flatness while minimizing the duration of oscillations in the time domain waveform.

We first illustrate the spectral shaping concept using a baseband impulsive waveform. In Figure 7.16a, we present an impulsive waveform tailored to produce an extremely flat 4 GHz bandwidth baseband RF spectrum. The impulsive waveform is approximately 450 ps (between first two nulls) and rides on a small DC pedestal (again, we are shaping the optical intensity). As shown in Figure 7.16b, the RF power spectrum (10 kHz RBW) is very broad and shows a nearly super-Gaussian shape with a bandwidth of 3.2 GHz. The flatness is within 6.78 dB over spectral range from 0.59 to 3.77 GHz. This fluctuation results from the squarelike DC pedestal in the time domain.

Although the DC offset of our signals is limited to the minimum necessary to achieve a particular oscillatory waveform, the steep rising and falling edges of the resulting DC pedestal result in RF spectral distortions. Several examples of this effect are the sinc-like ripples apparent in the power spectra of Figure 7.13c and Figure 7.16b. To remove this spectral ripple while maintaining the overall spectral content of our waveforms, proper apodization may be applied to the time domain waveform. This is analogous to using windowed FTs for spectral analysis applications [38]. As an example of this process, again consider the baseband impulse and corresponding RF power spectrum of Figure 7.16. To remove the steep rising and falling edges of the time domain waveform, we employ a 1.5 ns FWHM Gaussian apodization window. The resulting apodized impulse is shown in Figure 7.17a. Clearly, the steep rising and falling edges of the DC pedestal have been suppressed. As expected, in the frequency domain the ~6.8 dB ripple has been effectively

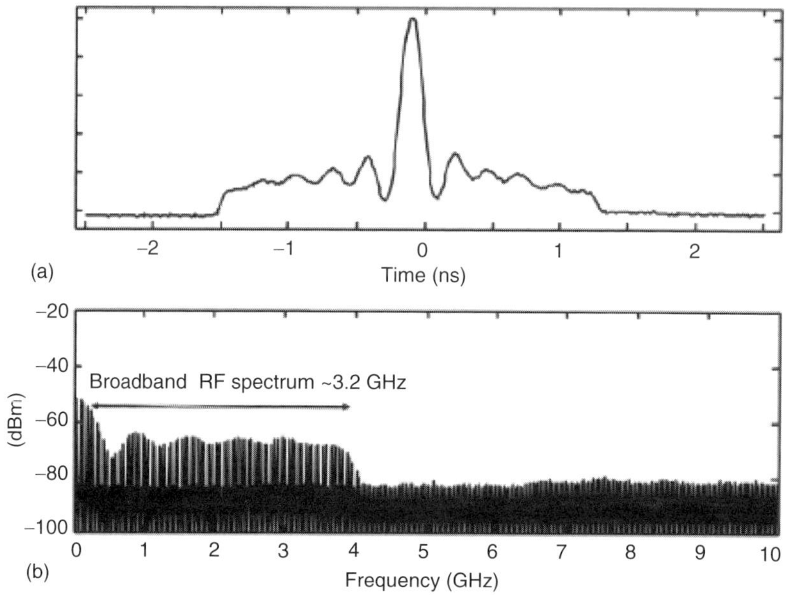


**FIGURE 7.16** Broadband RF impulse with a time domain shape designed to yield an extremely flat (super-Gaussian) RF power spectral density. (a) Time domain impulse. (b) Measured RF PSD. The ~6.8 dB maximum ripple is due to the steep rising and falling edges of the time domain waveform (~ $\pm$1.5 ns on either side of the central peak). (From Lin, I., McKinney, J.D., and Weiner, A.M., *IEEE Microw. Wireless Compon. Lett.*, 15, 226, 2005. With permission.)

removed as shown in Figure 7.17b. The measured RF spectrum is now smooth from DC to ~4 GHz (see Figure 7.16b) for comparison.

The ability to control or remove undesired low-frequency content in our waveforms enables synthesis of modulated waveforms with frequency content well isolated from DC. Such flexibility could prove quite useful in UWB systems by allowing frequency components falling within bands of concern (for example, those used for GPS applications) to be suppressed. An example of a broadband-modulated impulse and corresponding RF power spectrum is given in Figure 7.18. Here, a baseband impulse is designed to have an RF bandwidth of ~4 GHz and is subsequently amplitude-modulated at a frequency of 5 GHz to achieve an RF waveform exhibiting an 80% fractional bandwidth. It should be emphasized here that the desired time domain signal is applied as the optical filter function in our optical pulse–shaping apparatus—in this case, the modulated impulse. The shape of this impulse is again determined by first defining our target RF spectrum—here the target is a 4 GHz bandwidth super-Gaussian spectrum centered at 5 GHz. Thus, an additional local oscillator is not required for frequency up-conversion; the

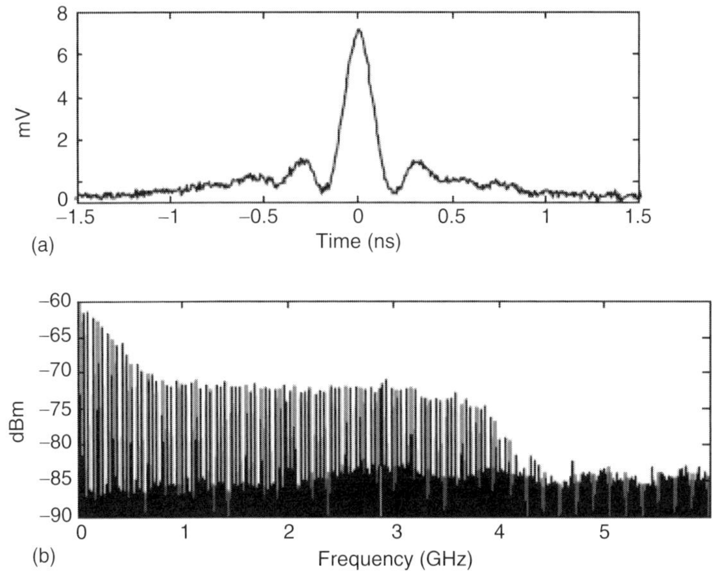


**FIGURE 7.17** The RF impulse (a) and measured RF PSD (b) when the impulse of Figure 7.16a is apodized with a 1.5 ns (FWHM) Gaussian window. Apodization removes the steep rising and falling edges of the time domain waveform, which leads to a smooth RF PSD.

center frequency of our modulated waveforms may be directly specified anywhere in the DC—~10 GHz range. In the measured RF power spectrum (Figure 7.18b), the 3–7 GHz oscillatory components are cleanly offset from the DC. Comparing the peak PSD at ~3.5 GHz (~ −45 dBm) with the peak noise floor of ~ −60 dBm (at 2 GHz, 30 kHz RBW) shows an extinction of 15 dB. We note the edges of the DC pedestal in the time domain waveform have been mostly, though not completely, suppressed (very small rising and falling edges at ~ −1.4 and +1.75 ns). The result is that there is some unintended low-frequency content in our waveform, most notably the spectral peak at ~1 GHz. This low-frequency content may again be suppressed through apodization with a slightly narrower windowing function. In this particular example, a total of 50 dB (power gain) electrical amplification was used to achieve the ~250 mV (peak) impulse.

Another example of an ultrabroadband-modulated impulse is shown in Figure 7.19a. Here, we appeal directly to UWB applications' requiring signals in the 3.1–10.6 GHz frequency band by synthesizing an approximately 135 ps (between first two nulls) modulated impulse. The temporal duration and modulation frequency of 6.85 GHz are chosen to yield a 7.5 GHz bandwidth filling the above-mentioned frequency band. The resulting RF power spectrum (30 kHz RBW) is shown in Figure 7.19b. The RF spectrum is observed

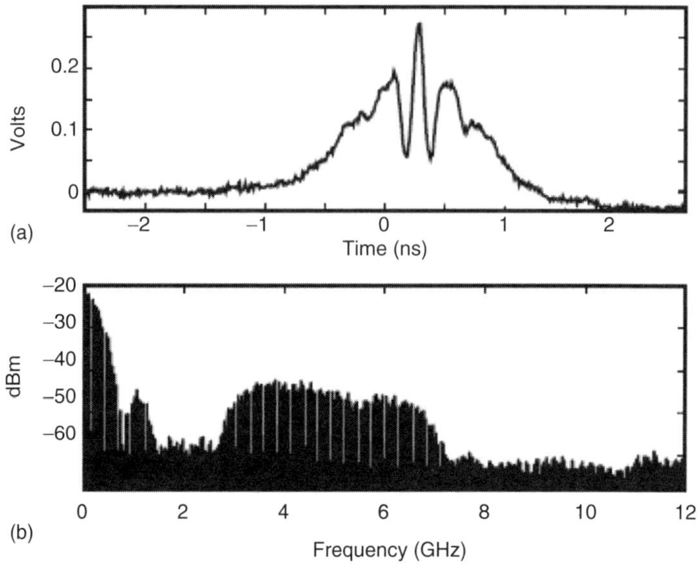


**FIGURE 7.18** Modulated impulsive waveform designed to exhibit an 80% fractional bandwidth at ~5 GHz. To achieve the modulated waveform, a baseband impulse designed to have a 4 GHz bandwidth is amplitude-modulated with a 5 GHz sinusoid; the resulting waveform is applied as the optical filter function in the optical pulse shaper. (a) Time domain impulse. (b) Measured RF power spectrum. Apodization of the time domain waveform allows suppression of low-frequency content in the waveform.

to be extremely flat over the range from ~3 to ~10 GHz; in the 3.5–9.5 GHz band the amplitude fluctuations are limited to approximately $\pm 2$ dBm. The significant low-frequency content (below 3 GHz) is a result of the fact that apodization was not employed for this waveform. We note the negative portion of the waveform occurring for times $\geq 1.25$ ns is a result of the high-pass filtering effect of our low-noise wideband electrical amplifier (20 dB power gain).

Given the operation of our system, that is, mapping of the optical intensity to a time domain electrical waveform, and assuming that the system is set for a fixed time aperture, the quantity that fundamentally determines the attainable RF energy and PSD in the output electrical waveform is the per-pulse energy from the optical source. Thus, the time domain electrical signal that most efficiently utilizes the available optical energy will result in the largest RF PSD and energy. Here, we show that the RF PSD may be increased on the order of 10 dB and the waveform energy by greater than a factor of $10\times$—for a given RF spectral bandwidth and fixed input optical energy—by utilizing smoothly chirped sinusoids instead of UWB impulses. This concept is similar

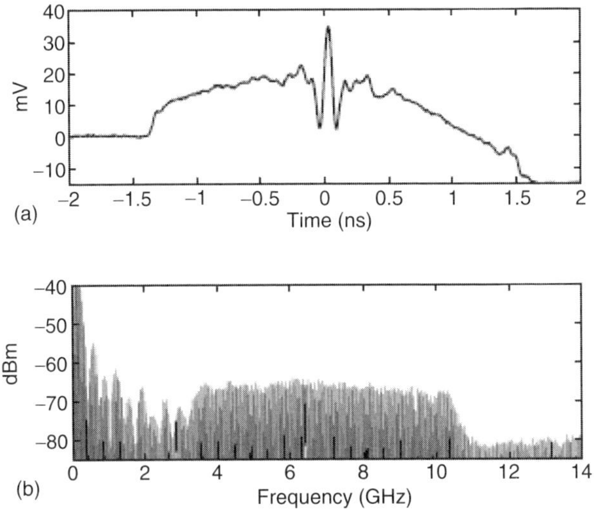


**FIGURE 7.19** ~135 ps Ultrawideband impulse (a) and the corresponding RF power spectrum (b). The impulse is tailored to possess spectral content filling the 3.1–10.6 GHz frequency band. The significant frequency content below 3 GHz may be removed with proper apodization of the time domain waveform. In (a), the negative portion of the waveform (times ≥1.25 ns) is due to the high-pass filtering effect of the electrical amplifier used in our experiment.

to the use of chirped waveforms to avoid peak power limitations in optical or microwave amplifiers (for example, optical chirped pulse amplification or chirp radar applications). For systems where the RF spectral content is of more importance than a minimal temporal duration, this provides a simple technique for increasing the RF energy without adding additional electronic amplification.

To illustrate this concept, the ultrabroadband impulse of Figure 7.19a is reproduced in Figure 7.20a. Superposed with the impulse is the electrical waveform produced when the optical filter function in our apparatus is set to unity for all wavelengths (dashed curve). This is an electrical representation of the unshaped optical power spectrum and provides an intuitive measure of the amount of the available optical energy used to synthesize this broadband impulse. The dashed curve also indicates the peak voltage level attainable in our apparatus—clearly this impulsive waveform spends only a fraction (~135 ps) of the available ~3 ns time aperture near this peak value. In contrast, the chirped waveform of Figure 7.20b spends a significantly longer time near the peak voltage level of our system. Here, the center frequency of the waveform sweeps linearly from 3.1 to 10.6 GHz to attain spectral content comparable to the impulse of Figure 7.20a. Under the assumption that both

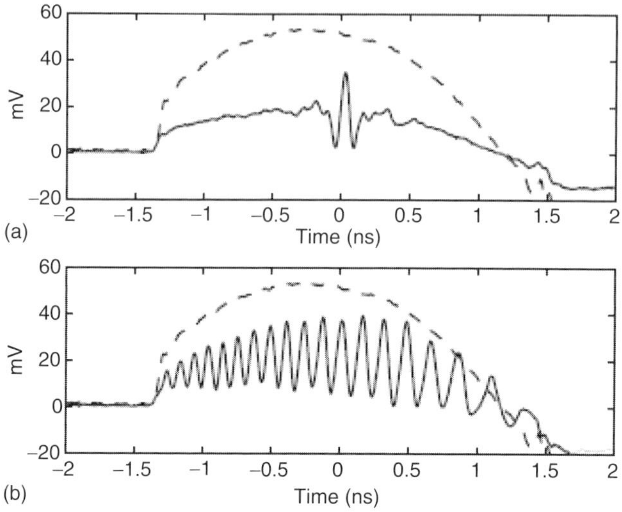


**FIGURE 7.20** The ultrawideband impulse (a) of Figure 7.19a and a linearly chirped RF waveform (b) designed to exhibit similar frequency content. The dashed curve represents the RF waveform achieved when the optical filter function in the shaper is set to unity; This provides an intuitive measure of how much optical energy is utilized to create each waveform, as well as indicating the peak voltage attainable in our system.

waveforms maintain the same RF bandwidth, the predicted increase in energy (or peak PSD in the frequency range of interest, assuming a flat RF power spectrum) for the chirped waveform depends on both the effective waveform amplitude as well as the waveform duration. The effective waveform amplitude accounts for the necessary DC component in our waveforms; for the UWB impulse the effective amplitude is ~13.1 mV and the effective amplitude of the chirped sinusoid is approximately 19 mV. The RMS durations of the impulsive and chirped waveforms are ~230 ps and ~1.68 ns, respectively. The predicted energy increase is then given by the ratio of the waveform durations multiplied by the square of the ratio of effective voltage amplitudes. In dB,

$$10\log_{10}\left[\frac{1.68}{0.23}\left(\frac{1.9}{13.1}\right)^2\right] = 11.65\,\text{dB} \tag{7.7}$$

Figure 7.21 shows the measured RF power spectra corresponding to the UWB impulse (Figure 7.21a) and linearly chirped waveform (Figure 7.21b). Both spectra are observed to be very broadband—both fill the frequency range of 3.1–10.6 GHz. The peak PSD increase achieved through the use of a chirped waveform is measured to be approximately 10 dBm (occurring at ~6 GHz)

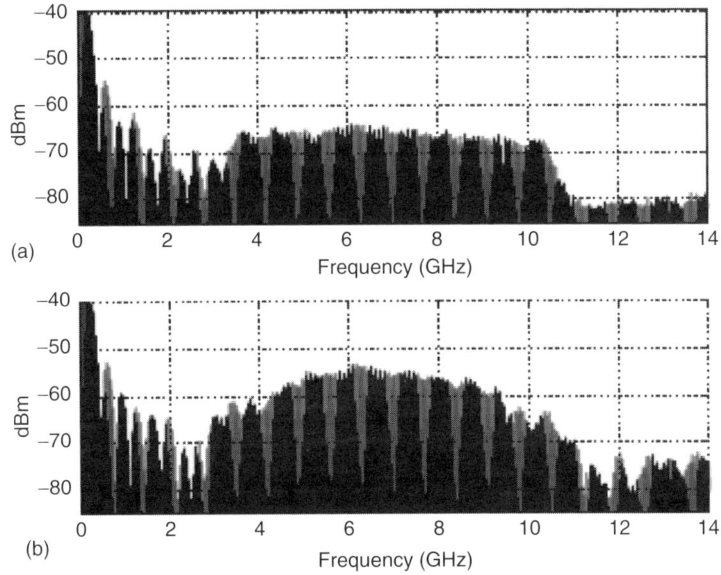


**FIGURE 7.21** Measured RF power spectra corresponding to the ultrawideband impulse (a) and chirped waveform (b) of Figure 7.19. The chirped waveform (Figure 7.19b) clearly spends a longer period of time near the peak voltage level than the impulse. As a result, the spectra of this chirped waveform show an increase in PSD of ~10 dBm compared to that of the impulse.

and agrees quite well with the predicted increase of 11.65 dBm. This clearly illustrates that impulsive waveforms may be utilized to achieve extremely flat RF power spectral densities and that the use of chirped waveforms enables a significant increase in RF energy given the available optical input energy. The choice of impulsive or chirped waveforms may then be based on the requirements of a particular application.

### 7.3.5 Amplification of Ultrawideband Waveforms

For applications, it is desirable to increase the energy of our waveforms. This may be accomplished either through wideband electrical amplification or by increasing the optical intensity incident on the photodiode (O-E converter) in our system. It is important to note that electrical amplification, even with low-noise amplifiers, leads to a decrease in the signal-to-noise ratio (SNR). Optical amplification, however, has shown the potential to increase the received electrical SNR in certain circumstances [39].

To compare these two methods of signal amplification, we focus on the UWB impulsive waveform previously illustrated in Figure 7.19 and

Figure 7.20a. This waveform is chosen for its applicability to UWB systems as its bandwidth fills the 3.1–10.6 GHz frequency band. The frequency content of this waveform also fills a large portion of the bandwidth of our electrical amplifiers allowing comparisons of these two techniques at both low and high RF frequencies. Here, the comparison is between a wideband low-noise amplifier (Miteq AMF-4D-001180-24-10P, 0.1–18 GHz, measured power gain $G_e$ ~33 dB, noise figure (NF) of ~1.85 dB) that amplifies the photodiode output prior to the electrical measurement and a femtosecond erbium-doped fiber amplifier (EDFA, PriTel FA-13) placed immediately before the optical pulse shaper in our system. To compare these two methods for an equivalent increase in electrical power, the gain of the EDFA was adjusted to yield a similar peak voltage value as that obtained via electrical amplification. Since optical power contributes to electrical voltage (current), the gain of the EDFA (increase in optical intensity) necessary to obtain an equivalent increase in electrical power is a factor of 2 lower (in dB) than the required electrical gain. Thus, the optical amplifier was adjusted for a gain of ~16.5 dBm.

Figure 7.22 compares the time domain impulses achieved via wideband electrical amplification and increased optical power. Here, all three waveforms are acquired with the sampling oscilloscope configured in a 64-trace averaging mode. In Figure 7.22a, we show the measured time domain impulse directly from our apparatus (no amplification). Here, the peak voltage of $V_{pk} \sim 1.75$ mV is determined by the extremely modest optical power incident on the photodiode (approximately 1.46 μW, average) and the RMS electrical noise ($V_n \approx 0.05$ mV) is dominated by the noise inherent in our sampling oscilloscope. The electrically amplified impulse and the impulse resulting from increased optical excitation are shown in Figure 7.22b and Figure 7.22c, respectively. A useful metric for impulsive signals (that would, for example, affect the decision threshold in impulsive RF systems) is the time domain SNR. Following the convention of work in the terahertz regime [40], we define the time domain SNR as the ratio of the peak signal power to the RMS noise power. Expressed in dB as a function of the peak signal voltage ($V_{pk}$) and RMS noise voltage ($V_n$)

$$\mathrm{SNR}_{\mathrm{TD}} = 20 \log 10 \frac{V_{pk}}{V_n} \,(\mathrm{dB}) \tag{7.8}$$

For the waveform of Figure 7.22a, the time domain SNR is approximately $\mathrm{SNR}_{\mathrm{TD}} = 30.9$ dB. The time domain SNR for the electrically amplified impulse ($V_{pk} = 81.15$ mV, $V_n = 0.66$ mV) is $\mathrm{SNR}_{\mathrm{TD}} \approx 41.8$ dB and that corresponding to the waveform obtained with increased optical excitation is $\mathrm{SNR}_{\mathrm{TD}} \approx 44.4$ dB ($V_{pk} = 77.9$ mV, $V_n = 0.47$ mV). The fact that the time domain SNR is a factor of 2.6 dB larger in the case of increased optical excitation is not surprising given the electrical amplifier not only amplifies

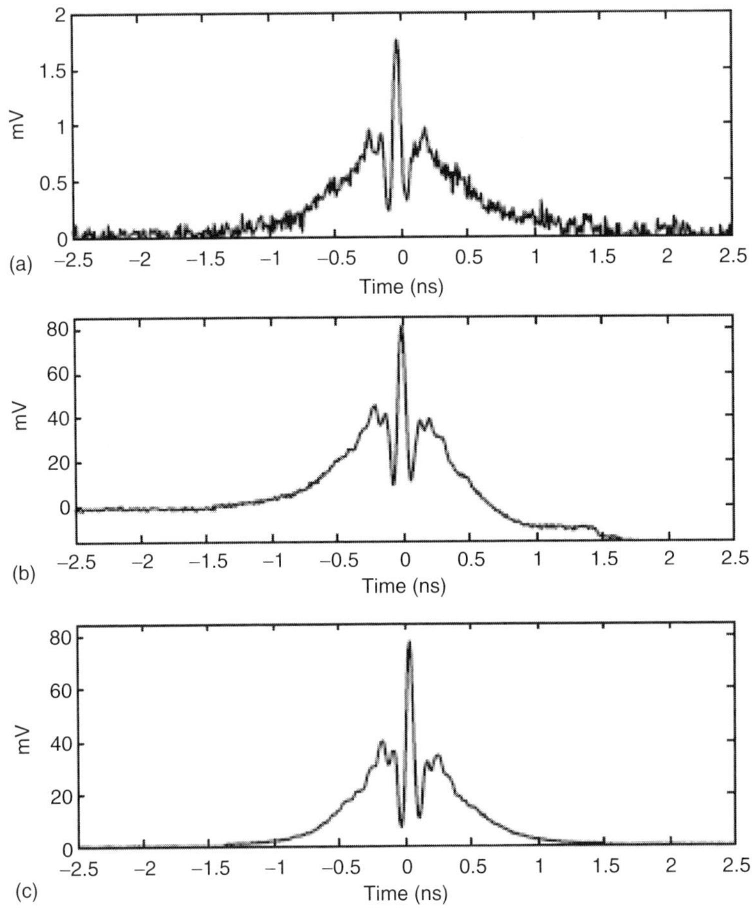


**FIGURE 7.22** Time domain comparison of the effects of electrical amplification versus increased optical excitation in our apparatus. (a) Ultrawideband impulsive waveform generated in our apparatus with an average optical power of ~1.46 μW. (b) Electrically amplified time domain impulse. (c) Ultrawideband impulse achieved by increasing the optical excitation power in our apparatus.

the incident thermal noise, but also exhibits a nonzero NF. It is interesting to note that, despite the fact that our system functions by shaping the optical intensity (a positive-definite quantity), the electrically amplified voltage waveform is negative for times greater than ~0.66 ns in Figure 7.22b; this effect arises from the high-pass filter response of the amplifier. In contrast the optical amplifier does not exhibit this behavior as its functionality is independent of RF frequency. From Figure 7.22c, it is apparent that the voltage waveform remains positive and that the high-pass filtering effects have been removed.

The effects of electrical amplification and increased optical excitation may also be observed in the frequency domain. As previously mentioned, our source laser operates with a repetition-rate of 50 MHz. Thus, the broadband RF spectral envelope of our waveforms is multiplied by a 50 MHz comb in the frequency domain. We focus on the frequency range of 5.0–5.1 GHz as this range lies near the central portion of the tailored RF spectrum. The 5.0–5.1 GHz portion of the RF spectrum (1 kHz RBW) for the unamplified impulse, electrically amplified impulse, and the impulse obtained via increased optical excitation are shown in Figure 7.23a through Figure 7.23c, respectively. In Figure 7.23a, the average noise floor of approximately −113 dBm corresponds to the displayed average noise level (DANL) of the spectrum analyzer for a 1 kHz RBW. Comparison of the comb amplitudes in Figure 7.23b and Figure 7.23c to that of Figure 7.23a clearly illustrates that

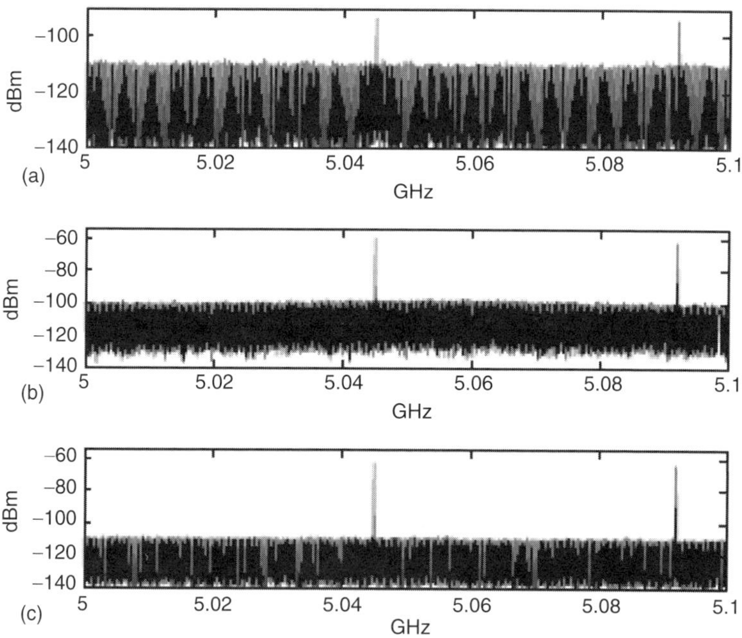


**FIGURE 7.23** Frequency domain comparison of the effects of electrical amplification versus increased optical excitation. Here, we compare the portion of the ultrawideband RF spectrum between 5 and 5.1 GHz for the (a) ultrawideband impulsive waveform generated in our apparatus with an average optical power of ~1.46 μW, (b) electrically amplified impulse, and (c) ultrawideband impulse achieved by increasing the optical excitation power in our apparatus.

the signal power is increased by ~33 dB by both electrical amplification and increased optical excitation. The noise floor in the case of electrical amplification, however, is seen to increase by roughly 10 dBm (~ −102.9 dBm/kHz). In the case of increased optical excitation, the noise floor remains essentially at the DANL of the spectrum analyzer (~ −112.4 dBm/kHz). As a measure of the frequency domain SNR of our waveforms, we compare the spectral amplitude of the 50 MHz comb teeth to the integrated noise in a 50 MHz bandwidth. For the electrically amplified impulse the comb amplitude of ~ −60.16 dBm yields SNR $\approx$ −4.26 dB, whereas increased optical excitation gives an SNR of approximately 2.74 dB (peak comb signal ~ −62.66 dBm, average noise ~ −112.4 dBm/kHz). Here, increased optical excitation shows better noise performance as the additional noise arising from optical amplification (limited by the beat noise between the optical signal and the amplified spontaneous emission (ASE) from the EDFA [39]) is significantly smaller than the thermal noise contribution from the electrical amplifier. We note that increased electrical amplitudes may also be achieved by using a higher power laser source—in this case, there would be no additional noise due to ASE and the system noise would be limited by thermal noise in the photodiode.

These measurements motivate the use of increased optical power, when feasible, over electronic amplification to achieve increased RF signal amplitudes. Fundamentally, this is limited by the maximum current rating of the photodiode. With our photodiode the peak voltage limit is ~1.5 V (current limit ~30 mA)—approximately 2× higher than available from our low-noise electrical amplifier which exhibits a 1 dB compression point (output) of 10 dBm (~700 mV into 50).

## 7.4 APPLICATION OF PHOTONICALLY SYNTHESIZED WAVEFORMS TO COMPENSATE FOR BROADBAND ANTENNAE PHASE DISTORTIONS

An area where photonic techniques could significantly impact RF systems, particulary UWB wireless systems, is in compensation of signal distortion—for example, distortions experienced by impulsive waveforms during radiation from broadband antennae. Given the broad frequency bandwidth available to UWB wireless systems, it is desirable for the antennae in these systems to exhibit an ideal frequency response over the 3.1–10.6 GHz UWB frequency band. Such an idealized response—consisting of a flat frequency response and linear phase delay—is, functionally, rarely achieved [41,42]. Although some antenna structures, such as the transverse electromagnetic (TEM) horn and Vivaldi antennae, do exhibit relatively flat frequency responses and relatively linear phase [27], most other structures show a nonlinear spectral phase response at a minimum. Therefore, methods to compensate for nonideal antenna behavior through design and implementation of phase-optimized UWB waveforms are necessary. Conceptually, this operation is analogous to dispersion compensation of ultrashort pulses in optical fiber. It is well known that chromatic dispersion in single-mode optical fiber, which contributes large quadratic and cubic spectral phase components to the input optical signal, leads to significant distortion of the optical pulse during fiber transmission. These pulse distortions may be removed by applying the conjugate spectral phase of that contributed by the optical fiber. In practice, this may be accomplished via precompensation (a length of dispersion compensating fiber or other device is placed before the fiber link) where the pulse is distorted prior to transmission, or through postcompensation where the pulse is recompressed at the output of the link.

Methods to compensate for phase-errors (both temporal and spatial) introduced by an antenna or array have previously been explored via theoretical techniques. Specifically, the concept of optimizing the antenna feed voltage to obtain desirable temporal properties in the received waveform—such as peak voltage amplitude or minimal received pulsewidth—has been explored in detail for dipole antennae [43]. This concept has also been extended to dipole arrays with extra constraints (for example, on the sidelobe-level of the far-field radiation pattern), and illustrates that the input excitation waveform may be tailored to optimize both temporal and spatial properties of the signal radiated from the array [44]. With regard to UWB systems, there has been significant theoretical interest in waveform optimization to combat antenna and channel distortions (specifically, dispersion) in recent years. Specifically with respect to antenna phase response, it has been shown that the transmit waveform may be designed to optimize the received waveform in UWB systems, for example, to maximize the received voltage amplitude or minimize the received voltage pulsewidth [26]. Though observation of antenna phase-distortions (dispersion) on UWB signals is readily observed and optimization techniques have been proposed to mitigate such distortions [26], to our knowledge there have been no experimental demonstrations of such techniques.

In this work, we utilize our photonic arbitrary electromagnetic waveform generator to precompensate UWB transmit waveforms to account for antenna dispersion. Through time domain impulse response measurements, we extract the RF spectral phase contributed by broadband ridged TEM horn antennae to signals transmitted over a wireless link and apply the conjugate spectral phase to our transmit waveform to achieve signal compression at the receiver. We demonstrate the received waveform in our system is indeed compressed and exhibits reduced duration and secondary oscillations as compared to the antenna link impulse response. This work constitutes the first experimental demonstration of antenna dispersion compensation of multigigahertz bandwidth UWB signals [28].

A schematic of our experimental apparatus is shown in Figure 7.24. Our antennae are ridged TEM horn antennae (Dorado International GH1-12N) with a specified bandwidth of 1–12 GHz. The output of our photonic time

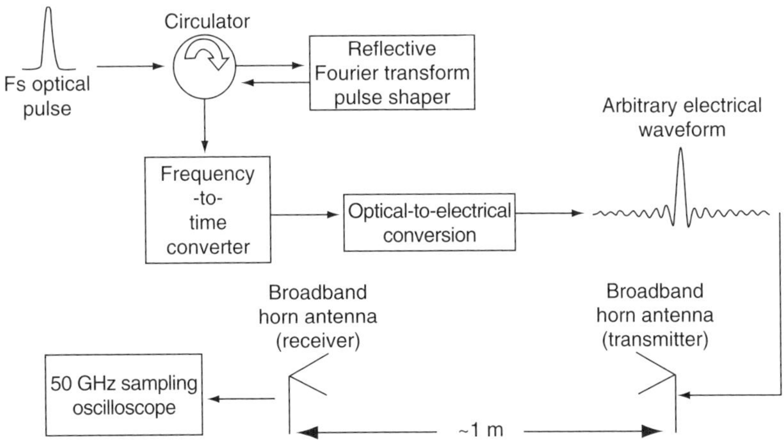


**FIGURE 7.24** Schematic representation of experimental apparatus.

domain electromagnetic signal generator is amplified with a broadband microwave amplifier (0.1–18 GHz, ~30 dB power gain) and used as the input to the transmit antenna. A second identical antenna is positioned 1 m away and functions as the receiver. The output of the second antenna is measured on a 50 GHz sampling oscilloscope. The distance of 1 m was chosen to satisfy the far-field constraint of $2D^2/\lambda$ at a frequency of 1 GHz and the antenna height (~1 m) was chosen to ensure any multipath components were clearly resolvable.

To characterize our antennae, we first obtain the impulse response $h(t)$ of the antenna link by exciting the transmit antenna with an ~18 ps, 400 mV peak voltage impulse derived directly from the laser output via photodiode. The measured impulse response and driving impulse (inset) are shown in Figure 7.25. Clearly, the short pulse input is dispersed; the received waveform shows clear oscillations over roughly 4 ns. Additionally, the received waveform is down-chirped with the low-frequency components near the antenna passband edge of ~900 MHz occurring roughly 4 ns after the initial monocycle at the leading-edge of the waveform. The large dispersion between the high-(~14 GHz) and low-(~900 MHz) frequency components is attributed to the strong variations in the RF spectral phase near cutoff in the antenna.

Because the bandwidth of the input pulse (approximately $1/(2 \times$ pulse-width) or ~27 GHz) is significantly greater than the bandwidth of our antennae, we can effectively regard the waveform of Figure 7.25 as the impulse response $h(t)$ of the transmit–receive antenna combination. From the FT of the impulse response we are then able to extract the additional spectral phase contributed by the antenna pair. The received waveform may then be compressed to a short

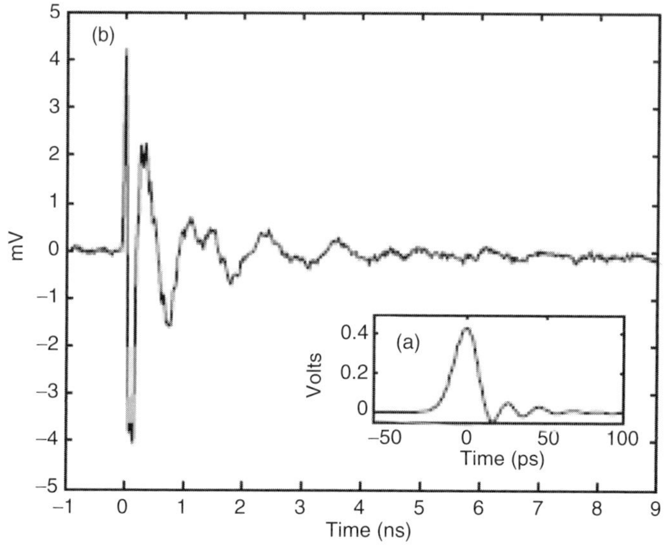


**FIGURE 7.25** (a) Photonically synthesized impulse excitation waveform. (b) Measured impulse response of the ~1 m antenna link.

pulse, provided the input signal exhibits the opposite RF spectral phase to that arising from the antenna.

So long as the input signal to the transmitting maintains this particular spectral phase, the spectral amplitude may also be tailored. For example, the spectral amplitude may be equalized to minimize the pulsewidth of the received voltage waveform. Intuitively, the simplest optimized input waveform $x(t)$ that will produce an impulse on reception is a signal "matched" to the antenna link, that is, a time-reversed version of the impulse response

$$x(t) = h(-t). \tag{7.9}$$

On reception then, the received voltage amplitude will be given by the autocorrelation of the link impulse response

$$r(t) = h(t)*h(-t), \tag{7.10}$$

where $*$ denotes convolution. To illustrate the capability of our technique to mitigate antenna phase distortions, we focus on this matched-filter solution [32] for the optimized input waveform here.

To achieve an approximation to the matched signal for our system, the antenna system impulse response is sampled and a time-reversed version of the sampled waveform is applied as the optical spectral filter function in our

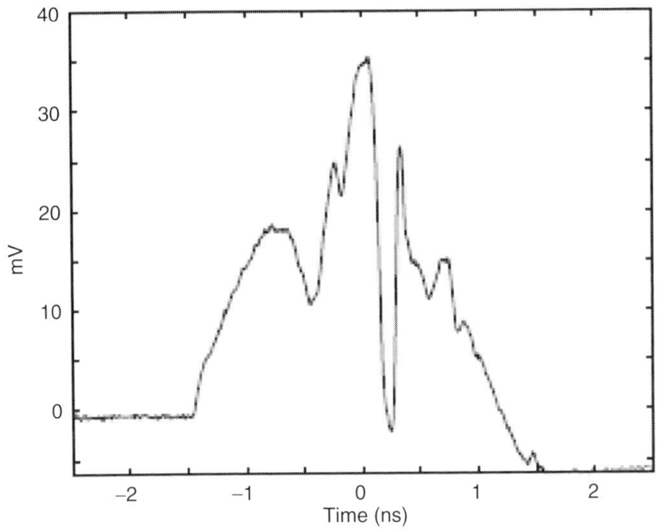


**FIGURE 7.26** Approximate matched input signal.

photonic time domain signal generator. The resulting electrical waveform is shown in Figure 7.26. Here, the approximate matched signal shows a duration of ~3 ns—roughly 75% of the duration of the impulse response—and is limited by the time aperture in our signal generator. Our system relies on the mapping of optical intensity to an electrical waveform; therefore, the resulting electrical signal is required to be a positive-definite quantity. This fact, along with the underlying shape of the optical power spectrum (manifested as the approximately Gaussian aperture of the electrical waveform) motivates us to choose the first 2 ns of the time-reversed impulse response as our desired electrical waveform. This ensures we retain the majority of the high-frequency content of the signal.

Figure 7.27 shows the received waveform when this matched signal is applied to the transmitting antenna. On comparison with the impulse response (Figure 7.25b), the received signal is visually more pulselike in nature and shows significantly less oscillation after the main central peak. It should be noted that the peak amplitude of the received signal is roughly 2× lower than the peak voltage in the impulsive drive case. However, the matched input signal used in our experiment is an order of magnitude lower in voltage than the impulsive input (40 vs. 400 mV). The small late-time oscillations (beginning at ~2 ns) are due to the fact that our matched signal only captures roughly the first 2 ns of the impulse response. Thus, the low-frequency components occurring outside this 2 ns aperture, though present in our matched waveform, are not phase-compensated. These oscillations decay on the timescale of the initial impulse response (~4 ns).

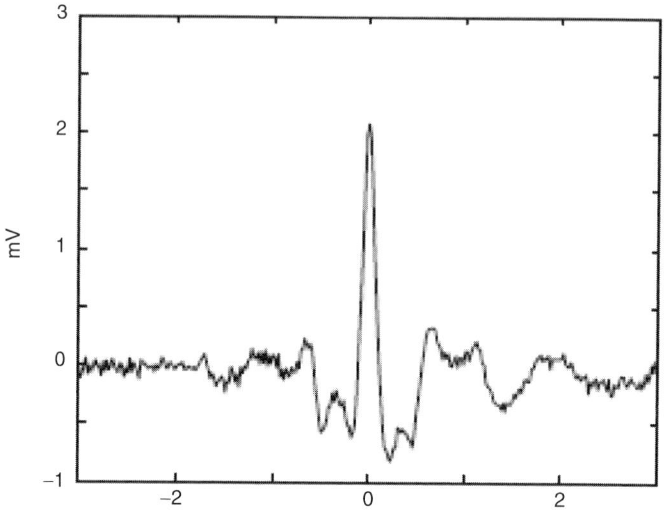


**FIGURE 7.27** Measured matched response.

To quantify the compression of the received waveform, we focus on comparing properties of the normalized power of the received waveforms, defined as

$$P_{\mathrm{N}} = \frac{|V_{\mathrm{rec}}|^2}{|V_{in_{\max}}|^2}, \tag{7.11}$$

where $V_{\mathrm{rec}}$ is the received voltage waveform and $V_{in_{\max}}$ is the peak value of the input voltage waveform. Our waveform generator is peak power limited—the maximum photodiode voltage is limited by the available optical energy and the maximum amplified electrical signal level is limited by gain saturation in our amplifier. Given this, the normalized power provides a direct measure of compression, or energy concentration, in the received signal for a fixed peak input signal power. The normalized powers for the impulse response and matched response are shown in Figure 7.28a and Figure 7.28b, respectively. The three metrics we choose to define compression of the received signal (in terms of the normalized power) are the peak normalized power, the duration, and the level of secondary oscillations in the received waveform compared to the main peak (sidelobe level).

Comparison of the normalized power of the impulse response (Figure 7.28a) to that of the matched response (Figure 7.28b) clearly illustrates the gain in peak normalized power. The matched response exhibits a 36-times larger peak power than the impulse response (recall, the peak received voltage for impulsive or matched excitation was approximately the same, whereas the

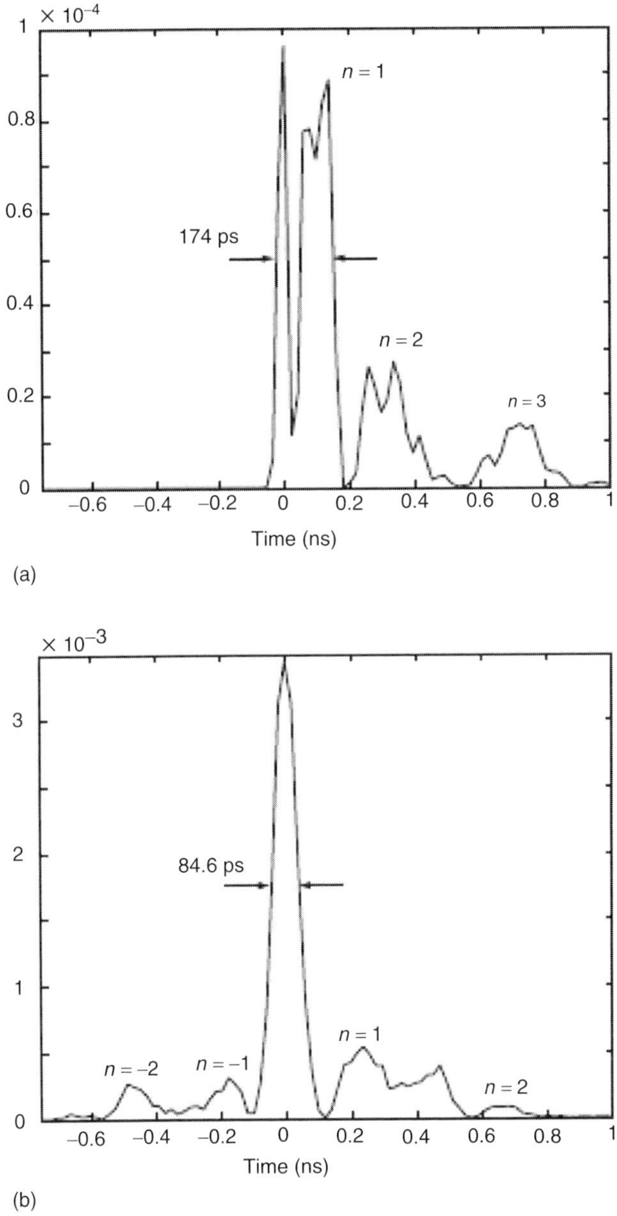


**FIGURE 7.28** Comparison of normalized power for the impulse (a) and matched (b) responses. The matched response is clearly compressed relative to the impulse response. The matched response exhibits a 36× larger peak normalized power, shorter duration, and reduced secondary oscillations (sidelobes) compared to the impulse response.

matched input had an order of magnitude lower peak voltage value). The pulsewidth—defined as the time interval over which any oscillations reach 50% of the peak normalized power—is also clearly shorter for the matched response. The duration of the matched response (~85 ps) is essentially half that of the impulse response (~174 ps).

Another (perhaps more insightful) metric that can be used to describe how well a signal is matched to a particular system is the ratio of the peak normalized power of the largest (main) signal oscillation to that of the secondary oscillations in the received signal. This is analogous to comparing the power in the main beam of an antenna array to the power in the sidelobes. We define the sidelobe level for our signals to be

$$S = 10\,\log_{10}\frac{|V_{\mathrm{n}}|^2}{|V_{\mathrm{p}}|^2}\ \mathrm{(dB)}, \tag{7.12}$$

where $|V_{\mathrm{n}}|$ is the voltage magnitude of the $n$-th secondary oscillation and $|V_{\mathrm{p}}|$ is the voltage magnitude of the largest oscillation. Table 7.2 summarizes this comparison. Comparison of the measured sidelobe levels for the matched signal with those of the impulse response shows a dramatic improvement. Though the experimental input matched signal is only an approximation to the ideal matched signal, the $n=1$ oscillation in the matched response has been pushed down to approximately the level of the $n=3$ sidelobe in the impulse response; all other oscillations in the matched response have been pushed well below the $n=3$ sidelobe in the impulse response.

Though we focus on the matched-filter solution to phase compensation in this work, our technique may be extended to optimize other parameters in the received waveform, for example, the pulsewidth or further reduction in secondary oscillations.

**TABLE 7.2**
**Comparison of Sidelobe Levels for Impulsive and Matched Excitation**

| Signal | Sidelobe # | Position (ns) | $S$ (dB) |
|---|---|---|---|
| Matched | $n=-2$ | −0.5 | −11.1 |
| | $n=-1$ | −0.18 | −10.5 |
| | $n=1$ | 0.23 | −8.1 |
| | $n=2$ | 0.68 | −16.1 |
| Impulse | $n=1$ | 0.135 | −0.3 |
| | $n=2$ | 0.331 | −5.4 |
| | $n=3$ | 0.71 | −8.5 |

## 7.5 CONCLUSION

In conclusion, we illustrate that optical pulse–shaping techniques, when applied to arbitrary electromagnetic waveform generation, enable cycle-by-cycle synthesis of a variety of ultrabroadband electromagnetic waveforms spanning the ~1–50 GHz range. Through proper choice of pulse-shaping geometry and O-E converter, our technique provides user-defined waveforms exhibiting arbitrary amplitude, phase, and frequency modulation, in addition to tailored temporal shapes and engineered spectral content. To date, and for the foreseeable future, these capabilities are unique to optical pulse shaper–based arbitrary electromagnetic waveform generators.

Several areas present opportunities for exciting research relative to system implementation, applications, and component development. In particular, novel optical pulse shapers exhibiting multinanosecond time apertures and increased time-bandwidth products would enable these systems to simultaneously achieve lower electrical frequencies and larger electrical bandwidths. Additionally, systems that employ optimization techniques to achieve tailored electrical frequency content [45] could be applied to address spectral occupancy constraints and potentially illustrate better performance than systems utilizing direct spectral specification. The use of photonic arbitrary waveform generators to address signal distortions in UWB systems could be applicable to mitigate multipath interference as well as mitigate antenna or other electrical system distortions. From the component point of view, high-speed parallel optical modulator arrays operating in the gigahertz to tens of gigahertz range could enable reprogrammable electrical (and optical) waveform synthesis on a subnanosecond timescale, thereby providing truly unprecedented agility. In addition, increased photodiode power-handling capability would enable high-fidelity electrical waveforms to be achieved at higher amplitudes via optical (instead of electrical) amplification. Though only a short list, these examples point to the wide variety of research opportunities and capabilities of our technique.

## ACKNOWLEDGMENTS

This work was supported by (or in part by) the US Army Research Office under contracts DAAD19-00-0497 and DAAD19-03-1-0275, the National Science Foundation under grant 0100949-ECS, Tektronix, and Intel. The authors also gratefully acknowledge the work of graduate student Ingrid S. Lin.

## REFERENCES

1. Weiner, A.M., Femtosecond pulse shaping using spatial light modulators, *Rev. Sci. Instrum.*, 71, 1929, 2000.
2. McKinney, J.D., Leaird, D.E., and Weiner, A.M., Millimeter-wave arbitrary waveform generation with a direct space-to-time pulse shaper, *Opt. Lett.*, 27, 1345, 2002.

3. McKinney, J.D., Seo, D.S., and Weiner, A.M., Photonically assisted generation of continuous arbitrary millimetre electromagnetic waveforms, *Electron. Lett.*, 39, 309, 2003.
4. McKinney, J.D., Seo, D.S., Leaird, D.E., and Weiner, A.M., Photonically assisted generation of arbitrary millimeter-wave and microwave electromagnetic waveforms via direct space-to-time optical pulse shaping, *J. Lightwave Technol.*, 21, 3020, 2003.
5. Lin, I., McKinney, J.D., and Weiner, A.M., Photonic synthesis of broadband microwave arbitrary waveforms applicable to ultra-wideband communication, *IEEE Microw. Wireless Compon. Lett.*, 15, 226, 2005.
6. Poinsot, S., Porte, H., Goedgebuer, J.P., Rhodes, W.T., and Boussert, B., Continuous radio-frequency tuning of an optoelectronic oscillator with dispersive feedback, *Opt. Lett.*, 27, 1300, 2002.
7. Yilmaz, T., DePriest, C.M., Turpin, T., Abeles, J.H., and Delfyett, P.J., Toward a photonic arbitrary waveform generator using a mode locked external cavity semiconductor laser, *IEEE Photon. Technol. Lett.*, 14, 1608, 2002.
8. Jalali, B., Kelkar, P., and Saxena, V., Photonic arbitrary waveform generator, in *Pro-c. 14th Annu. Meeting IEEE*, volume 1, 2001.
9. Chou, J., Han, Y., and Jalali, B., Adaptive rf-photonic arbitrary waveform generator, *IEEE Photon. Technol. Lett.*, 15, 581, 2003.
10. Lin, I.S., McKinney, J.D., Toong, F.S., and Weiner, A.M., Microwave arbitrary waveform generation via open-loop, reflective geometry fourier transform pulse shaper, in *Conference on Lasers and Electro-Optics (CLEO)*, volume 96 of OSA Trends in Optics and Photonics, San Francisco, CA, May 16–21, 2004, 427.
11. Ito, H., Kodama, S., Muramoto, Y., Furuta, T., Nagatsuma, T., and Ishibashi, T., High-speed and high-output InPInGaAs unitraveling-carrier photodiodes, *IEEE J. Sel. Top. Quantum Electron.*, 10, 709, 2004.
12. Liu, Y., Park, S.G., and Weiner, A.M., Enhancement of narrowband terahertz radiation from photoconducting antennas by optical pulse shaping, *Opt. Lett.*, 21, 1762, 1996.
13. Liu, Y., Park, S.G., and Weiner, A.M., Terahertz waveform synthesis via optical pulse shaping, *IEEE J. Sel. Top. Quantum Electron.*, 2, 709, 1996.
14. McKinney, J.D., Seo, D.S., and Weiner, A.M., Direct space-to-time pulse shaping at 1.5 μm, *IEEE J. Quantum Electron.*, 39, 1635, 2003.
15. Jiang, Z., Seo, D., Leaird, D.E., and Weiner, A.M., Spectral line-by-line pulse shaping, *Opt. Lett.*, 30, 2005.
16. Xiao, S., McKinney, J.D., and Weiner, A.M., Photonic microwave arbitrary waveform generation using a virtually-imaged phased-array (vipa) direct space-to-time pulse shaper, *IEEE Photon. Technol. Lett.*, 16, 1936, 2004.
17. Leaird, D.E. and Weiner, A.M., Femtosecond direct space-to-time pulse shaping, *IEEE J. Quantum Electron.*, 37, 494, 2001.
18. Sardesai, H.P., Chang, C.C., and Weiner, A.M., A femtosecond code-division multiple-access communication system test bed, *J. Lightwave Technol.*, 16, 1953, 1998.
19. McKinney, J.D. and Weiner, A.M., Engineering of the passband function of a generalized spectrometer, *Opt. Express*, 12, 5022, 2004.
20. Martinez, O.E., Pulse distortions in tilted pulse schemes for ultrashort pulses, *Opt. Commun.*, 59, 229, 1986.

21. Martinez, O.E., Grating and prism compressors in the case of finite beam size, *J. Opt. Soc. Am. B*, 3, 929, 1986.
22. Win, M.Z. and Scholtz, R.A., Ultra-wide bandwidth time-hopping spread-spectrum impulse radio for wireless multiple-access communications, *IEEE Trans. Commun.*, 48, 679, 2000.
23. Lee, J.S., Nguyen, C., and Scullion, T., A novel, compact, low-cost, impulse ground-penetrating radar for nondestructive evaluation of pavements, *IEEE Trans. Instrum. Meas.*, 53, 1502, 2004.
24. Fontana, R.J., Recent system applications of short-pulse ultra-wideband (uwb) technology, *IEEE Trans. Microw. Theory Tech.*, 52, 2087, 2004.
25. First report and order, ET Docket 98–153 (Revision of Part 15 of the Commission's Rules Regarding Ultra-Wideband Transmission Systems), Federal Communications Commission, adopted February 14, 2002, released April 22, 2002.
26. Pozar, D.M., Waveform optimizations for ultrawideband radio systems, *IEEE Trans. Antennas Propagat.*, 51, 2335, 2003.
27. Reed, J.H. (Ed.), *An Introduction to Ultra Wideband Communication Systems*, Prentice Hall, Upper Saddle River, NJ 2005.
28. McKinney, J.D. and Weiner, A.M., Compensation of the effects of antenna dispersion on uwb waveforms via optical pulse shaping techniques, submitted for publication to the *IEEE Trans. Microw. Theory Tech.*, Special Issue on Ultra-Wideband, 2005.
29. Han, J. and Nguyen, C., Ultra-wideband electronically tunable pulse generators, *IEEE Microwave Wireless Compon. Lett.*, 14, 112, 2004.
30. Immoreev, I.Y., Samkov, S., and Tao, T.H., Short-distance ultrawideband radars, *IEEE Aerosp. Electron. Syst. Mag.*, 20, 9, 2005.
31. Sun, F.G., Jiang, Z., and Zhang, X.-C., Analysis of terahertz pulse measurement with a chirped probe beam, *Appl. Phys. Lett.*, 73, 2233, 1998.
32. Franks, L.E., Signal Theory, Prentice Hall, Englewood Cliffs, NJ 1969.
33. Sorokin, E., Tempea, G., and Brabec, T., Measurement of the root-mean-square width and the root-mean-square chirp in ultrafast optics, *J. Opt. Soc. Am. B*, 17, 146, 2000.
34. Papoulis, A., Pulse compression, fiber communications, and diffraction: A unified approach, *J. Opt. Soc. Am. A*, 11, 3, 1994.
35. Han, J. and Nguyen, C., A new ultra-wideband, ultra-short monocycle pulse generator with reduced ringing, *IEEE Microw. Wireless Compon. Lett.*, 12, 206, 2002.
36. McKinney, J.D., Lin, I.S., and Weiner, A.M., Engineering of the radio-frequency spectra of ultrawideband electromagnetic waveforms via optical pulse shaping techniques (poster), in MWP 2004, *Int. Topical Meet. Microw. Photon.*, Ogunquit, ME, Oct. 4–6, 2004.
37. McKinney, J.D., Lin, I.S., and Weiner, A.M., Tailoring of the power spectral density of ultrawideband rf and microwave signals, in *IEEE/LEOS Summer Topical Meetings*, San Diego, CA, July 25–27, 2005.
38. Mallat, S., *A Wavelet Tour of Signal Processing*, Academic Press, San Diego, 1998.
39. Desurvire, E., *Erbium-Doped Fiber Amplifiers: Principles and Applications*, John Wiley and Sons, Inc., New York, 1994.
40. van Exter, M. and Grischkowsky, D.R., Characterization of an optoelectronic terahertz beam system, *IEEE Trans. Microw. Theory Tech.*, 38, 1684, 1990.

41. Lamensdorf, D. and Susman, L., Baseband-pulse-antenna techniques, *IEEE Antennas Propagat. Mag.*, 36, 20, 1994.
42. Shlivinski, A., Heyman, E., and Kastner, R., Antenna characterization in the time domain, *IEEE Trans. Antennas Propagat.*, 45, 1140, 1997.
43. Pozar, D.M., McIntosh, R.E., and Walker, S.G., The optimum feed voltage for a dipole antenna for pulse radiation, *IEEE Trans. Antennas Propagat.*, 31, 563, 1983.
44. Pozar, D.M., Kang, Y.W., Schaubert, D.H., and McIntosh, R.E., Optimization of the transient radiation from a dipole array, *IEEE Trans. Antennas Propagat.*, 33, 69, 1985.
45. Lin, I.S., McKinney, J.D., and Weiner, A.M., Optimization approach to ultrawide-band spectrum design via optical pulse shaping, in OSA Frontiers in Optics/Laser Science XXI Conference, Tucson, AZ, October 16–20, 2005.
46. Kawanishi, S., Takara, H., Uchiyama, K., Shake, I., and Mori, K., 3 tbit/s (160 gbit/s $\times$ 19 channel) optical tdm and wdm transmission experiment, *Electron. Lett.*, 35, 826, 1999.

# 8 Ultrafast Optoelectronic Switching Devices and Applications Based on Optically Injected Free Carriers

*Chi H. Lee*

## CONTENTS

8.1 Introduction ........ 260
8.2 Lightwave Switching in Semiconductor Microring Devices by Free Carrier Injection ........ 262
  8.2.1 Background ........ 262
  8.2.2 Theory ........ 263
  8.2.3 Experiment ........ 264
  8.2.4 Discussion ........ 268
  8.2.5 Phase Switching ........ 270
  8.2.6 Carrier Lifetime ........ 273
  8.2.7 Summary of the Lightwave Switching in Semiconductor Microring ........ 275
8.3 Polarization Switching Dynamics of Thin-Film Ferroelectric Capacitors ........ 275
  8.3.1 Background ........ 275
  8.3.2 Pulse Method and Simulation ........ 276
  8.3.3 Experiment and Results ........ 281
    8.3.3.1 Experiment ........ 281
    8.3.3.2 Experimental Results ........ 284
  8.3.4 Summary on the Polarization Switching Dynamics of Ferroelectric Thin-Film Capacitors ........ 287
8.4 Polymer High-Speed Photoconductive Effect in Thin Films ........ 289
  8.4.1 Background ........ 289
  8.4.2 Experiment ........ 290

8.4.2.1 Sampling Oscilloscope Measurement ..... 290
8.4.2.2 Photoconductive Sampling Using BAMH-PPV Switches ..... 297
8.4.2.3 Result of Photoconductive Sampling Measurements ..... 298
8.4.3 Experimental Results and Discussions ..... 299
8.4.4 Summary on Polymer High-Speed Photoconductive Effect in Thin Films ..... 301
8.5 Conclusion ..... 302
Acknowledgment ..... 302
References ..... 302

## 8.1 INTRODUCTION

The plasma or free carrier state in a semiconductor has received considerable attention in the past, owing to its applications in microwave control components. Essentially, both the dielectric and conductive properties of the plasma in a semiconductor can be used to control the transmission and propagation of an electromagnetic wave. Optical carrier injection can be achieved easily by single-photon absorption if the photon energy is larger than the bandgap energy of the material. The phenomenon that causes the transformation of material property is referred to as the photoconductivity effect. A photoconductivity effect that is initiated by ultrashort optical pulses is referred to as the ultrafast photoconductivity effect [1].

One of the major applications of ultrafast photoconductivity effect has been in ultrafast optoelectronic switching, which can be done with picosecond or even femtosecond precision. The use of ultrafast optoelectronic switching technique in microwave circuits and systems continues to be an area of intense research and development. Rapid advances in both microwave and optical technologies make it possible now to integrate them to form a new class of devices that can perform high-speed and high-frequency microwave photonic functions. With the advent of femtosecond lasers, we envision such devices to work in the picosecond time domain [2].

The ideas of using the photoconductivity effect for the control of microwave solid-state devices have been of great interest to the microwave community. For example, optical control of IMPATT [3], TRAPATT [4], and HEMT and MESFET [5] has been reported. The application of injection-locked oscillators for optically fed phased-array antennas has been reported [6]. The basic operation principle of all these devices is the photogeneration of free electron-hole pairs within the active region of the device. Photoexcited carriers modulate transconductance and other circuit parameters of the devices.

The optical technique for controlling microwave devices offers unique advantages in (1) near-perfect isolation between controlling and controlled devices, (2) elimination of RF feed in a large array system, (3) immunity from electromagnetic interference, (4) extremely fast response, (5) high power-handling capability, and (6) possibility of monolithic integration.

In our laboratory, we have taken advantage of the unique speed capability of the optical pulses. Electrical pulses are instantaneously generated via the ultrafast photoconductivity effect in a photoconductor [7]. Picosecond photoconductors can now be engineered to generate electrical pulses shorter than 1 ps [8] with jitter-free switching. All subsequent electrophotonic events can be slaved to this initial pulse. Using picosecond optical electronic techniques, we are able to generate microwave and millimeter-wave signals that are in complete time synchronization with the exciting optical pulses. Radio frequency (RF) waveform generation, ranging from one monocycle [9] to Continuous wave (CW), with peak power up to a few megawatts in the pulsed case and frequency up to 20 GHz in the CW case [10], has been achieved. Phase control of the millimeter-wave signals can be accomplished by simply controlling the arrival time of the exciting optical pulses, which offers the potential for a true time-delay phased-array radiation.

In this chapter we will present recent advances in the field of ultrafast optoelectronic switching devices and applications based on optically injected free carriers. In Section 8.2, we will present work on lightwave switching in semiconductor microring devices by free carrier injection. Using a conventional microwave photonic switching technique, we demonstrate lightwave switching in laterally coupled GaAs–AlGaAs microring resonators by free carrier injection. The ring waveguide is optically pumped just above its band gap energy, which results in a temporal tuning of the microring resonant wavelengths by the refractive index change due to the induced free carriers. Both the transmission and the phase function of the resonators are investigated and used to demonstrate all-optical switching. The switching time is 20 ps, limited by surface recombination of carriers. A switching energy of a few picojoules shifts the microring resonance by 1.2 nm.

In Section 8.3, we will present the application of an ultrafast photonic technique to study polarization switching dynamics of thin-film ferroelectric capacitors. We demonstrate how to use fentosecond laser pulses to generate and control ultrafast rise-time electric pulses. These very special electric pulses are then used to study the fast polarization switching process in fully integrated ferroelectric capacitors. Using these techniques, an experimentally measured switching time $t_s = 200$ ps was obtained and the characteristics switching time $t_o = 70$ ps was deduced. This is the shortest switching time reported to date.

Organic thin-film devices have played an important role in electronics and photonics. In Section 8.4, we will present our current study of ultrafast photoconductive effect in poly 2,5-*bis*[*N*-methyl-*N*-hexylamino] phenylene vinylene (BAMH-PPV) polymer thin films. The carrier lifetime in BAMH-PPV is measured by photoconductive sampling. The results of a systematic study of the influence of both external bias electric field and illumination intensity on the photoconductance of the polymer will be reported. Electric pulses with a pulse width of 2 ps were generated by the polymer photoconductor. The transient carrier mobility is found to be much higher than that of the steady-state value. For 93 nm thick film, the lower limit is estimated to be 0.24 ($cm^2/V$ s) when assuming unity quantum efficiency. This transient mobility value could be 480 ($cm^2/V$ s) when taking quantum efficiency to be 0.05%.

## 8.2 LIGHTWAVE SWITCHING IN SEMICONDUCTOR MICRORING DEVICES BY FREE CARRIER INJECTION

### 8.2.1 BACKGROUND

All-optical switching devices operating at the 1.55 μm wavelength are necessary components for future high-speed optical telecommunication systems. Therefore, the need for a fast, low threshold, compact, and all-optical switch has drastically increased in the last few years. In general, semiconductor optical switching devices have been the main focus of the ongoing research, because semiconductors have shown large nonlinearities and they can be integrated with other optoelectronic devices, such as microlasers and amplifiers. One main factor would limit the performance of a semiconductor straight waveguide optical switch is its limited interaction length [11,12]. This problem has been partially solved by the introduction of semiconductor microring devices, because the interaction length of the device is effectively enhanced by the finesse of the microring cavity [13]. In this type of device, the phase-shift of the propagating lightwave in the microring is caused by the free carrier generation, resulting from the absorption of photons in the optical pump beam. The increased density of the free carriers will cause a refractive index change for the propagating wave, leading to a phase-shift. This effect is similar to that in the millimeter-wave case, as demonstrated in Ref. [14]. Both the lightwave signal to be controlled and the controlling pump beam have to be coupled into the microring first. This can be achieved by tuning the wavelengths of both the signal and the pump laser to the resonant wavelengths of the microring modes. The subsequent optical absorption of the pump beam has to be via two-photon absorption in order that the free carriers can be uniformly generated throughout the entire microring. These requirements for the precise tuning of wavelengths for coupling purpose are disadvantages in the application of optical switching. One simple solution is to illuminate the pump beam from above the microring. It does not require wavelength tuning of the pump beam as long as the photon energy of the pump beam is greater than the band gap energy of the semiconducting material of the microring. The photoabsorption process is by single-photon absorption. If the semiconductor is a direct band gap material, the absorption can be very efficient, with a possibility of 100% quantum efficiency. The free carriers generated in this case will be near the surface facing the incoming pump beam. The carrier density will be much higher than that generated by the two-photon absorption process. It does not matter that the carriers are confined to the surface within the absorption depth of the pump beam because the increased carrier density will cause a refractive index change that will subsequently lead to the phase-shift of the signal wave propagating inside the microring. If the pump beam is a femtosecond laser beam, the optical switching can be ultrafast. This optoelectronics technique for free carrier injection has become a conventional microwave photonic method for optically controlled microwave devices [1]. The optical technique for controlling microwave devices and circuits offers unique advantages in: (1) near-perfect isolation between the controlling and controlled devices, (2) elimination of RF feed in a large array system, (3) immunity from electromagnetic interference, (4) high power-handling capability, (5) light weight and compact size, (6) extremely fast response, and (7) possibility of monolithic integration. Therefore, it has been studied extensively in the millimeter-wave region. It is a simple matter to extend the theory to the optical region. I shall review the theory and experiment of all-optical switching by injection of free carriers by optical means.

### 8.2.2 Theory

Using Figure 8.1, one can analyze the optical fields at the throughput and drop ports. The electric field circulating inside the microring resonator can be expressed as

$$E_1(t) = -j\kappa_1 E_0 + r_1 r_2 a E_1(t-\tau)\mathrm{e}^{j\phi}, \tag{8.1}$$

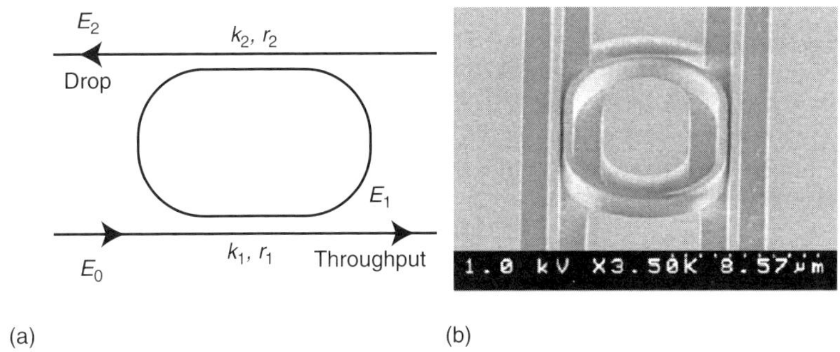


**FIGURE 8.1** (a) Schematic diagram and (b) SEM picture of the microring resonator. (From T.A. Ibrahim et al., *IEEE J. Lightwave Tech.* 21, 2997–3003, 2003. With permission.)

where $E_0$ is the input cw probe beam, which represents the lightwave to be switched, $r_1$ and $r_2$ are the field transmission coefficients at the input and output waveguides, respectively, $a$ is the field round-trip loss, $\kappa_1$ is the field coupling coefficient between the input waveguide and the ring waveguide such that $\kappa_1^2 + r_1^2 = 1$, and $\tau$ is the round-trip time of the resonator. The resonator round-trip phase $\phi$ is related to the refractive index of the waveguide. The refractive index is a function of density of the injected free carriers in the semiconductor waveguide. By injecting free carriers with optoelectronic switching technique, one can control the phase of the circulating field. This phase $\phi$ is given by

$$\phi = \frac{2\pi}{\lambda} nL, \tag{8.2}$$

where $\lambda$ is the probe beam wavelength, $n$ is the refractive index of the ring waveguide, and $L$ is the circumference of the ring resonator. The ratio of the field at the drop port to the input field at steady state is given by the following equation:

$$\frac{E_2}{E_0} = \frac{r_1 - ar_2 e^{j\phi}}{1 - ar_1 r_2 e^{j\phi}}. \tag{8.3}$$

It is clear that the field at the drop port is a nonlinear function of the phase $\phi$ and hence the free carrier density. The dynamic behavior of the output field can be solved numerically by substituting Equation 8.1 and Equation 8.2 into Equation 8.3. The device used for the demonstration of optical switching is a single race track laterally coupled to two straight waveguide buses with a straight coupling section of 10 μm, as shown in Figure 8.1b [15]. This is the same device used by Van et al. [16] to demonstrate the nonlinear optical switching via two-photon absorption [17].

As the input wavelength of the probe beam (the lightwave signal that is to be switched) is tuned, the spectral response of the ring resonator is obtained. One such mode is shown in Figure 8.2. It displays the measured spectral response of the microresonator at both the throughput and drop ports at its transverse electric (TE) resonance mode at 1543.4 nm. By fitting both the throughput and drop port spectral responses with the assumption of symmetrical couplings at the input and output waveguides, the microring linear parameters could be extracted. The round-trip power loss is 24% and the coupling coefficient is 11%. The ring has a 3 dB bandwidth of 1.3 nm, an intrinsic quality factor of 1200, a free spectral range of 18 nm, and a finesse of 14.

### 8.2.3 Experiment

All-optical switching is demonstrated by using the pump–probe scheme, in which a high-intensity pump pulse is used to control the transmission of a CW

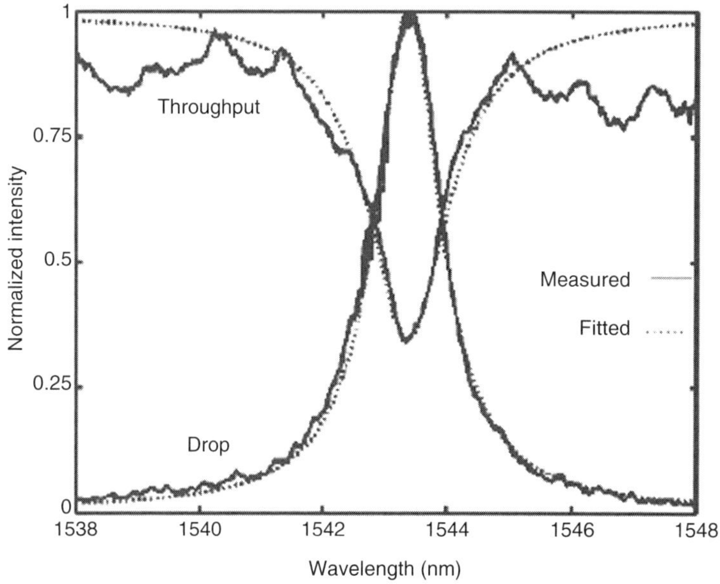


**FIGURE 8.2** Measured (solid lines) and fitted (dotted lines) spectral response at the throughput and drop ports of the microring resonator around 1543.4 nm. The discrepancy between the measured and fitted throughput spectrums is because of the resonances of the Fabry–Perot cavity formed by the cleaved facets of the throughput waveguide. (From T.A. Ibrahim et al., *IEEE J. Lightwave Tech.* 21, 2997–3003, 2003. With permission.)

probe beam tuned to the vicinity of one of the microring resonances. For example, if the initial probe beam is tuned to the edge of the resonant line, the pump beam induced refractive index change may sweep the line dynamically into resonance, resulting in a "switch-on" function. Conversely, it may lead to a "switch-out" effect. By sweeping the probe beam wavelength across the microring resonance and detecting the time variation of the dropped signal, the resonance wavelength shift of the microring and hence its refractive index change can be measured. The experimental setup is shown in Figure 8.3. The pump beam is a train of mode-locked Ti:sapphire pulses at 800 nm. The pulse width is typically 100 fs with 1 kHz repetition-rate. The average power of the beam is attenuated to less than 1 $\mu$W at the device surface. A lens system is used to focus the beam to about 50 $\mu$m spot size and to pump the microring from the top. The probe beam is provided by an external cavity semiconductor laser diode that is tunable around 1550 nm. The laser diode is tuned to the microring resonance wavelength at 1543.4 nm with an average power of a few milliwatts. The probe beam is fed into the input port of the resonator using a conically tipped fiber. The insertion loss is typically 7 dB per facet.

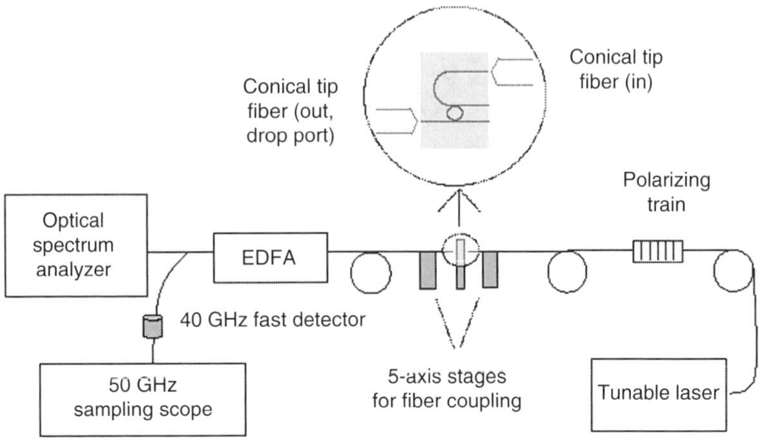


**FIGURE 8.3** Experimental setup for all-optical switching of the microring resonator. The femtosecond laser beam (not shown in the figure) is illuminated from above the microring to actuate the switching function.

The output of the drop port is also collected with another conically tipped fiber, optically amplified, fed to a 40 GHz detector and monitored by a 50 GHz scope.

Since the bandgap wavelength of GaAs is around 800 nm, the pump pulse is almost fully absorbed in the microring waveguide and high-density free carriers are generated. These carriers decrease the refractive index of the microring waveguide and cause a temporarily blue shift of the microring resonance wavelength. When the free carriers diffuse to the waveguide walls and recombine, the effect diminishes. If a probe beam is tuned to one of these microring resonant wavelengths, the output probe signal will be dynamically modulated. The depth of modulation depends on the magnitude of the resonance-wavelength shift. In addition, the steeper the resonator profile, or in other words, the higher the cavity finesse, the larger the modulation of the output signal for a given wavelength shift.

Figure 8.4a shows a three-dimensional plot of the time variation of the dropped probe signal versus its wavelength. When the probe beam was initially tuned to the shorter wavelength side of the microring resonant wavelength, the carrier-induced blue shift in the resonance brings the probe beam into resonance and is temporarily dropped at the output, thus it experiences a rapid increase in transmission. When the probe beam was initially tuned to the resonant wavelength, the carrier-induced index change shifts the resonance away from the probe beam, resulting in a rapid attenuation of the dropped signal.

Simulation results shown in Figure 8.4b are in good agreement with measured data. By measuring the switching contrast of the dropped beam at

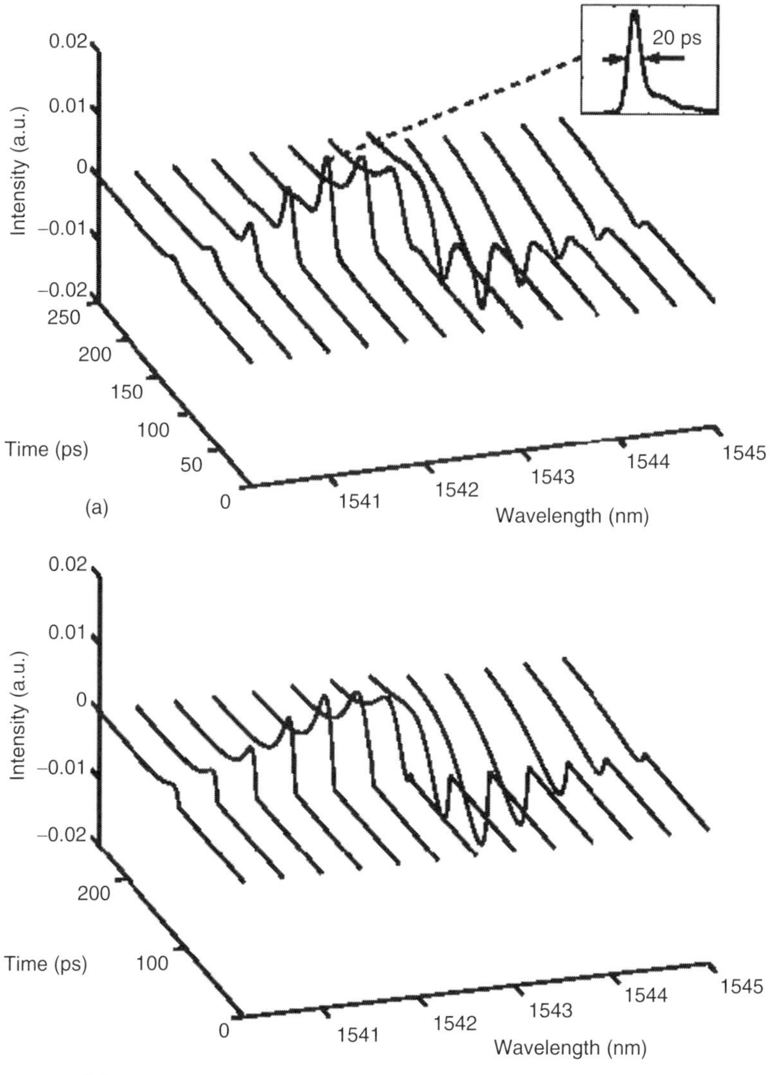


**FIGURE 8.4** (a) Measured and (b) simulated modulation of the dropped signal for different input probe beam wavelength (DC level has been removed in both (a) and (b) for clarity). (From T.A. Ibrahim et al., *IEEE J. Lightwave Tech.* 21, 2997–3003, 2003. With permission.)

the different probe wavelengths, one can deduce the amount of resonance wavelength shift. Our simulation model yields a maximum refractive index change of $2.6 \times 10^{-3}$ that leads to a resonance wavelength shift of 1.2 nm.

The switching contrast is approximately 7 dB. Higher switching contrast can be obtained by injecting more carriers or by using a critically coupled resonator [18]. A rise-time of 10 ps is measured for the dropped probe beam and is limited by the detection system. A switching window of 20 ps is obtained and shown in the inset of Figure 8.4a.

### 8.2.4 Discussion

When carriers are injected into a GaAs waveguide, its refractive index has the following expression [14,19]:

$$\eta, k = \frac{1}{2}\left\{ \pm\left(\varepsilon_{\mathrm{L}} - \frac{\omega_{\mathrm{p}}^2}{\omega^2 + \upsilon^2}\right) + \left[\left(\varepsilon_{\mathrm{L}} - \frac{\omega_{\mathrm{p}}^2}{\omega^2 + \upsilon^2}\right)^2 + \left(\frac{\upsilon}{\omega}\frac{\omega_{\mathrm{p}}^2}{\omega^2 + \upsilon^2}\right)^2\right]^{\frac{1}{2}}\right\}^{\frac{1}{2}} \tag{8.4}$$

in which $\eta$ and $k$, the real and imaginary parts of the complex refractive index, are obtained by selecting the "+" or "−" sign of the leading term of the right-hand-side of Equation 8.4, respectively. Equation 8.4 shows that the refractive index depends on the dielectric constant of the material $\varepsilon_{\mathrm{L}}$, the plasma frequency $\omega_{\mathrm{p}}$, the collision frequency of the carriers $\upsilon$, and the frequency of the propagating probe beam $\omega$. The plasma frequency is given by

$$\omega_{\mathrm{p}}^2 = \frac{n_{\mathrm{p}} e^2}{\varepsilon_0 m^*}, \tag{8.5}$$

where $n_{\mathrm{p}}$ is the free carrier density, $e$ is the free electron charge, $\varepsilon_0$ is the free space permittivity, and $m^*$ is the effective mass of the carrier. The effective pump energy that hits the microring is 20 pJ/pulse and this corresponds to $8 \times 10^7$ photons/pulse. Since the absorption coefficient of GaAs at 800 nm is 104 $\mathrm{cm}^{-1}$, a carrier density of $2.5 \times 10^{18}\ \mathrm{cm}^{-3}$ is expected. In Figure 8.5a we plot Equation 8.4 and in Figure 8.5b we plot the real part of refractive index change versus the injected free carrier density at a probe beam wavelength of 1543.4 nm.

For a carrier density of $2.5 \times 10^{18}\ \mathrm{cm}^{-3}$, a refractive index change of $3 \times 10^{-3}$ is estimated, which is consistent with the value obtained from the experiment. Figure 8.5a shows that the change in the imaginary part of the refractive index at that value of carrier density is in the order of $10^{-5}$. Thus the additional loss due to the free carriers can be neglected.

One aspect to be addressed here is the pump signal wavelength. Two-photon absorption requires that the pump beam signal to be tuned exactly to the microring resonance wavelength so that it can be coupled into the ring. This restricts the switching speed of the device to that of the device bandwidth. Moreover, as the resonance is tuned in the presence of the pump signal,

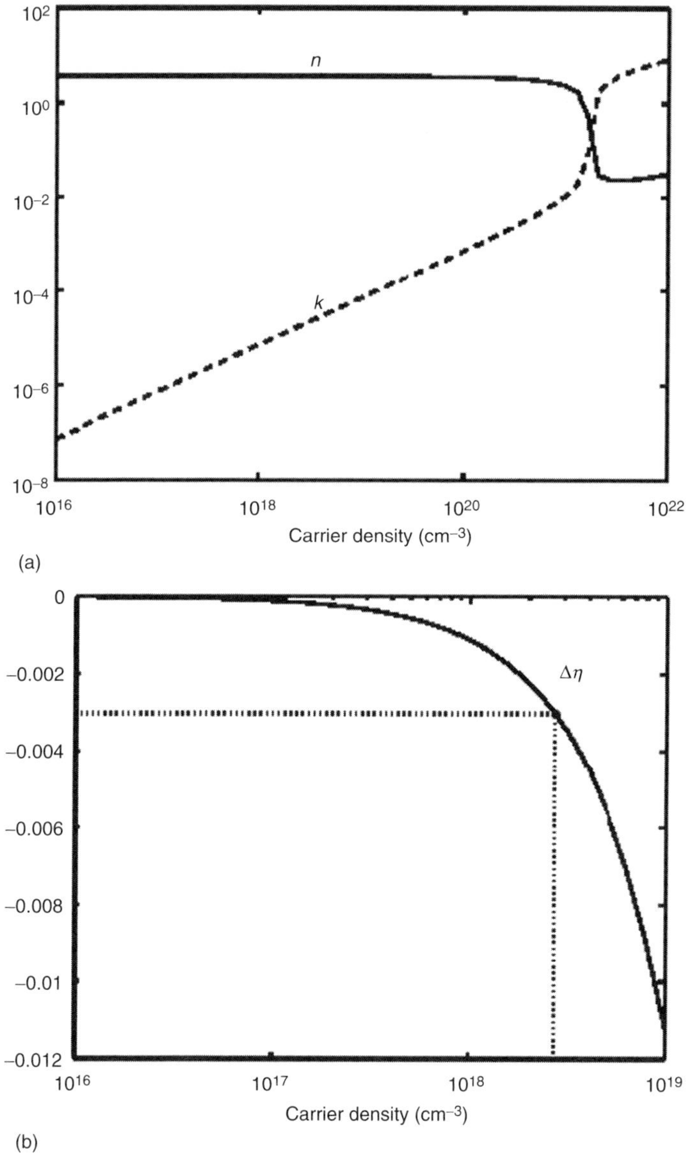


**FIGURE 8.5** (a) Real (solid) and imaginary (dotted) parts of the refractive index and (b) real part of the refractive index change versus carrier density for the microring GaAs waveguide at the probe wavelength. (From T.A. Ibrahim et al., *IEEE J. Lightwave Tech.* 21, 2997–3003, 2003. With permission.)

the pump signal itself becomes out of resonance and the dynamic performance of the device is degraded. Single-photon absorption mechanism with the beam illuminated from the top of the ring does not suffer from the above disadvantage because the pump beam wavelength is not tuned to the device resonance, which allows ultrafast switching to take place and gives more flexibility in terms of the pump source wavelength.

### 8.2.5 Phase Switching

So far we have only discussed the use of the transmission-transfer function of the microring resonator for switching. However, an examination of its phase-transfer function reveals a similar nonlinear enhancement. A microring coupled to a single waveguide, as shown in Figure 8.6a, forms a phase filter. The phase-shift of a signal after passing through the ring is given by [20]

$$\phi_{\text{eff}} = \pi + \phi + \tan^{-1}\frac{r\sin\phi}{a - r\cos\phi} + \tan^{-1}\frac{ar\sin\phi}{1 - ar\cos\phi}, \tag{8.6}$$

where $r$ is the field transmission coefficient and $\phi$ is the round-trip phase of the resonator, defined by Equation 8.2. In Figure 8.6b, we plot Equation 8.6 for different values of the field coupling coefficient, $\kappa = (1 - r^2)^{1/2}$, for a lossless resonator ($a = 1$). Near resonance ($\phi \approx 0$), the slope of the effective phase becomes very steep, and hence the device output phase is very sensitive to changes in the single-pass phase-shift of the microring. This phase enhancement can be observed by introducing the microring resonator into one arm of a Mach–Zehnder interferometer (MZI). Although an MZI requires a $\pi$ phase-shift for switching, Figure 8.6b shows that such an amount of phase-shift can be obtained with a drastically reduced single-pass phase-shift of a microring resonator introduced on one arm of the MZI. In other words, the amount of index change required to switch the MZI is dramatically reduced. It is worth mentioning here that the propagation loss of the microring has to be minimum for high interference visibility between the two arms of the MZI and hence efficient switching. Figure 8.7 shows a SEM picture of an MZI loaded with a microring resonator. The spectral response of the device is shown in Figure 8.8. The device has an intrinsic quality factor of 2060 and an estimated finesse of 18.

We used the same experimental setup mentioned earlier to pump the device from the top. In Figure 8.9, we plot the off–on switching behavior of the device, when the probe beam was initially tuned to the microring resonance at 1556 nm. A switching window of 24 ps is obtained, allowing 40 GHz data processing. An effective energy of 4 pJ/pulse is used to switch the device, which is one order of magnitude less than that required by two-photon absorption [16].

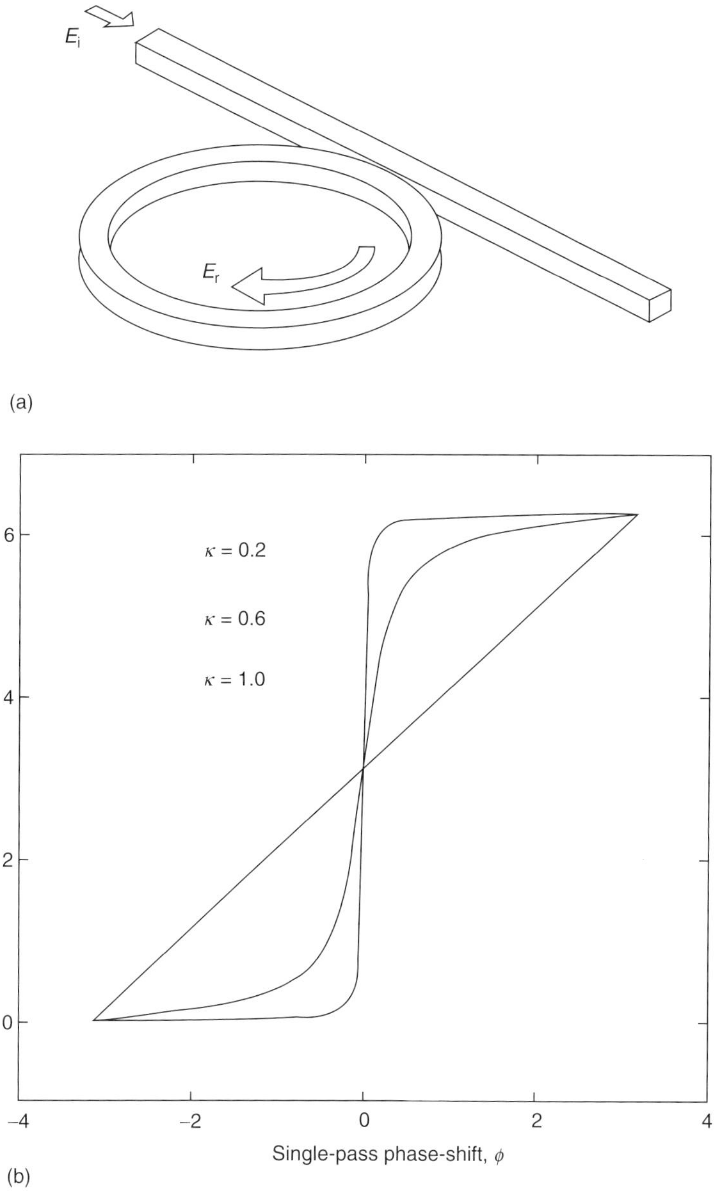


**FIGURE 8.6** (a) Schematic diagram of a ring resonator coupled to a straight waveguide, and (b) phase enhancement of a lossless resonator for different coupling coefficient, $\kappa$. (From T.A. Ibrahim et al., *IEEE J. Lightwave Tech.* 21, 2997–3003, 2003. With Permission.)

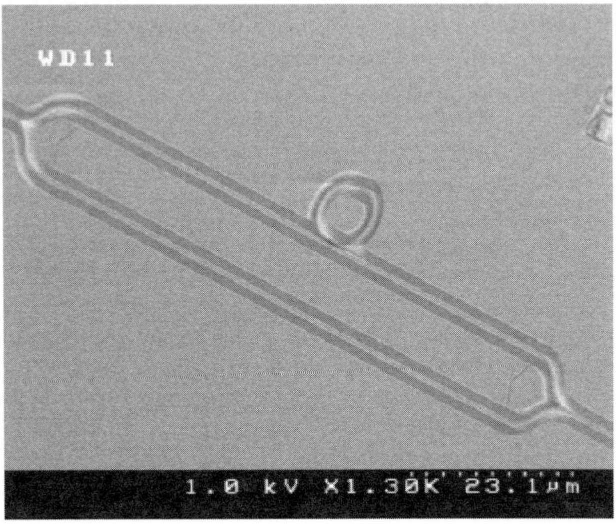


**FIGURE 8.7** SEM picture of the MZI loaded with a microring resonator device. (From T.A. Ibrahim et al., *IEEE J. Lightwave Tech.* 21, 2997–3003, 2003. With Permission.)

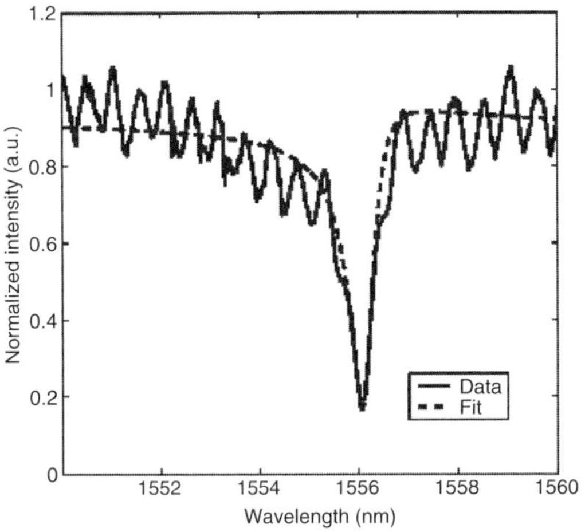


**FIGURE 8.8** Measured (solid line) and fitted (dashed line) spectral response of the MZI loaded ring device around its resonance wavelength at 1556 nm. The ringing in the measured spectrum is due to the high reflectivity of the uncoated input and output cleaved facets of the device that acts as a Fabry–Perot cavity. (From T.A. Ibrahim et al., *IEEE J. Lightwave Tech.* 21, 2997–3003, 2003. With permission.)

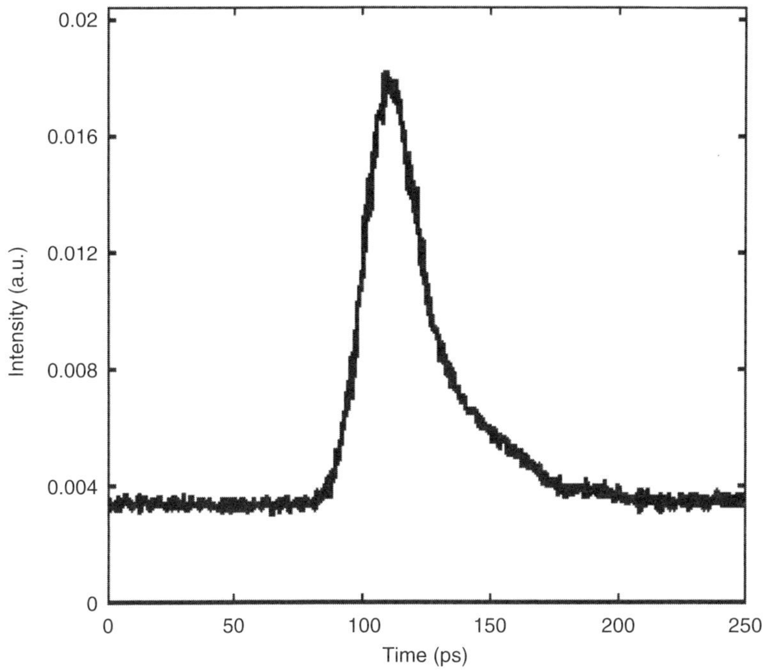


**FIGURE 8.9** The off–on switching behavior of the MZI loaded ring device when the probe beam was initially tuned to the microring resonance at 1556 nm. (From T.A. Ibrahim et al., *IEEE J. Lightwave Tech.* 21, 2997–3003, 2003. With permission.)

### 8.2.6 Carrier Lifetime

The speed of these devices is mainly limited by the carrier lifetime of the guiding material. To measure the carrier lifetime independently, we optically pumped a tapered straight waveguide at different width sections. We increased the pumping energy so that enough carriers were injected into the waveguide and absorption was observed in a probe beam. By detecting the intensity variation of the probe beam, we could measure the recovery time of the transmitted signal and hence the carrier lifetime. In a tightly confined waveguide, diffusion to the waveguide walls dominates, where fast surface recombination occurs. In Figure 8.10a, we plot the normalized transmitted probe intensity when the waveguide was pumped at different cross sections and in Figure 8.10b, we plot the carrier lifetime versus its diffusion length, which is approximately half the waveguide width. As can be seen, the carrier lifetime varies quadratically with the diffusion length as expected because of the diffusion processes. Since the guiding layer is made of intrinsic bulk material, ambipolar diffusion takes place where both electrons and holes contribute to the diffusion process. Using the Einstein relation, we fitted the

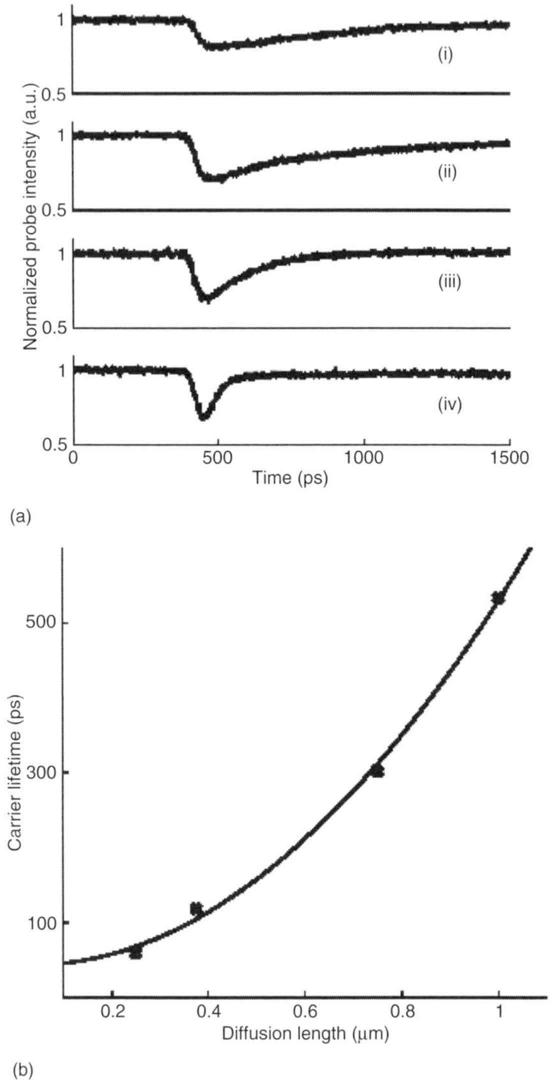


**FIGURE 8.10** (a) Normalized transmitted probe intensity when the tapered straight waveguide was optically pumped at its width of (i) 2.0 fm, (ii) 1.5 μm, (iii) 0.8 μm, and (iv) 0.5 μm. (b) Measured and fitted carrier lifetime of the GaAs waveguide versus the carrier diffusion length. (From T.A. Ibrahim et al., *IEEE J. Lightwave Tech.* 21, 2997–3003, 2003. With Permission.)

carrier lifetime curve and extracted the ambipolar diffusion constant to be 19.2 $cm^2/s$. The ambipolar diffusion constant calculated from published values of electrons and holes mobilities and diffusion constants of GaAs [21] is found to be 19.4 $cm^2/s$, which is very close to the measured value. Faster response time can be obtained by DC-biasing the microresonator waveguide, where a static electric field normal to the epitaxial layer is induced and rapidly sweeps the free carriers out of the waveguide core [22].

### 8.2.7 Summary of the Lightwave Switching in Semiconductor Microring

In summary for this section, we have shown all-optical switching by carrier injection induced by single-photon absorption in laterally coupled GaAs-AlGaAs microring resonators. Simulation results have been shown to be in good agreement with measured data. We investigated both the transmission and phase functions of the resonator using two different device configurations. The switching energy is found to be at least one order of magnitude less than that required by two-photon absorption. These devices have great potential for switching, routing, multiplexing and demultiplexing, and reshaping. Optical pumping can be accomplished by integrated vertical cavity surface emitting semiconductor lasers (VCSELs) underneath the microrings or by an array of laser diodes mounted on top of the microrings. The switching speed is only limited by the carrier lifetime and can be enhanced by DC-biasing the microring waveguide to sweep out carriers.

## 8.3 POLARIZATION SWITCHING DYNAMICS OF THIN-FILM FERROELECTRIC CAPACITORS

### 8.3.1 Background

Ferroelectric materials exhibit a hysterisis loop of polarization as a function of electric field. The application of ferroelectric (FE) capacitors to FE nonvolatile random accesses memories (FeRAMs) is based on their two remnant polarization states and the capability of switching from one polarization state to the other by applying a short voltage pulse [23,24]. The $+P_r$ and $-P_r$ polarizations can be used as logic "1" and "0", respectively. The word nonvolatile indicates that the memory can preserve the stored data information even when the power is off. The study of the intrinsic polarization switching time of a FE material is important from both the physics and application points of view. For physics, it is a fundamental question on how fast the polarization can be switched. For applications, a faster polarization switching time leads to a faster *write* and *read* operation of a memory device. To obtain the intrinsic polarization switching time, we have to understand the microscopic mechanisms of the polarization switching process. One method used to study the polarization switching process is the so-called "pulse method," in which the temporal waveform of the displacement current flowing through the FE capacitor is measured. However, the measurement of the polarization switching time in the pulse method is hindered by a variety of circuit-related delays such as the rise-time of the input electric pulse, the timing jitter, and the *RC* time constant of the circuit. To obtain the intrinsic polarization switching time, the rise-time of the input electric pulse and the $RC_{FE}$ time constant must be at least on the same order of magnitude as the intrinsic switching time. Experimentally, the rise-time of the input electrical pulse, which is a critical limiting factor of the pulse method, is of great concern. The faster the rise-time is, the higher the temporal resolution. The ideal input electric pulse for the experiment is a step-function pulse with an extremely fast rise-time and constant amplitude thereafter. Such an electric pulse is not readily attainable by electronic techniques. However, the microwave photonic technique using a photoconductive switch, actuated by an ultrashort laser pulse, is perfectly suited for producing such a pulse [1,2]. In this chapter, we demonstrate the utilization of jitter-free, ultrafast rise-time electrical pulses generated by a semiconductor photoconductive switch with femtosecond laser illumination, to study the fast polarization switching process in the fully integrated ferroelectric $Pb(Nb,Zr,Ti)O_3$ capacitors. A polarization switching time $t_s$ of $\cong$220 ps is obtained for a $4.5 \times 5.4\ \mu m^2$ PNZT capacitor. Prior to our work, the shortest polarization switching time reported was 390 ps by Larsen et al. [25,26]. Modeling of the switching transients using the Ishibashi–Merz model gives a characteristic switching time $t_0$ of $\cong$70 ps, approaching the theoretically predicted intrinsic polarization switching time of 50 ps.

### 8.3.2 Pulse Method and Simulation

The pulse method is used to study the polarization switching process in FE materials. The polarization switching speed can be measured directly. In this section, we describe in detail the pulse method and present a numerical analysis of the measurement circuit. The equivalent circuit of the pulse method is shown in Figure 8.11. It consists of a pulse generator, an FE capacitor under test, and a sampling oscilloscope with a 50 $\Omega$ input impedance. The pulse generator provides an input positive–positive or positive–negative pulse train, as shown in Figure 8.12, where the first pulse is the *write* pulse and the second pulse is the *read* pulse. The pulses go through the FE capacitor and the oscilloscope is used to monitor the displacement current flowing through the capacitor. Figure 8.13 shows an expected output signal on the oscilloscope. With a positive–negative or positive–positive input, current transients corresponding to the polarization switched case ($P^*$) and the polarization nonswitched case ($P^\wedge$) are observed, respectively. By subtracting the $P^\wedge$ output from the $P^*$ output, we obtain $\Delta P$, which is only related to the domain switching and is of great interest.

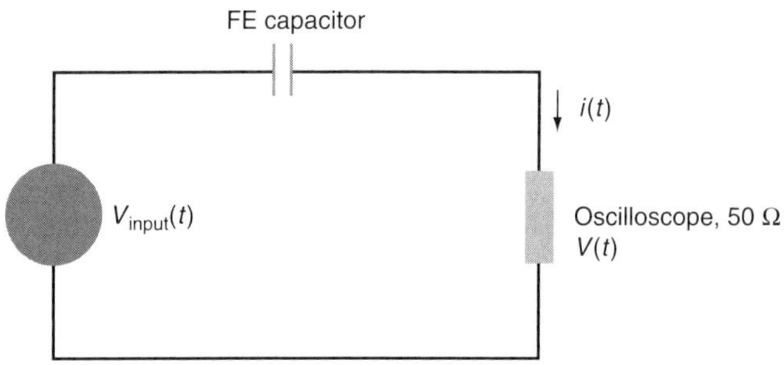


**FIGURE 8.11** Equivalent circuit of the "pulse method." (From J. Li et al., *IEEE J. Lightwave Tech.* 21, 3282–3291, 2003. With permission.)

To study the polarization switching process using the pulse method, we must first answer the following question. What circuit parameters are critical for accurate measurement of the intrinsic polarization switching time? This is the motivation for conducting circuit simulations of the pulse method, which will be presented in the following.

Assuming that the capacitor is a linear dielectric one, the ordinary differential equation of the circuit is

$$i(t)R + \frac{\int_0^t i(t)\,\mathrm{d}t}{C} = V_{\text{input}}(t), \tag{8.7}$$

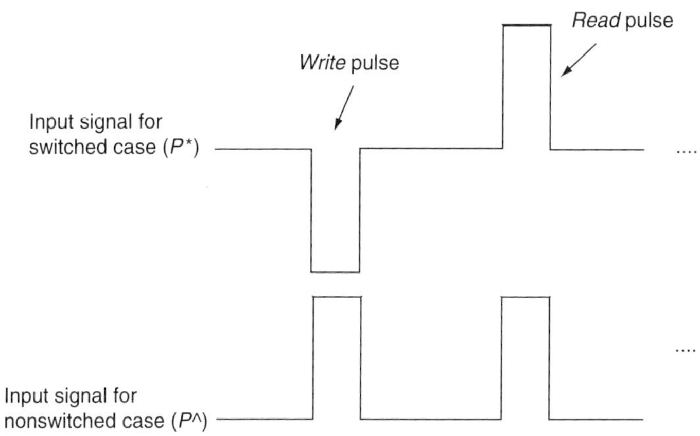


**FIGURE 8.12** Illustration of the input positive–positive or positive–negative pulse trains of the pulse method.

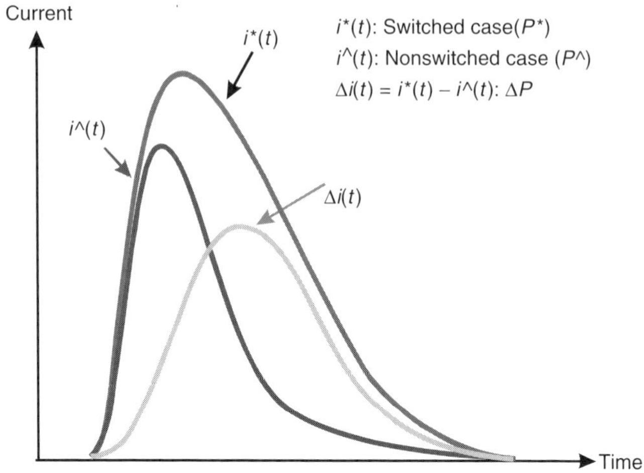


**FIGURE 8.13** The expected current transients shown on the oscilloscope using the pulse method.

where $i(t)$ is the current flowing through the oscilloscope and also the displacement current flowing through the capacitor. The oscilloscope is represented by a 50 Ω resistor $R$. $C$ is the capacitance of the dielectric capacitor. $V_{\text{input}}(t)$ is the step-like input electric pulse to the capacitor. The displacement current $i(t)$ is directly related to $V(t)$, which is the output response displayed on the oscilloscope, or specifically the voltage drop across the 50 Ω node, by

$$i(t) = \frac{V(t)}{R}. \tag{8.8}$$

The voltage drop across the capacitor is given by

$$V_{\text{cap}}(t) = V_{\text{input}}(t) - V(t) = \frac{\int_0^t i(t)\,\mathrm{d}t}{C}. \tag{8.9}$$

Equation 8.7 can be solved using a Laplace transform if $V_{\text{input}}(t)$ is an ideal step function. Realistically, the rise-time of $V_{\text{input}}(t)$ can never be zero. A numerical calculation is required to solve this equation. Simulations are carried out using the Runga–Kutta method. In simulations, the input step-like electric pulse ($V_{\text{input}}(t)$) is mathematically modeled as

$$V_{\text{input}}(t) = A * \exp\left[-\exp\left(-k(t - t_{\text{c}})\right)\right], \tag{8.10}$$

where $A$ controls the amplitude, $k$ controls the rise-time, and $t_c$ controls the position. The selection of this model is based on the fact that our experimentally applied input electric pulse can be best fitted with this model.

Figure 8.14 illustrates an experimentally obtained hysteresis loop of a thin-film PNZT capacitor along with parameters of concern. The simulation of a nonlinear FE capacitor is accomplished in the following way. From Figure 8.14, the polarization reversal process ($P^*$) consists of both the domain polarization switching process ($\Delta P$) and the linear dielectric polarization variation process ($P^\wedge$), which is the charging or discharging of the capacitor. To simplify our analysis, the $P^*$ process is modeled using a simple dielectric capacitor with capacitance $C_{\text{nonlinear}}$,

$$C_{\text{nonlinear}} = \frac{P^* \cdot A_{\text{FE}}}{V_{\text{applied}}}, \tag{8.11}$$

while the $P^\wedge$ process is modeled by another linear dielectric capacitor with capacitance $C_{\text{linear}}$,

$$C_{\text{linear}} = \frac{P^\wedge \cdot A_{\text{FE}}}{V_{\text{applied}}}. \tag{8.12}$$

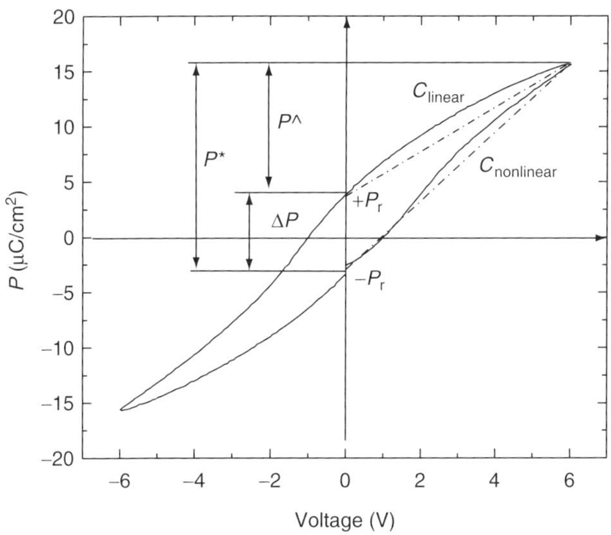


**FIGURE 8.14** An experimentally obtained hysteresis loop of a thin-film PNZT capacitor. $P^\wedge$ and $P^*$ correspond to the nonswitched and switched polarization, respectively. (From J. Li et al., *IEEE J. Lightwave Tech.* 21, 3282–3291, 2003. With permission.)

where $A_{FE}$ is the area of the capacitor. The values of $P^*$ and $P^\wedge$ can be obtained directly from Figure 8.14. Giving $V_{input}(t)$, $C_{nonlinear}$, and $C_{linear}$, voltage responses corresponding to the $P^*$ and $P^\wedge$ processes, which are denoted by $V(t)_{P*}$ and $V(t)_{P\wedge}$, respectively, can be simulated. The $\Delta V(t)$ response, which corresponds only to the domain polarization switching $\Delta P$ process, is given by

$$\Delta V(t) = V(t)_{P*} - V(t)_{P\wedge}. \tag{8.13}$$

$\Delta V(t)$ is a very important quantity because it ultimately determines the polarization switching speed, dictating how fast an FE memory device can be. It may also provide information of the domain polarization switching dynamics with the hclp of a certain theoretical model (i.e., the Ishibashi–Merz model, which will be introduced in detail in a later section).

The area of the capacitor and the rise-time of the input electric pulse are two parameters characterizing the measurement circuit. Therefore, simulations are carried out to study their effects on the polarization switching, specifically, on the output response $V(t)$. Figure 8.15a and Figure 8.15b show the output responses as a function of time for various capacitances and rise-times, respectively. An output response with a longer tail is expected for a larger capacitance due to the increase of the circuit $RC_{FE}$ time constant, indicating the need for small-size capacitors to realize the measurement of the ultrafast polarization switching process. Similarly, a slow rise-time of the input pulse results in a slow response. Even if there does exist an ultrafast polarization switching process, the slow rise-time makes it impossible to resolve this fast process that is embedded within. This is the exact reason why a step-like electric pulse with a fast rise-time is necessary. Figure 8.15a

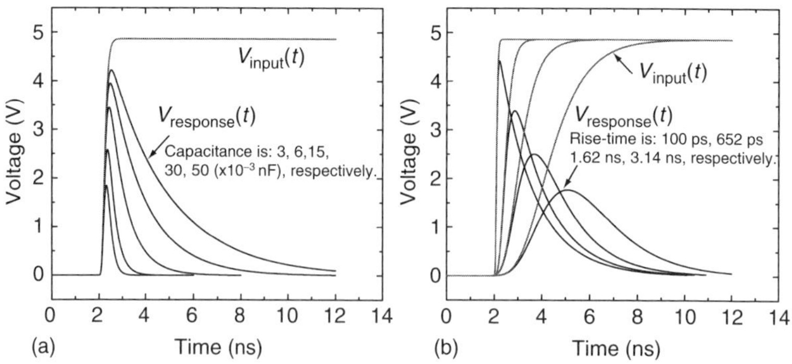


**FIGURE 8.15** Simulated output responses $V(t)$ for various (a) capacitances and (b) rise-times of the input electric pulse $V_{input}(t)$. In (a), the rise-time of the input electric pulse is kept constant at 300 ps. In (b), the capacitance is kept constant at $3 \times 10^{-2}$ nF. (From J. Li et al., *IEEE J. Lightwave Tech.* 21, 3282–3291, 2003. With permission.)

and Figure 8.15b also show that the shape of the output response changes with the capacitance and the rise-time. With an increase in the rise-time (keeping capacitance constant) or a decrease in the capacitance (keeping rise-time constant), the shape of the output response becomes more symmetric. It is important to understand this shape variation because it explicitly shows whether the $RC_{FE}$ or the rise-time effect dominates under a given condition.

### 8.3.3 Experiment and Results

#### 8.3.3.1 Experiment

To measure the 50 ps theoretically estimated polarization switching time [27–31], one requires a setup with a time resolution of at least of the same order of magnitude. This requires the generation of a step-like pulse with less than 50 ps rise-time. Such an electric pulse is best generated by a photoconductive switch made of semiconductor with long carrier lifetime and actuated by a femtosecond laser pulse. In our experiment, a silicon semiconductor photoconductive switch is used as a fast "pulse generator" to produce step-like pulses with rise-times as short as ≅50–100 ps and amplitudes up to 10 V. Si is chosen because of its long carrier lifetime, leading to a step-function-like electric pulse. Figure 8.16 shows the structure of the Si photoconductive switch.

Photoconductive switches with interdigitated fingers are patterned into the coplanar transmission line structure with a 50 Ω characteristic impedance to transmit the high-speed broadband signal with negligible dispersion and attenuation. Without laser illumination, the photoconductive switch is in its dark state, or high resistance state, with a resistance of ≅100 kΩ to MΩ. On laser illumination, if the laser photon energy is greater than the semiconductor bandgap energy, electron-hole pairs are generated, transforming the semiconductor gap into a quasiconducting state, with its on-state resistance dropping immediately to a few ohms. Consequently, a voltage pulse is produced. Theoretically, the rise-time of this voltage pulse could be as fast as a few picoseconds. However, in practice, the dispersion of the cable and the connectors will slow down the rise-time to a few tens of picoseconds when the electric pulse propagates from the photoconductive switch to the FE capacitor under test. A typical electric pulse produced for this work is shown in Figure 8.17. The rise-time of this pulse is 60 ps. To further reduce the rise-time of the step-function pulse, one needs to minimize the dispersion of all circuit elements and monolithically integrate the photoconductive switch and the FE capacitor on the same chip.

The constant amplitude of the voltage pulse will enable us to systematically study the polarization switching process of the FE capacitors of different sizes. The detailed experimental setup is shown in Figure 8.18. A conventional pulse generator (pulse generator 1 in the figure) was used to produce the *write* pulse that sets the FE device to a stable positive or negative

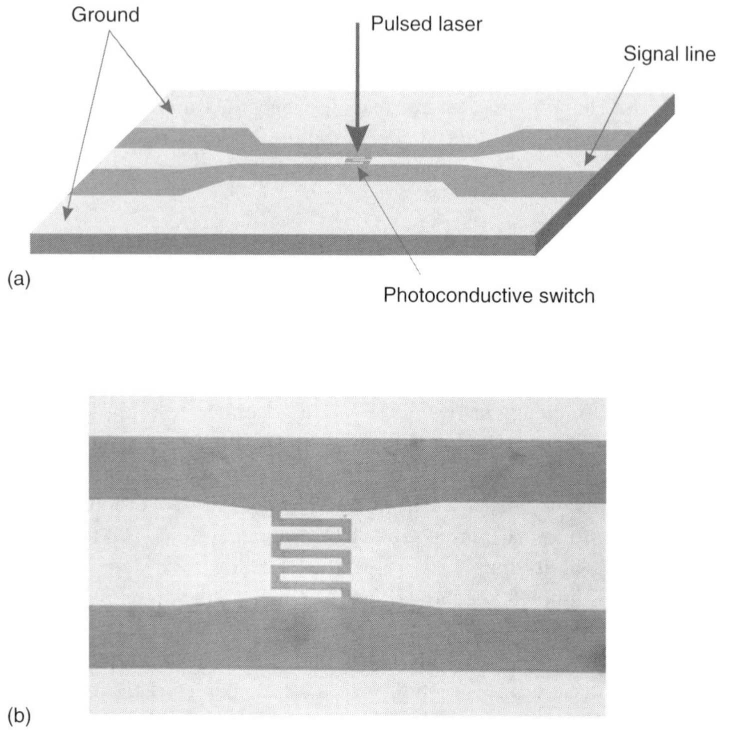


**FIGURE 8.16** (a) Schematic and (b) real image of a semiconductor photoconductive switch device.

polarization state by applying a positive or negative square pulse. An electric pulse with a ≅7 ns rise-time, 8 V amplitude, and 1 ms duration was used as this *write* pulse. Femtosecond laser pulses from the Ti:sapphire regenerative amplifier (repetition-rate = 10 kHz, laser pulse width ≅100 fs) illuminated the Si switch to produce the *read* pulses with fast rise-time, sufficiently high amplitude, and long duration. The bias of the Si switch was provided by another conventional pulse generator (pulse generator 2 in the figure). The delay between the two pulse generators was carefully tuned so that the *write* pulses are in the middle of the *read* pulses. The output pulse train to the FE device is a train of pulses consisting of a series of the *write* and *read* pulses. The new pulse train can be a sequence of negative–positive pulses or positive–positive pulses. A negative–positive pulse sequence applied to the FE device produces a switched response $V(t)_{P*}$, while a positive–positive sequence gives the nonswitched response $V(t)_{P\wedge}$. By subtracting the nonswitched response from the switched response, the $\Delta V(t)$ response is obtained, which gives information depending only on the domain polarization switching $\Delta P$. An HP54750A sampling oscilloscope with a time resolution of ≅20 ps was used

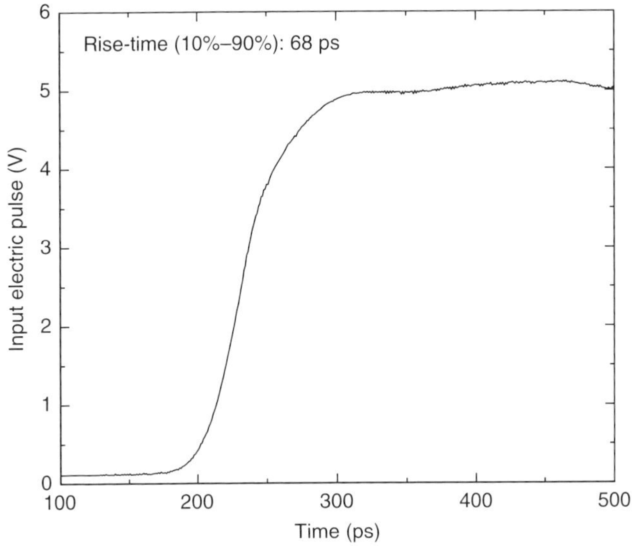


**FIGURE 8.17** An input 5 V electric pulse generated by a Si photoconductive switch. (From J. Li et al., *IEEE J. Lightwave Tech.* 21, 3282–3291, 2003. With permission.)

to monitor the output polarization switching response. The input impedance of the oscilloscope is 50 Ω, which is impedance-matched to the transmission line to prevent the signal from reflecting back to the setup. Figure 8.19 shows the timing sequence of each pulse train. All instruments are synchronized by a low jitter trigger signal generated by a photodiode trigger switch. A unique

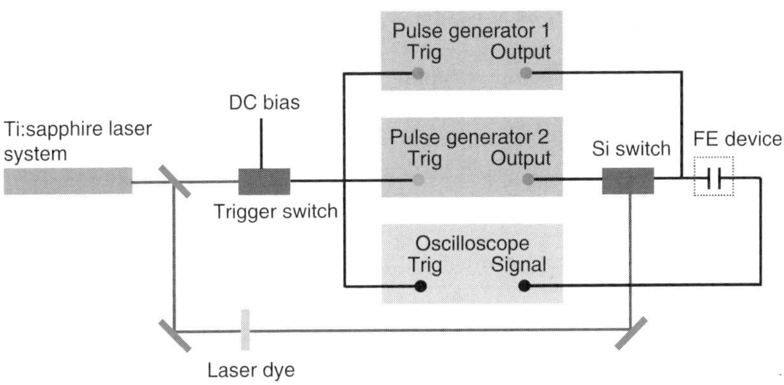


**FIGURE 8.18** Experimental setup of the pulse method using a Si photoconductive switch as a fast pulse generator. (From J. Li et al., *IEEE J. Lightwave Tech.* 21, 3282–3291, 2003. With permission.)

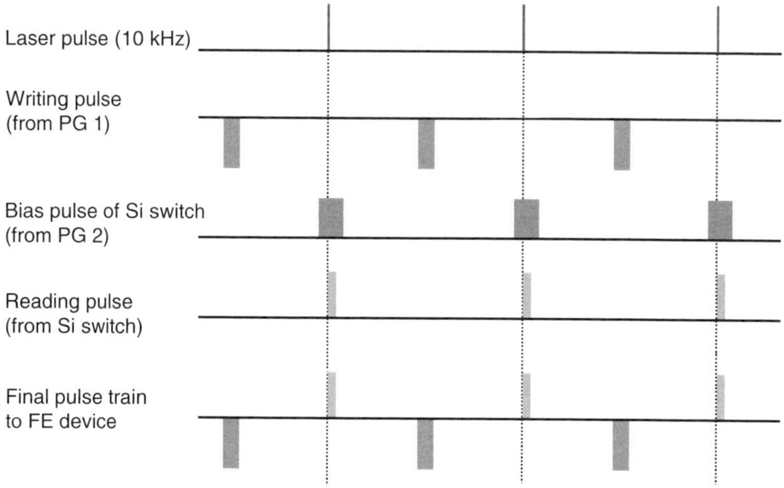


**FIGURE 8.19** Time sequences of pulse trains. (From J. Li et al., *IEEE J. Lightwave Tech.* 21, 3282–3291, 2003. With permission.)

advantage of using the photoconductive switch is that all events are timing jitter free. All events are perfectly time-synchronized to each other. Figure 8.18 shows that there is a dye inserted in the path of the laser beam traveling toward the Si switch. This dye solution is used as a saturable absorber to eliminate the prepulse of each laser pulse. For our ≅800 nm working wavelength of the laser, the dye material selected is *IR*-140 from Exciton Inc. Its central absorption wavelength is 810 nm. The excited state lifetime is 600 ps. The corresponding solvent is ethylene glycol/propylene carbonate (1:1). The density of the dye solvent is ≅0.38 mg/mL. The selection of the dye density depends on laser power and laser pulse width.

Because of the high-speed transmission lines connected to the FE capacitors and the high-bandwidth sampling oscilloscope, the time resolution of our experimental setup depends only on the rise-time of the input electric pulse generated by the Si switch, which is ≅50–100 ps. Therefore, our experimental system is capable of detecting the fast polarization switching for FE capacitors.

#### 8.3.3.2 Experimental Results

In this section we present the systematic study of the ultrafast polarization switching dynamics of the PNZT FE capacitors. Figure 8.20a and Figure 8.20b show the corresponding voltage responses of $P^*$, $P^\wedge$, and $\Delta P$ processes for a $4.5 \times 5.4\ \mu m^2$ PNZT capacitor. The polarization switching time, $t_s$, which is defined as the time from the onset of switching to the point 90% down from the maximum, is ≅220 ps.

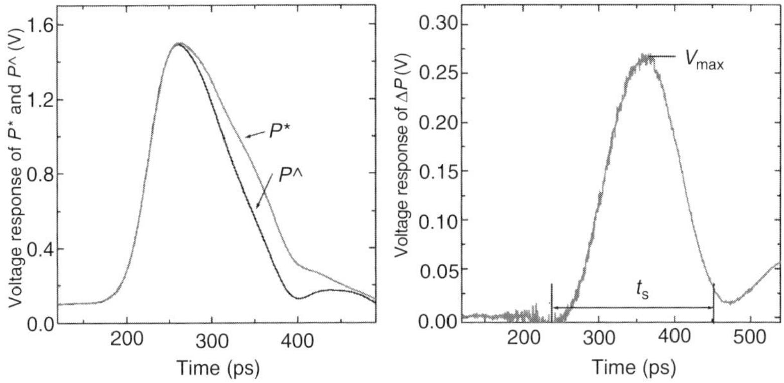


**FIGURE 8.20** Voltage responses of $P^*$, $P^\wedge$, and $\Delta P$ processes for a $4.5 \times 5.4\ \mu m^2$ PNZT capacitor.

The polarization switching time, $t_s$, is clearly dependent on the circuit parameters. To extract from the experimental data obtained by the pulse method as shown in Figure 8.20, one needs to calculate the fraction of the switched domains as a function of time, $f(t)$. This will be illustrated as follows.

The displacement current, $\Delta i(t)$, is related solely to the domain polarization switching process and can be expressed as

$$\Delta i(t) = \frac{V(t)_{p^*} - V(t)_{P^\wedge}}{50}, \tag{8.14}$$

where $V(t)_{P^*}$ and $V(t)_{P^\wedge}$ are the voltage transients displayed on the oscilloscope corresponding to the polarization-switched and nonswitched processes, respectively. $\Delta P(t)$, which represents the switched polarization as a function of time, can be obtained by integrating the current transient $\Delta i(t)$,

$$\Delta P(t) = \frac{\int_0^\infty \Delta i(t)\,\mathrm{d}t}{A_{FE}}, \tag{8.15}$$

where $A_{FE}$ is the area of the capacitor. Given the total switched polarization $P_{total}$, $f(t)$ can be expressed as

$$f(\text{switched}) = \frac{\Delta P(t)}{P_{total}}. \tag{8.16}$$

In other words, $f(t)$ is the normalized $\Delta P(t)$ and represents the fraction of the switched domains as a function of time.

Mathematical modeling of the polarization switching process in FEs has been studied for many years. Through this kind of model one can connect the experimental data obtained by the pulse method to the microscopic domain polarization switching mechanism in ferroelectrics.

One of the most popular model is the so-called Ishibashi–Merz model [27,32–34], as defined by

$$f(t) = 1 - \exp\left[-\exp\left(-\frac{\alpha}{E}\right)\left(\frac{t}{t_0}\right)^n\right], \tag{8.17}$$

where $\alpha$ is the activation field. The exponent $n$ is referred to as the domain growth dimensionality of the nucleating domain, and $t_0$ is the characteristic switching time, which is the time at which 63% of polarization has been switched after an infinitely long applied electric pulse is applied. Being roughly equal to 63% of the intrinsic polarization switching time, $t_0$ is also an intrinsic parameter. By integrating Equation 8.16 and using Equation 8.17 we can plot $f(t)$ for the PNZT capacitor with a dimension of $4.5 \times 5.4\ \mu m^2$.

Figure 8.21 shows the plots of $f(t)$ and the corresponding fitting curves using the Ishibashi–Merz model for different capacitor sizes. Using the above model we extracted $t_0$ and dimensionality $n$ from $f(t)$. Using a 500-kV/cm activation field one can extract $t_0$ and $n$ value. For the

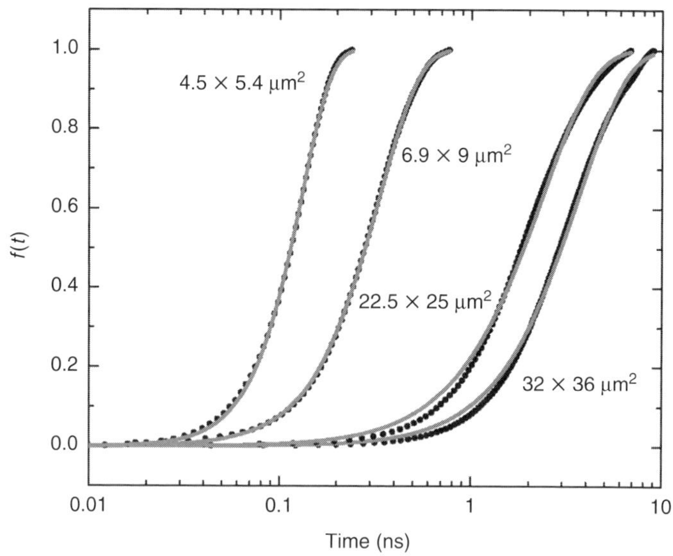


**FIGURE 8.21** Plots of measured (dot line) and fitted (solid line) $f(t)$ curves for various capacitor areas. (From J. Li et al., *IEEE J. Lightwave Tech.* 21, 3282–3291, 2003. With permission.)

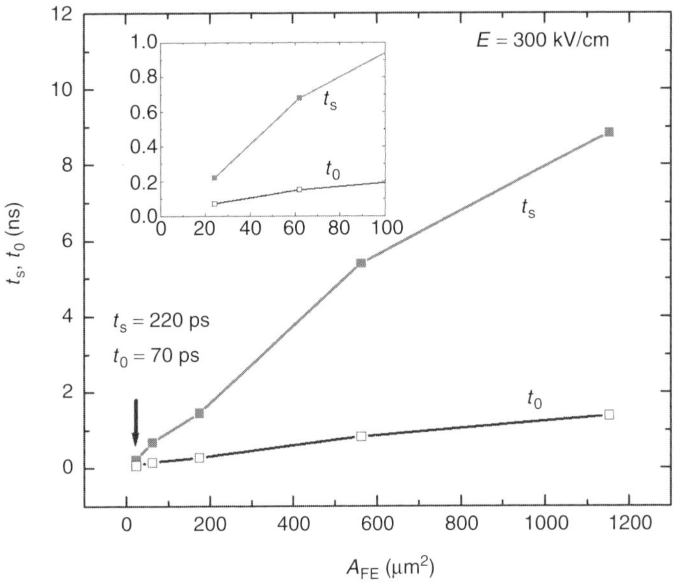


**FIGURE 8.22** Polarization switching time $t_s$ and characteristic switching time $t_0$ as functions of capacitor area. The inset shows the front part of the curve. (From J. Li et al., *IEEE J. Lightwave Tech.* 21, 3282–3291, 2003. With permission.)

$4.5 \times 5.4\ \mu m^2$ capacitor, $t_0$ is 70 ps for the experimentally obtained polarization switching time, $t_s$, of 220 ps. Figure 8.22 shows the dependence of $t_s$ and $t_0$ on capacitor size. Both $t_s$ and $t_0$ decrease linearly with area, illustrating the $RC_{FE}$ time limitation. The fitting parameter $t_0$, which is assumed to be intrinsic, should not depend on capacitor area. However, our data clearly show that $t_0$ varies with capacitor size, although this change is less than that of $t_s$. The reason for this discrepancy could be that circuit factors in the polarization switching process are not considered by the Ishibashi–Merz model. Dimensionality $n$ varying with the capacitor area is shown in Figure 8.23.

### 8.3.4 Summary on the Polarization Switching Dynamics of Ferroelectric Thin-Film Capacitors

The approach we present in this section is the widely applied pulse method, which can directly measure the domain switching speed. Numerical simulations of the circuit of this method were carried out to investigate the impact of circuit parameters on the experimental results. It has been shown that to obtain the accurate intrinsic polarization switching time, the rise-time of the input electric pulse and the capacitor area, or equivalently the $RC_{FE}$ time

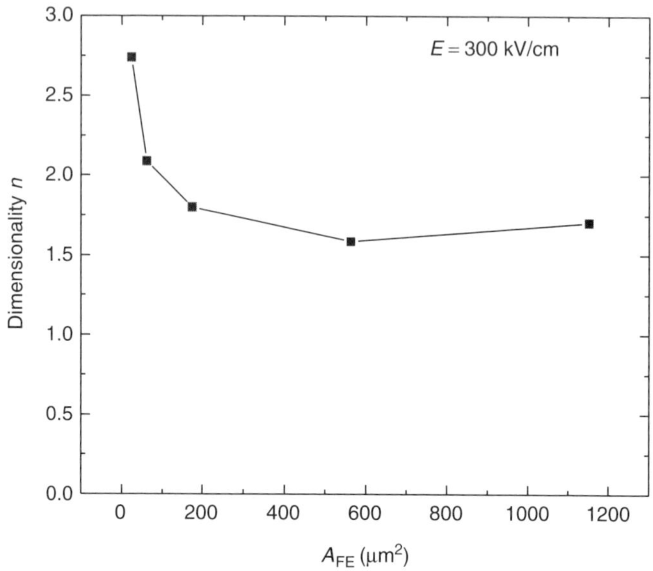


**FIGURE 8.23** Dimensionality $n$ as a function of capacitor area. (From J. Li et al., *IEEE J. Lightwave Tech.* 21, 3282–3291, 2003. With permission.)

constant, must be as small as possible or at least on the same order as the intrinsic switching time, which is estimated theoretically ≅50 ps. In this work, we designed and realized a novel experimental setup with a time resolution as small as ≅50–100 ps to study the fast polarization switching processes in FE capacitors. In this setup a semiconductor photoconductive switch actuated by an ultrafast laser pulse is utilized as a fast pulse generator to produce jitter-free, step-like electric pulses with fast rise-times. The input pulse rise-time limitation on obtaining the intrinsic polarization switching time is mostly eliminated.

The experiments were conducted on fully integrated FE PNZT capacitors by varying either the capacitor size or the rise-time of the input electric pulse. For a $4.5 \times 5.4\ \mu m^2$ PNZT capacitor, a switching time $t_s$ of ≅220 ps was obtained, which is to our knowledge, the fastest switching time ever reported. From the experimental data we confirmed that to obtain the real intrinsic polarization switching time both the rise-time of the input electric pulses and the capacitor size have to be sufficiently small. The polarization switching time measurement is limited by the larger of the two parameters, namely the $RC$ time constant and the rise-time of the input electric pulse. In other words, the rise-time becomes important only when the capacitor size is small enough. For larger size capacitors, a fast rise-time will still result in a large switching time because it is now limited by the $RC$ time constant of the capacitor.

Conversely, for smaller size capacitors, a longer rise-time will still result in a larger switching time. Only when the above two parameters work together can one obtain a switching time that is close to the intrinsic switching time.

The experimentally obtained switching transients were fitted with the widely accepted Ishibashi–Merz model to extract the inherent parameters associated with the domain switching process, namely the characteristic switching time $t_0$ and the domain growth dimensionality $n$. A $\cong$70 ps characteristic switching time was obtained for a $4.5 \times 5.4\ \mu m^2$ PNZT capacitor. In the Ishibashi–Merz model, $t_0$ and $n$ are supposed to be intrinsic parameters. However, it was shown that both parameters are extremely sensitive to the capacitor size and the rise-time of the input electric pulse. This circuit parameter dependence of $t_0$ and $n$ is due to the inevitable distortion of the experimental data by circuit effects in the pulse method, while the Ishibashi–Merz model is a theoretical model solely based on the physical process of the polarization switching.

## 8.4 POLYMER HIGH-SPEED PHOTOCONDUCTIVE EFFECT IN THIN FILMS

### 8.4.1 BACKGROUND

Organic devices promise the possibility of revolution in electronics by providing inexpensive, lightweight, and easy-to-process polymer alternatives for inorganic semiconductors. Since the first report of metallic conductivities in "doped" polyacetylene in 1977 [35], the research on electrically conducting polymers has attracted considerable attention and has led to rapid progress in understanding the physics of conducting polymers and their many applications. Recently, as high-purity polymers have become available, a series of semiconducting polymer devices have been realized, ranging from thin-film transistors, photodiodes, light-emitting diodes (LEDs) to plastic lasers [36–39]. Although considerable progress has been made toward understanding the properties of semiconducting polymers, some issues still remain controversial. One of the long-standing debates is whether the photocarriers are generated via direct interband excitations and relaxations [40] or via exciton formation and dissociation afterwards [41]. Some earlier work proved the presence of free carriers in the process of photocarrier generation [42], but the lack of time resolution made it difficult to resolve the excitation process. More recently, researchers employed terahertz emission spectroscopy to study the carrier lifetime [43,44], which gives better time resolution but is an indirect measurement. Detailed analysis has to be carried out to get the desired parameters. Photoconductive switching has long been used as a method to generate ultrafast electric pulses, and photoconductive sampling can be used to measure the pulse width with a resolution only restricted by the laser pulse width and RC constant of the switches employed. With a carefully designed structure, a resolution of subpicosecond can be obtained.

In this section we report the generation of few picoseconds electrical pulses from BAMH-PPV photoconductive switches and its application in the study of dynamic photoconductive properties of BAMH-PPV polymers and carrier mobilities. Photoconductive carrier lifetimes of less than two picoseconds have been observed. The transient mobility was estimated to be 400 $cm^2/V$ s.

## 8.4.2 Experiment

### 8.4.2.1 Sampling Oscilloscope Measurement

The polymer sample under study is a conjugated polymer BAMH-PPV. BAMH-PPV was synthesized at China Lake Naval Weapons Center and the monomer structure is shown in Figure 8.24. Most of the conjugated polymers have little or no absorption at longer wavelengths or infrared. To study the optical response of these polymers, UV or the short wavelength part of the visible spectrum needs to be employed. Photoconductive switches were fabricated on a fused quartz substrate in the configuration of coplanar waveguides. BAMH-PPV polymer solution was spin-coated on top of the metal electrodes to form the active layer. Polymer films of different thickness were prepared: (1) 280 nm film from 2% solution, (2) 780 nm film from 6%

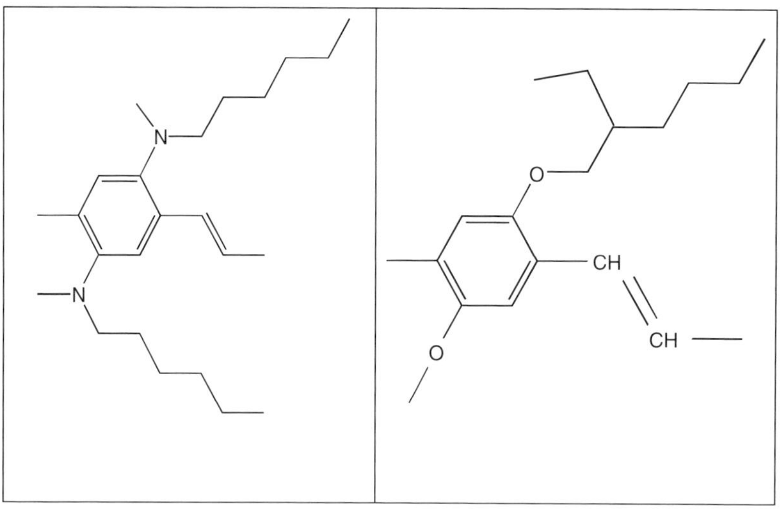


(a) (b)

**FIGURE 8.24** Monomers of conjugated polymers: (a) BAMH-PPV and (b) MEHPPV. (From H.Y. Liang et al., *Chemical Physics Lett.*, 419, 292–296, 2006. With permission.)

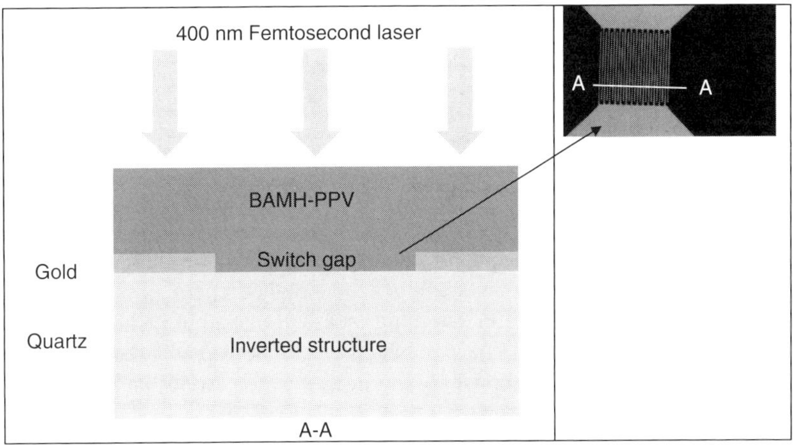


**FIGURE 8.25** Inverted structure photoconductive switch, interdigitated, with gaps of 2 μm.

solution, and (3) 93 nm from 0.67% solution. Using an inverted structure (electrodes under a semiconducting polymer medium) the fabrication process can be much easier than making electrodes on top of polymer films, which are generally not as smooth as semiconductor wafers. Meanwhile, to spin-coat the polymer after the electrodes have been fabricated, avoid exposing the film to UV light illumination in the photolithography process, to reduce the possibility of polymer photooxidation and degradation. This configuration has a limit on the film thickness to allow light to penetrate through the polymer film to transfer the carriers onto the metallic electrodes. An interdigitated metal–polymer–metal (MPM) switch structure was adopted. Gold was deposited on the other side of the quartz to form a 50 Ω impedance line, as shown in Figure 8.25.

The gold finger width was 1 μm with a 2 μm gap between fingers. The finger length was 50 μm and the capacitance of the switch was 20 fF. Contacting pads were added at the end of each line to accommodate the GGB pico-probes in the probe station. Interdigitated electrodes were employed in the active region to increase the sensitivity because polymers are far less sensitive to optical excitation than semiconductors. The capacitance of the gap increases logarithmically as the gap becomes smaller, while the sensitivity increases as $l^{-2}$ ($l$ is gap width); hence, using a small gap length to increase sensitivity will not significantly affect the speed of the device.

The BAMH-PPV photoconductive switch was illuminated with 400 nm, 100 fs laser pulses from a frequency-doubled Ti:sapphire laser amplifier. The transient photoconductive signal from the polymer switch was then recorded with a *Tektronix* sampling oscilloscope (Figure 8.26) to make systematic

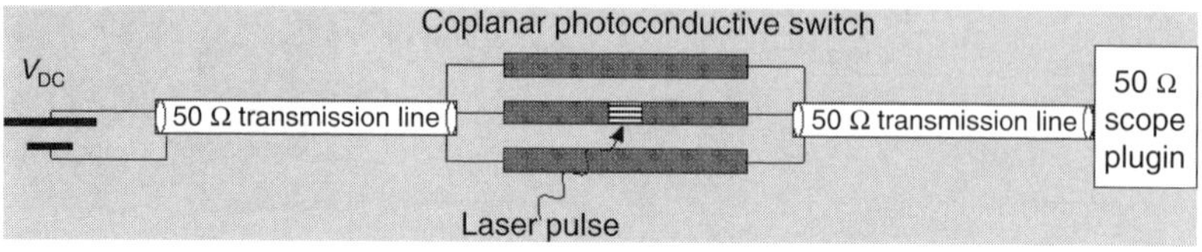


**FIGURE 8.26** Test setup for photoconductivity of MPM devices.

measurements of the transient current dependence on the illuminating laser intensity and applied bias voltage. In general a higher repetition-rate will give better signal-to-noise ratio for sampling measurements, but if the repetition-rate is too high, the BAMH-PPV polymer will be burned away quickly from the quartz surface. The following experiments were carried out with 1 kHz laser repetition-rate.

A typical transient voltage waveform is shown in Figure 8.27, measured at room temperature with laser pulse energy 0.5 nJ/pulse and bias 200 kV/cm. The rise-time and fall-time were less than 20 ps, with a full width at half maximum FWHM ~ 20 ps. The sharp falling edge indicates the polymer might have a very short carrier lifetime. The circuit in Figure 8.26 can be represented by the equivalent circuit as shown in Figure 8.28. The switch is represented by

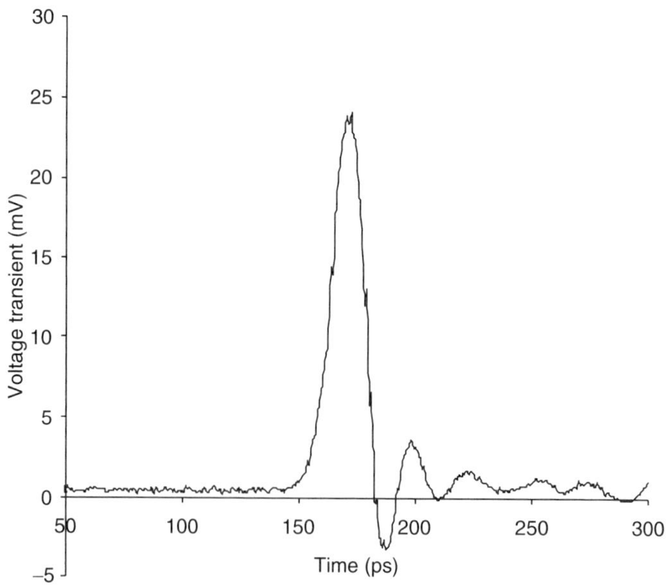


**FIGURE 8.27** Voltage transient wave form from the MPM switch probed by oscilloscope.

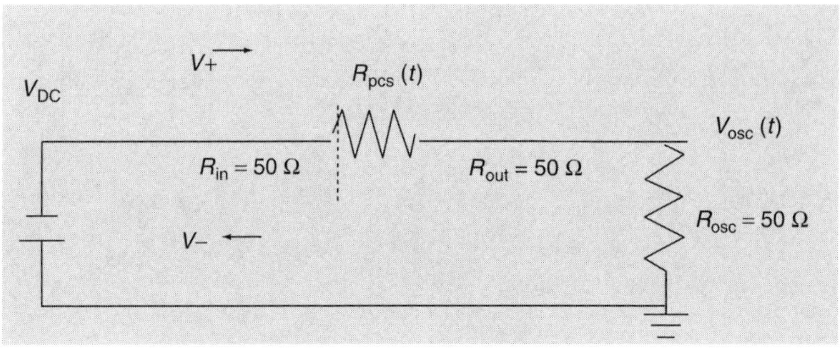


**FIGURE 8.28** Equivalent circuit of the oscilloscope measurement.

a time varying resistance $R_{pcs}(t)$. A parallel capacitance is omitted because the usual value of 20–50 fF will not have a significant affect on the waveform.

The value of the photoconductance can be calculated from the measured value of $V_{osc}(t)$, the voltage across the input impedance of the sampling oscilloscope as shown below:

$$G_{psc}(t) = \frac{1}{R_{pcs}(t)} = \frac{V_{osc}(t)}{(V_D + V_{off} - 2 \bullet V_{osc}(t))50\,\Omega}. \tag{8.18}$$

Since the sampling oscilloscope does not have sufficient time resolution, the photoconductance obtained is actually an average photoconductance, averaged over the time resolution of the sampling oscilloscope. The intrinsic peak photoconductance, which is related to the mobility, is a factor of $F$ larger than that measured by the oscilloscope. F is the ratio of the oscilloscope resoultion time to the pulse width of the photoconductive signal, measured by the pump–probe method. We shall elaborate on this point later.

Figure 8.29 shows the dependence of transient photoconductance on laser illumination energy. The time-varying photoconductance was calculated from the oscilloscope traces with Equation 8.18. During the laser pulse illumination, the resistance of the photoconductive switch varies with time; therefore, photoconductance is a function of time [45]. Here we are interested in the peak photoconductance. It can be seen that the peak transient photoconductance increases linearly with the laser pulse energy up to 2 nJ/pulse.

From Figure 8.30, it can be seen that the transient photoconductance remained roughly constant at different external bias fields up to $6 \times 10^5$ V/cm. This is typical of photoconductivity in semiconductors where the electronic structure can be described by band theory [40]. While increasing the applied voltage on the switch, no noticeable change in pulse width was observed during the experiment (Figure 8.31). This is an indication that no carrier sweep-out occurs during the pulse formation. The electrical pulse width is usually determined by one of the two processes: carrier lifetime and carrier sweep-out,

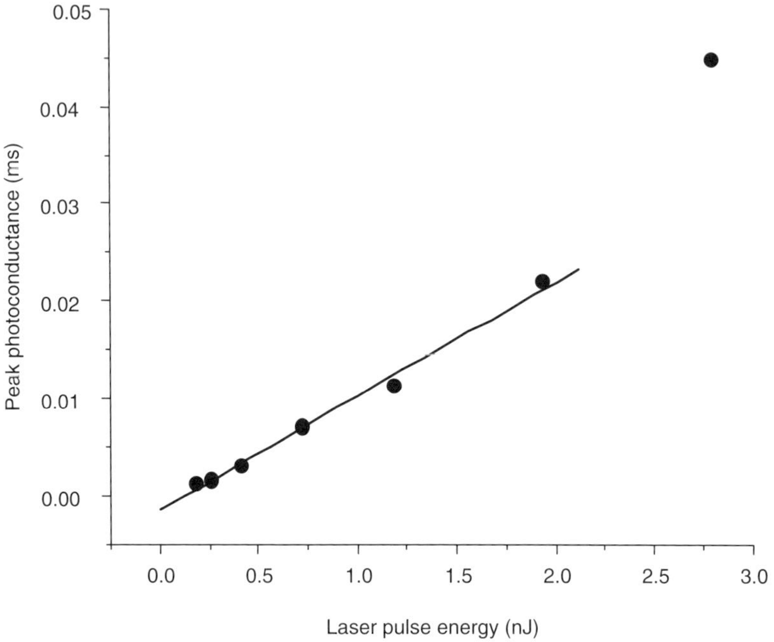


**FIGURE 8.29** Peak transient photoconductance with different laser illumination, for electric field of 250 kV/cm. Photoconductivity is estimated based on the temporal resolution of the sampling oscilloscope of 20 ps. The actual peak transient photoconductivity will be higher if the bandwidth of the measuring system is higher. (From H.Y. Liang et al., *Chemical Phys. Lett.*, 419, 292–296, 2006. With permission.)

whichever is shorter. Because of the low mobility of the polymers, carrier sweep-out is not a main factor that influences the formation of the picosecond electric pulse. The determining factor is therefore the carrier lifetime. Hence we can assume that during the formation of the electric pulse, carrier recombination is the dominating process that can be characterized by the photoconductive carrier lifetime. The external electric field is limited by breakdown to $6 \times 10^5$ V/cm.

Transient photoconductance for different thicknesses of polymer films and different gap sizes is shown in Figure 8.32. Switches fabricated on thicker film yield lower photoconductance. This is because the absorption coefficient $\alpha$ (at 400 nm) of these BAMH-PPV polymer films was found to depend on the film thickness (Figure 8.33). Thicker films tend to exhibit crystallization, hence the absorption is higher than that of thinner ones. By examining polymer switches of different thickness, we observed that the 280 nm thick film was more even and smooth, while the 789 nm film was coarse due to crystallization. The absorption layer on the surface is thin and cannot penetrate

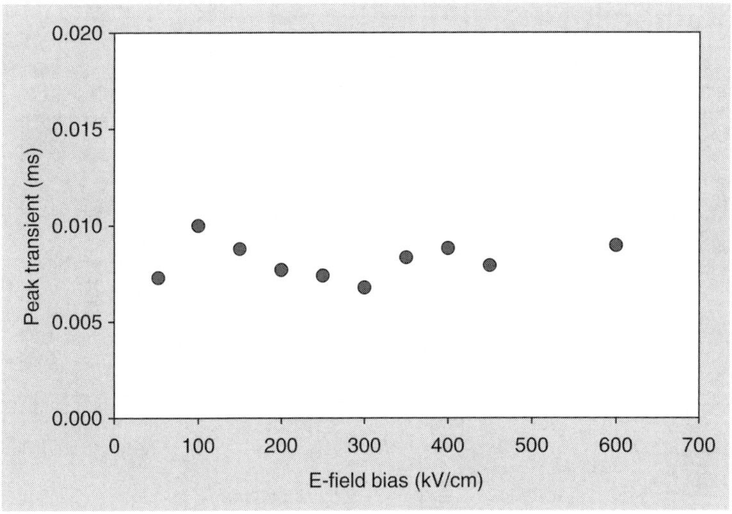


**FIGURE 8.30** Peak transient photoconductance under different bias, laser power 0.72 nJ/pulse. (From H.Y. Liang et al., *Chemical Phys. Lett.*, 419, 292–296, 2006. With permission.)

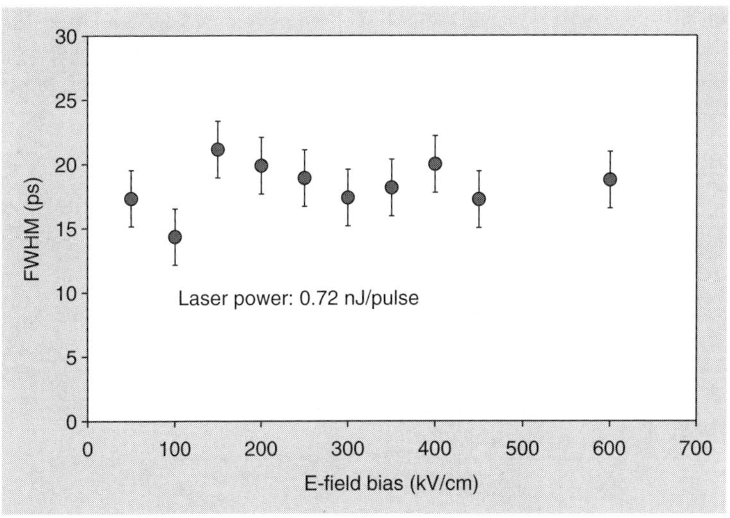


**FIGURE 8.31** Photocurrent pulse width dependence on external field with illumination of laser pulse energy 0.72 nJ/pulse.

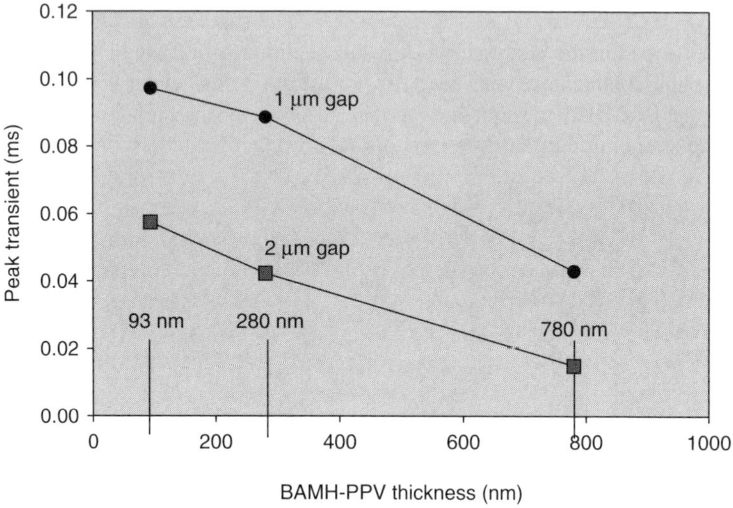


**FIGURE 8.32** Gap-width dependence of transient photoconductance, with laser pulse energy 1.5 nJ and external electric field 250 kV/cm.

deep enough to the metal electrodes in the case of thick film (780 nm here); this is because the device has an inverted structure, with electrodes lying under the polymer film. When the film is thin, more carriers have a chance to reach the electrodes beneath the polymer film. Switches with a larger finger gap width also have lower photoconductance. To obtain better photoconductive switch behavior, we need to use thinner films with small finger gaps.

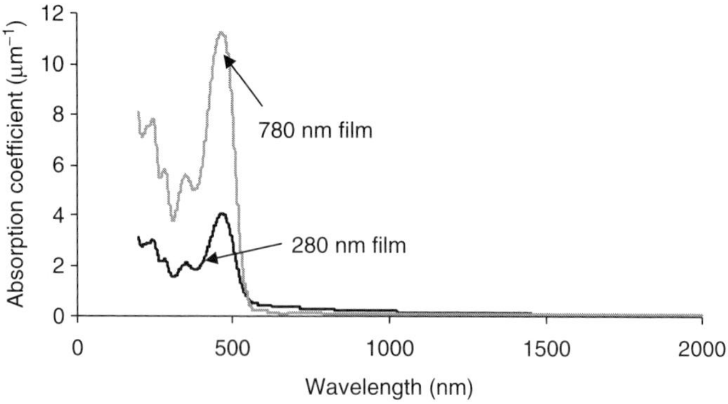


**FIGURE 8.33** Absorption coefficient spectrum of BAMH-PPV. At 400 nm, $\alpha = 2.1$ $\mu m^{-1}$, which is consistent with that measured by frequency-doubled femtosecond laser pulses.

#### 8.4.2.2 Photoconductive Sampling Using BAMH-PPV Switches

As mentioned in the discussion in the last section, to find the actual intrinsic peak photoconductance one needs to know the actual electric pulse width from the BAMH-PPV switch. This can be done by the standard photoconductive sampling method using the pump–probe technique.

Two interdigitated, MPM photoconductive switches were fabricated on fused quartz substrate in a cross configuration. Polymer solution was spin-coated on top of the metal electrodes to form the active layer, as shown in Figure 8.34. The polymer used was BAMH-PPV for the pump–probe measurement. The two MPM switches were at a distance of less than 50 μm right across the center line to minimize the possible dispersion caused by the transmission line. The gold finger width was 1 μm with a 2 μm gap between fingers. The finger length was 50 μm and the capacitance of each switch was estimated to be 20 fF.

To measure the duration of the electric pulse from the BAMH-PPV photoconductive switch, which is a representative of the carrier lifetime of the material, a pump–probe scheme was employed. Instead of using the sampling oscilloscope, the second photoconductive switch was used to probe the transient current waveform. Both switches were illuminated with 400 nm light from a femtosecond laser amplifier running at a repetition of 40 kHz. The experimental layout is shown in Figure 8.35.

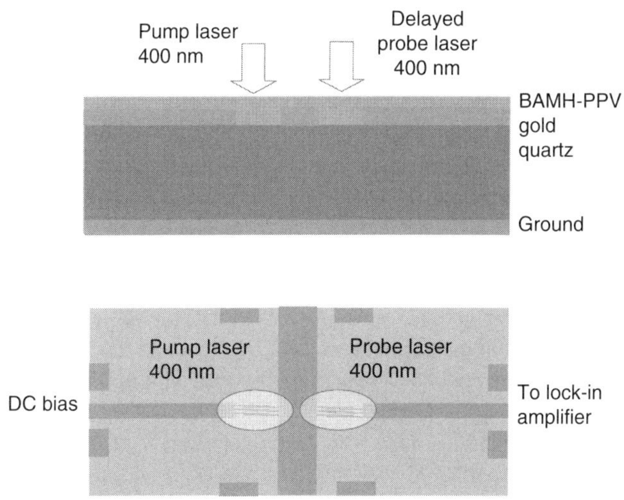


**FIGURE 8.34** Device for the pump–probe measurement. The switches are interdigitated, with gaps of 2 μm. The pads at each end of transmission line are designed to accommodate the pico-probes for applying bias and taking signal. (From H.Y. Liang et al., *Chemical Phys. Lett.*, 419, 292–296, 2006. With permission.)

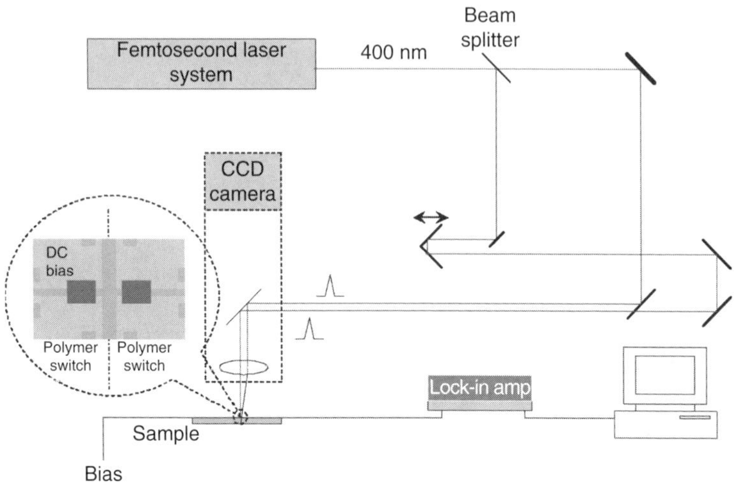


**FIGURE 8.35** Pump–probe measurement layout for photoconductive sampling.

The light from the femtosecond Ti:sapphire laser amplifier was frequency-doubled so that the photon energy could reach the absorption region of the polymer. A beam splitter divided the laser beam into two with almost equal power: one for pump, generating the transient electrical pulse to be measured; the other beam for probe, which was delayed with respect to the pump pulse by a translational stage driven by a stepping motor. After passing the optical delay line, the probe beam was combined with the pump beam by a beam splitter to copropagate to the focusing element. A Nectar upright microscope was used to focus both beams on the sample surface. At the same time, the CCD camera was used to observe the sample and focal points on the sample surface to ease alignment. When using a long working distance 50× objective, a focal point as small as 2 μm can be obtained, and it can easily be located on any part of the switch. For general measurement, a 10× objective was used. Because of the low pulse amplitude from the polymer photoconductive switch, the sampled signal was extremely weak and synchronous signal detection was employed by using a SRS 830 digital lock-in amplifier. A chopper was used in the pump beam, at a frequency of 1.6 kHz. The signal sampled by the probe switch was sent to the lock-in amplifier, which used the chopper frequency as a reference. The lock-in reading was transferred to the control computer, which was running Labview programs that both control the delay line translational stage and collect data.

#### 8.4.2.3 Result of Photoconductive Sampling Measurements

By delaying the probe beam with respect to the pump beam and taking the sampled data from the lock-in amplifier as a function of delay time, the pulse

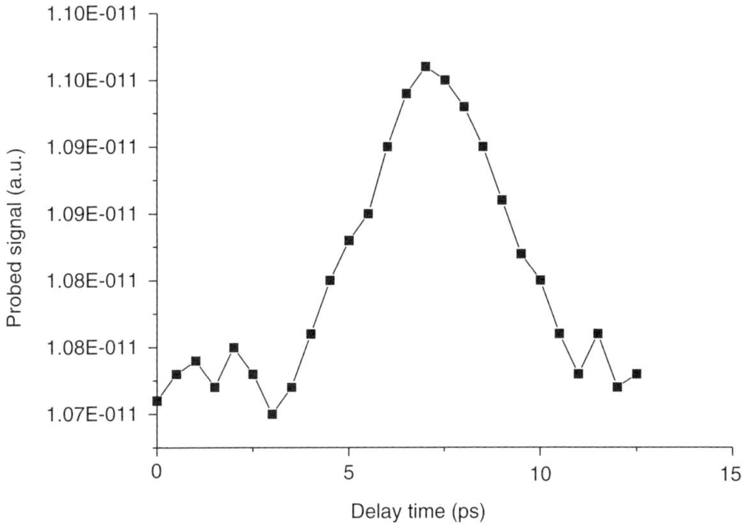


**FIGURE 8.36** Autocorrelation measurement of electric pulses from BAMH-PPV MPM switches, FWHM = 4 ps. (From H.Y. Liang et al., *Chemical Phys. Lett.*, 419, 292–296, 2006. With permission.)

width can be measured, which is only limited by the gap capacitance and laser pulse width. As seen in Figure 8.36, the measured autocorrelation curve has a FWHM of ~4 ps. If a Gaussian pulse shape is assumed, the electrical pulse from the BAMH-PPV photoconductive switch has a pulse duration of 2 ps, which is much shorter than the oscilloscope results.

### 8.4.3 Experimental Results and Discussions

One may estimate the carrier mobility in the BAMH-PPV film based on the transient photoconductance and the parameters of the laser and the MPM device. The absorption coefficient, $\alpha$, of BAMH-PPV polymer films depends on the film thickness and is greater for thicker films, as mentioned previously. The carrier density $n$ can also be estimated using the laser power, the absorption coefficient $\alpha$, and the geometry of the laser spot. As a result, the photoconductivity $\sigma$ and the mobility $\mu$ of the BAMH-PPV thin film can be estimated according to the following equations:

$$\sigma = \frac{GL}{A},$$
$$\mu = \frac{\sigma}{ne}, \tag{8.19}$$

where $G$ is the peak transient photoconductance, $A$ is the cross section of the switch (can be estimated by finger length multiplying by film thickness), $L$ is the width of finger gap, and $n$ is carrier density, given by Equation 8.20. At laser pulse energy of 1.5 nJ/pulse, the carrier density can be calculated:

$$n = \phi(1 - R)\frac{(1 - e^{-\alpha d})}{d}\frac{I_{pulse}\Delta t}{\hbar\omega}, \tag{8.20}$$

where $\phi$ represents quantum efficiency, $R$ represents film reflection, $d$ is film thickness, $I_{pulse}$ is laser intensity, and $\Delta t$ is laser pulse width. The results are shown in the Table 8.1, assuming a quantum efficiency of 1.

From Table 8.1, a transient carrier mobility of 0.2 $cm^2/V$ s was estimated for a 93 nm thick film of BAMH-PPV when the measuring system time resolution is 2 ps. This value is 10 times higher than the data in Ref. [40], in which the measurement was based on a system resolution of 50 ps. One obvious fact seen from Equation 8.19 is that the transient carrier mobility is time-dependent. When the measuring system resolution is lower than the actual mobility change, only an averaged value is detected, which is lower than the true value. A quick measuring facility is the key to obtaining a result close to reality.

When calculating the photocarrier density, we made an assumption that all the photons absorbed by the polymer contribute to carrier generation, that is, the quantum efficiency equals 1. In reality, only a portion of the photons creates carriers. From a recent paper by Hendry et al. [43], it was estimated that the quantum efficiency for charge generation in PPV polymers could be less than 0.05%. When taking this into account, the value of the transient carrier mobility in our experiment could be as high as 480 $cm^2/V$ s, based on the 2 ps time resolution.

In Hendry's work [43], terahertz emission spectroscopy was used to study charge carrier generation and time-dependent carrier mobility in PPV polymer. When analyzing the terahertz emission waveform from a polymer antenna, the carrier mobility showed strong time-dependence, with a very

**TABLE 8.1**
**Summary of Experimental Data on Transient Photoconductivity and Mobility Measurements (Assuming the Quantum Efficiency to Be 1)**

| Thickness (nm) | $G_{peak}$ (ms) | $L$ (μm) | $\alpha$ ($\mu m^{-1}$) | $N$ ($\times 10^{18}$ $cm^{-3}$) | $\sigma$ $(\Omega\ cm)^{-1}$ | $\mu$ ($cm^2/V$ s) |
|---|---|---|---|---|---|---|
| 93 | 0.10 | 1 | 1.9 | 2.6 | $8.3 \times 10^{-2}$ | 0.2 |
| 93 | 0.06 | 2 | 1.9 | 2.6 | $9.8 \times 10^{-2}$ | 0.24 |
| 280 | 0.09 | 1 | 2.1 | 2.3 | $2.6 \times 10^{-3}$ | 0.07 |
| 280 | 0.04 | 2 | 2.1 | 2.3 | $2.3 \times 10^{-3}$ | 0.06 |
| 780 | 0.05 | 1 | 5.6 | 1.8 | $5.1 \times 10^{-3}$ | 0.02 |
| 780 | 0.015 | 2 | 5.6 | 1.8 | $3.1 \times 10^{-3}$ | 0.01 |

*Source*: From H.Y. Liang et al., *Chemical Phys. Lett.*, 419, 292–296, 2006. With permission.

large initial value (within first 1 ps), decreasing rapidly toward a small steady-state value (~$10^{-5}$ cm$^2$/V s). The time resolution is on the order of 200 fs. When the time resolution is lower, the mobility will be an average value over the system's smallest resolvable time duration. Considering the measuring system time resolution to be 2 ps, by using the time-dependent mobility curve in Hendry's paper and integrating it from origin to 2 ps, we obtained an average mobility of ~540 cm$^2$/V s, which is extremely close to the results of our experiment.

From Table 8.1, it is seen that mobility is larger at lower film thickness. This is mainly device limited. As discussed in the previous section, thicker films have a higher absorption coefficient, leading to a lower photoconductivity. It can be noted that for 93 nm film the photoconductivity is around 15–30 times that of 780 nm films. An alternative method to avoid the intrinsic shortcomings of the inverted structure could be by illuminating from the bottom of the device.

Based on the values of carrier lifetime (2 ps), mobility (480 cm$^2$/V s), and applied electric field (250 kV/cm), we can calculate the mean free path of the carriers:

$$l = v \cdot \tau = 250 \times 10^3 \times 480 \times 2 \times 10^{-12} = 1.2 \times 10^{-4}\,\text{cm}.$$

### 8.4.4 Summary on Polymer High-Speed Photoconductive Effect in Thin Films

In this section the dynamic photoconductive properties of BAMH-PPV polymers have been investigated. The photocarrier generation process was studied by photoconductive sampling measurement. This has been done for the first time for the semiconducting polymer. This type of on-wafer measurement confirms the existence of free charge carriers after photoexcitation. However, they may be the products of exciton dissociation. Using an MPM photoconductive switch device configuration we observed that electric pulses as short as 1~2 ps were generated in MPM switches, the shortest on-wafer measurement result obtained on PPV devices to date. A carrier lifetime of less than 2 ps was estimated based on the generated picosecond electric pulses. The carrier mobility can be estimated if one knows the quantum efficiency of the carrier generation process. A low value of the mobility, $\mu = 0.4$ cm$^2$/V s, is obtained if one assumes the quantum efficiency to be 1. A high value of the mobility, 480 cm$^2$/V s, is estimated when a quantum efficiency of 0.05% is assumed. Furthermore our study also indicates that carrier mobility is a time-dependent parameter. It also shows that the carrier mobility is independent of the bias field, up to $6 \times 10^5$ kV/cm.

By using optimized device design and experimental setup, the achieved photoconductive switching performance is the best so far for conventional two-gap sampling. Signal-to-noise ratio needs to be improved to get a clearer picture of the carrier generation and transport mechanisms. This will enable us to measure the photoconductive effect on wafer with the fastest time response. If one can add a terahertz radiation as a probe for the femtosecond photoinduced carriers, one may learn more about the physics of photocarrier transport in semiconducting polymers.

## 8.5 CONCLUSION

In this chapter we have presented three different aspects of studies based on the ultrafast switching technique using femtosecond optical pulses and photoconductive response of material media. In the first application, we have demonstrated that it can be used to perform all-optical switching of micro ring resonators by carrier injection into the ring via single-photon absorption mechanism. The ring waveguide is optically pumped resulting in a temporal tuning of the microring resonant wavelengths by the refractive index change due to the induced free carriers. Switching speed of 20 ps has been achieved. In the second application we use a silicon photoconductive switch to generate step-function electrical pulse to study the polarization switching dynamic of FE capacitor. An intrinsic, characteristic switching time $t_0 = 70$ ps has been deduced from a small-size FE capacitor. This is the shortest switching time reported to date. The success of our measurement is credited to our ability to generate the step-function-like electrical pulses with a rise-time of 60 ps. Finally we have studied the ultrafast photoconductive phenomena in BAMH-PPV polymer film. We measured a photoconductive lifetime of 2 ps, the shortest reported to date. We also found that the transient mobility of the photoinduced carriers is much higher than that of the steady-state value, by many orders of magnitude. In general, to study ultrafast electro-optic effect, one needs ultrafast optoelectronic switching techniques.

## ACKNOWLEDGMENT

I wish to thank my colleagues and former students for contributing to this work. They are W. Herman, P.T. Ho, J. Goldhar, R. Ramesh, W.L. Cao, M. Du, T.A. Ibrahim, J.J. Li, and H.Y. Liang.

## REFERENCES

1. C.H. Lee, Picosecond optoelectronic devices based on optically injected electron-hole plasma, Chapter 5, in *Picosecond Optoelectronic Devices*, C.H. Lee, ed., Academic Press, Orlando, FL, 1984.
2. C.H. Lee, Picosecond optics and microwave technology, *IEEE Trans. Microw. Theory Tech.* 38, 596–607, 1990.
3. H.W. Yen, M.K. Barnoski, R.G. Hansperger, and R.T. Melville, Switching to GaAs IMPATT diode oscillator by optical illumination, *Appl. Phys. Lett.* 31, 120–122, 1977.

4. R.A. Keihl, Behavior and dynamics of optically controlled TRAPATT oscillators, *IEEE Trans. Electron Devices*, ED-25, 703–710, 1978.
5. R.N. Simons and K.B. Bhasin, Microwave performance of an optically controlled AlGaAs/GaAs high electron mobility transistor and GaAs MESFET, in *IEEE MTT-S Int. Microw. Symp. Dig.* pp. 815–818, June 1987.
6. T. Berceli, A.S. Daryoush, P.R. Herczfeld, W.D. Jemison, and A. Paolella, A MMIC based injection locked oscillator for optically fed phased array antennas, in *IEEE MTT-S Symp. Dig.* pp. 131–134, June 1989.
7. S. Jayaraman and C.H. Lee, Observation of two-photon conductivity in GaAs with nanosecond and picosecond light pulse, *Appl. Phys. Lett.* 20, 392–395, 1972.
8. M.B. Ketchen et al., Generation of subpicosecond electrical pulses on coplanar transmission lines, *Appl. Phys. Lett.* 48, 751–753, 1986.
9. H.A. Sayadin, M.G. Li, and C.H. Lee, Generation of high-power broad-band microwave pulses by picosecond optoelectronic technique, *IEEE Trans. Microw. Theory Tech.* 37, 43–50, 1989.
10. D. Butler et al., A cw 20-GHz opto-electronic source with phased-array applications, *Microw. Opt. Technol. Lett.* 1, 119–123, 1988.
11. S.T. Ho, C.E. Soccolich, M.N. Islam, W.S. Hobson, A.F.J. Levi, and R.E. Slusher, Large nonlinear phase shifts in low-loss AlxGa1-xAs waveguides near half-gap, *App. Phys. Lett.* 59, 2558–2560, 1991.
12. C. Coriasso, D. Campi, C. Cacciatore, L. Faustini, C. Rigo, and A. Stano, All-optical switching and pulse routing in a distributed-feedback waveguide device, *Opt. Lett.* 23, 183–185, 1998.
13. S. Blair, J.E. Heebner, and R.W. Boyd, Beyond the absorption-limited nonlinear phase shift with microring resonators, *Opt. Lett.* 27, 357–359, 2002.
14. Chi H. Lee, Paul S. Mak, and A.P. DeFonzo, Optical control of millimeter-wave propagation in dielectric waveguides, *IEEE J. Quantum Electron.* 16, 277–287, 1980.
15. J.V. Hryniewicz, P.P. Absil, B.E. Little, R.A. Wilson, and P.-T. Ho, Higher order filter response in coupled micro-ring resonators, *IEEE Photon. Technol. Lett.* 12, 320–322, 2000.
16. V. Van, T.A. Ibrahim, K. Ritter, P.P. Absil, F.G. Johnson, R. Grover, J. Goldhar, and P.T. Ho, All-optical nonlinear switching in GaAs-AlGaAs microring resonators, *IEEE Photonic Tech. Lett.* 14, 74–76, 2002.
17. V. Van, T.A. Ibrahim, P.P. Absil, F.G. Johnson, R. Grover, and P.T. Ho, Optical signal processing using nonlinear semiconductor microring resonators, *IEEE J. Select. Topics Quantum Electron.* 8, 705–713, 2002.
18. P.P. Absil, J.V. Hryniewicz, B.E. Little, R.A. Wilson, L.G. Joneckis, and P.-T. Ho, Compact microring notch filters, *IEEE Photon. Technol. Lett.* 12, 398–400, 2000.
19. A.M. Yurek, C.D. Striffler, and C.H. Lee, Optoelectronic devices for millimeter waves, *Infrared and Millimeter Waves*, Vol. 14, Chapter 4, pp. 249–290, K.J. Button, ed., Academic Press, Inc., 1985.
20. J.E. Heebner and R.W. Boyd, Enhanced all-optical switching by use of a nonlinear fiber ring resonator, *Opt. Lett.* 24, 847–849, 1999.
21. S.M. Sze, *Physics of Semiconductor Devices*, 2nd Ed., Wiley, New York, 1981.
22. M. Yairi and D. Miller, Equivalence of diffusive conduction and giant ambipolar diffusion, *J. Appl. Phys.* 91, 4374–4381, 2002.
23. J.F. Scott and C.A.P. Dearaujo, Ferroelectric memories, *Science* 246, 1400–1405, 1989.

24. A.I. Kingon and E.R. Myers, eds., *Ferroelectric Thin Films*, Vol. 200, 1990, Material Research Society Proceedings.
25. P.K. Larsen, G.L.M. Kampschoer, M.B. van der Mark, and M. Klee, Ultrafast polarization switching of lead zirconate titanate thin films, in *Eighth IEEE Int. Symp. Appl. Ferroelectrics*, 1992.
26. P.K. Larsen, R. Cuppens, and G.J.M. Dormans, Pulse switching characterization of ferroelectric thin films, in *Science and Technology of Electronic Thin Films*, O. Auciello and R. Waser, eds., Kluwer Academic, Dodrecht, pp. 201–221, 1995.
27. E. Fatuzzo and W.J. Merz, *Ferroelectricity*, John Wiley & Sons, New York, 1967.
28. W.J. Merz, Switching time in ferroelectric $BaTiO_3$ and its dependence on crystal thickness, *J. Appl. Phys.* 27, 938–943, 1956.
29. H.L. Stadler and Z. Pj, Nucleation and growth of ferroelectric domains in $BaTiO_3$ at Fields from 2 to 450 kV/cm, *J. Appl. Phys.* 34, 3255, 1963.
30. J.F. Scott, Limitations on ULSI-FeRAMs, *IEICE T Electron*, E81C, 477–487, 1998.
31. J.F. Scott, The physics of ferroelectric ceramic thin films for memory applications, *Ferroelectrics Rev.* 1, 1–129, 1998.
32. W.J. Merz, Domain formation and domain wall motions in ferroelectric $BaTiO_3$ single crystals, *Phys. Rev.* 95, 690–698, 1954.
33. Y. Ishibashi, A model of polarization reversal in ferroelectrics, *J. Phys. Soc. Jpn.* 59, 4148–4154, 1990.
34. Y. Ishibash and Y. Takagi, Ferroelectric domain switching, *J. Phys. Soc. Jpn.* 31, 506, 1971.
35. C.K. Chiang, C.R. Fincher, Jr., Y.W. Park, A.J. Heeger, H. Shirakawa, E.J. Louis, S.C. Gau, and A.G. MacDiarmid, Electrical conductivity in doped polyacetylene, *Phys. Rev. Lett.* 39, 1098, 1977.
36. A. Dodabalapur, Z. Bao, A. Makhija, J.G. Laquindanum, V.R. Raju, Y. Feng, H.E. Katz, and J. Rogers, Organic smart pixels, *Appl. Phys. Lett.* 73, 142, 1998.
37. C.J. Drury, C.M.J. Mutsaers, C.M. Hart, M. Matters, and D.M. Leeuw, Low cost all polymer integrated circuits, *Appl. Phys. Lett.* 73, 108, 1998.
38. S. Scheinert, T. Doll, A. Scherer, G. Paasch, and I. Horselmann, Organic field-effect transistors with nonlithographycally defined submicrometer channel length, *Appl. Phys. Lett.* 84, 4427, 2004.
39. J.R. Lawrence, G.A. Turnbull, and I.D.W. Samuel, Polymer laser fabricated by a simple micromolding process, *Appl. Phys. Lett.* 82, 4023, 2003.
40. D. Moses, H. Okumoto, C.H. Lee, and A.J. Heeger, Mechanism of carrier generation in poly(phenylene vinylene): Transient photoconductivity and photoluminescence at high electric field, *Phys. Rev. B* 54, 4748–4754, 1996.
41. D.M. Basco and E.M. Conwell, Hot exciton dissociation in conjugated polymers, *Phys. Rev. B* 66, 155210, 2002.
42. C.H. Lee, G. Yu, and A.J. Heeger, Persistent photoconductivity in poly(*p*-phenylenevinylene): Spectral response and slow relaxation, *Phys. Rev. B* 47, 15543, 1993.
43. E. Hendry, M. Koeberg, J.M. Schins, L.D.A. Siebbeles, and M. Bonn, Ultrafast charge generation in a semiconducting polymer studied with THz emission spectroscopy, *Phys. Rev. B* 70, 033202-1-4, 2004.

44. E. Hendry, J.M. Schins, L.P. Candeias, L.D.A. Siebbeles, and M. Bonn, Efficiency of exciton and charge carrier photogeneration in a semiconducting polymer, *Phy. Rev. Lett.* 92, 196601, 2004.
45. W.L. Cao, M. Du, Y. Kim, W.N. Herman, and C.H. Lee, Ultrafast switching of poly(2,5-*bis*[*N*-methyl-*N*-hexyl amino] phenylene-vinylene) (BAMH-PPV) polymer switches, *OSA/ACS FIOLS Organic Thin Film Symposium*, Tucson, Oct. 2003.

# 9 Tera Sample-per-Second Time-Stretched Analog-to-Digital Conversion

*Bahram Jalali and Yan Han*

## CONTENTS


9.1 Overview ... 308
9.2 Time–Wavelength Mapping ... 311
9.3 Discrete Time–Wavelength Processing ... 314
9.4 Impact of Time-Stretch on Signal-to-Noise Ratio ... 315
9.5 Effect of Optical Dispersion on Fidelity of the Time-Stretched Electrical Signal ... 318
9.5.1 Mathematical Framework ... 318
9.5.2 Dispersion Penalty and Harmonic Distortion ... 319
9.5.3 Residual Phase Errors ... 324
9.6 Time-Bandwidth Product ... 327
9.7 Techniques to Overcome Dispersion-Induced Bandwidth Limitation ... 330
9.7.1 Single-Sideband Modulation ... 330
9.7.2 Practical Consideration on SSB Modulation ... 333
9.7.3 Phase Diversity and Maximum Ratio Combining Algorithm ... 336
9.7.4 Tera-Sample Transient Digitizer ... 339
9.8 Effect of Optical Nonlinearity on Fidelity of the Time-Stretched Electrical Signal ... 342
9.9 Removal of Distortion Caused by Spectral Nonuniformities of the Optical Source ... 344
9.10 Continuous-Time Operation ... 349
9.11 Conclusion ... 356
9.12 Appendix ... 358
References ... 359


The technique of time-to-wavelength mapping followed by spectral domain processing represents a unique and powerful method for detection, generation, and processing of ultrawideband waveforms. This chapter presents the fundamentals of time–wavelength signal processing by a comprehensive analysis of its important applications, the ultrawideband time-stretched analog-to-digital conversion. By reducing the signal bandwidth prior to digitization, this technique offers revolutionary enhancements in the performance of electronic converters. The chapter starts with a fundamental-physics analysis of the time–wavelength transformation including both continuous-time and discrete-time implementations, and the implication of time-dilation on the signal-to-noise ratio (SNR). A detailed mathematical description of the time-stretch process is then constructed. It elucidates the influence of optical dispersion and nonlinearity on the fidelity of the electrical signal. Two techniques for ultrawideband operation, single-sideband (SSB) modulation and phase diversity, are examined. Further, problems arising from the nonuniform optical power-spectral-density are explained, and two methods for overcoming them are described. Finally, design issues and performance features of a time-stretch analog-to-digital converter array are discussed. The experimental results include a recent demonstration of a record 1 tera-sample/s analog-to-digital conversion of a transient waveform.

## 9.1 OVERVIEW

Digital signal processing (DSP) has revolutionized modern communication and radar systems by offering unprecedented performance and adaptivity. For broadband systems, however, the application of DSP is hindered by the difficulty in capturing the wideband signal [1]. A sampling oscilloscope is not an option because it requires the signal to be repetitive in time. It only provides information about the average signal behavior; hence, it does not operate in real time. Real-time capture of ultrafast electrical signals is a difficult problem that requires wideband analog-to-digital converters (ADCs).

A detailed discussion on the limits of electronic ADCs is beyond the scope of this chapter. An excellent review by R.H. Walden can be found in Ref. [2]. In general, the performance is limited by one or more of the following problems: (i) jitter in the sampling clock, (ii) settling time of the sample-and-hold circuit, (iii) speed of the comparator (comparator ambiguity), and (iv) mismatches in the transistor thresholds and passive component values. The limitations imposed by all these factors become more severe at higher frequencies. By lowering the carrier frequency and the bandwidth, the time-stretch technique addresses all these limitations simultaneously.

The standard approach to deal with this problem is to employ parallelism through the use of the time-interleaved ADC architecture, shown in Figure 9.1 [3,4]. Here the signal is captured by a parallel array of slow digitizers, each clocked at a fraction of the Nyquist rate. The Nyquist criterion is only satisfied when the signal is reconstructed, sample by sample, in the digital domain. In this chapter, we use the term "sample-interleaved" to refer to this architecture. It is well known that mismatches between digitizers limit the dynamic range and hence the resolution of sample-interleaved ADC systems

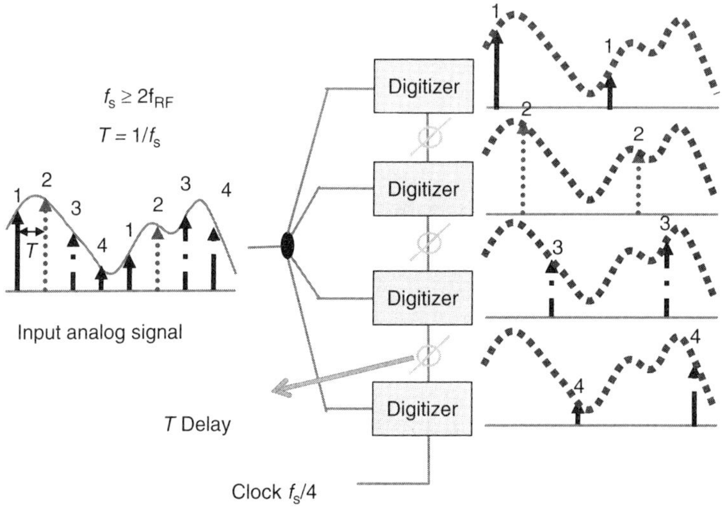


**FIGURE 9.1** Conventional sample-interleaved analog-to-digital converter architecture. The signal is captured by a parallel array of slow digitizer, each clocked at a fraction of the Nyquist rate. The Nyquist criterion is only satisfied when the signal is reconstructed, sample-by-sample, in the digital domain.

[3,4]. A state-of-the-art electronic ADC, embodied by real-time digitizing oscilloscopes (Tektronix TDS7404, or Agilent 54854A), boasts 20 GSa/s, 4 GHz analog bandwidth, and 8 nominal bits of resolution. In practice, depending on the input signal amplitude, these instruments exhibit an effective number of bits (ENOB) of approximately 4.5 to 5.5 bits when measured over the full bandwidth.

An entirely new analog-to-digital (A–D) architecture is the so-called time-stretched ADC, shown in Figure 9.2 [5–8]. Here the analog signal is slowed down prior to sampling and quantization by an electronic digitizer. For a time-limited input, a single channel, shown in Figure 9.2a, suffices, whereas a continuous input can be captured with a multichannel system shown in Figure 9.2b. The partitioning of the continuous-time signal into parallel segments does not necessarily require electronic switches. As shown later in this chapter, it can be performed passively in the optical domain.

Slowing down the signal prior to digitization has several advantages. For a stretch factor of $M$, the effective sampling rate of the electronic digitizer, $f_s$, is increased to $M \cdot f_s$. The effective input bandwidth of the electronic digitizer is also increased by $M$. The error associated with the jitter in the sampling clock of the digitizer is reduced due to a reduction in the signal slew rate, a key feature not shared by the conventional sample-interleaved architecture. For a time-limited input, only a single digitizer is needed (Figure 9.2a).

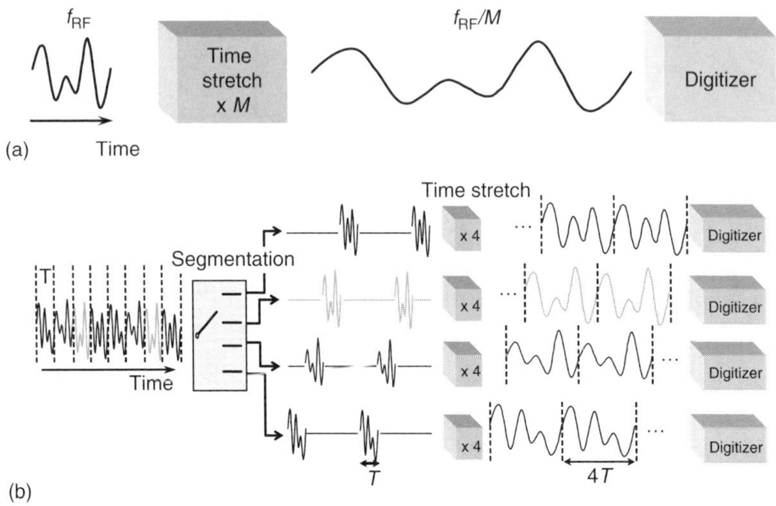


**FIGURE 9.2** Conceptual block diagram for the time-stretched analog-to-digital converter. (a) Single-channel system for capturing a time-limited signal. (b) Multichannel system for capturing a continuous-time signal. In contrast to the sample-interleaved system (Figure 9.1), each channel here is sampled at or above the Nyquist rate.

Hence, interchannel mismatch problems are avoided as there is no need to use multiple digitizers. For the continuous-time system (Figure 9.2b), one would expect the interchannel mismatches similar to those in the conventional sample-interleaved ADC. However, there is a fundamental difference between the two systems. In the time-stretched ADC (TS-ADC), the signal in each channel is sampled at or above the Nyquist rate, whereas in the sample-interleaved ADC it is sampled at a fraction of the Nyquist rate. It has recently been shown that, because of this advantage, mismatch errors in the time TS-ADC can be corrected, in real time, using the information available in the signal itself [9].

A practical method for implementing the time-stretch function is to use the photonic system shown in Figure 9.3 [10–12]. A linearly chirped optical pulse is generated by propagating the broadband (nearly transform-limited) pulses generated by a supercontinuum (SC) source [13] in a chromatic dispersive medium, such as an optical fiber. The electrical input signal modulates the intensity of the chirped optical pulses in an electro-optic modulator. The envelope is subsequently stretched in a second spool of fiber before photodetection and digitization by a moderate-speed electronic ADC. The same system can be used to perform time compression [14] and time reversal [15]. Time compression has application in ultrawideband signal generation, whereas time reversal represents wideband phase conjugation.

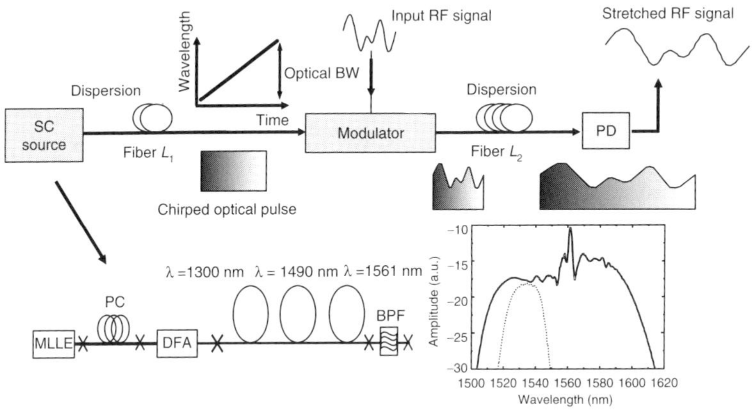


**FIGURE 9.3** Functional block diagram for the photonic time-stretch preprocessor. A linearly chirped optical pulse is obtained by dispersing an ultrashort ultrawideband pulse (the SC). Time-to-wavelength mapping is achieved when this pulse is intensity-modulated by the electrical signal. The inset is the diagram of three-stage SC generation. SC: Supercontinuum, PD: Photo-Detector.

## 9.2 TIME–WAVELENGTH MAPPING

An intuitive approach to understanding the time-stretch process is through the time–wavelength mapping representation. The stretch process consists of two steps. Step 1 is the time-to-wavelength ($t$-to-$\lambda$) mapping performed by the combination of the chirped optical pulse and the electro-optic modulator. As shown in the subset of Figure 9.3, 1 ps pulse generated by a passively mode-locked fiber laser at 1560 nm is used to generate broadband SC in three stages of fibers after the amplification [13]. An optical bandpass filter is placed after SC generation to slice a suitable portion for time stretching. The so-obtained nearly transform-limited optical pulse is chirped after propagating through a dispersive medium. When this chirped pulse is modulated by an intensity modulator, the time domain signal is mapped into the wavelength domain. Step 2 is the wavelength-to-time ($\lambda$-to-$t$) mapping performed by the second fiber. The modulated chirped pulse propagates through a second dispersive medium. As a result, the signal modulated onto the chirp pulse is stretched in the time as the pulse is further chirped.

The validity of time–wavelength representation can be understood through the following argument. Suppose that the chirped pulse is obtained by linearly dispersing a transform-limited pulse of the width, $\Delta t_o$. This value represents the system's ultimate time resolution; in other words, events cannot be localized in time beyond $\Delta t_o$. Hence, the wavelength-to-time mapping picture is only valid when the timescale of interest is sufficiently larger than this time ambiguity. The timescale of interest depends on the speed of the electrical signal. The well-known Nyquist sampling theorem states that it is enough to sample a band-limited signal with the maximum frequency $\omega_{RF}$ at the sample interval of $\Delta T = \pi/\omega_{RF}$, without the loss of information. To map the timescale of an electrical signal with frequency $\omega_{RF}$ to optical wavelength, it is required that $\Delta t_o \ll \pi/\omega_{RF}$. Fundamentally, $\Delta t_o$ is related to the optical bandwidth, $\omega_{opt}$, by the uncertainty principle, $\omega_{opt}\, \Delta t_o \geq 1/2$. Relating optical and electrical bandwidths, the combination of uncertainty and Nyquist principles implies that the time–wavelength mapping representation is valid as long as $\omega_{RF} \ll 2\pi\omega_{opt}$. Typically, $\omega_{opt}$ is in the terahertz range, whereas $\omega_{RF}$ is in the gigahertz range. Therefore, the assumption is readily justified.

The time-stretch system, as described above, relies on the linear group velocity dispersion (GVD) characterized by the GVD parameter, $\beta_2$, to linearly stretch the radio frequency (RF) waveform in time. However, it is well known that the optical fiber possesses nonlinear GVD, i.e., GVD is a function of wavelength. The wavelength dependence of GVD is characterized by the higher order terms in the expansion of $\beta$ as a function of frequency, namely $\beta_3$, $\beta_4$, etc. Wavelength-dependent GVD may distort the RF signal since different time segments may be stretched by different amounts. In other words, the stretch factor can be time-dependent. In this section, we analyze this phenomenon and demonstrate that, surprisingly, the nonlinear dispersion has no effects on the stretch factor in the time-stretch system. This fortuitous effect occurs because the effect in the two fibers counteract.

The GVD in optical fibers is conveniently described in terms of the dispersion parameter, $D(\lambda)$. As an example, we consider $D(\lambda)$ for the SMF-28 fiber, which in units ps/km/nm is given by (from Corning SMF-28 product information sheet)

$$D(\lambda) = \frac{S_0}{4}\left(\lambda - \frac{\lambda_0^4}{\lambda^3}\right), \tag{9.1}$$

where $\lambda_0$ is the zero-dispersion wavelength and $S_0$ is dispersion slope at $\lambda_0$. After traveling through a unit length of fiber, the time–wavelength transformation is governed by GVD. The group delay per unit length is described by

$$\tau_g(\lambda) = \int_{\lambda_{REF}}^{\lambda} D(\lambda)\,d\lambda = \frac{S_0}{8}\left(\lambda^2 + \frac{\lambda_0^4}{\lambda^2} - \lambda_{REF}^2 - \frac{\lambda_0^4}{\lambda_{REF}^2}\right),$$

where $\lambda_{REF}$ represents the reference wavelength. Then, at the input to the modulator, the input timescale is related to the wavelength as

$$t_{in} = L_1 \cdot \tau_{g1}(\lambda), \tag{9.2}$$

where $L_1$ is the length of the first fiber and $\tau_{g1}$ is the $\tau_g$ for fiber 1. At the output of fiber 2, the output timescale and wavelength are related by

$$t_{out} = L_1\tau_{g1}(\lambda) + L_2\tau_{g2}(\lambda), \tag{9.3}$$

where $L_2$ is the second fiber length and $\tau_{g2}$ is the $\tau_g$ for fiber 2. It can be clearly seen that as long as the two fibers have identical dispersion characteristics ($\tau_{g1} = \tau_{g2}$), the time transformation from the input to output is linear and can be simply described by

$$t_{out} = (L_1 + L_2)/L_1 t_{in} = Mt_{in}, \tag{9.4}$$

where we define the magnification factor (or the stretch factor) $M = (L_1 + L_2)/L_1$. Wavelength dependence of GVD (described by $\beta_3$ and higher order terms) is canceled and hence does not cause a nonuniform stretch. This property is independent of the length of fibers (hence independent of the stretch factor) as long as the two fibers have identical dispersion behaviors. As will be shown by a more detailed model in Section 9.5, $\beta_3$ causes a form of distortion through its interaction with the optical modulator. In addition, small differences in the dispersion property (GVD per unit length) of the two fibers result in some amount of wavelength dependence (and hence dependence) in the stretch factor. We now summarize the key parameters of the TS-ADC. Through Equation 9.1 to Equation 9.4, it was shown that the stretch factor is given by $M = 1 + L_2/L_1$, or more generally, $M = 1 + D_{t,2}/D_{t,1}$, where $D_{t,1}$ and $D_{t,2}$ represent the total dispersion in each dispersive element. The time aperture in a single-channel system, or the segment length in the case of continuous-time operation (see Figure 9.2), is equal to the chirped pulse duration at the modulator input. This is given by $T_A = \Delta\lambda D_1 L_1$, where $D_1$ is the dispersion parameter of fiber 1 and $\Delta\lambda$ is the optical bandwidth. Here, it is assumed that the chirped pulse width is much larger than initial nearly-transform-limited pulse width. As an example, assuming 20 nm optical bandwidth and 500 ps/nm dispersion (roughly equivalent to a spool of dispersion compensating fiber (DCF) fiber with a 3 dB total loss) the time aperture, $T_A$, is 10 ns. Another important system parameter is the 3 dB electrical bandwidth. As will be shown in Section 9.5, in the case of double sideband (DSB) intensity modulation, dispersion limits the 3 dB bandwidth to

$$\Delta f_{RF} = [1/(8\pi|\beta_2|L_2/M)]^{1/2}.$$

Finally, we point out the difference between the time-stretch system described here and the time-lens concept [5,16,17]. Shown in Figure 9.4, the latter exploits the analogy between paraxial diffraction and temporal dispersion to magnify an optical waveform. The input signal is first dispersed,

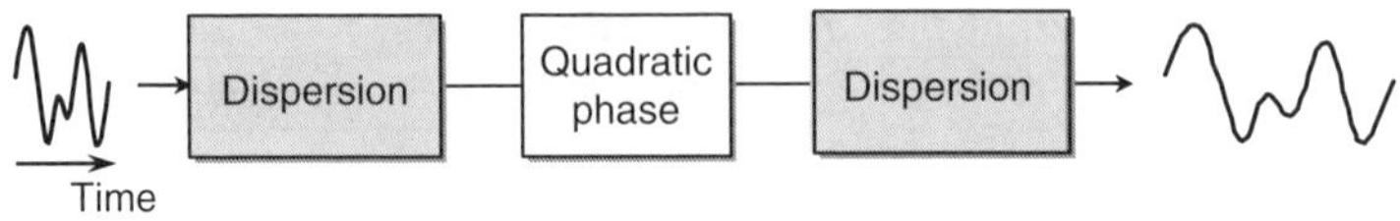


**FIGURE 9.4** Conceptual diagram of time-lens system exploiting the analogy between paraxial diffraction and temporal dispersion. In the time-stretch system, the input signal is not dispersed prior to receiving the linear chirp (or quadratic phase) due to the lack of low-loss wide-band electrical dispersive devices. The time-stretch system with application of magnifying electrical signal is different from the time lens with application of magnifying optical signal.

then linearly chirped (quadratic phase-shifted), and finally dispersed again. For applications in analog-to-digital conversion, the input signal is electrical. The lack of low-loss wide-band electrical dispersive devices renders the time lens inapplicable to analog-to-digital conversion. In the time-stretch system, a linear chirp is imposed onto the signal by the combination of the chirped optical pulse and the electro-optic modulator. The input signal is not dispersed prior to receiving the linear chirp. Hence the system is not a time lens. The main consequence is a limitation on the system's RF bandwidth, which can be removed by exploiting the phase diversity inherent in electro-optic modulation, a topic discussed in Section 9.7.3.

## 9.3 DISCRETE TIME–WAVELENGTH PROCESSING

Time-stretch processing described in the previous section uses optical fiber, a continuous-GVD device, resulting in continuous time–wavelength mapping. It is also possible to perform the time stretch in a discrete-time format by replacing the fiber with discrete-GVD components [6,8]. Possible implementations include an array waveguide grating (AWG) in the loopback configuration (Figure 9.5a) [18] or fiber bragg grating (FBGs) (Figure 9.5b). Here, the discrete-GVD convert SC pulses to a multiwavelength pulse train that samples the electrical signal in the electro-optic modulator. A second discrete-GVD then stretches the interpulse spacing, leading to slowing down of the modulation envelope. We note that a fiber in place of the first discrete-GVD device yields the same results at the output.

A major difference between continuous- and discrete-GVD implementation is the latency in a time-stretch preprocessing. For example, in the former, around 20 km fiber (DCF with a total dispersion of $-2000$ ps/nm) has to be used to achieve a stretch factor of 20, if a 2 ns time aperture and 20 nm optical bandwidth are used. This suggests a latency as large as 100 μs, dictated by the propagating time. On the other hand, the latency in a discrete time-stretch preprocessing is minimal. The main contribution to latency is the feedback delays used to reduce the repetition rate of the sampled pulse stream. Specifically, the latency is equal to the pulse-propagating time in AWGs and the

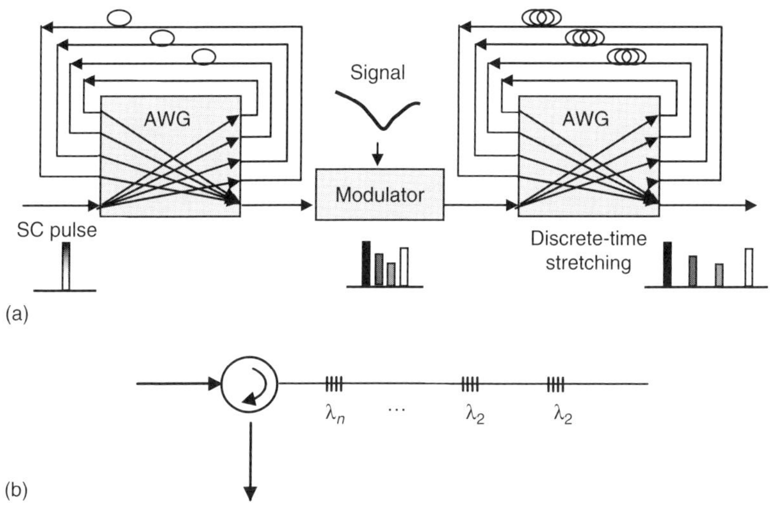


**FIGURE 9.5** Discrete-time implementation of the time-stretch system by (a) the re-circulating array waveguide grating or (b) the fiber Bragg gratings.

longest delay in the feedback paths. For example, to realize the same 20× stretch factor with a 2 ns time aperture, the latency will be approximately 40 ns. This is a 2500 times reduction in latency. The discrete time-stretch implemented using FBGs has the similar latency as the implementation using AWGs.

Another difference is that, in the discrete-GVD system, the signal is slowed down by propagating through a short delay line. The impact of chromatic dispersion in the delay line on optical pulse is negligible. In contrast, optical pulses in the continuous-GVD system need to propagate through tens of kilometers of fiber. As will be shown in the Section 9.5, the chromatic dispersion in fiber imposes a limit on the RF of the bandwidth. This limitation can be overcome by using SSB modulation or, preferably, using the phase-diversity technique described in Section 9.7.

## 9.4 IMPACT OF TIME-STRETCH ON SIGNAL-TO-NOISE RATIO

Intuitively, one would expect a reduction in the signal power on time stretching. This would be rather undesirable if it impacts the SNR. In this section, we analyze this phenomenon and show that, serendipitously, there is no impact on the shot- and thermal-noise-limited SNR. In addition, we show that time stretching alleviates the SNR reduction due to the jitter in the sampling clock.

We also explain the impact of optical amplifiers when used to mitigate losses in the time-stretch system.

The shot-noise-limited SNR in a given electrical bandwidth, $B_e$, can be written as

$$\mathrm{SNR}_{\mathrm{shot}} = \frac{\langle i \rangle}{2eB_{\mathrm{e}}}, \tag{9.5}$$

where $\langle i \rangle$ is the average photocurrent, and $e$ is the electron charge. For a lossless system, when the signal is stretched by the factor of $M$, the average signal power, and hence $\langle i \rangle$, are reduced by $M$ to preserve the total energy. At the same time, since the signal is slowed down, the required receiver bandwidth is reduced by $M$. In other words, both the numerator and the denominator in Equation 9.5 are reduced by the same amount factor, rendering the $SNR_{\mathrm{shot}}$ unchanged.

With respect to thermal noise, the corresponding SNR is given by

$$\mathrm{SNR}_{\mathrm{thermal}} = \frac{\langle i \rangle^2 R_{\mathrm{equ}}}{4kTB_{\mathrm{e}}} \propto \frac{\langle i \rangle^2}{4kTB_{\mathrm{e}}^2 C}, \tag{9.6}$$

where $k$ is Boltzman constant, $T$ is the temperature, $R_{\mathrm{equ}}$ and $C$ are the equivalent input resistance and the total input capacitance of the optical-to-electrical (O–E) receiver. After time stretching, both $\langle i \rangle$ and $B_{\mathrm{e}}$ are reduced by $M$, hence the thermal-noise-limited SNR also remains unchanged.

When the analog signal is sampled prior to quantization, an additional noise is produced by the jitter in the sampling clock. This so-called aperture uncertainty error is the main obstacle in achieving a high resolution in a high-speed A–D converter [2]. The resulting SNR is related to the root-mean-square (RMS) jitter, $\tau_j$, as

$$\mathrm{SNR}_{\mathrm{jitter}} = \frac{1}{8\pi^2 f_{\mathrm{RF}}^2 \tau_j^2}, \tag{9.7}$$

where $f_{\mathrm{RF}}$ is the maximum input RF signal frequency [2]. Since time stretch reduces $f_{\mathrm{RF}}$ into $f_{\mathrm{RF}}/M$, it renders the ADC less sensitive to jitter-induced noise in the digitizer. Quantitatively, the jitter-limited SNR improves by 6 dB per octave of time stretch. This is an important distinguishing property of the time stretch compared to time-interleaved architectures (Figure 9.1).

It should be noted that, in the time-stretch system, two potential types of timing jitters exist: optical jitter from pulsed laser and electronic jitter from digitizer's sampling clock. In the time-limited application where a single digitizer is sufficient, optical jitter does not influence the system in that only a single optical pulse is used, therefore, the RF signal is not influenced by the jitter between optical pulses. Optical jitter changes the absolute-time instant of samples. In a parallel time-stretch system, where multiple segments are concatenated, optical pulse jitter is important. The influences of interpulse timing errors on the RF signal fidelity are discussed in detail in Refs. [9,19].

Since the shot-noise-limited and thermal-noise-limited SNR are unaffected by the time-stretch process, the system, from the SNR point of view, can be treated as an RF fiber-optic link [20]. The losses in a link generally consist of E–O and O–E conversion losses and insertion loss of optical components. A practical time-stretch system uses optical amplifiers to compensate for the optical losses. Similar to an amplified optical link, the amplified spontaneous emission (ASE) noise is a concern.

The signal-ASE beat-noise-limited SNR can be expressed as

$$\mathrm{SNR}_{\mathrm{ASE}} = \frac{(mR_{\mathrm{PD}}P_{\mathrm{s}})^2}{2(R_{\mathrm{PD}}P_{\mathrm{s}})(R_{\mathrm{PD}}\rho_{\mathrm{ASE}})B_{\mathrm{e}}} = \frac{m^2 P_{\mathrm{s}}}{2\rho_{\mathrm{ASE}}B_{\mathrm{e}}}, \tag{9.8}$$

where $P_{\mathrm{s}}$ is the optical signal power, $\rho_{\mathrm{ASE}}$ is the ASE power spectral density, and $R_{\mathrm{PD}}$ is responsivity of the photodetector (PD). Also, $m$ is the modulation index defined as $m = \pi V_{\mathrm{RF}}/V_{\pi}$, and $V_{\mathrm{RF}}$ and $V_{\pi}$ are the RF voltage amplitude and the half-wave voltage, respectively. After time stretching, both $P_{\mathrm{s}}$ and $B_{\mathrm{e}}$ are reduced by the same factor $M$, hence the signal-ASE beat-noise-limited SNR remains unchanged.

Assuming a single optical amplifier is used, Equation 9.8 can be rewritten in terms of the input power of optical amplifier, $P_{\mathrm{s,in}}$:

$$\mathrm{SNR}_{\mathrm{ASE}} = \frac{m^2 P_{\mathrm{s},in} G}{2[2n_{\mathrm{sp}}h\nu(G-1)]B_{\mathrm{e}}} \approx \frac{m^2 P_{\mathrm{s},in}}{4n_{\mathrm{sp}}h\nu B_{\mathrm{e}}}, \tag{9.9}$$

where $h$ is plank constant, $\nu$ is the optical frequency, $G$ is the gain of optical amplifier, and $n_{\mathrm{sp}}$ is the population inversion parameter with a value close to 2 for a typical amplifier. Compared to the thermal noise in Equation 9.6, $4n_{\mathrm{sp}}h\nu$ plays a similar role of $4kT$. In a typical situation, $T = 300$ K, $n_{\mathrm{sp}} = 2$, $\lambda = 1.55$ μm, and $4n_{\mathrm{sp}}h\nu = -150$ dBm/Hz. This is much larger than $4kT = -168$ dBm/Hz and hence, the signal-ASE beat noise will limit the SNR of the time-stretch preprocessor. As an example, in our current setup, the power spectral density of the SC source at the input of the first erbium-doped fiber amplifier (EDFA) is around $-18$ dBm/nm with the repetition rate of 20 MHz. Assuming an optical bandwidth of 15 nm, the average input power is 0.24 mW. With $m = 40\%$, $n_{\mathrm{sp}} = 2$, $\lambda = 1.58$ μm, $B_{\mathrm{e}} = 4$ GHz (after stretching), the calculated signal-ASE beat-noise-limited SNR is 39.7, or 6.3 SNR bits. Here, the stretched signal is assumed to occupy the pulse repetition period, and hence the signal power is equal to the average power. The effective bandwidth of TS-ADC is 4 GHz times the stretch factor. Clearly, an SC source with a higher power spectral density is important to achieve a higher resolution. Every 6 dB increase in power spectral density increases the ASE limited resolution by 1 bit. In practice, since the optical bandwidth of time-stretch system is much larger than the electrical bandwidth, ASE–ASE beat noise should also be considered.

The SNR analysis can be extended to the case where multiple optical amplifiers are present. The treatment and conclusions will be similar to those in long-haul optical links with cascaded optical amplifiers [21]. Since $\rho_{ASE}$ accumulates after cascading, this suggests that proper location and gain of amplifiers are critical in maximizing the SNR. This noise should become the limiting factor if a large fiber propagating loss needs to be compensated. One can conclude that future development of low-loss highly dispersive devices will be highly valuable to the TS-ADC system.

## 9.5 EFFECT OF OPTICAL DISPERSION ON FIDELITY OF THE TIME-STRETCHED ELECTRICAL SIGNAL

While dispersion performs the desired function of time stretch, it has other influences on the electrical signals; it has other, potentially adverse, effects as well. It is therefore imperative to fully understand and be able to quantify the influence of dispersion on electrical signal fidelity. In Section 9.5.1 below, we present the mathematical framework for this purpose. Section 9.5.2 investigates dispersion penalty and harmonic distortion caused by dispersion. Section 9.5.3 reveals the higher order effects, which we term "residual phase errors."

### 9.5.1 MATHEMATICAL FRAMEWORK

In this section, we develop a comprehensive mathematical model for the time-stretch systems. The model is capable of predicting fundamental and higher order phenomena that modify the electrical signal during the time-stretch process.

We assume the optical SC pulse is transform-limited and has a Gaussian envelope. In the frequency domain, its electric field can be represented as

$$E_1(\omega) = E_0(2\pi T_0^2)^{1/2} \exp(-\omega^2 T_0^2/2), \tag{9.10}$$

where $T_0$ is the pulse half-width and $E_0$ is the pulse amplitude. $E_1$ indicates the electrical field at the output of the SC source in Figure 9.3. After propagating through the first spool of fiber, the field becomes [22]

$$E_2(\omega) = E_1(\omega) \exp\left(\frac{j}{2}\beta_2\omega^2 L_1 + \frac{j}{6}\beta_3\omega^3 L_1\right), \tag{9.11}$$

where $\beta_2$ and $\beta_3$ are the second-order and third-order GVD, respectively. $E_2$ indicates the electrical field before the modulator in Figure 9.3. Assuming a push–pull Mach–Zehnder modulator biased at the quadrature point, after modulation by the sinusoid RF signal of angular frequency $\omega_{RF}$, the field can be represented (in time domain) as

$$E_3(t) = E_2(t)\cos\left(\frac{\pi}{4} + \frac{m}{2}\cos\omega_{RF}t\right), \tag{9.12}$$

where $E_3$ is the electrical field at the output of modulator in Figure 9.3. After propagation through the second fiber, the field becomes

$$E_4(\omega) = E_3(\omega)\exp\left(\frac{j}{2}\beta_2\omega^2 L_2 + \frac{j}{6}\beta_3\omega^3 L_2\right), \tag{9.13}$$

where $E_4$ is the electrical field in the PD in Figure 9.3. Finally, the photocurrent is given by

$$I(t) = KE_4(t)E_4^*(t), \tag{9.14}$$

where $K = (c\varepsilon_0 n/2)R_{PD}A_{eff}$, $R_{PD}$ is the detector responsivity, $n$ is the refractive index, $c$ is the speed of light, $\varepsilon_0$ is the vacuum permittivity, and $A_{eff}$ is the fiber effective mode area.

Next, we analyze the signal in the electrical domain using analytical approximations to the above model, and verify their accuracy by comparison with numerical simulations. In Section 9.5.2, we consider the influence of the linear dispersion on the spectrum of the electrical system. We show that linear dispersion results in a frequency-dependent attenuation as well as frequency-dependent harmonic distortion in the time-stretched electrical signal. In Section 9.5.3, we consider the influence of nonlinear dispersion and describe the resulting residual phase error.

### 9.5.2 Dispersion Penalty and Harmonic Distortion

In this section, we describe and quantify dispersion-induced power penalty and harmonic distortion in the time-stretch system. Since this effect originates from the linear dispersion, $\beta_2$, we temporarily neglect the nonlinear dispersion term, $\beta_3$.

Using Bessel functions, Equation 9.12 can be expanded as follows:

$$E_3(t) = \frac{\sqrt{2}}{2}E_2(t)\left[J_0\left(\frac{m}{2}\right) + \sum_{n=1}^{\infty} 2c(n)J_n\left(\frac{m}{2}\right)\cos(n\omega_{RF}t)\right], \tag{9.15}$$

$$\text{where } c(n) = \begin{cases} (-1)^{n/2}, & n = \text{even} \\ (-1)^{n+1/2}, & n = \text{odd.} \end{cases}$$

Combining Equation 9.15 with Equation 9.10, Equation 9.11, and Equation 9.13, the electrical field entering the PD is written as

$$E_4(\omega) \approx J_0\left(\frac{m}{2}\right)E_{\text{env}}(\omega)$$
$$+\sum_{n=1}^{\infty} c(n)J_n\left(\frac{m}{2}\right)\exp\left(jn^2\phi_{\text{DIP}}\right)\left[E_{\text{env}}\left(\omega-\frac{n\omega_{\text{RF}}}{M}\right)+E_{\text{env}}\left(\omega+\frac{n\omega_{\text{RF}}}{M}\right)\right],$$
$$E_{\text{env}}(\omega)=\frac{\sqrt{2}}{2}E_0(2\pi T_0^2)^{1/2}\exp\left[-\frac{\omega^2T_0^2}{2}+\frac{j}{2}\beta_2\omega^2(L_1+L_2)\right], \tag{9.16}$$

where $E_{\text{env}}(\omega)$, the envelope function, corresponds to the electrical field when $m-0$.

$$\phi_{\text{DIP}}=\frac{1}{2}\beta_2\frac{L_2}{M}\omega_{\text{RF}}^2$$

is the dispersion-induced phase (DIP). As will be shown in this paper, DIP plays a central role on the distortion in the stretched signal, whether in the DSB or SSB modulation. In obtaining Equation 9.16, the following relation was used:

$$\frac{1}{2}\beta_2L_1(\omega-\omega_{\text{RF}})^2+\frac{1}{2}\beta_2L_2\omega^2=\frac{1}{2}\beta_2(L_1+L_2)\left(\omega-\frac{\omega_{\text{RF}}}{M}\right)^2+\phi_{\text{DIP}}.$$

In the expression for $\phi_{\text{DIP}}$, terms related to $T_0$ are neglected. This is equivalent to assuming an infinite optical bandwidth, or more precisely, $\omega_{\text{opt}} \gg \omega_{\text{RF}}$. This approximation is removed in the next section, and it is shown that having a finite optical bandwidth results in a subtle phenomenon, which we term the "aperture effect."

We now invoke the following property of Bessel functions:

$$J_n(m)\approx\frac{1}{2^n n!}m^n.$$

Hence,

$$J_p(m)J_q(m)\approx\frac{1}{2^{p+q}p!q!}m^{p+q}.$$

We substitute Equation 9.16 into Equation 9.14 and, for each harmonic, we keep terms with the lowest $(p+q)$. For example, for the DC term, this requires

$$J_0^2\left(\frac{m}{2}\right) \gg 2\sum_{n=1}^{\infty} J_n^2\left(\frac{m}{2}\right).$$

The photocurrent, so obtained, can be expressed as

$$I(t) = I_{\text{Envelope}} \times \begin{bmatrix} J_0^2\left(\frac{m}{2}\right) - 4J_0\left(\frac{m}{2}\right)J_1\left(\frac{m}{2}\right)\cos\phi_{\text{DIP}}\cos\frac{\omega_{\text{RF}}}{M}t \\ +\left[2J_1^2\left(\frac{m}{2}\right) - 4J_0\left(\frac{m}{2}\right)J_2\left(\frac{m}{2}\right)\cos 4\phi_{\text{DIP}}\right]\cos\frac{2\omega_{\text{RF}}}{M}t \\ +\left[4J_0\left(\frac{m}{2}\right)J_3\left(\frac{m}{2}\right)\cos 9\phi_{\text{DIP}} + 4J_1\left(\frac{m}{2}\right)J_2\left(\frac{m}{2}\right)\cos 3\phi_{\text{DIP}}\right]\cos\frac{3\omega_{\text{RF}}}{M}t \\ +\cdots \end{bmatrix}, \tag{9.17}$$

where

$$I_{\text{Envelope}} = \frac{1}{2}KE_0^2\frac{1}{[1+(L_1+L_2)^2/L_{\text{D}}^2]^{1/2}}\exp\left(-\frac{t^2}{T_0^2[1+(L_1+L_2)^2/L_{\text{D}}^2]}\right)$$

is the static ($m = 0$) photocurrent and $L_D = T_0^2/|\beta_2|$ is the dispersion length. In the case, when the modulation index $m$ is small, Equation 9.12 can be simplified into

$$I(t) = I_{\text{Envelope}} \times \begin{bmatrix} 1 - m\cos\phi_{\text{DIP}}\cos\frac{\omega_{\text{RF}}}{M}t + \frac{m^2}{8}(1-\cos 4\phi_{\text{DIP}})\cos 2\frac{\omega_{\text{RF}}}{M}t + \\ \frac{m^3}{96}(\cos 9\phi_{\text{DIP}} + 3\cos 3\phi_{\text{DIP}})\cos 3\frac{\omega_{\text{RF}}}{M}t + \cdots \end{bmatrix}. \tag{9.18}$$

The second term in Equation 9.18 is the desired time-stretched signal, and the following terms describe its harmonics. The desired signal contains a frequency-dependent attenuation described by the $\cos\phi_{\text{DIP}}$ term. For baseband signals, the system has a low-pass transfer function where high RF frequencies are attenuated. This is caused by interference between two beating terms: the beating of carrier with the upper sideband and the beating with the lower sideband [6]. Qualitatively, this penalty is similar to that observed in a conventional RF fiber-optic link [23]. However, its magnitude is reduced by the stretch factor, $M$, appearing in $\phi_{\text{DIP}}$ [6]. The reduction in the dispersion penalty occurs because, as the signal travels along the length of the fiber, its modulation frequency is reduced. Recognizing the similarity of this effect to that in an analog optical link, SSB modulation has been proposed as a means to mitigate the effect in TSADC [24].

The above results were obtained for a zero-chirp modulation achievable with a push–pull Mach–Zehnder (MZ) modulator, as described in Equation 9.12. The RF bandwidth is higher when chirped modulation is employed. When the chirp introduced by the modulator is opposite to that introduced by fiber dispersion, the dispersion penalty can be mitigated. Chirped modulation can be achieved using a single-arm MZ modulator biased at the quadrature point. In this case, the $\cos\phi_{\mathrm{DIP}}$ in Equation 9.18 should be replaced by $\cos(\phi_{\mathrm{DIP}} \pm \pi/4)$, where the sign depends on the polarity of the bias. When the sign of the first term is opposite to the second term, the 3 dB RF bandwidth of the time-stretch system is equal to $\Delta f_{\mathrm{RF}} = [1/(4\pi\,|\beta_2|L_2/M)]^{1/2}$, which is 41.4% larger than in the zero-chirp case, $\Delta f_{\mathrm{RF}} = [1/(8\pi|\beta_2|L_2/M)]^{1/2}$ [25,26].

The third and forth terms in Equation 9.18 are responsible for second- and third-order harmonic distortion, respectively. The amplitudes are $(m^2/8)(1 - \cos 4\phi_{\mathrm{DIP}})$ for the second harmonic and $(m^3/96)(\cos 9\phi_{\mathrm{DIP}} + 3\cos 3\phi_{\mathrm{DIP}})$ for the third harmonic. Figure 9.6a shows the fundamental and second harmonic power in the time-stretch system with a stretch factor, $M = 5$ (dashed line) and the same for a conventional fiber-optic link (solid line). The latter describes a case where the RF signal is modulated onto a single-wavelength carrier and travels through a fiber of length $L_2$. For an MZ modulator biased at the quadrature bias point, second harmonic is suppressed. This occurs because the second harmonics in the optical field spectrum are canceled in the photodetector by those arising from the beating of the upper and lower fundamental sidebands. As the input signal frequency is increased, the phase-shift induced by dispersion renders this cancelation incomplete, resulting in residual second harmonic tones in the RF spectrum. The frequency dependence is periodic as suggested by Equation 9.18. Second-order distortion is not a concern in systems operating in the suboctave bandwidth regime. In addition, while DIP shift results in second-order distortion in the time-stretch system, its magnitude is far less than that in a conventional RF fiber-optic link. Figure 9.6b shows the same for the third harmonic distortion. At very low frequency, the third harmonic is due to the nonlinear transfer function of the modulator (biased at quadrature). Dispersion-induced phase-shift results in a periodic behavior for the third-order distortion, as predicted by Equation 9.18. The frequency dependence of harmonic magnitudes indicates that time stretch is a nonlinear system with memory. A more general description of such a system needs to use the Volterra series technique.

We note that in the limit of $L_1$ approaching infinity, the time-stretch system reduces to a conventional RF fiber-optic link, consisting of an externally modulated narrow band source (e.g., DFB laser) that propagates over a fiber of length $L_2$. In this situation, over the limited-system time aperture, the source in the time-stretch system appears to be narrowband as in the conventional RF fiber-optic link. Mathematically, the stretch factor, $M$, approaches unity with

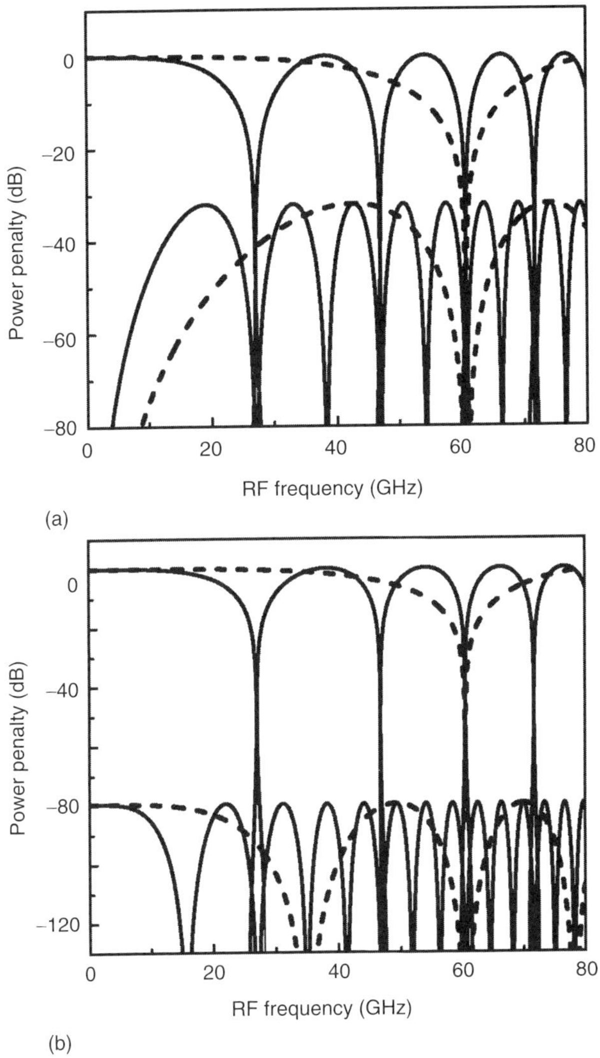


**FIGURE 9.6** (a) Electrical power transfer function for the fundamental frequency and second harmonic distortion. (b) Same for fundamental frequency and third harmonic distortion. Dashed line is for the time-stretched system, solid line is for a conventional RF fiber-optic link. Calculations are for zero-chirp modulation. $L_2 = 5$ km, $M = 5, m = 5\%$.

$$\phi_{\mathrm{DIP}} = \frac{1}{2}\beta_2 L_2 \omega_{\mathrm{RF}}^2.$$

Equation 9.18 then becomes the same as the results obtained for the dispersion penalty in a conventional RF fiber-optic link [23,27].

### 9.5.3 Residual Phase Errors

In this section, we remove two assumptions made in the derivation of Equation 9.18, i.e., ignoring third-order dispersion, $\beta_3$, and also terms related to $T_o$ in the expression for $\phi_{DIP}$. At the same time, the modulation is assumed linear to simplify the mathematical analysis.

Under the assumption of linear modulation, Equation 9.12 can be simplified into

$$E_3(t) \approx \frac{\sqrt{2}}{2} E_2(t) \left[ J_0\left(\frac{m}{2}\right) - 2J_1\left(\frac{m}{2}\right) \cos\left(\omega_{RF} t\right) \right], \tag{9.19}$$

where $E_2(t)$ is given by Equation 9.11. Combining Equation 9.19 with Equation 9.10, Equation 9.11, and Equation 9.13, the signal entering the PD can be written as

$$\begin{aligned} E_4(\omega) = {} & J_0\left(\frac{m}{2}\right) E_{env}(\omega) - J_1\left(\frac{m}{2}\right) E_{env}\left(\omega - \frac{\omega_{RF}}{M}\right) \exp\left[j\phi(-\omega_{RF})\right] \\ & - J_1\left(\frac{m}{2}\right) E_{env}\left(\omega + \frac{\omega_{RF}}{M}\right) \exp\left[j\phi(\omega_{RF})\right], \end{aligned} \tag{9.20}$$

where the envelope function is

$$E_{env}(\omega) = \frac{\sqrt{2}}{2} E_0 (2\pi T_o^2)^{1/2} \exp\left( -\frac{\omega^2 T_0^2}{2} + \frac{j}{2}\beta_2 \omega^2 (L_1 + L_2) + \frac{j}{6}\beta_3 \omega^3 (L_1 + L_2) \right).$$

Compared to the same name function in Equation 9.16, $\beta_3$ terms are included here. As shown in Equation 9.20, the optical field spectrum consists of DSBs at the stretched RF. This is the desired time-stretch signal. The nonideal effects, caused by dispersion, are contained in the phase of the upper and lower stretched sidebands. For the lower ($-\omega_{RF}$) and upper ($+\omega_{RF}$) sidebands,

$$\begin{aligned} \phi(\mp \omega_{RF}) &= \phi_{DIP} + \left[ 3\omega\omega_{RF}^2 \left(1 - \frac{1}{M}\right) \mp \omega_{RF}^3 \left(1 - \frac{1}{M^2}\right) \right] \frac{1}{6}\beta_3 L_1 \\ &\quad - \frac{j}{2} \left[ \pm 2\frac{\omega\omega_{RF}}{\omega_{opt}^2} \left(1 - \frac{1}{M}\right) - \frac{\omega_{RF}^2}{\omega_{opt}^2} \left(1 - \frac{1}{M^2}\right) \right] \\ &\approx \phi_{DIP} + \frac{1}{2}\beta_3 \omega\omega_{RF}^2 \frac{L_2}{M} \mp j \frac{\omega\omega_{RF}}{\omega_{opt}^2} \left(1 - \frac{1}{M}\right), \end{aligned} \tag{9.21}$$

where $\omega_{opt} = 1/T_o$ is the optical bandwidth (half width at $1/e$ intensity point).

The phase $\phi(\mp \omega_{RF})$ in Equation 9.21 is composed of three terms. The first term is the phase distortion due to linear dispersion, $\beta_2$, which causes attenuation at high frequencies, discussed in the previous section. The second term is the phase distortion caused by $\beta_3$, which combined with the first term, creates a wavelength-dependent (and hence time-dependent) attenuation, for a given frequency. The third term is error caused by the finite optical bandwidth.

The photocurrent in time domain can be expressed as (see Appendix)

$$I(t) = I_{\text{Envelope}} \times \begin{bmatrix} J_0^2\left(\frac{m}{2}\right) - 4J_0\left(\frac{m}{2}\right)J_1\left(\frac{m}{2}\right) \\ \left(\cos\phi_{\text{DIP}}\cos\frac{\omega_{\text{RF}}}{M}t + \text{Err}_{\beta_3} + \text{Err}_{\text{BW}}\right) \end{bmatrix}, \tag{9.22}$$

$$\text{Err}_{\beta_3} = -\frac{\beta_3\omega_{\text{RF}}^2(1-1/M)\sin\phi_{\text{DIP}}}{2\beta_2 M}t\cos\left(\frac{\omega_{\text{RF}}}{M}t\right), \tag{9.23}$$

$$\text{Err}_{\text{BW}} = \frac{\omega_{\text{RF}}(1-1/M)\sin\phi_{\text{DIP}}}{\omega_{\text{opt}}^2\beta_2(L_1+L_2)}t\sin\left(\frac{\omega_{\text{RF}}}{M}t\right). \tag{9.24}$$

Equation 9.23 and Equation 9.24 are the errors due to $\beta_3$ and finite optical bandwidth, respectively.

We now show the individual effects of these sources of error on a single-tone RF signal. Figure 9.7a is a plot of

$$\left(\cos\phi_{\text{DIP}}\cos\frac{\omega_{\text{RF}}}{M}t + \text{Err}_{\beta 3}\right)$$

and shows the influence of $\beta_3$ on the stretched signal, in time domain. To highlight the $\beta_3$ impact, we use a value $\beta_3 = 10\ \text{ps}^3/\text{km}$ that is two orders of magnitude higher than that in the single-mode fiber (SMF-28). As suggested by Equation 9.23, this introduces a deleterious linear amplitude modulation. In the previous section, it is shown that $\beta_2$ introduces attenuation at high RFs. The influence of $\beta_3$ is to make $\beta_2$ wavelength sensitive and hence, via wavelength-to-time mapping, cause a time-dependent attenuation. To verify the validity of approximations made and conclusion of the analytical model, numerical simulations were performed. Figure 9.8 shows the error introduced by including $\beta_3$ in the simulations. The functional dependence is in full agreement with Equation 9.23, verifying the validity of the approximations made in the derivation.

Figure 9.7b is a plot of $\cos\phi_{\text{DIP}}\cos(\omega_{\text{RF}}t/M) + \text{Err}_{\text{BW}}$ and shows the effect of the error introduced by the finite optical bandwidth. For the purpose of illustration, an unusually small optical bandwidth, 0.5 nm, is used to accentuate this effect. The envelope of this function behaves as $\sqrt{1+ct^2}$. In an ideal situation, where the optical bandwidth is much larger than the RF bandwidth ($\omega_{\text{opt}} \gg \omega_{\text{RF}}$), this error diminishes, as suggested by Equation 9.24.

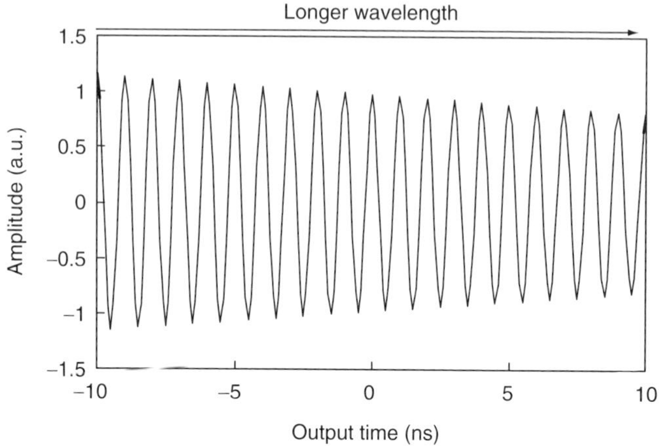


(a)

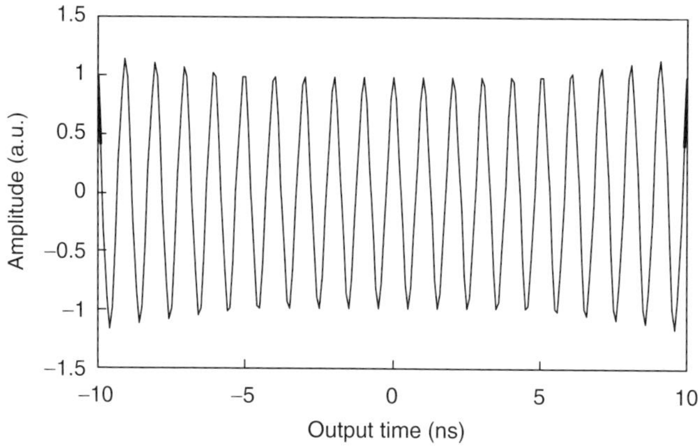


(b)

**FIGURE 9.7** (a) The influence of nonlinear group velocity dispersion, characterized by $\beta_3$, on the time-stretched signal. 100 times larger $\beta_3$ than SMF-28 is used. The behavior can be understood in terms of a wavelength-dependent dispersion penalty, causing higher attenuation at longer wavelengths. (b) Time-stretched output signal when the effect of finite optical bandwidth is included. An unusual small optical bandwidth, 0.5 nm is used to accentuate this effect. $\omega_{RF} = 10$ GHz, $M = 10$, $\beta_2 = -20$ ps$^2$/km, $\beta_3 = 10$ ps$^3$/km, $L_2 = 50$ km, $\Delta\lambda = 0.5$ nm.

Typically, these errors are relatively small to be observed in experiments as they are masked by other sources of noise and distortion. However, when bandwidth limits are removed by using techniques described in Section 9.7, the ultrabroadband operation generally intensifies these high-order distortions and they need to be corrected.

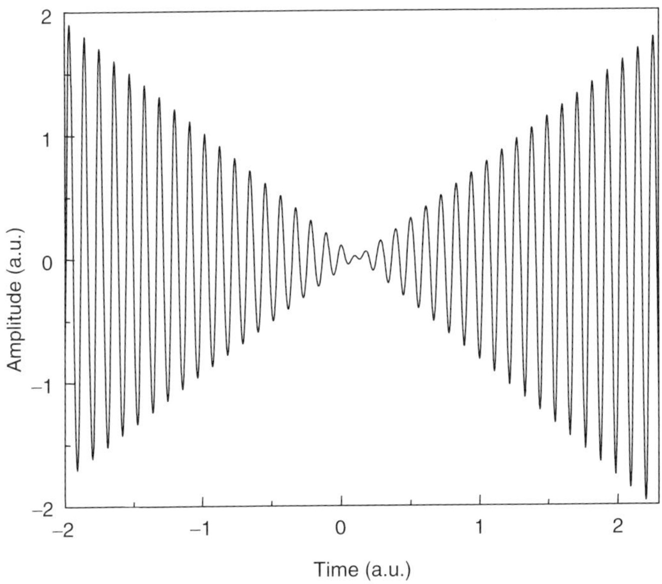


**FIGURE 9.8** Numerical simulation result showing the error caused by nonlinear group velocity dispersion. The functional dependence is in full agreement with Equation 9.23, verifying the validity of the approximations made in the analytical model.

## 9.6 TIME-BANDWIDTH PRODUCT

A photonic time-stretch system could be described by the following three parameters: the stretch factor, time aperture, and RF bandwidth. The stretch factor is given by $M = 1 + L_2/L_1$. The time aperture is given by $T_A = \Delta\lambda D_1 L_1 = 2\pi|\beta_2|\Delta f_{\text{opt}} L_1$, where $\Delta f_{\text{opt}} = \Delta\omega_{\text{opt}}/2\pi$ is the optical bandwidth. The 3 dB RF bandwidth in the case of DSB modulation is equal to $\Delta f_{\text{RF}} = [1/(8\pi|\beta_2|L_1)]^{1/2}$ for $M \gg 1$. Naturally, it is highly desirable to maximize the time aperture. For a fixed optical bandwidth, it is possible to do so and maintain a constant stretch factor by increasing $L_1$ while keeping the $L_2/L_1$ ratio constant. However, a larger $L_1$ increases the dispersion penalty and reduces the RF bandwidth. This trade-off renders neither the time-aperture nor the bandwidth capable of assessing the performance alone. In this section, the product of time aperture and RF bandwidth is identified as an optimum metric to evaluate the overall performance and is analyzed for DSB-modulated time-stretch systems. This product also has profound significance as it reflects amounts of information contained in a segment. In addition, the limiting factor on the stretch factor is also discussed.

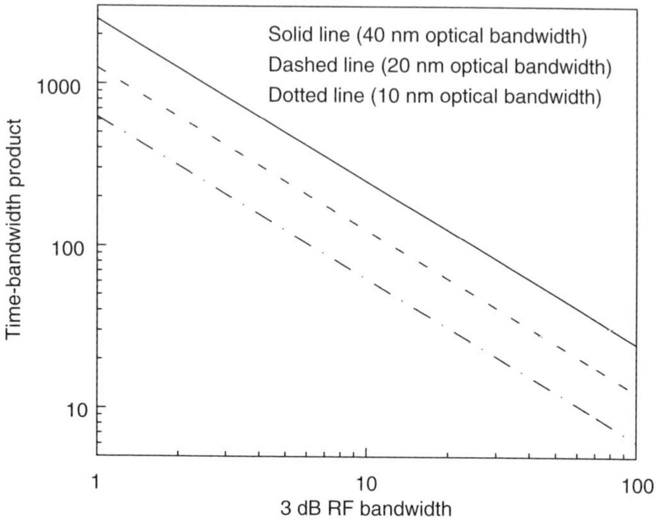


**FIGURE 9.9** The maximum achievable time-bandwidth product for a given 3 dB bandwidth, in a DSB-modulated time-stretched system. Zero-chirp modulation, achieved with a push–pull MZ modulator, is assumed.

Using the above expressions of time aperture and RF bandwidth, the time-bandwidth product (TBP) of the DSB-modulated system can be obtained:

$$\mathrm{TBP_{DSB}} = 2T_{\mathrm{A}}\Delta f_{\mathrm{RF}} = \frac{\Delta f_{\mathrm{opt}}}{2\Delta f_{\mathrm{RF}}}. \tag{9.25}$$

In Equation 9.25, the TBP only depends on the ratio of the O–E bandwidth and this puts the fundamental limits on the achievable TBP. Figure 9.9 shows the calculated TBP for a center wavelength of 1.55 μm and optical bandwidths of 10, 20, and 40 nm, respectively. Clearly, there is a trade-off between TBP and RF bandwidth, while a broadband DSB modulation is relatively easy to accomplish.

In a DSB-modulated time-stretch system, the TBP could be 41.4% larger than the one in Equation 9.25 if chirped modulation with proper bias, discussed in section 9.5.2, is used. In addition, for pass-band RF signals, one may be able to operate at one of the high-frequency lobes of the transfer function. Hence, a longer $L_1$ could be used while maintaining a low dispersion penalty and the time aperture of the system could be increased for a given operation frequency. However, this comes at the expense of RF bandwidth. A detailed analysis demonstrates that it does not render a larger TBP [26].

Equation 9.25 indicates that a larger optical bandwidth is required to obtain a larger TBP in DSB modulation system. In practice, the maximum SC bandwidth that can be used will depend on the wavelength response of the optical modulator. For example, in an MZ modulator, deviation from the center wavelength results in a bias shift away from the quadrature point and a concomitant increase in second-order distortion [28]. In the multioctave system, this effect can limit the spur free dynamic range (SFDR). As an example, for an SFDR = 80 dB, the optical bandwidth is approximately 24 nm for an internally biased modulator. By properly choosing the physical path length mismatch and external bias, the first-order wavelength dependence of the modulator can be eliminated, removing the limitation on the optical bandwidth [28]. Another possible solution is to use the wavelength-insensitive internal bias. This can be accomplished by replacing the Y-junction in the MZ modulator with the $1 \times 2$ MMI (multimode interferometer) coupler. Since two outputs of the $1 \times 2$ MMI coupler have a natural $90^{\circ}$ phase-shift independent of the operation wavelength, no physical optical path difference is required and, hence, the dependence on wavelength can be minimized. The wavelength dependence of the PD response is of lesser concern as it can be accurately characterized and calibrated.

The dispersion-induced bandwidth limitation can be overcome by using techniques described in the next section. In the context of the TBP, these techniques eliminate the trade-off between the large time aperture and the large bandwidth. The length of fiber 1, or equivalently the time aperture, could be freely optimized without sacrificing the RF bandwidth. When the trade-off between time aperture and bandwidth is removed, propagation loss of the fiber may become the dominant limit to the time-bandwidth product. The loss is proportional to the fiber length and, hence, is proportional to the total amount of dispersion needed in the system. Using the definition of time aperture and TBP, the total fiber loss $L_{\mathrm{TS}}$ that is required for a given stretch factor, TBP, and RF bandwidth, is given by

$$L_{\mathrm{TS}} \approx \frac{l}{D}\frac{MT_{\mathrm{A}}}{\Delta\lambda} = \frac{l}{D}\frac{MTBP}{2\Delta\lambda\,\Delta f_{\mathrm{RF}}} \tag{9.26}$$

for $M \gg 1$. Here, $D = (-2\pi c/\lambda^2)\beta_2$ is the dispersion parameter and $l$ is the fiber loss per unit length. As an example, for a stretch factor $M = 10$, an RF bandwidth $\Delta f_{\mathrm{RF}} = 50$ GHz, an optical bandwidth $\Delta\lambda = 40$ nm and $TBP = 1000$, the total fiber loss is 29.4 dB if a standard single-mode fiber (SMF-28) is used ($D = 17$ ps/km/nm, $l = 0.2$ dB/km). Better performance can be obtained if DCF is used ($D = -90$ ps/km/nm, $l = 0.6$ dB/km). With the DCF, the loss is reduced to 16.7 dB. According to Equation 9.26, the loss of the dispersive elements is also the main limitation in obtaining a larger stretch factor. A large fiber loss limits the achievable SNR. From this point of view, dispersive devices with superior dispersion-to-loss ratio than that obtainable with the DCF are highly desirable. Since optical amplifiers are needed to compensate the loss, proper design of an amplified photonic time-stretched system is important. SC sources with high power spectral densities are also desired.

## 9.7 TECHNIQUES TO OVERCOME DISPERSION-INDUCED BANDWIDTH LIMITATION

We have shown that the bandwidth of a time-stretch system is limited by the DIP fading. This bandwidth limit together with time aperture sets the fundamental limit on the amount of information in a time-stretch segment, which can be quantified by the TBP. In this section, two techniques to overcome this bandwidth limitation, SSB modulation and phase diversity are discussed. In Section 9.7.1, the theoretical model in Section 9.5 is revised for SSB modulation. Then, the challenge of performing SSB modulation over a wide bandwidth is examined. Section 9.7.3 presents the principle of phase diversity and maximum ratio combining algorithm—perhaps a superior technique to overcome the bandwidth limitation. Using this technique, a transient digitizer with a record 1 TSa/s sampling rate is demonstrated in Section 9.7.4.

### 9.7.1 SINGLE-SIDEBAND MODULATION

In this section, we provide a complete analysis of the dispersion-induced impairments in the case of SSB modulation. It extends the analysis reported in Ref. [24] by including the harmonic distortion (as in Section 9.5.2), the residual phase errors, including third-order dispersion and finite optical bandwidth (as in Section 10.5.3). SSB modulation can be achieved by using a dual-arm modulator biased at the quadrature point with the $\pi/2$ phase-shifted RF signals applied to the two arms [29]. An SSB-modulated time-stretch system is shown in Figure 9.10. Under this modulation scheme, Equation 9.12 is replaced by

$$E_3(t) = E_2(t)\left\{\exp\left[j\left(\frac{\pi}{2} + \frac{m}{2}\cos\omega_{\mathrm{RF}}t\right)\right] + \exp\left(j\frac{m}{2}\sin\omega_{\mathrm{RF}}t\right)\right\}/2. \quad (9.27)$$

Harmonic distortions in the SSB system can be obtained through the same procedure as in Section 9.5.2 by replacing Equation 9.15 with

$$E_3(t) = \frac{E_2(t)}{2}\sum_{n=-\infty}^{\infty}\left[(1 + j^{n+1})J_n\left(\frac{m}{2}\right)\exp\left(jn\omega_{\mathrm{RF}}t\right)\right]. \quad (9.28)$$

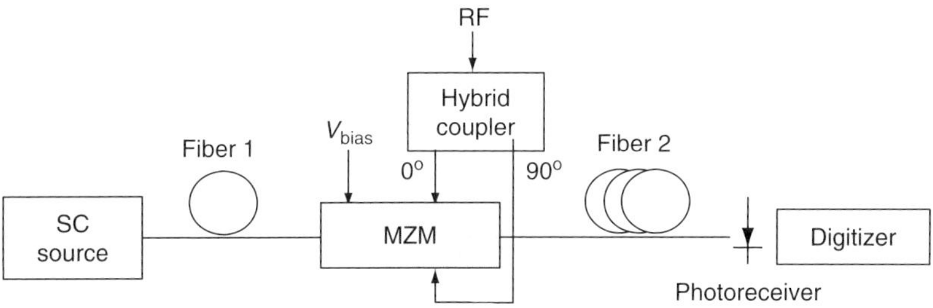


**FIGURE 9.10** Block diagram of the SSB-modulated photonic time-stretched analog-to-digital converter. The hybrid coupler provides a 90° RF phase-shift between its two output ports. MZM: Mach–Zehnder Modulator. SC: Supercontinuum.

In case $m$ is small, the photocurrent can be expressed as

$$I = I_{\text{Envelope}} \times \begin{bmatrix} 1 - \dfrac{m}{\sqrt{2}} \cos\left(\dfrac{\omega_{\text{RF}}}{M} t - \phi_{\text{DIP}} + \dfrac{\pi}{4}\right) \\ + \dfrac{m^2}{8} \sin 4\phi_{\text{DIP}} \cos 2 \dfrac{\omega_{\text{RF}}}{M} t \\ + \dfrac{\sqrt{2} m^3}{192} \cos\left(3 \dfrac{\omega_{\text{RF}}}{M} t + 9\phi_{\text{DIP}} - \dfrac{\pi}{4}\right) \\ - \dfrac{\sqrt{2} m^3}{128} \cos\left(3 \dfrac{\omega_{\text{RF}}}{M} t - 3\phi_{\text{DIP}} + \dfrac{\pi}{4}\right) + \cdots \end{bmatrix}. \tag{9.29}$$

The first two terms describe the results reported in Ref. [24]. The remaining terms are the harmonic distortion in the SSB modulation, reported here for the first time. Clearly, there is no $\beta_2$-induced attenuation in the SSB system. However, a nonlinear phase distortion is introduced through the term $\phi_{\text{DIP}}$. In the context of TS-ADC, the phase distortion could be removed in the digital domain using the known dispersion behavior of the fiber. This process is facilitated by the fact that, because interference of the sidebands is avoided, signal power at the ADC input is preserved, and hence this side effect is tolerable. Therefore, the usefulness of the SSB technique largely depends on the effectiveness of digital correction of the phase distortion. In addition, harmonics distortion observed in the DSB modulation also appears in SSB modulation.

Similarly, the residual phase distortions in the SSB system can be obtained through the same procedure as in Section 9.5.3 by replacing Equation 9.19 by

$$E_3(t) \approx E_2(t) \left[ \frac{1+j}{2} J_0\left(\frac{m}{2}\right) - J_1\left(\frac{m}{2}\right) \exp(-j\omega_{\text{RF}} t) \right]. \tag{9.30}$$

The photocurrent is then given by

$$I(t) = I_{\text{Envelope}} \times \left\{ \begin{array}{l} J_0^2\left(\frac{m}{2}\right) - 2\sqrt{2}J_0\left(\frac{m}{2}\right)J_1\left(\frac{m}{2}\right) \\ \left[\cos\left(\frac{\omega_{\text{RF}}}{M}t - \phi_{\text{DIP}} + \frac{\pi}{4}\right) + \text{Err}_{\beta_3} + \text{Err}_{\text{BW}}\right] \end{array} \right\}, \quad (9.31)$$

$$\text{Err}_{\beta_3} = \frac{\beta_3\omega_{\text{RF}}^2(1-1/M)}{2\beta_2 M} t \sin\left(\frac{\omega_{\text{RF}}}{M}t - \phi_{\text{DIP}} + \frac{\pi}{4}\right), \quad (9.32)$$

$$\text{Err}_{\text{BW}} = \frac{\omega_{\text{RF}}(1-1/M)}{\omega_{\text{opt}}^2\beta_2(L_1+L_2)} t \cos\left(\frac{\omega_{\text{RF}}}{M}t - \phi_{\text{DIP}} + \frac{\pi}{4}\right). \quad (9.33)$$

Although the attenuation of the fundamental frequency due to $\beta_2$ is eliminated, the errors due to $\beta_3$ and due to the finite optical bandwidth remain. In fact, the amplitude of these errors is larger in SSB modulation, by the factor $\sin \phi_{\text{DIP}}$ compared to DSB modulation. This is because, in the DSB modulation, the interference between two carrier-sideband beating terms partially cancels such a distortion.

$\beta_3$ error could also be visualized as a weak dependence of stretch factor on the RF signal frequency. When $\beta_3 = 0$, the stretch factor, $M$, is decided by the following identity as in the derivation of Equation 9.16:

$$\frac{1}{2}\beta_2 L_1(\omega - \omega_{\text{RF}})^2 + \frac{1}{2}\beta_2 L_2\omega^2 = \frac{1}{2}\beta_2(L_1+L_2)\left(\omega - \frac{\omega_{\text{RF}}}{M}\right)^2 + \phi_{\text{DIP}}. \quad (9.34)$$

When $\beta_3$ is nonzero, Equation 9.34 can be revised as

$$\begin{aligned} &\frac{1}{2}\beta_2 L_1(\omega - \omega_{\text{RF}})^2 + \frac{1}{6}\beta_3 L_1(\omega - \omega_{\text{RF}})^3 + \frac{1}{2}\beta_2 L_2\omega^2 + \frac{1}{6}\beta_3 L_2\omega^3 \\ &\approx \frac{1}{2}\beta_2(L_1+L_2)\left(\omega - \frac{\omega_{\text{RF}}}{M}\right)^2 + \frac{1}{6}\beta_3(L_1+L_2)\left(\omega - \frac{\omega_{\text{RF}}}{M}\right)^3 + \phi_{\text{DIP}} + \frac{1}{2}\beta_3\omega\omega_{\text{RF}}^2\frac{L_2}{M} \\ &\approx \frac{1}{2}\beta_2(L_1+L_2)\left(\omega - \frac{\omega_{\text{RF}}}{M'}\right)^2 + \frac{1}{6}\beta_3(L_1+L_2)\left(\omega - \frac{\omega_{\text{RF}}}{M'}\right)^3 + \phi_{\text{DIP}}, \end{aligned} \quad (9.35)$$

where $M' = M[1 + (\beta_3/2\beta_2)\omega_{\text{RF}}]$. The approximation is the same as that in Equation 9.21. Equation 9.35 shows that the residual phase error $\beta_3\omega\omega_{\text{RF}}^2 L_2/2M$ can also be interpreted as an RF-dependent stretch factor. This dependence may cause a nonuniform time stretch and should be minimized by limiting the magnitude of $\beta_3$. In Equation 9.35, the frequency of signal sideband is assumed to be lower than the carrier. When the sideband is higher than the carrier, Equation 9.35 should be modified as $M' = M[1 - (\beta_3/2\beta_2)\omega_{\text{RF}}]$. Therefore, in the DSB modulation, the signal corresponding to one of the two sidebands is slightly overstretched, while the other is slightly understretched. In the PD, the interference between two carrier-sideband beating tones alleviates the overall signal distortion. This agrees with the results in Equation 9.23 and Equation 9.32.

From a dispersion penalty point of view, SSB may be the preferred modulation method. However, the merits of DSB and SSB should be carefully examined in a particular system and for specific applications.

### 9.7.2 Practical Consideration on SSB Modulation

The challenge in SSB time stretch is the ability to perform SSB modulation over a large RF bandwidth. Because the optical carrier in the time-stretched system is broadband, traditional filtering techniques [30,31] for SSB modulation cannot be used. Phase discrimination methods for SSB modulation require a 90° RF phase-shifter. Such devices exhibit amplitude and phase ripples over a broad bandwidth [32]. This practical consideration is discussed in this section.

In a practical broadband SSB modulator, only finite suppression of one of the sidebands can be accomplished. The quality of SSB modulation can be quantified by the sideband field suppression ratio, which is defined here as the ratio of amplitudes of the two sidebands. Assuming that the SSB modulator shown in Figure 9.10 is used, the output optical field $E_{\text{out}}(t)$ can be related to the input optical field $E_{\text{in}}(t)$ by

$$\begin{aligned} E_{\text{out}}(t) &= \frac{E_{\text{in}}(t)}{2}\left\{\exp\left[i\left(\frac{\pi}{2}+\frac{m}{2}\cos\omega_{\text{RF}}t\right)\right]+\exp\left[i\frac{rm}{2}\sin(\omega_{\text{RF}}t+\Delta\phi)\right]\right\} \\ &= \frac{E_{\text{in}}(t)}{2}\sum_{n=-\infty}^{\infty}\left[J_n\left(\frac{m}{2}\right)i^{n+1}+J_n\left(\frac{rm}{2}\right)\exp\left(in\Delta\phi\right)\right]\exp(in\omega_{\text{RF}}t), \end{aligned} \tag{9.36}$$

where the modulator is biased at the quadrature point, $r$ is the interelectrode signal-amplitude ratio (ideally equal to unity), and $\Delta\phi$ is the deviation of interelectrode phase difference from 90°. The quality of hybrid coupler and symmetry of dual-electrode modulator are the main factors that determine $r$ and $\Delta\phi$.

From Equation 9.34, the sideband field suppression ratio, $R$ can be expressed as

$$R=\left|\frac{J_1(rm/2)\exp\left(i\Delta\phi\right)-J_1(m/2)}{J_1(rm/2)\exp\left(-i\Delta\phi\right)+J_1(m/2)}\right|\approx\left|\frac{r\exp\left(i\Delta\phi\right)-1}{r\exp\left(-\,i\Delta\phi\right)+1}\right|, \tag{9.37}$$

where the approximation is valid when the modulation index, $m$, is small. Equation 9.37 shows that, for small modulation depths, $R$ is independent of the modulation depth and depends only on the phase and amplitude imbalances between two arms. Over its operating bandwidth, typical imbalance parameters for a commercially available hybrid coupler are $r=\pm 0.6$ dB and

$\Delta\phi = \pm 7^\circ$ (Anaren 90° hybrid coupler, model# 10029-3). Substituting these values into Equation 9.37, we obtain $R = -10.4$ dB.

The interference between the two sideband-carrier beat signals results in the PD current of the form $I(t) \propto \cos(\omega_{RF}t - \phi_{DIP}) + R\cos(\omega_{RF}t + \phi_{DIP})$. A detailed derivation could follow Section 9.5.2 by replacing Equation 9.15 with Equation 9.36. For a DSB-modulated system, where $R = 1$, it reduces to the dispersion penalty equation, which suggests that the RF power is scaled by $\cos^2 \phi_{DIP}$ (as in Section 9.5). More generally, however, the power transfer function depends on $R$ based on this interference equation and is given by

$$H = \left(\frac{1 - R}{1 + R}\right)^2 \sin^2 \phi_{DIP} + \cos^2 \phi_{DIP}. \tag{9.38}$$

When $\phi_{DIP} = \pi/2$, dispersion penalty reaches its maximum value $[(1 - R)/(1 + R)]^2$. Figure 9.11a shows the dispersion penalty for different sideband amplitude suppression ratios, $R$, for a system with $L_2 = 50$ km and $M = 10$. It is evident that, even with an imperfect SSB modulation, i.e., one with a modest SSB, suppression ratio greatly reduces the dispersion penalty. As an example, when $R = -10.4$ dB, the maximum power penalty is 1.6 dB, which is acceptable for most applications.

An important consideration is the sensitivity of dispersion penalty to physical system parameters. The contour plot in Figure 9.11b shows the maximum power penalty as a function of amplitude and phase imbalances in the dual-arm SSB modulator. As can be observed, to obtain a power penalty of less than 3 dB, the maximum amplitude imbalance of 1.5 dB, or a maximum phase imbalance of 19° can be tolerated. This information is of critical importance in the design of the SSB modulated TS-ADC system. Further, with $M \longrightarrow 1$, the expression for DIP term, $\phi_{DIP}$, reduces to the same for an analog optical link of length $L_2$. Therefore, the analytical model developed here is equally applicable to RF fiber-optical links with the same modulation.

To investigate the performance of the SSB modulator used in the TS-ADC (Figure 9.10), an external-cavity diode laser was used as the carrier so that the modulation sidebands can be examined directly in the optical spectrum analyzer. The typical modulated optical spectrum is shown in Figure 9.12a. Figure 9.12b shows the sideband power suppression ratio (SPSR $= R^2$), obtained from the measured optical spectrum. Results show that a minimum suppression of $-18.8$ dB is achieved across the bandwidth 8–18 GHz. On the basis of the worst-case amplitude and phase imbalances of the hybrid coupler (0.6 dB and 7° across 4–18 GHz), we calculate an SPSR $= R^2 = -20.8$ dB. The difference between this and the measured value can be attributed to the imbalance in the modulator and cables, which is not considered in the theoretical value.

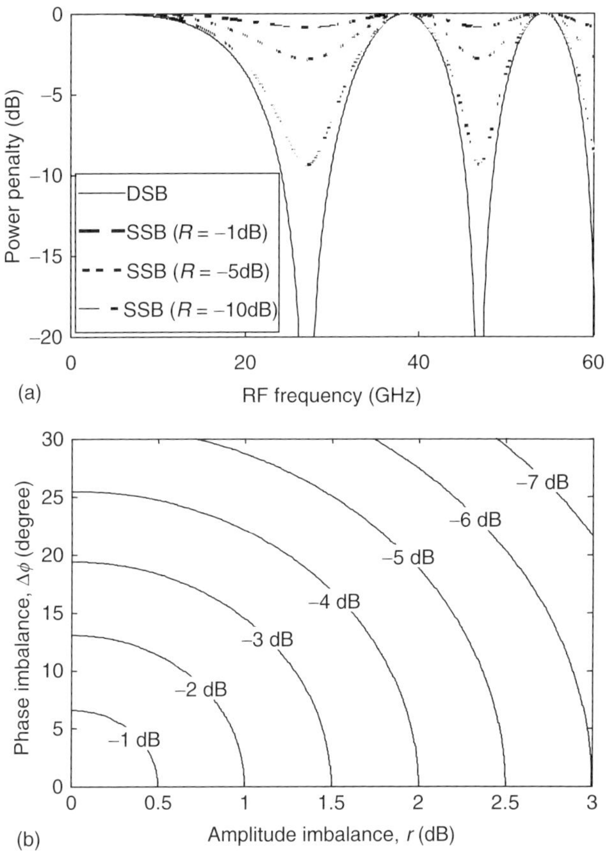


**FIGURE 9.11** (a) Dispersion penalty curves for DSB and SSB modulation, with different sideband suppression ratios, $R$. (b) Contour map of the maximum dispersion penalty (marker values) as a function of amplitude and phase imbalances in the SSB modulator.

Also shown in Figure 9.12b is the maximum dispersion penalty corresponding to the measured SPSR based on Equation 9.38. Results indicated that the penalty can be kept below 2.5 dB, from 8 to 20 GHz. The lowest frequency that could be measured was limited to 8 GHz by the optical spectrum analyzer. However, since the hybrid coupler maintains the specified amplitude and phase imbalance down to 4 GHz, we conclude that the maximum penalty is 2.5 dB across 4–20 GHz.

Figure 9.13 shows a 20 GHz signal captured at 120 GSa/s with the SSB TS-ADC and the corresponding sine wave curve fit test. The noise bandwidth was limited to 4 to around 20 GHz. DCF was used as the dispersive element.

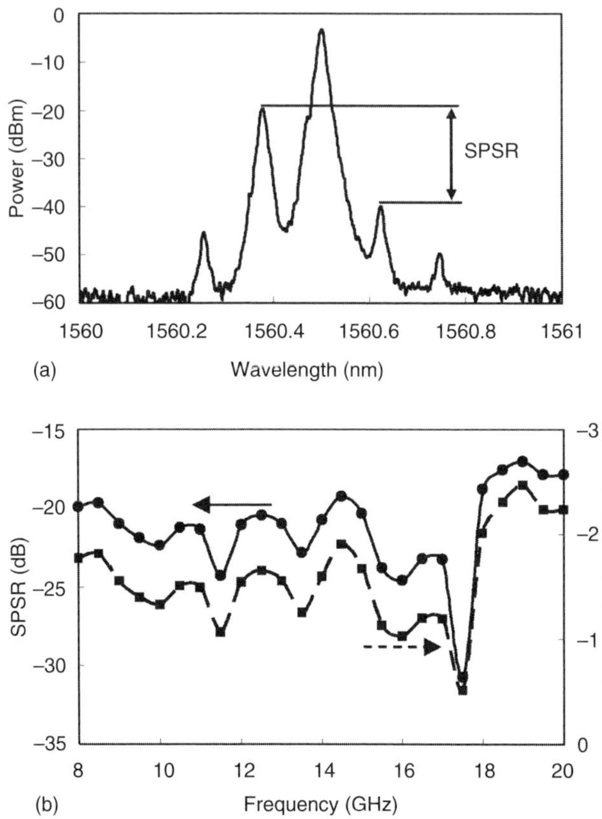


**FIGURE 9.12** (a) Measured optical spectrum of the SSB-modulated signal. SPSR denotes the sideband power suppression ratio. (b) The measured frequency dependence of SPSR (solid line) and the corresponding maximum power penalty (dash line).

The stretch factor was measured to be 5.94, the optical modulation index was 18.6%, and the input aperture was 2.9 ns. The electronic ADC was the Tektronix TDS7404 digital oscilloscope with 4 GHz input bandwidth and 20 GSa/s sample rate. Therefore, the effective input sampling rate was approximately 120 GSa/s.

### 9.7.3 Phase Diversity and Maximum Ratio Combining Algorithm

As discussed in Section 9.7.1, SSB modulation is able to mitigate dispersion penalty; however, it introduces phase distortion, which must be equalized in the digital domain. Additionally, the bandwidth over which SSB modulation

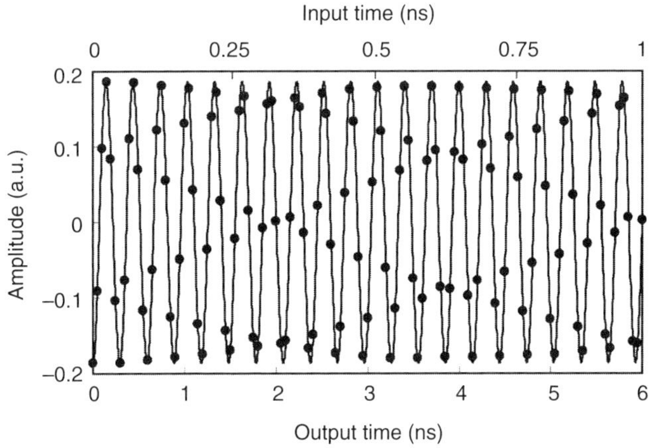


**FIGURE 9.13** 120 GSa/s real-time capture of a 20 GHz RF signal by the SSB TS-ADC system. Measured data (symbols) and fitted sine curve (solid line) of a time-stretched ADC. Effective noise bandwidth is 4 GHz.

can be achieved is limited by the microwave hybrid used in the setup. In this section, we propose another technique that overcomes the dispersion penalty in a TS-ADC. We exploit natural phase diversity that exists between the two outputs of a dual-output MZ modulator to overcome the RF fading problem caused by dispersion. We also present an extension of the maximum ratio combining (MRC) algorithm, imported from the world of wireless communication, for application to this system. The MRC not only eliminates the fading but also offers maximum SNR. This technique is also applicable to traditional fiber communication links.

The RF fading phenomenon in the time-stretch system is somewhat similar to the channel fading problem in wireless communication. There, multipath propagation causes interference resulting in spatial fading [33]. In time stretching, frequency fading arises from the interference between two RF sidebands when they beat with the carrier after propagation through the fiber. Spatial diversity, achieved using multiple antennas, has proven to be an effective method to overcome fading in the wireless systems [33]. In the same spirit, the proposed technique makes use of phase diversity, inherent in a dual-output modulator, to overcome the frequency fading in the time-stretch system. In contrast to the wireless system, the fading in time stretch is deterministic, greatly simplifying the system design and signal processing.

Figure 9.14 shows a single-arm dual-output MZ modulator. The output optical fields $E_{out,1}$ and $E_{out,2}$ are related to the input optical field $E_{in}$ by

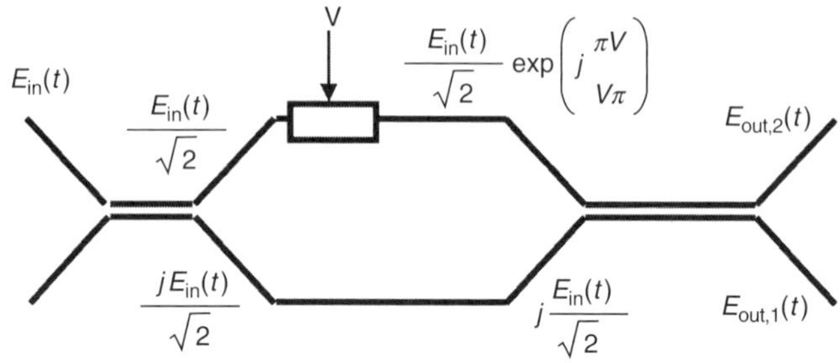


**FIGURE 9.14** Signal propagation in a single-arm dual-output MZ modulator.

$$\begin{aligned} E_{out,1} &= j\frac{E_{in}}{2}\left(\exp\left(j\frac{\pi V}{V_\pi}\right)+1\right) = jE_{in}\exp\left(j\frac{\pi V}{2V_\pi}\right)\cos\left(\frac{\pi V}{2V_\pi}\right), \\ E_{out,2} &= \frac{E_{in}}{2}\left(\exp\left(j\frac{\pi V}{V_\pi}\right)-1\right) = jE_{in}\exp\left(j\frac{\pi V}{2V_\pi}\right)\sin\left(\frac{\pi V}{2V_\pi}\right). \end{aligned} \tag{9.39}$$

The outputs contain both amplitude and phase modulation. The phase modulation is converted to amplitude modulation after dispersion in the fiber. The key observation is that the two ports differ by 90° in the phase; hence, their fading characteristics are different. The RF power spectrum of two outputs, using the same procedure in Section 9.5, is given by

$$H_1(f) = \cos^2\left(\phi_{DIP} - \frac{\pi}{4}\right), \qquad H_2(f) = \cos^2\left(\phi_{DIP} + \frac{\pi}{4}\right). \tag{9.40}$$

The plot of power penalty versus the signal frequency is shown in Figure 9.15. When one output experiences a null, the other reaches the maximum. This complementary diversity behavior is used to overcome the frequency fading when both ports are collected. Specifically, $H_1(f) + H_2(f) = \cos^2(\phi_{DIP} - \pi/4) + \cos^2(\phi_{DIP} + \pi/4) = 1$, free of any fading. We note that both outputs can be stretched in the same fiber and captured by the same digitizer by delaying one, by half the pulse repetition period, and combining them before the stretch fiber.

Owing to the similarity between the RF fading in time stretch and the spatial fading in the wireless system, antenna combining algorithms used to maximize the SNR in wireless may also be applicable here. There are some differences, however. In particular, the time stretch is a broadband (multi-octave) and deterministic system, while wireless is narrowband (suboctave) and stochastic. In general, the algorithm used in time stretch is a broadband extension of algorithm used in wireless. The fact that the system is deterministic greatly simplifies the problem. A simple solution is to use a switch to select the output having a higher SNR. A more interesting approach is MRC. The MRC algorithm sums the channels with a weight that reflects the voltage

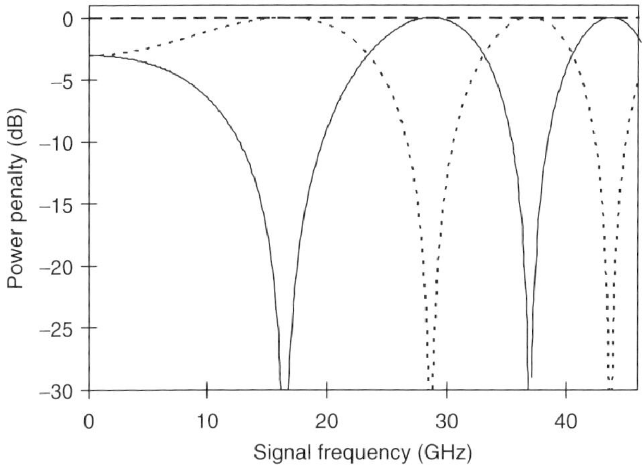


**FIGURE 9.15** Calculated dispersion penalty for two outputs of a single-arm MZ modulator. $H_1(f)$: the dot line; $H_2(f)$: the solid line; $H_1(f)+H_2(f)$: the dashed line.

SNR in that channel [33]. In time stretch, the weights are $W_1 = \cos(\phi_{DIP} + \pi/4)$ and $W_2 = \cos(\phi_{DIP} - \pi/4)$. It can readily be seen that

$$W_1 \cdot \cos(\phi_{DIP} + \pi/4) + W_2 \cdot \cos(\phi_{DIP} - \pi/4) = 1. \tag{9.41}$$

Clearly, the weights are proportional to the signal voltage-to-noise ratios. The complementary characteristics of phase diversity ensure that the combined output is independent of frequency, as described in Equation 9.41. In other words, frequency fading is fully eliminated. The broadband requirement implies that the MRC weights are frequency-dependent, as reflected in the expression for $\phi_{DIP}$. MRC algorithm can also be used to remove the $\beta_3$ distortion, if the deterministic wavelength dependence caused by $\beta_3$ is included in the weights in Equation 9.41. The function in Equation 9. 41 is best implemented in digital domain after A/D conversion.

### 9.7.4 Tera-Sample Transient Digitizer

Using the phase diversity technique, a time-stretch transient digitizer with a record sampling rate of 1 TSa/s and intrinsic bandwidth of 200 GHz is demonstrated in this section. The TSa/s digitizer consists of a 50× time-stretch preprocessor and a 20 GSa/s electronic digitizer. A practical difficulty in accomplishing a 50× stretch factor with a time aperture of more than 1 ns is to overcome the large loss of dispersive fibers. This is mitigated by the judicious use of optical amplification in such a manner as to optimize the overall SNR, while avoiding the distortion from optical nonlinearity.

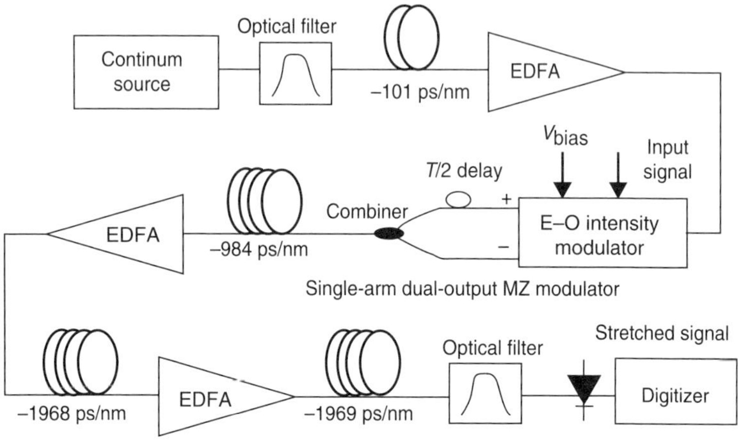


**FIGURE 9.16** Experiment setup of TSa/s digitizer with a 50× photonic time-stretch preprocessor.

The block diagram of TSa/s TS-ADC setup is shown in Figure 9.16. A similar SC source is used, but with the reduced repetition rate (by a factor of 5) to avoid the optical pulse overlap after stretching. This is performed by using a pulse picker circuit and an optical switch. Around 15 nm of the optical spectrum (centered at 1590 nm) is sliced using an optical bandpass filter and is used in the experiment. After being linearly chirped by propagation in a spool of dispersive fiber, a given pulse enters the electro-optic modulator where it captures the electrical transient. The chirped optical pulse has finite amplitude variations resulting in distortion of the captured signal. The distortion is removed by using a reference waveform and/or digital filtering, as described in Section 9.9. The modulator has two outputs that have the complementary fading characteristics, and, after delaying one by $T/2$, the outputs are time-interleaved and stretched together in the second dispersive fiber. The first fiber has a total dispersion of $D_1 = -101$ ps/nm, creating a chirped optical pulse with around 1.5 ns duration. The second fiber has a total dispersion of $D_2 = -4921$ ps/nm. The stretch factor is given by $(D_1+D_2)/D_1 = 50$. To minimize the noise contributed by the optical amplifiers, the waveform is filtered by an optical bandpass filter before photodetection. The detected waveform is subsequently captured by a Tektronix TDS7404 (4 GHz analog bandwidth, 20 GSa/s) real-time digitizer. The effective sampling rate of this system is 20 GSa/s × 50 = 1 TSa/s and the intrinsic input analog bandwidth is 4 GHz × 50 = 200 GHz.

Figure 9.17 shows the real-time capture of a 48 GHz tone at 1 TSa/s. To measure the SNR, similar measurements using 200 optical pulses are obtained. Over a 10 GHz digitally filtered bandwidth, the average SNR (over 200 measurements) is 22.7 ± 1.08 dB corresponding to 3.5 ± 0.2

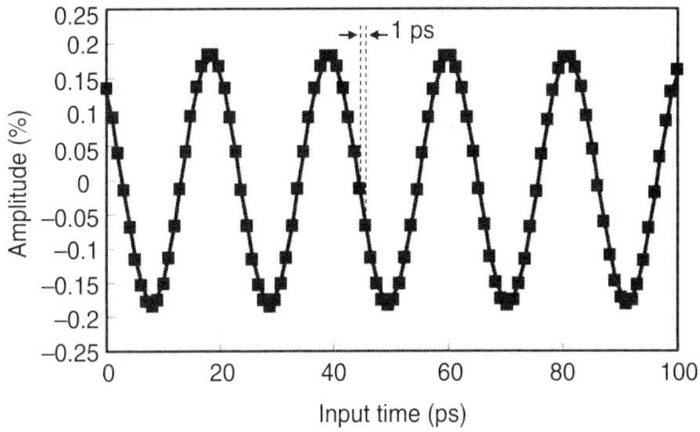


**FIGURE 9.17** Digitization of a 48 GHz sinusoid at 1 TSa/s. Lines are obtained using standard sine curve fitting.

ENOB. Measurements at other input frequencies resulted in up to 4.2 ENOB. The measured stretch factor and voltage full width half maximum (FWHM) time aperture are 50.5 and 1.1 ns, respectively.

The digitizer performance is also measured in the frequency domain. Figure 9.18 shows the digital spectrum of a 26 GHz tone and a 48 GHz tone. The Hanning window was used before discrete Fourier transform (DFT). In Figure 9.18, the spurious tone at 125 GHz (2.5 GHz after 50× stretch) is due to the digitizer and is not caused by the photonic time stretcher. We note that no digital filter was used to obtain the data in Figure 9.18. Hence, the input analog bandwidth of these spectrums is 0–200 GHz. A 2nd harmonic

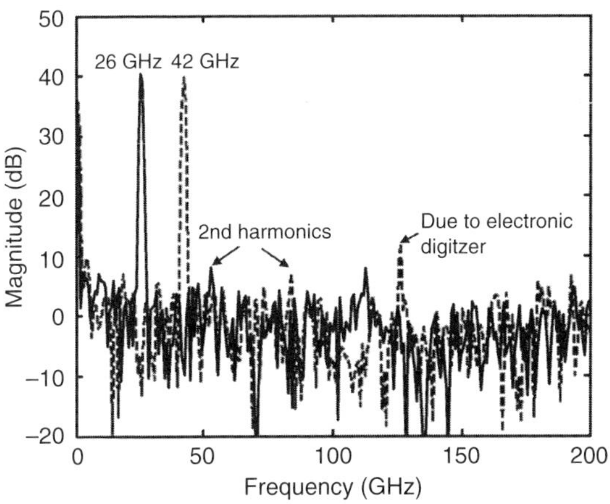


**FIGURE 9.18** The digital magnitude spectrum of captured signals at 26 GHz (solid line) and 48 GHz (dashed line).

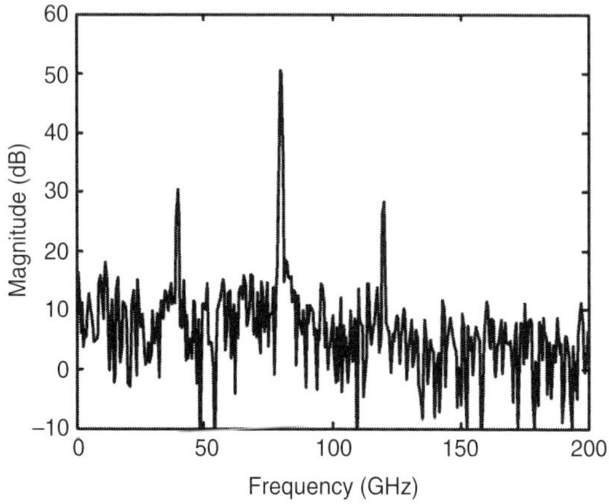


**FIGURE 9.19** The digital magnitude spectrum of captured signals at 80 GHz.

distortion tone is observed in Figure 9.18. This has been theoretically predicted in Section 9.5. Postdistortion performed in the digital domain can potentially mitigate this type of distortion.

The TSa/s digitizer has an intrinsic bandwidth of 200 GHz, as shown in Figure 9.18. Practically, the input bandwidth is limited by the bandwidth of the electro-optic modulator to approximately 50 GHz. One method to effectively increase the modulator bandwidth is to use harmonic modulation. The MZ modulator is conventionally biased at the quadrature point, at which even harmonics are suppressed. If an MZ modulator is biased at the peak or null of transmission, odd harmonics, instead, are suppressed. The lowest frequency component at the output of modulator is the second harmonic. For a sinusoid input, the bandwidth of MZ modulator is effectively doubled. An 80 GHz sinusoid modulated tone is generated from a 40 GHz input by using this technique. The digital magnitude spectrum of captured signal is shown in Figure 9.19. The clear 80 GHz tone demonstrates the effectiveness of the time-stretch processor beyond 50 GHz. The spurious tones at 40 and 120 GHz are from the undersuppressed fundamental frequency and the 3rd harmonic.

## 9.8 EFFECT OF OPTICAL NONLINEARITY ON FIDELITY OF THE TIME-STRETCHED ELECTRICAL SIGNAL

If the optical power is low, the chirped pulse propagation in fiber is linear and can be accurately described by Equation 9.11 and Equation 9.13. However, higher optical power is desired, as shown in Section 9.4, so that every 6 dB increase in power spectral density increases the ASE-signal beat-noise-limited resolution by 1 bit. On the other hand, a larger stretch factor requires a longer fiber to be used. Both the high power and longer propagation distance may cause the pulse propagation to deviate from the linear regime. Therefore, the model used in Section 9.5 needs to be modified to include fiber nonlinearity, otherwise known as the "Kerr effect."

In the nonlinear regime, the optical pulse propagation is governed by the nonlinear Schrödinger Equation 9.22,

$$\frac{\partial A}{\partial z} + \frac{\alpha}{2}A + \frac{i}{2}\beta_2\frac{\partial^2 A}{\partial t^2} - \frac{1}{6}\beta_3\frac{\partial^3 A}{\partial t^3} = i\gamma|A|^2 A, \tag{9.42}$$

where $\alpha$ is the attenuation constant and the nonlinear coefficient $\gamma$ is related to the nonlinear index coefficient $n_2$ by $\gamma = (n_2\omega_0)/(cA_{\text{eff}})$. $A$ is the slowly varying amplitude and is related to the optical field used in Section 9.5 by $E(\vec{r},t) = \hat{x}\frac{1}{2}\{F(x,y)A(z,t)\exp[i(\beta_0 z - \omega_0 t)] + c.c.\}$.When the nonlinear term $\gamma|A|^2A$ is neglected, Equation 9.42 reduces to Equation 9.11 and Equation 9.13.

Generally, no analytical solution exists when dealing with the nonlinear propagation Equation 9.42. The numerical approach of the split-step Fourier method [22] is used in the simulations presented here. The result is shown in Figure 9.20. The parameters used in the simulation are summarized here. $\beta_2 = -21.67$ ps$^2$/km, $\gamma = 1.3$ rad/W/km, $L_1 = 7.25$ km, $M = 17.5$, $\Delta\lambda = 15$ nm, $m = 0.1$, $f_{\text{RF}} = 48$ GHz. The energy of Gaussian optical pulse is 50 pJ. The single-arm MZ modulator is assumed and fiber loss, ASE noise, and $\beta_3$ are neglected. Since the electrical signal does not interact with the optical pulse until the electro-optic modulation, only fiber nonlinearity in the second fiber has notable effects on the fidelity of stretched signal.

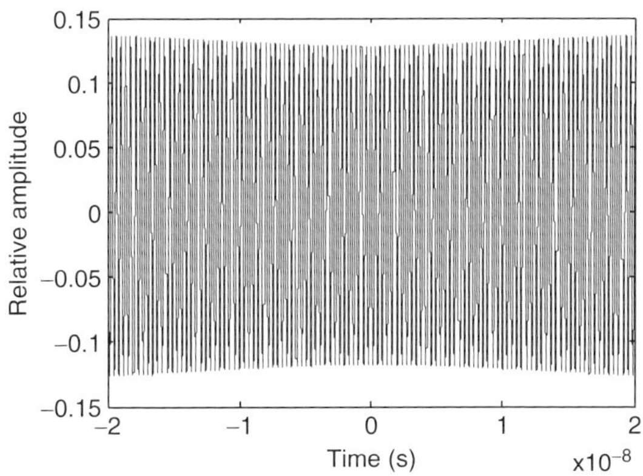


**FIGURE 9.20** The simulated time-stretched sinusoid signal where the fiber nonlinearity is considered. The distortion is the result of interaction between dispersion and fiber nonlinearity.

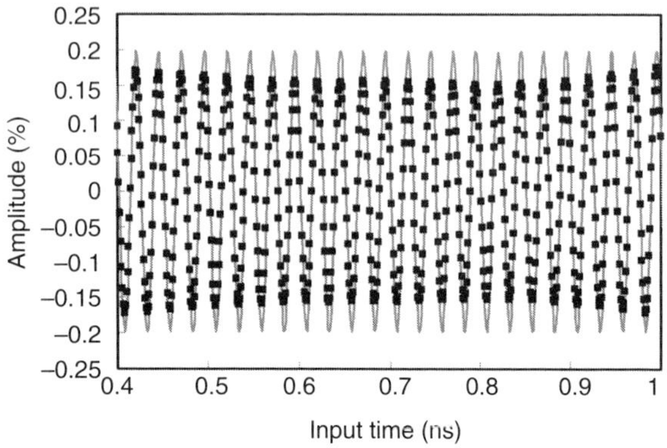


**FIGURE 9.21** Distortion caused by nonlinear optical effect in the dispersive fiber.

The central portion of the Gaussian optical pulse has higher instantaneous power. Figure 9.20 shows that there is a larger attenuation of the RF signal in this region. This amplitude dependent attenuation is the result of interaction between dispersion and fiber nonlinearity. In Section 9.5, it was shown that the amount of dispersion-induced attenuation depends on the initial modulation chirp. When the modulated optical pulse is propagating through the second fiber, optical nonlinearity introduces an intensity-dependent phase-shift (self-phase modulation). In case of sinusoid intensity modulation, the nonlinear phase-shift is also a sinusoid function of the same frequency. This phase modulation due to fiber nonlinearity mixes with the amplitude modulation signal and varies the effective modulation chirp. As a result, the amount of attenuation caused by dispersion changes. Because the instantaneous power is not constant along the pulse, the extent of nonlinear phase-shift varies and introduces the distortion pattern shown in Figure 9.20. As a result, the optical power in the time-stretch system should be carefully controlled to avoid distortion from fiber nonlinearity.

The distortion pattern predicted in simulation is also observed in experiments. An example of a severely distorted signal is shown in Figure 9.21. The setup is the same as the one in the TSa/s experiment, but with a higher optical power. The distortion (amplitude reduction) is most severe at the center of pulse, where the pulse has highest peak power. Distributed Raman amplifiers are potentially useful in eliminating this effect.

## 9.9 REMOVAL OF DISTORTION CAUSED BY SPECTRAL NONUNIFORMITIES OF THE OPTICAL SOURCE

Another challenge in the TS-ADC system is the nonuniform spectral density of the SC source and its pulse-to-pulse variation. The flatness is determined primarily by that of the source as well as the influence from optical filters and the EDFA gain profile. Owing to the wavelength-to-time mapping occurring in the first fiber, spectral nonuniformities appear as a temporal modulation of the chirped pulse entering the modulator and, hence, mix with and distort the desired electrical signal. Moreover, the spectral nonuniformity is not static due to pulse-to-pulse variation of the SC source [34,35]. To correct these, a dynamic calibration method that is able to track the pulse-to-pulse variation is needed. In this section, a technique using a dual-output differential MZ modulator is discussed [36]. As will be shown, this is an effective method for removing this form of distortion.

In a differential modulated system, the RF signal, $s(t)$, is obtained by performing the following operation,

$$s(t) = \frac{1}{m} \frac{I^{+}(t) - I^{-}(t)}{I^{+}(t) + I^{-}(t)}, \tag{9.43}$$

where

$$I^{+}(t) = \frac{env(t)}{2} \cdot [1 + m \cdot s(t)]$$

and

$$I^{-}(t) = \frac{env(t)}{2} \cdot [1 - m \cdot s(t)].$$

Here, $env(t)$ is the intensity of the chirped carrier, which contains unwanted temporal variations and represents the input of modulator, and $I^{+}(t)$ and $I^{-}(t)$ are the differential outputs of the MZ modulator. One of the differential outputs follows the typical modulation equation, such as Equation 9.12, with the small signal linear approximation. The other complement output could be easily obtained by the law of power conservation. Unwanted temporal variations, $env(t)$, arise when nonuniform spectrum of the SC source and ripples in the optical filter are mapped into time through wavelength-to-time mapping. The subtraction and division operation remove the additive and multiplicative distortion, respectively. Since the correction is performed for each pulse, the system is immune to pulse-to-pulse variations of the SC source. In addition, Equation 9.43 can be used to linearize the modulation if the arcsin() function is applied to $ms(t)$. We note that the differential signal processing technique described above has been used by others for noise and distortion suppression in other types of optical systems that make use of MZ modulators [37].

The experimental setup is shown in Figure 9.22. The SC is obtained by amplifying and passing the output of a 20 MHz passively mode-locked laser

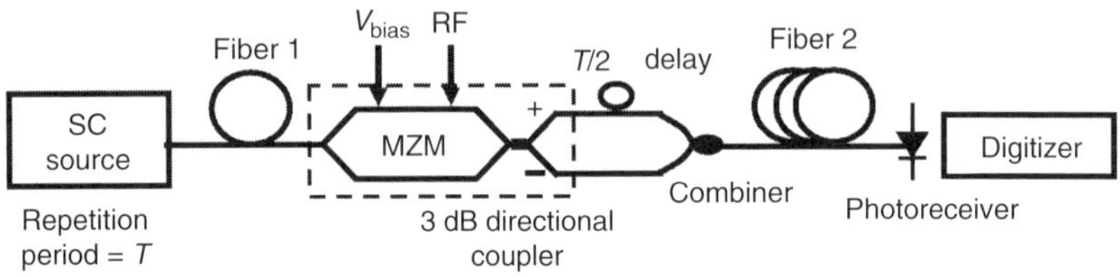


**FIGURE 9.22** Block diagram of the differential photonic time-stretched analog-to-digital converter. Repetition rate of the optical source is $T$. One of the outputs is delayed by $T/2$ before it is combined with the other output. Both outputs can now be captured with a single digitizer.

through a dispersion decreasing fiber. A bandpass filter selects a portion of the SC in the L band centered at 1587 nm. The L band is chosen because the source has a higher signal to ASE noise ratio away from the C band. DCF is used as the dispersive medium. An optical amplifier is inserted before the modulator to compensate for the system loss. The differential outputs of modulator are time-interleaved by a 3 dB coupler after one channel is delayed by 25 ns (half the repetition period). The electronic ADC is the Tektronix TDS7404 digital oscilloscope with 4 GHz input bandwidth and 20 GSa/s sampling rate. The stretch factor, $M = 1 + L_2/L_1$, of the system is 6. Therefore, the effective sampling rate of this system is 120 GSa/s.

The differential time-stretch system block diagram, Figure 9.22, appears similar to the phase diversity time-stretch system, Figure 9.16, described in Section 9.7. Both techniques use the dual-output modulator. However, there is a fundamental distinction between two systems. The differential technique uses the push–pull dual-electrode MZ modulator. Here the output is pure amplitude modulation, as shown in Equation 9.12, with the two ports experiencing the same fading characteristics as those discussed in Section 9.5. In contrast, the phase-diversity technique uses the single-electrode MZ modulator, which produces both amplitude and phase modulation, as shown in Equation 9.39, and the fading characteristics are different.

In the experiment, a 12 GHz sinusoid signal is captured by the time-stretched ADC. Figure 9.23 shows the signals from both modulator arms, i.e., $I^{+}(t)$ and $I^{-}(t)$ along with the average value $[I^{+}(t) + I^{-}(t)]/2$. The average represents the nonuniform SC envelope. Its slow variations are due to the SC source, and the fast variations originate from the ripples of the bandpass filter used to select the suitable SC portion. The gain mismatch and delay between two outputs have been calibrated. The distortion caused by the SC envelope appears as common-mode distortion of both outputs. Performing the operation described in Equation 9.43 provides the waveform shown in Figure 9.24, the samples and the corresponding sine wave curve fit test. The sampling rate is 120 GSa/s and the time aperture is 2.1 ns. The average SNR over 100 pulses is $22.7 \pm 0.6$ dB, which corresponds to $3.5 \pm 0.1$ bits, with the

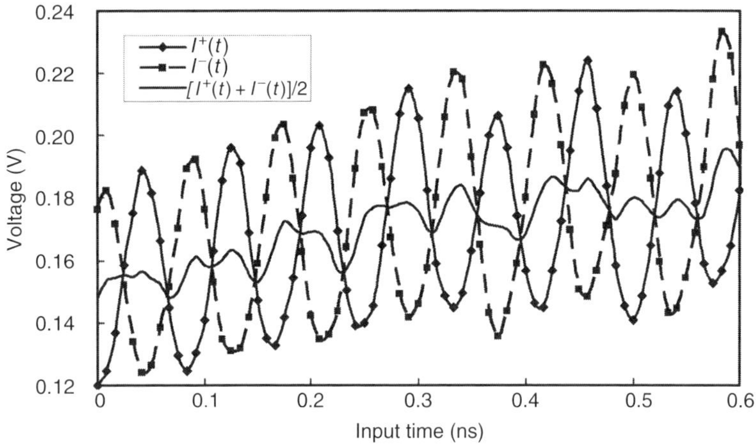


**FIGURE 9.23** Differential outputs, $I^+$ and $I^-$ of the modulator and their average value: $(I^+ + I^-)/2$. The average represents the spectral shape of the optical source including the influence of the bandpass filter.

24 GHz effective noise bandwidth. We separately characterized the digitizer with a test tone of the same amplitude as the stretched signal and equal to the digitizer's full scale range. The measured SNR over the full bandwidth

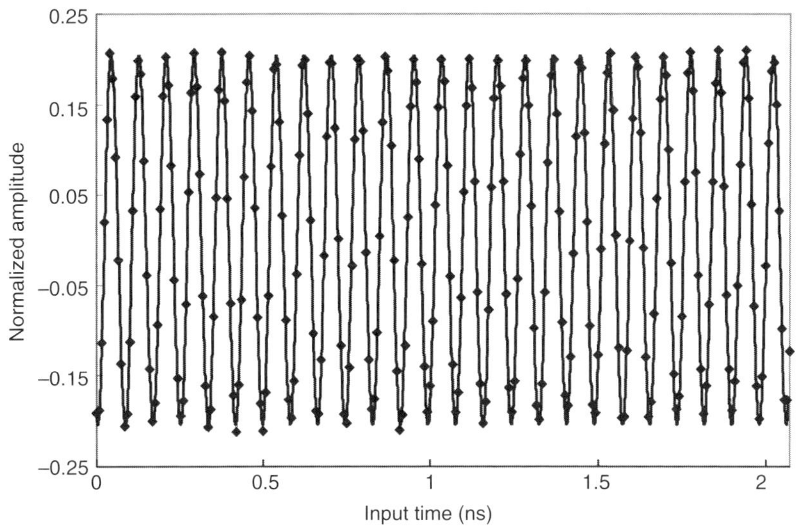


**FIGURE 9.24** 120 GSa/s real-time capture of a 12 GHz RF signal by the differential TS-ADC system. The nonuniform envelope is removed by using Equation 9.37. Effective noise bandwidth is 24 GHz.

(4 GHz) was 28.8 dB, which is approximately one bit higher than that of TS-ADC.

The differential architecture removes the distortion caused by the nonuniform SC spectrum. The remaining fundamental sources of error are the ASE-signal-beat noise and the digitizer's quantization noise. In this experiment, we observe a reduction in SNR when the full scale voltage, $V_{FS}$, of the quantizer is increased. This is consistent with a quantization noise-limited performance. If $A$ is the stretched signal amplitude, then the quantization-limited SNR is given by [38]:

$$\mathrm{SNR} = 6.02 \cdot \mathrm{ENOB} + 1.76 + 20 \log_{10} \frac{A}{V_{FS}}. \tag{9.44}$$

Equation 9.44 is also the definition of ENOB. The highest SNR is obtained when $A = V_{FS}$. If $V_{FS}$ is increased, the quantization-limited SNR decreases. By contrast, the ASE-signal beat noise is independent of the ratio $A/V_{FS}$. In addition, as can be seen in Figure 9.23, due to the nonuniform SC envelope, signals $I^{+}(t)$ and $I^{-}(t)$ could not occupy the full-scale of the digitizer over the entire time aperture. Hence, $A < V_{FS}$ in the experiment, resulting in a reduction in the SNR due to the increase of the quantization noise. This is the part of reason for the 1-bit difference between the measured ENOB of TS-ADC and back-to-back experiment. The fact that the SNR is not limited by the ASE represents an improvement over our earlier experiment [10]. This has been achieved by reducing the system losses (hence a reduction in the number of optical amplifiers) and by operating in the L-band portion of the SC where ASE noise level is lower.

The following explanation is necessary with respect to system bandwidth in this particular experiment. In an SSB-modulated system [24,26], the effective input bandwidth would be 4 GHz $\times$ 6 = 24 GHz. In the above experiments, because of the unavailability of a dual-output SSB modulator, the experiments were performed with DSB modulation. The dispersion penalty in this system limits the input of 3 dB bandwidth to 19.3 GHz. However, for quantization noise and ASE noise, the effective bandwidth is 24 GHz, independent of the modulation format (DSB or SSB).

The differential technique is one method of removing the distortion caused by spectral nonuniformities. Another method is filtering in the digital domain. If the variation caused by distortion is slow compared to the RF input signal, it can be separated from the signal in the frequency domain by low-pass filtering. The envelope so obtained can then be used to correct for the distortion. This method will place a low-frequency cutoff on the system, but is otherwise able to compensate the spectral nonuniformity and its pulse-to-pulse dependence. If frequency components corresponding to nonuniform spectrum coincide with the stretched signal, it is often necessary to separately measure a reference pulse shape when no RF signal is fed into the time-stretch system. Since high-frequency ripples mostly come from fixed optical components and do not change with time, they can be effectively characterized ahead of the measurement and be calibrated. In the general approach, static nonuniformities are first removed by comparing to the reference. Then, the self-background correction technique is used to remove the residual time-varying spectral nonuniformities. This method was employed in the SSB and phase diversity experiments described in Section 9.7.

The observation that spectral variations are mapped to similar variations in time (after dispersion) has led us to propose a new method for generating ultrawideband electrical waveforms with arbitrary modulation [39,40]. In this approach, the spectrum of a broadband optical pulse is shaped according to the waveform of interest. Following dispersion, this spectral waveform is directly transformed into an identical time waveform, which is subsequently converted to an electronic domain using a PD. In a recent demonstration, intelligent digital control of the spectrum has been used to render the system insensitive to the nonuniform power spectral density of the optical source [40]. The system directly converts digital data to ultrawideband analog waveforms.

We have extended the concept of time–wavelength transformation to spectroscopy and chemical detection [41]. Here, an SC pulse passing through a gas is intensity modulated by the absorption spectrum of the pulse. After dispersion, the spectrum is mapped into time and can be measured using an oscilloscope, or directly using an electronic ADC. We then apply a digital time domain filter, resulting in binary detection of the gas constituents [41]. This technique has several advantages: (1) it eliminates the need for ultrawideband tunable lasers, (2) it eliminates the need for bulky optical spectrum analyzers or monochromators, and (3) it measures the instantaneous changes in the spectrum, allowing real-time study of molecular dynamics.

## 9.10 CONTINUOUS-TIME OPERATION

To fully understand the operation of the continuous-time TS-ADC, shown in Figure 9.2b, it is instructive to compare it with the conventional sample-interleaved ADC shown in Figure 9.1. The sample-interleaved approach is the standard approach for achieving high sampling rate and is used in all such electronic digitizers. The input analog signal is simultaneously applied to a bank on $M$ digitizers. Digitizers are then sequentially clocked at a rate of $f_s/M$, where $f_s$ is the desired aggregate sampling rate. The signal is then reconstructed in the digital domain by combining sequential samples from the array of digitizers. There are two important differences between this architecture and the continuous-time time-stretch system shown in Figure 9.2b. First is the fact that each digitizer in Figure 9.1 sees the input signal at its full bandwidth. Although the requirement on the sampling rate for each digitizer is reduced from $f_s$ to $f_s/M$, there is no such reduction on the required input bandwidth. Therefore, the input bandwidth of the system is not improved in this architecture. In contrast, the signal is slowed down before it enters the digitizers in the TS-ADC. The second important difference between the two systems is the relative (to input frequency) sampling rate of each digitizer. In the sample-interleaved system, sampling by each digitizer is done at a fraction of the required Nyquist rate. The original signal can only be recovered *after* sample-by-sample reconstruction in the digital domain. In contrast, because the signal is slowed down prior to digitization, each digitizer samples the signal above the Nyquist rate in the TS-ADC. In other words, a time-segment of the input signal is available from each digitizer *before* reconstruction. This is a significant feature of the TS-ADC and permits real-time cancelation of mismatch errors, as discussed below.

It is well established that the performance of a sample-interleaved ADC is limited by the mismatches between different channels [3]. Interchannel gain, offset, and clock skew create spurious tones in the frequency domain and limit the dynamic range. Owing to its parallel architecture, TS-ADC also suffers from interchannel mismatch, because different segments have to be combined to reconstruct the original signal. However, the fact that sampling in each channel meets the Nyquist criteria can be used to correct the mismatch errors. This can be achieved by allowing a finite overlap between adjacent segments, as shown in Figure 9.25. By comparing signals from two adjacent digitizers in the overlap region, mismatch errors can be calculated and corrected in the digital domain [9]. Because the error is obtained from the signal itself, not from a test signal, one is able to perform real-time error correction, addressing amplitude and frequency dependencies and dynamic variations of the error. The overhead associated with the overlap decreases with the increase in segment length and is not significant [9].

An interesting mode of operation for the continuous-time TS-ADC is one in which reconstruction of the signal from different segments is entirely avoided. If the segment length is sufficiently long, each segment can provide the necessary spectral information. The minimum segment length is then determined by the required resolution bandwidth. In this mode of operation,

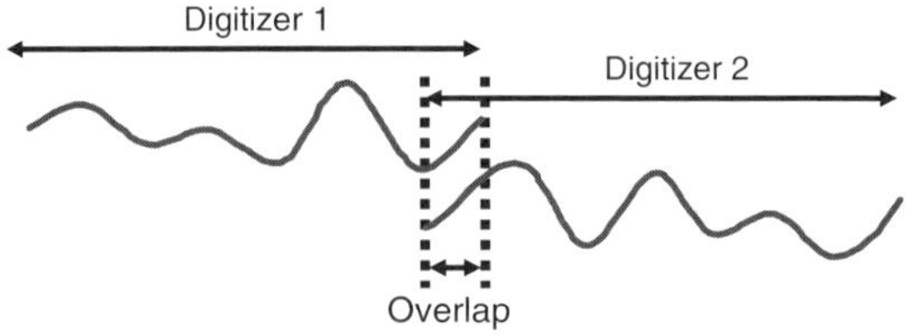


**FIGURE 9.25** Two adjacent channels of the input signal. By allowing for a finite overlap of the two channels, interchannel mismatch errors can be estimated and corrected.

subsequent segments simply update the captured signal spectrum at the segment arrival rate. Offset mismatch manifests itself as uncertainty in the DC component between each segment. Gain mismatch affects the signal amplitude, resulting in an uncertainty in the absolute value of the power spectrum. However, this does not affect the relative power of the spectral components, which is the quantity of interest in most applications. Clock (segment-to-segment timing) skew has no impact on the resulting spectrum—it will introduce a time offset. Since the segment length might not always be an integer multiple of the RF signal period, the segmentation might introduce errors similar to the leakage errors in DSP. However, the error can be minimized with windowing and/or sufficiently long segments.

The continuous-time TS-ADCs rely on a means to segment-interleave the incoming signal. The use of an active gating switch is the most obvious approach. However, the time–wavelength mapping offers a clever alternative to do this. The gating can be performed using a passive optical filter. The principle of passive virtual time-gating is shown in Figure 9.26 and it greatly simplifies the implementation of continuous-time TS-ADC. To illustrate the process, a two-dimensional time–wavelength representation is used. Figure 9.26a shows a pulse train consisting of ultrashort broadband optical pulses. For clarity, the optical spectrum is divided into four passbands (PBs). After propagation through the first dispersive elements, pulses are linearly chirped and broadened in time by an amount equal to the interpulse spacing (Figure 9.26b). Following modulation of the electrical signal onto this waveform (not shown), the waveform is dispersed again leading to time-dilation (Figure 9.26c). Since the electrical signal is superimposed onto the optical carrier, it is also stretched in the process. At this point, there is significant temporal overlap (cross talk) between adjacent segments. Fortuitously, however, the segments that overlap in time are separated in an optical wavelength. This permits their separation using a passive optical filter that carves the spectrum

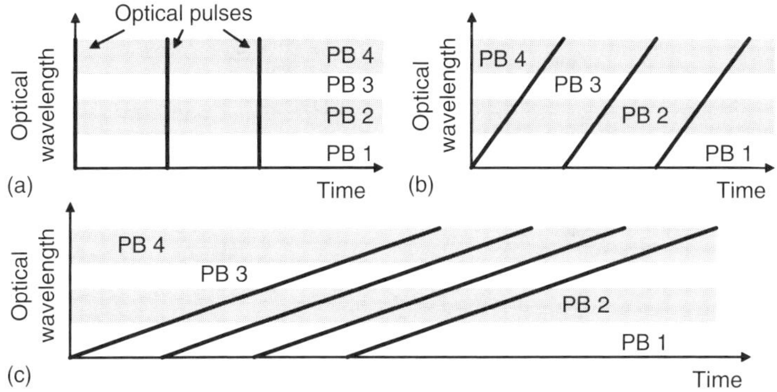


**FIGURE 9.26** Virtual time-gating principle.

into individual optical PBs. Because of the linear time-wavelength mapping, the spectrum carving is equivalent to temporal gating. In other words, each optical PB corresponds to a distinct time segment.

An important feature of the virtual time-gating is that temporal segmentation of the electrical signal is performed using a passive optical filter and without the need for an electronic switch or demultiplexer. Another important feature is that a single electro-optic modulator and a single set of dispersive elements are used for all channels. This avoids interchannel mismatch problems and offers hardware simplicity. Again, the resulting overlap in time can be tolerated because of separation in optical wavelength. This time–wavelength gating scheme is adopted in the experiment described below.

A two-channel continuous-time time-stretch experiment that uses the virtual time-gating scheme is shown in Figure 9.27. A high-pass filter (1570 nm and above) is used to select the long wavelength portion of the SC spectrum. The so-obtained nearly-transform-limited pulses propagate through a length of DCF with a total dispersion of $-1467$ ps/nm. An optical amplifier is inserted in the middle of fiber to compensate for fiber propagation loss. The location is chosen to minimize the detrimental effect of optical nonlinearities in the fiber, while maintaining a good SNR. The chirped pulse is modulated by electrical signal in an MZ modulator. The modulated pulse then propagates through a length of DCF with a –750 ps/nm dispersion to stretch the modulation. The stretch factor is $1 + D_2 L_2/D_1 L_1 = 1.5$. At the end of fiber, a 3 dB coupler is used to split two channels and each channel is filtered by a PB optical filter. The two ~20 nm PBs are centered at 1576 nm and 1594 nm. We note that this configuration is not ideal because of the splitting loss of the 3 dB coupler. To eliminate the splitting loss, which would be required in a system with large number of channels, a coarse wavelength division multiplexing (WDM) filter is used. The two channels are subsequently photodetected and digitized concurrently, by two inputs of Tektronix real-time oscilloscope TDS7404 (4 GHz input bandwidth and 10 GSa/s per channel). In principle,

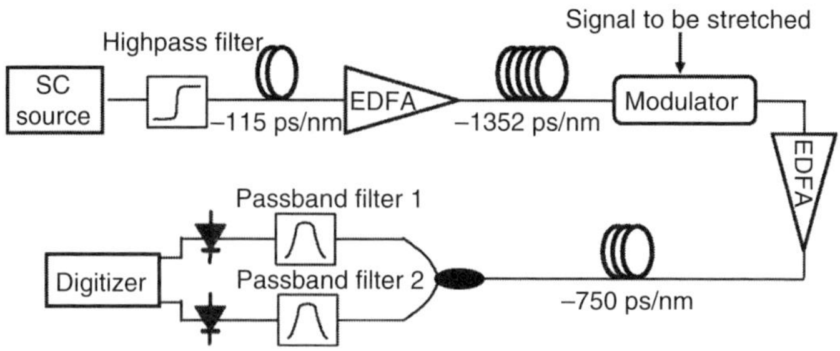


**FIGURE 9.27** Experiment setup for a two-channel continuous-time-stretch analog-to-digital converter. High-pass filter: 1570 nm above. Passband filter 1: 1576 nm with 20 nm bandwidth. Passband filter 2: 1594 nm with 20 nm bandwidth.

the digitizer and the SC source must be synchronized, although this was not implemented in the present experiments since the TDS7404 does not provide external synchronization. The distortion caused by the nonuniform spectrum of the chirped pulse is suppressed using the digital filtering technique described in Section 9.9. We note that the two channels are stretched in the same piece of fiber and are not separated until the digitization step. Also, the time-gating is performed using passive optical filters eliminating the need for electronic switches.

The measured signal quality is evaluated using a two-tone test. A 2 GHz and a 3.1 GHz signal are combined before the modulator. Each segment is subsequently stretched, digitized, and recorded. The magnitude spectrum of one segment is shown in Figure 9.28, obtained using DFT. A Hanning window is applied to limit the frequency leakage due to the finite window length. This also gives a larger weight to the center of each segment, where the SNR is highest. The resulting spectrum shows the stretched tones with more than 30 dB SFDR. Intermodulation tones corresponding to the second- and third-order distortion are marked. The third-order distortion is due to the nonlinear transfer function of the MZ modulator, which imposes a limit on dynamic range given by SFDR $= -20 \log (m^2/8) = 34$ dB, where $m = 0.4$ is the optical modulation index of each tone (defined as half the peak-to-peak optical power divided by the average). The second-order distortion is due to the dispersion in the fiber, which distorts the phase balance between modulation sidebands and generates a frequency-dependent second-order nonlinearity (Section 9.5.2). Broadband linearization techniques in the form of postdistortion performed in the digital domain can potentially be useful.

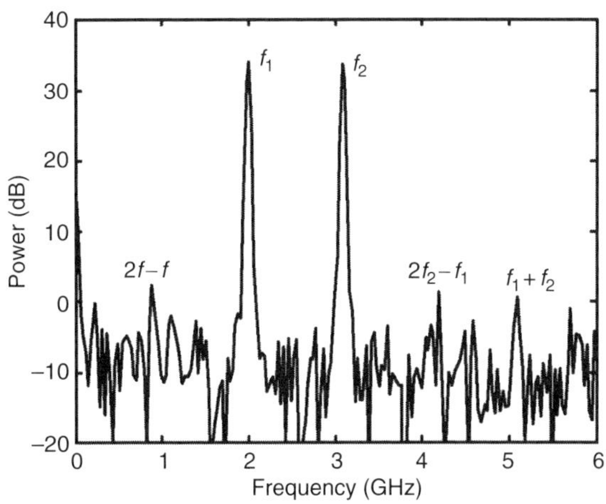


**FIGURE 9.28** Measured digital magnitude spectrum using two-tone test.

The virtual time-gating concept is clearly demonstrated in Figure 9.29, which shows a captured 4.7 GHz tone. The tone is slowed down to $4.7/1.5 = 3.13$ GHz, allowing it to be captured by the 4 GHz digitizer. The continuous-time signal is passively interleaved into two channels using two optical filters. In the experiment, a clock skew exists between the two channels due to the propagation length mismatch after the 3 dB coupler. This is partially calibrated by measuring the propagation delay with oscilloscope using the 1 ps pulse laser. The calibration accuracy is equal to the minimum

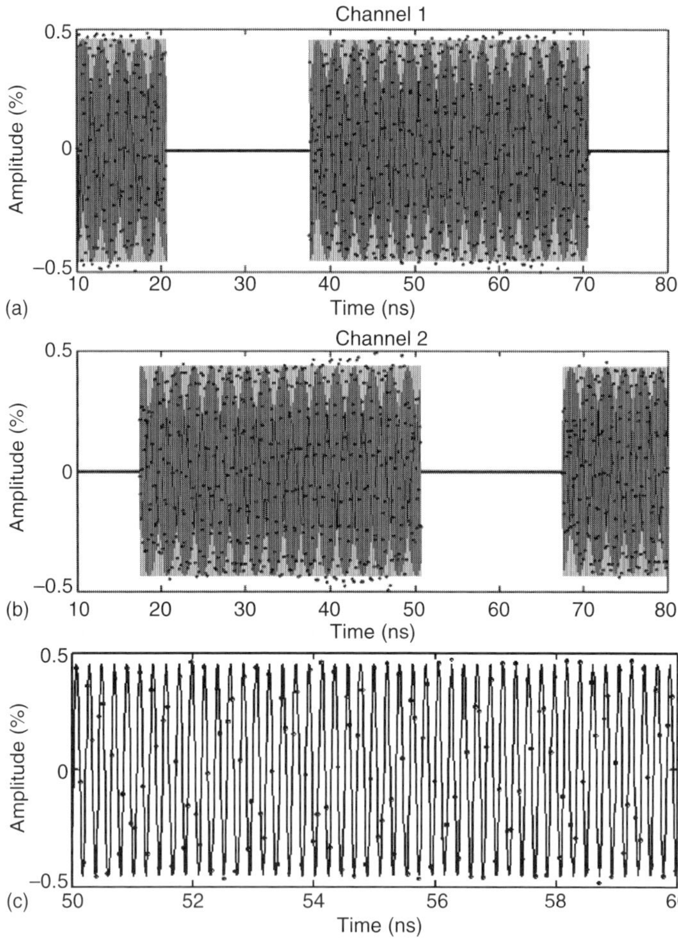


**FIGURE 9.29** Measured 4.7 GHz signal after interleaving. The symbols are sample points and the line is the fitted sine curve. (a) Channel 1 at 1576 nm, (b) channel 2 at 1594 nm, (c) portion of (a).

resolution of oscilloscope (~50 ps). The overlap between two channels corresponds to the slightly spectral overlap and can be used to estimate interchannel mismatches. The slight amplitude variation across the time window is caused by the nonlinear GVD (dependence of $D$ on wavelength) of the fiber (Section 9.5.3).

The overlap section of two adjacent channels in Figure 9.29 is shown in Figure 9.30a, where gain mismatch and clock skew are clearly visible. The digital filtering technique used to correct for the spectral nonuniformity removes the DC component, hence no DC offset is present. The gain mismatch and clock skew are estimated by an adaptive online calibration technique, which was proposed and numerically studied in Ref. [9]. If $S_1(k)$ and $S_2(k)$

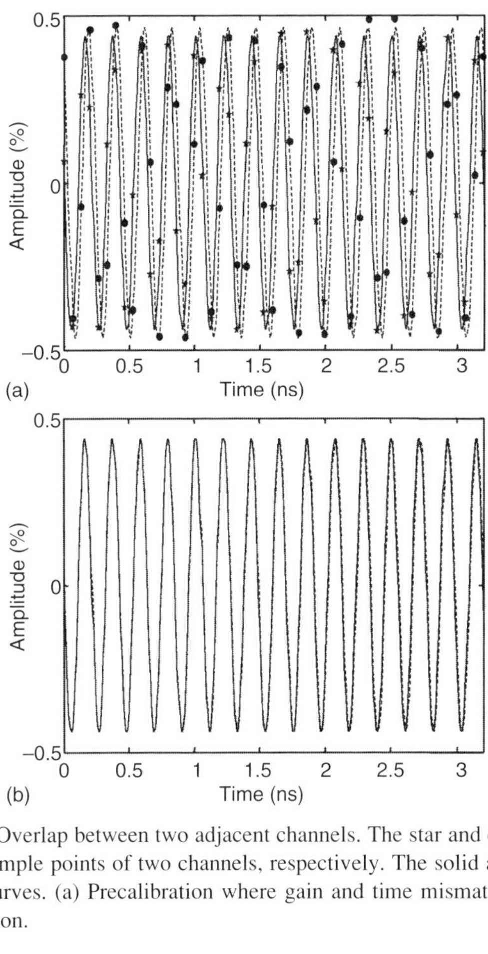


**FIGURE 9.30** Overlap between two adjacent channels. The star and circular symbols represent the sample points of two channels, respectively. The solid and dashed lines are the fitted curves. (a) Precalibration where gain and time mismatches are visible, (b) postcalibration.

represent the DFT of samples in the overlapped section between adjacent segments, the DC mismatch, $a$, gain mismatch, $b$, and clock skew, $r$, can be calculated as

$$
\begin{aligned}
a &\cong \left[ \sum_{k=1}^{W-1} |S_2(k)|^2 / \sum_{k=1}^{W-1} |S_1(k)|^2 \right]^{1/2}, \\
b &\cong [|S_2(0)| - a|S_1(0)|]/W, \\
r &\cong \frac{W}{2\pi} \frac{\Delta}{\Delta k} [\angle S_2(k) - \angle S_1(k)],
\end{aligned}
\tag{9.45}
$$

where $r$ is normalized to the sample interval, and $W$ is the number of sample points in the overlapped section. $\angle$ denotes the phase angle, and $\Delta/\Delta k$ operator takes the slope of its argument. Figure 9.30b shows the overlapped sections after correction using the estimated gain mismatch and clock skew obtained from Equation 9.45. The estimated gain mismatch was 1.05 and the clock skew (relative to the sample interval) was $-0.398$.

As demonstrated, the mismatches between optical PB channels can be corrected using the above algorithm. Within each channel, the accuracy between segments needs to be examined as well. Since each segment in a channel corresponds to an optical pulse of source laser, the quality of signal in a channel is mainly decided by the pulse laser stability. In principle, the laser repetition frequency must also be phase-locked to the clock of digitizer. The lack of phase-locking causes the clock walk-off between segments in a channel. In this experiment, the lack of external synchronization function of the used digitizer and laser prevents us from evaluating the ADC performance over a long time aperture.

## 9.11 CONCLUSION

Time-stretch preprocessing prior to digitization can mitigate many of the problems that plague ultrawideband analog-to-digital conversion. Currently, photonics represent the most promising approach to manipulating the timescale of an electrical signal. Since the digitization is performed in the electronic domain, this approach does not compete with electronics; it enhances it. Enhancements include increase in the sampling rate and the input bandwidth, reduction in the digitizer sampling jitter noise, and interchannel mismatch errors.

A detailed analytical model that predicts the fidelity of the electrical output signal has been developed, and its accuracy has been verified by comparing its results with those obtained using numerical simulations. The model suggests that the fundamental performance improves with a larger O–E bandwidth ratio. TBP has been identified as the proper figure of merit for comparing different architectures. With the present technology, TBP is limited to a few hundred when DSB modulation is employed. Because this limit is set by dispersion in the fiber, one can overcome it with SSB modulation if the remaining phase distortion can be corrected by an equalization filter in the digital domain. Such a system will become loss-limited, and TBP with values of several thousands can be achieved for losses can be readily overcome by using optical amplifiers. Assuming that the quantization noise of the electronic digitizer is low enough, the SNR of an amplified time-stretch system will be limited by the ASE-signal beat noise. To maximize the SNR, a proper design with optimum amplifier gains and locations and an SC source with high-power spectral density are critical. Raman amplification is expected to be beneficial.

An important practical challenge is the nonuniform power spectral density, and pulse-to-pulse variations of the broadband optical source used for time-to-wavelength transformation. We have presented two techniques for overcoming these challenges, both of which rely on modest amount of signal processing in the digital domain.

To capture a continuous-time input signal, a multichannel time-stretch system is required. This architecture overcomes the TBP limitation detailed above. In principle, however, it can suffer from interchannel mismatch problems that degrade the performance of conventional sample-interleaved converters. However, there exists a fundamental difference between a parallel time-stretched array and a sample-interleaved converter array. In the latter, the signal in each channel is sampled at a fraction of the required Nyquist rate, whereas, in the former, it is sampled at, or above, the Nyquist rate. This permits an effective on-line calibration of the mismatch errors.

Finally, comparison must be made with another alternative wideband ADC architecture where the RF signal is channelized before entering an array of narrow band ADCs. This approach also slows down the RF signal, similar to what happens in the continuous-time TS-ADC. The difference is that the channelized ADC operates in frequency domain compared to time domain operation in the TS-ADC. It places the burden on analog filter design and exacerbates the chronic problem in RF system design. Further, it is unable to provide real-time information about the RF spectrum, but rather provides the spectrum averaged over a time window corresponding to the inverse of the single channel bandwidth. In contrast, the TS-ADC provides real-time spectral information that is critical when the input signal contains fast chirps. In fact, the system performs short-time Fourier Transform [42]. The dispersed SC pulse entering the electro-optic modulator captures a short duration of the input RF waveform. The waveform is slowed down, digitized, and then transformed into the frequency domain. Successive chirped pulses capture adjacent segments, leading to complete temporal coverage of the input waveform. As a demonstration of this feature of the TS-ADC, measurements of RF chirp rates as high as 350 GHz/ns have already been reported [42].

## 9.12 APPENDIX

Here we describe the derivation for Equation 9.22–Equation 9.24.

Using Equation 9.21, when the distortion is relatively small, the exponential terms, $e^{j\phi(\mp\omega_{RF})}$ in Equation 9.20 can be written in a linear form as

$$e^{j\phi(\mp\omega_{\rm RF})} \approx \exp(j\phi_{\rm DIP})\left[1+\frac{j}{2}\beta_3\omega\omega_{\rm RF}^2\frac{L_2}{M}\pm\frac{\omega\omega_{\rm RF}}{\omega_{\rm opt}^2}\left(1-\frac{1}{M}\right)\right]. \tag{A1}$$

Use Equation A1 and transform Equation 9.20 into the time domain. The lower sideband signal can be obtained as

$$\begin{aligned}E_{4,\rm lower}(t) = &-J_1\left(\frac{m}{2}\right)\exp(j\phi_{\rm DIP})\\ &\left[\begin{array}{l}E_{\rm env}(t)\exp\left(-j\frac{\omega_{\rm RF}}{M}t\right)-\frac{1}{2}\beta_3\omega_{\rm RF}^2\frac{L_2}{M}\frac{\rm d}{{\rm d}t}\left[E_{\rm env}(t)\exp\left(-j\frac{\omega_{\rm RF}}{M}t\right)\right]\\ +j\frac{\omega_{\rm RF}}{\omega_{\rm opt}^2}\left(1-\frac{1}{M}\right)\frac{\rm d}{{\rm d}t}\left[E_{\rm env}(t)\exp\left(-j\frac{\omega_{\rm RF}}{M}t\right)\right]\end{array}\right].\end{aligned} \tag{A2}$$

In Equation A2, the second and third terms correspond to the $\beta_3$ distortion and the limited optical bandwidth, respectively. To further simplify the derivative term in A2, the third-order dispersion term in the envelope is ignored and we obtain

$$\begin{aligned}\frac{{\rm d}E_{\rm env}/{\rm d}t}{E_{\rm env}} &\approx \frac{{\rm d}\exp\{-t^2[T_{\rm o}^2-j\beta_2(L_1+L_2)]/2\}/{\rm d}t}{\exp\{-t^2[T_{\rm o}^2-j\beta_2(L_1+L_2)]/2\}}\\ &=\frac{-t}{T_{\rm o}^2-j\beta_2(L_1+L_2)}\approx\frac{t}{j\beta_2(L_1+L_2)}.\end{aligned} \tag{A3}$$

Using Equation A3, it can be proved that

$$\frac{\rm d}{{\rm d}t}\left[E_{\rm car}(t)\exp\left(-j\frac{\omega_{\rm RF}}{M}t\right)\right]\approx\frac{{\rm d}E_{\rm car}(t)}{{\rm d}t}\exp\left(-j\frac{\omega_{\rm RF}}{M}t\right)$$

holds. The upper sideband can be similarly handled. Substituting all the results into Equation 9.14, the photocurrent can be obtained as

$$\begin{aligned}I(t) = &\,I_{\rm Envelope}\times\\ &\left[J_0^2\left(\frac{m}{2}\right)-4J_0\left(\frac{m}{2}\right)J_1\left(\frac{m}{2}\right)\left(\cos\phi_{\rm DIP}\cos\frac{\omega_{\rm RF}}{M}t+{\rm Err}_{\beta_3}+{\rm Err}_{\rm BW}\right)\right],\end{aligned} \tag{A4}$$

$${\rm Err}_{\beta_3} = -\frac{\beta_3\omega_{\rm RF}^2(1-1/M)\sin\phi_{\rm DIP}}{2\beta_2M}t\cos\left(\frac{\omega_{\rm RF}}{M}t\right), \tag{A5}$$

$${\rm Err}_{\rm BW} = \frac{\omega_{\rm RF}(1-1/M)\sin\phi_{\rm DIP}}{\omega_{\rm opt}^2\beta_2(L_1+L_2)}t\sin\left(\frac{\omega_{\rm RF}}{M}t\right). \tag{A6}$$

## REFERENCES


1. J.A. Wepman, Analog-to-digital converters and their applications in radio receivers, *IEEE Communications Magazine*, 33(5), 39–45, 1995.
2. R.H. Walden, "Analog-to-digital converter survey and analysis," *IEEE Journal on Selected Areas in Communications*, 17(4), 539–550, 1999.
3. W.C. Black and D.A. Hodges, "Time interleaved converter arrays," *IEEE Journal of Solid-State Circuits*, SC-15, 1022–1029, 1980.
4. A. Montijo and K. Rush, "Accuracy in interleaved ADC systems," *Hewlett-Packard Journal*, 44(5), 38–46, 1993.
5. W.J. Caputi, "Stretch: a time-transformation technique," *IEEE Transactions on Aerospace and Electronics Systems*, AES-7(2), 269–278, 1971.
6. F. Coppinger, A.S. Bhushan, and B. Jalali, "Photonic time stretch and its application to analog-to-digital conversion," *IEEE Transactions on Microwave Theory and Techniques*, 47(7), 1309–1314, 1999.
7. Y. Han and B. Jalali, "Photonic time-stretched analog-to-digital converter: fundamental concepts and practical considerations," *Journal of Lightwave Technology*, 21(12), 3085–3103, 2003.
8. B. Jalali and F. Coppinger, "Data conversion using time manipulation," U.S. Patent 6 288 659, Sep 11, 2001.
9. Y. Han, B. Rezaei, V.P. Roychowdhury, and B. Jalali, "Adaptive online calibration in time stretched ADC arrays," *Instrumentation and Measurement Technology Conference*, 2, 1212–1216, 2003.
10. A.S. Bhushan, P.V. Kelkar, B. Jalali, O. Boyraz, and M. Islam, "130 GSa/s photonic analog-to-digital converter with time stretch preprocessor," *IEEE Photonics Technology Letters*, 14, 684–686, 2002.
11. Y. Han, O. Boyraz, and B. Jalali, "Ultra wideband photonic time-stretch A/D converter employing phase diversity," *IEEE Transactions on Microwave Theory and Techniques*, 53(4), 1404–1408, 2005.
12. Y. Han, O. Boyraz, and B. Jalali, "Tera-sample-per-second Real-time Waveform Digitizer," *App. Phys. Lett.*, 87, 241116, 2005.
13. O. Boyraz, J. Kim, M.N. Islam, F. Coppinger, and B. Jalali, "10 Gb/s Multiple wavelength, Coherent short pulse source based on spectral carving of supercontinuum generated in fibers," *Journal of Lightwave Technology*, 18(12), 2167–75, December 2000.
14. B. Jalali, F. Coppinger, and A.S. Bhushan, "Time-stretch preprocessing overcomes ADC limitations," *Microwave & RF Magazine*, 38(3), 57–66, Mar 1999.
15. F. Coppinger, A.S. Bhushan, and B. Jalali, "Time reversal of broadband microwave signals," *Electronics Letters*, 35(15), 1230–1232, 1999.
16. B.H. Kolner and M. Nazarathy, "Temporal imaging with a time lens," *Optics Letters*, 14(12), 630–632, 1989.
17. C. Bennett and B. Kolner, "Upconversion time microscope demonstrating magnification of femtosecond waveforms," *Optics Letters*, 24, 783–785, 1999.
18. A.S. Bhushan, F. Coppinger, S. Yegnanarayanan, and B. Jalali, "Nondispersive Wavelength-Division-Sampling," *Optics Letters*, 24, 738–740, 1999.
19. B. Asuri, Y. Han, and B. Jalali, "Time-stretched ADC arrays," *IEEE Transactions on Circuits and Systems II*, 49(7), 521–524, 2002.

20. E.I. Ackerman and C.H. Cox, "RF fiber-optic link performance," *IEEE Microwave Magazine*, 2(4), 50–58, 2001.
21. R. Ramaswami and K.N. Sivarajan, *Optical Networks*, Morgan Kaufman Publishing, San Francisco, 2nd ed., 2002, Chapter 5.
22. G.P. Agrawal, *Nonlinear Fiber Optics*, Academic Press, New york, 3rd ed., 2001, Chapter 3.
23. H. Schmuck, "Comparison of optical millimeter-wave system concepts with regard to chromatic dispersion," *Electronics Letters*, 31(21), 1848–1849, 1995.
24. J.M. Fuster, D. Novak, A. Nirmalathas, and J. Marti, "Single-sideband modulation in photonic time-stretch analogue-to-digital conversion," *Electronics Letters*, 37(1), 67–68, 2001.
25. J. Han, B. Seo, Y. Han, B. Jalali, and H.R. Fetterman, "Reduction of fiber chromatic dispersion effects in fiber-wireless and photonic time-stretching system using polymer modulators," *Journal of Lightwave Technology*, 21(6), 1504–1509, 2003.
26. Y. Han and B. Jalali, "Time-bandwidth product of the photonic time-stretched analog-to-digital converter," *IEEE Transactions on Microwave Theory and Techniques*, 51(7), 1886–1892, 2003.
27. A. Hilt, "Microwave harmonic generation in fiber-optical links," *Conference Proceedings 13th International Conference on Microwaves, Radar and Wireless Communications*, 2, 693–698, 2000.
28. S. Dubovitsky, W.H. Steier, S. Yegnanarayanan, and B. Jalali, "Analysis and improvement of Mach–Zehnder modulator linearity performance for chirped and tunable optical carriers," *IEEE Journal of Lightwave Technology*, 20(5), 886–891, 2002.
29. G.H. Smith, D. Novak, and Z. Ahmed, "Technique for optical SSB generation to overcome dispersion penalties in fibre-radio systems," *Electronics Letters*, 33(1), 74–75, 1997.
30. K. Yonenaga and N. Takachio, "A fiber chromatic dispersion compensation technique with an optical SSB transmission in optical homodyne detection systems," *IEEE Photonics Technology Letters*, 5, 949–951, 1993.
31. J. Park, W.V. Sorin, and K.Y. Lau, "Elimination of the fibre chromatic dispersion penalty on 1550 nm millimetre-wave optical transmission," *Electronics Letters*, 33(6), 512–513, 1997.
32. K. Chang, I. Bahl, and V. Nair, *RF and Microwave Circuit and Component Design for Wireless Systems*, John Wiley, New Jersey, 2002, Chapter 6.
33. T. Rappaport, *Wireless Communications: Principles and Practice*, Prentice Hall, New Jersey, 1996, pp. 177–254.
34. M. Nakazawa, K. Tamura, H. Kubota, and E. Yoshida, "Coherence degradation in the process of supercontinuum generation in an optical fiber," *Optical Fiber Technology*, 4(2), 215–223, 1998.
35. O. Boyraz, J. Kim, M.N. Islam, F. Coppinger, and B. Jalali, "10 Gb/s multiple wavelength, coherent short pulse source based on spectral carving of supercontinuum generated in fibers," *IEEE Journal of Lightwave Technology*, 18, 2167–2175, 2000.
36. Y. Han and B. Jalali, "Differential photonic time stretch analog-to-digital converter," *16th Annual Meeting of Lasers and Electro-Optics Conference*, 2003, (LEOS 2003), Tucson, Arizona, Paper CWH2.

37. J.C. Twichell and R. Helkey, "Phase-encoded optical sampling for analog-to-digital converters," *IEEE Photonics Technology Letters*, 12(9), 1237–1239, 2000.
38. J.G. Proakis and D.G. Manolakis, *Digital Signal Processing: Principles, Algorithms, and Applications*, Prentice Hall, New Jersey, USA, 3rd ed., 1996, Chapter 1.
39. B. Jalali, P. Kelkar, and V. Saxena, "Photonic arbitrary waveform generator," *14th Annual Meeting of the IEEE Lasers and Electro-Optics Society*, 1, 253–254, 2001.
40. J. Chou, Y. Han, and B. Jalali, "Adaptive RF-photonic arbitrary waveform generator," *IEEE Photonics Technology Letters*, 15(4), 581–583, 2003.
41. J. Chou, Y. Han, and B. Jalali, "Time–wavelength spectroscopy for chemical sensing," *IEEE Photonics Technology Letters*, 16(4), 1140–1142, 2004.
42. A. Nuruzzaman, O. Boyraz, and B. Jalali, "Time stretch short time Fourier transform," IMTC Transactions on Instrumentation and Measurement, Como, Italy, 18–20 May 2004.

# 10 Terahertz Photonics

*Albert Redo-Sanchez and Xi-Cheng Zhang*

## CONTENTS

10.1 Introduction ... 363
10.2 Pulsed Systems ... 367
  10.2.1 Generation ... 368
  10.2.2 Detection ... 371
10.3 Continuous-Wave Systems ... 373
  10.3.1 Generation ... 374
  10.3.2 Detection ... 377
10.4 Sensing Techniques and Applications ... 379
  10.4.1 Time Domain Spectroscopy ... 379
  10.4.2 Explosive Identification ... 383
  10.4.3 Other Spectroscopic Applications ... 386
10.5 Imaging Techniques and Applications ... 388
  10.5.1 Passive and Active Imaging ... 388
  10.5.2 CCD-Based Two-Dimensional Imaging ... 391
  10.5.3 Imaging Geometries and Beam Guidance ... 391
  10.5.4 Compact CW THz Imaging Systems ... 393
  10.5.5 Space Shuttle Foam Inspection ... 395
  10.5.6 Security Scan and Content Inspection ... 398
  10.5.7 Tomography ... 399
10.6 Conclusion ... 405
Acknowledgments ... 408
References ... 408

## 10.1 INTRODUCTION

Terahertz (THz) wave radiation is defined as the far-infrared (FIR) electromagnetic radiation between 0.1 and 10 THz (1 THz $= 10^{12}$ Hz), and it has become a rich field for fundamental physics and technological application. THz wave science and technology is expected to become one of the key fields in the twenty-first century. The interest in THz wave started as early as 1930s, but the lack of good emission and detection techniques and devices left THz less mature compared to its neighbor technologies (microwave and photonics). At lower frequencies, microwave technology has been developed on the basis of the electronics industry and most of the physics involved can be explained by classical or semiclassical transport theories. The main applications are in communications, radar, and food industry. On the other hand, at higher frequencies, visible radiation generation and detection mechanisms and related technologies (known as photonics) are strongly based on quantum physics and have found several different applications such as in the medical field, machining, lithography, and entertainment. The region between microwave and IR technologies has been known as the THz gap for many years. This THz gap (Figure 10.1) provides opportunity and exploration as a result of the development of new generation and detection technologies. This gap can also be seen as a bridge that can use both technologies to create its own and the fusion of these technologies is one of the reasons why THz has emerged as a completely new field. Two main approaches exist: pulsed and continuous wave (CW) and both can be used as sensing and imaging tools. Table 10.1 displays the most common generation and detection techniques and devices for pulsed and CW systems.

The generation performance of a source is mainly described by its power, but for pulsed system, the energy, pulse duration and repetition rate are also important. In pulsed systems, the output is characterized by the average power, peak power, and pulse energy. They are related through the timing parameters such as pulse duration, and repetition rate. A related parameter is the duty cycle, which is equal to the product of the repetition rate and pulse duration. The pulse energy is the total energy emitted during the time period considered as the pulse duration. The average power is obtained by multiplying the pulse energy with the repetition rate and the peak power is obtained by multiplying the pulse energy with the pulse duration. All these parameters are used either to characterize an optical laser or THz pulses. For an optical laser pulse which is very narrowband, it is important to report the wavelength, but for a THz pulse the central frequency is important. The central frequency is defined as the frequency where the spectrum shows a maximum. The central

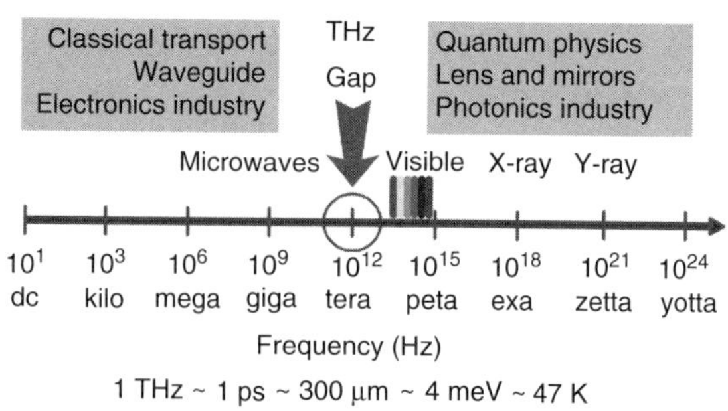


**FIGURE 10.1** THz gap between electronics (lower frequencies) and photonics (higher frequencies), and some typical values for a THz photon.

**TABLE 10.1**
**Main Generation and Detection Techniques and Devices for Pulsed and CW Systems**

| | THz Wave Source | THz Wave Detector |
|---|---|---|
| PULSED | Electro-optical rectification | Electro-optical sampling |
| | Photoconductive antenna | Photoconductive antenna |
| | Bulk semiconductor surface | Deuterated triglycine sulfate (DTGS) |
| | Superlattices and heterostructures | Pyroelectric detector |
| | THz optical parametric generation | Bolometer |
| CW | Gas vapor laser | |
| | Photomixing | |
| | Gunn oscillator | Schottky diode |
| | Free electron laser (FEL) | Golay cell |
| | Quantum-cascade laser (QCL) | Bolometer |
| | Arc lamp/globars | Pyroelectric detector |
| | Backward-wave oscillator (BWO) | |
| | Smith–Purcell | |

frequency is related to the optical pulse duration in a way that, the shorter the pulse, the higher is the central frequency. CW systems are better described by using the power, and they can be considered as the limit of a pulsed system with infinite pulse duration. This power is delivered continuously and then the stability, which is defined as the drift of the power with respect to time, is also a very important parameter. Usually a good CW source should be as powerful as possible and should be highly stable, which means that the oscillation of the power with time should be as small as possible.

With regard to detection, the main parameters are the signal-to-noise ratio (SNR), dynamic range, noise equivalent power (NEP), and responsivity (Figure 10.2). They can be used either for a CW or a pulsed system. SNR is defined as the power ratio between the signal and the noise measured within the system bandwidth and is usually expressed in dB. Dynamic range is defined as the ratio between the smallest and the largest signal. The smallest is usually related to the noise floor and the largest to the maximum power that the source can handle without damage or saturation.

The NEP is defined as the input power that produces an SNR equal to 1 at the output of a detector at a given modulation frequency, operating wavelength, and noise bandwidth. The unit is expressed in power, but sometimes it is defined as the minimum detectable power per square root bandwidth. This last definition is usually used for broadband detectors and involves the bandwidth of the detector. The NEP gives the theoretical detection limit of

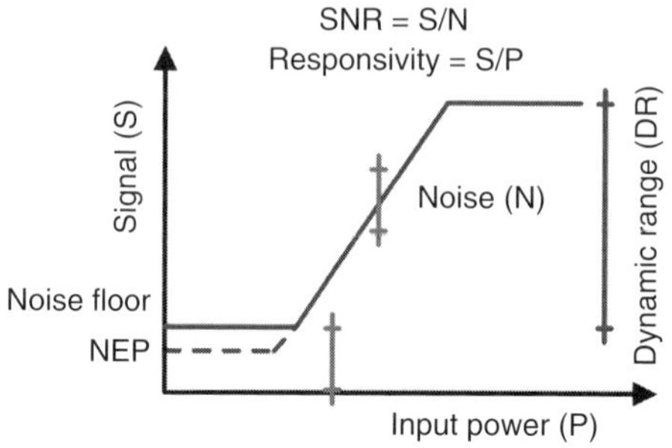


**FIGURE 10.2** Detection parameter relation. In this example, the noise floor is higher than the NEP and it limits the dynamic range. If the system is optimized, the noise floor should be closer or lower than the NEP.

the system and can be higher or lower than the actual noise floor. In an optimal situation, the noise floor should be lower than the NEP.

Responsivity and sensitivity are confused very often and sometimes they are used indistinctly to mean the same. The responsivity is the ratio of the electrical output to the excitation signal and it is usually expressed in units of voltage/current divided by input power. The sensitivity is defined as the minimum input signal required to produce a specified output signal having a specified SNR. They are both used to indicate how a system is sensible to a particular input signal. They usually depend strongly on the frequency of the radiation, modulation frequency, and bandwidth of the detector.

The noise floor is the measurement of the signal generated by all noise sources and unwanted signals within a detection system. There are several noise sources, which can come from the sample and environment, but others come from the device itself due to the electronics, contacts resistance, wiring induction, temperature, etc. The most important electronic noise sources are Johnson noise, Shot noise, and pink noise. Johnson noise is generated by the collisions of the electrons with the lattice, and the typical value for a resistor of 1 KΩ in a bandwidth of 1 Hz is ~4 $nV/Hz^{1/2}$. Shot noise arises from the discrete nature of the electrons crossing a potential barrier. The fluctuation follows a Poisson distribution and the strength of this noise increases as the average current increases, but often the SNR also increases; thus, shot noise is usually a problem at small currents. A typical value for a 1 Hz bandwidth and a current of 1 mA is ~0.5 $nA/Hz^{1/2}$. Pink noise is a signal or process with a frequency spectrum such that the power spectral density is proportional to the reciprocal of the frequency and is also known as $1/f$ noise or flicker noise.

Because of their proximity to optical technology, some parameters used to characterize optical beams are also used to characterize THz beams. THz wave propagation can be considered, most of the times, as a quasioptical rather than radio-frequency technology. Thus, the equations and principles

**TABLE 10.2**
**Strengths and Weaknesses of Pulsed and CW Systems**

| | Strengths | Weaknesses |
|---|---|---|
| PULSED | Broadband | Complex setup |
| | Amplitude and phase information (coherent) | Low power |
| | | Expensive |
| CW | Narrowband | |
| | Less complex than a pulsed setup | No spectral information |
| | Can be compact and portable | No phase information |
| | Less expensive than pulsed systems | |

used in laser optics can be applied with minor modifications. In general, laser-beam propagation can be approximated by assuming that the laser beam has a Gaussian intensity profile, corresponding to the theoretical $TEM_{00}$ mode or fundamental mode. This approximation can also be assumed for THz beams. A Gaussian beam is formed by electromagnetic radiation whose electric field intensity transverse distribution follows a Gaussian function along the radius. Important optical parameters are the Rayleigh range and resolution. The Rayleigh range is defined as the distance where the beam is considered collimated. Distances close to the Rayleigh range are considered the near-field region, and distances that are far away are considered the far-field region. Resolution is affected by diffraction effects of the system and it is evaluated according to the Rayleigh criteria, which states that the imaging process is diffraction-limited when the first diffraction minimum of the image of one point source coincides with the maximum of another point source (Table 10.2).

## 10.2 PULSED SYSTEMS

Pulsed systems refer commonly to those that generate and detect a transient electromagnetic pulse. The pulse information is taken in a time domain and it is transformed in the frequency domain by a Fourier transform. To have the electromagnetic energy distributed through the THz range, the duration of the electromagnetic oscillation must be within the picoseconds range. Early work began in late 1970s and early 1980s with the study of the response of photoconductors to laser pulses [1–4], which have attracted many academic and industrial interests since then. In the late 1980s, the use of photoconductive antennas generated the first prototype of pulsed THz wave emitter and detector [5,6], and it is still a widely used device for emission and detection of THz radiation.

In recent years, THz wave-pulsed system and applications have been oriented toward spectroscopy (THz wave time domain spectroscopy) and imaging (T-ray imaging). Terahertz wave time domain spectroscopy, or THz-TDS, is a time-resolved technique with coherent detection measurement and it has a frequency covering from the FIR region to the mid-infrared (MIR) region. T-ray imaging uses similar generation and detection techniques, but uses the pulse information to get a spectroscopic picture of the sample. The most common way that pulsed systems work is by splitting a beam generated by a femtosecond laser into two: the probe and the pump beams. The pump beam is used to generate the THz pulse, while the probe beam is used to obtain the pulse profile of the pump beam. This is performed by modulating the probe beam's amplitude with the pump beam's amplitude. The probe beam is then delayed with respect to the pump with a time delay and the entire waveform is obtained point by point. This pulse information acquired in the time domain is transformed in the frequency domain with a Fourier transform and thus, spectral information can be obtained.

### 10.2.1 Generation

The broadband pulsed generation is based on the excitation of different materials with ultrashort laser pulses. Different mechanisms have been exploited to generate THz radiation, including photocarrier acceleration in photoconducting antennas, second-order nonlinear effects in electro-optic crystals (optical rectification), plasma oscillations, and electronic nonlinear transmission lines. Photoconduction and optical rectification are the two most common approaches for generating broadband, pulsed THz beams. Actual conversion efficiencies for these techniques are very low and THz powers tend to be in the nanowatts to microwatts range, while the average power of the femtosecond optical source is in the region of 100 mW to 1 W.

Optical rectification is based on the inverse process of the electro-optic effect [7]. The electro-optic effect is a change in the optical properties of a material as a result of applying an electric field with a lower frequency than the frequency of light. The most important electro-optic effects are the Pockels effect and the Kerr effect, in which birefringence is induced or modified by the electric field applied onto the material. The Pockels effect is a change in the refractive index or birefringence that depends linearly on the field, while the Kerr effect is a change in the index or birefringence that is quadratic with the field and is generally weaker than the Pockels effect. All materials display the Kerr effect, but only certain crystalline solids show the Pockels effect because it requires the crystal to have inversion symmetry. THz wave generation by optical rectification can be coupled into free space. It is essentially a difference-frequency generation process in which the frequency difference is close to zero. Femtosecond laser pulses generated by a Ti:sapphire laser are used to generate THz radiation from an electro-optical (EO) crystal because the femtosecond pulse contains many frequency components and all the differences between the components contribute to the difference-frequency generation. This technique was first demonstrated for generating FIR radiation using $LiNbO_3$ [8], and a lot of research has focused on optimizing THz generation through investigating the electro-optic properties of different materials including traditional semiconductors such as GaAs and ZnTe (Figure 10.3), organic crystals such as the ionic salt 4-dimethylamino-*N*-methyl-4-stilbazolium-tosylate (DAST), and many others [9–14].

The main factors that affect the effectiveness of optical rectification are as follows: the pulse duration of the laser, the phase-matching conditions, and the absorption of the EO crystal. The pulse duration is the main input for the crystal and determines the bandwidth of the THz pulse. Shorter pulses are expected to extend the bandwidth and, actually, with the development of sub 10 femtosecond range laser, bandwidth of 100 THz could be generated. With such broadband pulses, it is impossible to select EO materials that fulfill the condition of group velocity matching for all frequency components. It can be seen that the thickness must be small to obtain broad-bandwidth emission. On the other hand, if the sample is very thin, it may happen that secondary pulses or echoes generated by multiple reflections mix together with the main pulse and make signal analysis more difficult to analyze. The time between echoes should be much larger than the pulse duration itself. Finally, the absorption in the EO crystal also affects the performance; usually, the thicker the crystal, the higher is the absorption, resulting in a lower output. Generally, materials with a large second-order nonlinear coefficient are good candidates for the THz source [15]. Currently, ZnTe is the material of choice due to its high nonlinear coefficient, high damage threshold, and best phase-match condition with the Ti:sapphire laser [16].

Photoconductive antennas are biased THz wave emitters [17,18] and are capable of relatively large average THz powers in excess of 40 $\mu$W [19] and bandwidths as high as 10 THz [20]. The photoconductive antenna uses

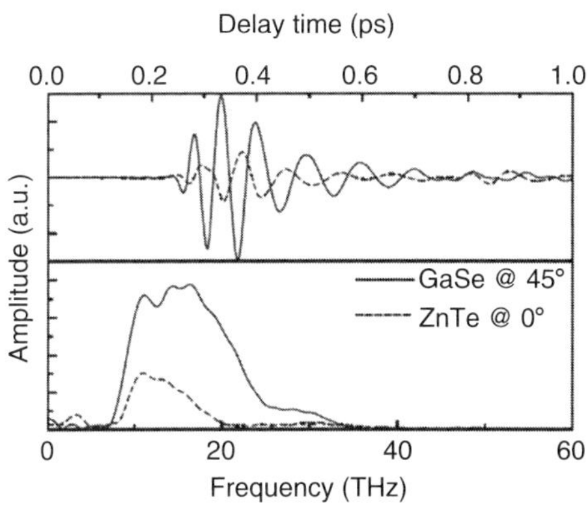


**FIGURE 10.3** Pulse generation and detection with GaSe and ZnTe crystals.

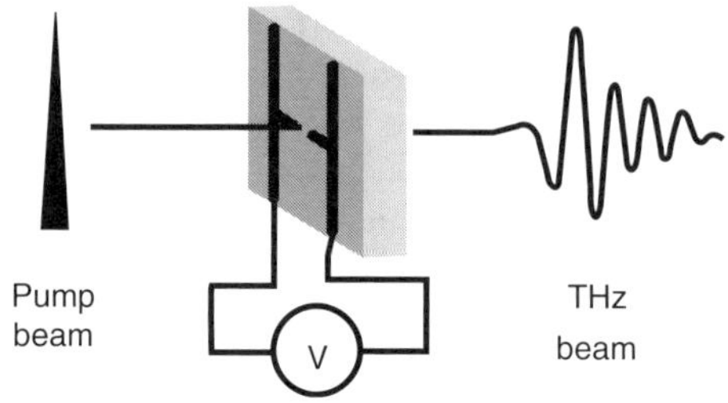


**FIGURE 10.4** THz pulse generation by a photoconductive antenna. A femtosecond pulse duration laser is sent to the photoconductive structure that is under a DC voltage. The carriers generated by the laser pulse are accelerated in the material and THz pulse emission occurs.

high-speed photoconductors as transient current sources that generate the radiation, and metallic electrodes are used to bias the photoconductive gap and form the antenna (Figure 10.4). The physical mechanism for THz wave generation in photoconductive antennas is the generation of electron–hole pairs when an ultrafast laser pulse with photon energy larger than the bandgap of the material strikes the photoconductor. The free carriers accelerate in the static bias field and form a transient photocurrent, which generates electromagnetic waves. The intensity and bandwidth of the THz radiation are largely affected by the material parameters such as the bandgap, carrier's lifetime, and resistivity. The most widely used materials for photoconductive antennas are semiinsulating GaAs [19,21] and low-temperature-grown (LTG) GaAs [22,23]. Compared to other traditional THz wave emitters such as the ZnTe crystal, GaAs-based photoconductive antennas do not have phonon absorption around 5 THz, which is quite advantageous for gap-free broadband spectroscopy up to 7 THz. However, optical rectification is advantageous in broadband THz radiation because the bandwidth of the THz spectra is determined only by the laser duration pulse, and not by the response time of the material [24]. With an ultrashort laser pulse, the bandwidth of the generated THz radiation could be as high as 100 THz [25] (Table10.3).

Other types of unbiased THz wave emitters are based on the THz surface emission by the ultrafast transport of charged carriers. THz transients arise either due to the charge separation by the built-in electrical field at the semiconductor surface [26] or the difference in the electron and hole diffusion coefficients (photo-Dember effect) [27]. The study of the THz radiation mechanism from bulk materials promoted a new category of THz science: THz emission spectroscopy [28]. The analysis of THz pulses irradiated from bulk semiconductors can help us understand the materials themselves. For efficient THz radiation, it is desirable to have rapid photocurrent rise and decay times. Thus, semiconductors with small effective electron masses such as InAs and InP are as attractive as GaAs. The maximum drift velocity is also

**TABLE 10.3**
**Generation Performance for Optical Rectification (OR) and Photoconductive Antennas (PCA)**

| | Laser Beam/THz Beam |
|---|---|
| Energy per pulse [J] | $10^{-3}$–$10^{-8}$/$10^{-9}$–$10^{-14}$ |
| Peak power [W] | $10^{10}$–$10^{5}$/$10^{3}$–$10^{-2}$ |
| Average power [W] | $10^{-6}$ (typical) |
| Pulse duration [ps] | ~0.100/~1 |
| Repetition rate [Hz] | $10^{8}$–$10^{3}$ |
| Bandwidth [THz] | ~3 (typical); 100 (max) |

an important material parameter; it is generally limited by the intraband scattering rate or intervalley scattering in direct semiconductors such as GaAs [29–31]. Since the radiating energy mainly comes from stored surface energy in the form of the static bias field, the THz radiation energy scales up with the bias and optical fluency [32,33]. The breakdown field of the material is another important parameter as this determines the maximum bias that may be applied [34]. Other materials such as semiconductor quantum wells [35–37], superlattices [38,39], superconductors [40–42], dye molecules [43], and even magnetomolecules [44,45] have also been studied for THz radiation. The acceleration of electron bunches from free electron lasers is a brand new source of pulsed THz wave [46–48]. This method provides a powerful pulsed source, which offers the feasibility of nonlinear THz research.

Recently, it has been demonstrated that it is possible to generate THz radiation using air [49]. Air does not show a second nonlinear coefficient, but it has a third-order nonlinear coefficient instead. By generating plasma with a pulsed laser, it is possible to mix a fundamental wave and its second harmonic through the third-order nonlinearity and generate efficient THz wave generation. This process opens the possibility of using THz for remote sensing by sending a laser beam far away, generating THz locally, and transforming the reflected THz radiation into an optical beam again and returning to the origin.

### 10.2.2 Detection

The traditional detection techniques in pulsed THz wave technology are EO sampling and the use of photoconductive antennas (Table 10.4). FIR interferometry techniques using incoherent detectors such as bolometers [50] have also been used to perform detection, although they lose the coherency that EO sampling and photoconductive antennas provide. The bolometer requires a liquid helium–cooled environment but, more recently, the use of Golay cell

**TABLE 10.4**
**Detection Performance for Optical Rectification (OR) and Photoconductive Antennas (PCA)**

| | |
|---|---|
| Dynamic range | 10–100 K |
| SNR | 100–10 K |
| NEP [W/Hz1/2] | $\sim 10^{-15}$ |
| Bandwidth [THz] | ~5 (typical); 100 THz (max) for EO and 60 (max) for PCA |

detector provides a new and convenient approach for interferometry detection because it does not require a low temperature environment [51]. Both the bolometer and the Golay cell detect power rather than the electric field and, thus, cannot provide a coherent measurement.

Photoconductive antennas are widely used for pulsed THz detection and an identical structure to the photoconductive antenna emitter may be used. Rather than applying a bias voltage to the electrodes of the antenna, a current amplifier and ammeter are used to measure the transient current generated by an optical pulse and biased by the instantaneous THz field (Figure 10.5). The biased current is proportional to the applied THz field [5,52–54]. It is possible to use a photoconductive antenna for broadband THz wave detection by properly selecting the substrate materials. Ultrahigh bandwidth detection has been demonstrated using photoconductive antenna detectors using LTG-GaAs with detectable frequencies in excess of 60 THz [54]. The apertures of

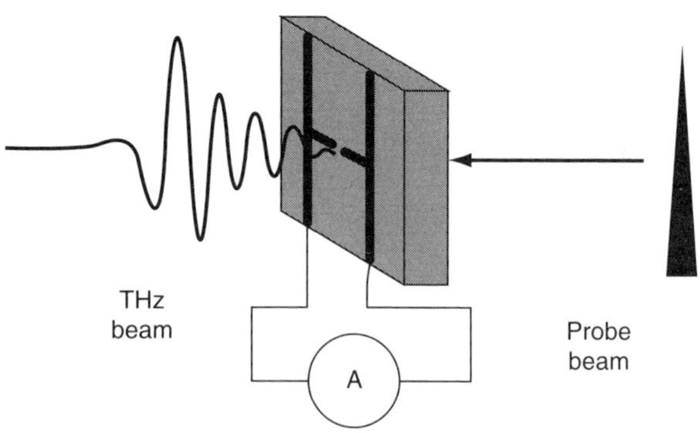


**FIGURE 10.5** Detection of a THz pulse with a photoconductive antenna. The electric field generated by the laser pulse is biased by the THz field. This biased current is measured with an ammeter.

the photoconductive antennas are usually in the micron range, and the optical alignment is more difficult compared to the free-space EO sampling.

Free-space EO sampling utilizes second-order nonlinear crystals to detect the polarization variation of the probe pulse interfered by a THz pulse. Fundamentally, the EO effect is a coupling effect between a low-frequency electric field (THz pulse) and a laser beam (optical pulse) in the sensor crystal. The THz electric field modulates the birefringence of the sensor crystal, thus modulating the polarization ellipticity of the optical probe beam passing through the crystal. The ellipticity modulation of the optical beam is then polarization-analyzed to provide information on both the amplitude and phase of the applied electric field [55–57]. The frequency response function for detection is the same as that for generation, and the phase-matching conditions and crystal absorption also play a role in the detector's performance. Thus, thinner crystals provide broader frequency response, but then the interaction distance is shorter, resulting in a smaller sensitivity. Among many EO crystals such as GaP [58], GaAs [59], organic crystals [12], and EO polymers [13], ZnTe [60] is the best candidate for THz wave detection below 4 THz. The use of an extremely short laser pulse ($<15$ fs) and a thin sensor crystal ($<30$ $\mu$m) allows EO detection of signals into the MIR range. Extremely high detection bandwidths in excess of 100 THz have been demonstrated using thin sensors [61].

## 10.3 CONTINUOUS-WAVE SYSTEMS

CW can be considered as the extreme case of pulsed emission when the duration of the pulse is infinite. Generally, CW systems can work only at a single frequency at a time but, in some particular cases, broadband sources may be used. Regarding the detectors, either broadband or narrowband are available. The frequency range that CW systems offer is less than in those operating in pulsed mode. On the other hand, CW systems offer higher power than pulsed systems and they are less affected by water vapor absorption. CW systems have been traditionally evolved from microwave and electronics technology, but some techniques come from the photonics technology as well. CW systems are good for field applications because they are more suitable than pulsed systems to achieve small, reliable, and portable devices. The main foreseen applications for CW systems are related to imaging and nondestructive evaluation (NDE), but some research is being done in sensing and identification when the spectrum features requiring identification are known previously. Narrowband sources are very important for high-resolution spectroscopy applications and also have broad potential applications in telecommunications. A lot of effort has been made to develop such narrowband sources and detectors over the past century. A multitude of techniques are under development, including up-conversion of electronic RF sources, down-conversion of optical sources, lasers, and backward-wave oscillator (BWO) tubes.

### 10.3.1 Generation

The most commonly employed technique for generating low power (from 100 μW to 20 mW) CW THz radiation up to 0.6 THz is through up-conversion of lower frequency microwave oscillators such as voltage-controlled oscillators (VCO) and dielectric-resonator oscillators. Up-conversion is typically achieved using a chain of planar GaAs Schottky diode multipliers after a Gunn or impact ionization avalanche transit-time (IMPATT) diode. Other technologies available to generate CW THz radiation are backward-wave oscillator (BWO), gas laser, free electron lasers, quantum-cascade lasers (QCLS), and photomixing.

The Gunn effect was discovered while testing the behavior of *n*-GaAs samples under high electric fields applied within the bulk of the material. These samples showed microwave radiation emission for a certain range of DC bias voltage, which did not depend on the junction properties, but on the properties of the semiconductor itself. This effect only happens in materials where conduction electrons can exist in more than one stable state with different mobility. The most common material used to fabricate such diodes is *n*-GaAs. The semiconductor is located within a resonant cavity to achieve a narrow bandwidth and oscillation stability [62]. A Gunn diode consists of two heavily *n*-doped regions on each terminal and a thin lightly doped layer in between. This structure generates two bands with different electron mobility, which results in a negative differential resistance behavior. If the device is biased near the threshold voltage, it shows an oscillating current that generates an RF output. The oscillation frequency is determined by the properties of the middle layer such as the thickness, doping concentration, and the applied bias voltage. An IMPATT consists of two regions: the avalanche region or injection region, and the drift region. The avalanche or injection region creates the carriers (holes or electrons) and the drift region is where the carriers move across the diode. Most of the structures available are based on a basic PN junction with an intrinsic region between the P-type and N-type regions. The IMPATT diode is operated under reverse bias conditions, which are set so that avalanche breakdown occurs very close to the P region. Under these circumstances, any of the carriers are accelerated very quickly and emit in a particular frequency range. They are typically made with silicon carbide because of its high breakdown fields. One of the limitations of the current Gunn and IMPATT diode technology is that output power decreases rapidly above 0.1 THz. To overcome this problem, a series of multipliers are used, the most common are doublers and triplers with a conversion efficiency up to 10%. Typical output powers lie between tens of milliwatts for lower frequencies and hundreds of microwatts for higher frequencies. A typical frequency limit for commercially available devices is about 0.6 THz, but frequencies as high as 2.7 THz have been reported [63]. Important research is directed toward increasing the frequency of Gunn and IMPATT diodes by using alternate semiconducting structures and improved fabrication techniques [64].

In gas laser sources, a $CO_2$ laser pumps a low-pressure gas cavity, which emits radiation at a frequency that lie within the THz range [65]. This gas is usually methanol or hydrogen cyanide. Gas lasers can provide a broad range of frequencies, and are tunable to distinct lines by varying the composition and pressure of the gas in the laser cavity and the frequency of the $CO_2$ pump laser. The $CO_2$ laser is usually grating-tuned and its emission lies between 9 and 11 $\mu$m. This radiation is introduced inside the low-pressure gas cavity through a small coupling hole in one end mirror and the THz radiation produced is emitted through another coupling hole in the other end mirror [66]. The THz laser process operates on the molecular rotational transitions of the low-pressure gas. An infrared photon generated by the $CO_2$ with energy close to a transition from a particular rotational ground state to a rotational excited state is absorbed by a gas molecule. If the population inversion is achieved, the molecule decays through different rotational excited states to the ground state and emits laser radiation according to the energy difference of these excited states.

The basic structure of a BWO is a magnetically focused electron beam passing through a corrugated structure carrying an RF signal. The electrons in the structure are periodically decelerated, forming bunches that interact with the RF field and producing electromagnetic radiation in the opposite direction to the motion of the electrons. This radiation is coupled into a waveguide and sent to the free space. The frequency of the emitted radiation is determined by the speed of the electrons, which can be adjusted via the potential difference between the cathode and the anode of the electron gun. BWO sources are capable of output powers of tens of milliwatts in the subterahertz range and available devices can operate above 1 THz [67].

Free electron lasers generate extremely high-power THz radiation [46]; Williams [68] demonstrated this using free electron lasers with energy-recovering linear accelerators by forcing bunches of high-speed electrons in vacuum to go through regions where magnetic field variations are very strong. The magnetic field causes the electron bunches to oscillate and emit photons. Mirrors are used to confine the photons to the electron beam line, which forms the gain medium for the laser. The frequency of the photons depends on the electrons' speed, magnetic field intensity, and spatial variation. To generate this radiation, an energy-recovering linear accelerator is necessary, which requires a dedicated facility and imposes prohibitive cost and size constraints. However, they may operate CW or pulsed waves and provide average brightnesses more than six orders of magnitude higher than typical photoconductive antenna emitters. Free electron lasers have significant potential in applications where high-power sources are essential or in the investigation of nonlinear THz spectroscopy.

More recently, QCLs have evolved very fast and appear to be a promising technique to generate THz. The first QCL was demonstrated in 1994 at Bell Laboratories on a structure of a series of coupled quantum wells constructed using molecular beam epitaxy (MBE) [69]. Some previous work was done with semiconductor lasers over 20 years ago using lightly doped p-type germanium with hole population inversion induced by crossed electric and magnetic fields [70]. These lasers are tunable by adjusting the magnetic field or external stress. THz lasing in germanium has also been demonstrated by applying a strong uniaxial stress to the crystal to induce hole population inversion [71]. Recent advances in semiconductor deposition techniques such as MBE allow the construction of multiple quantum well semiconductor structures and QCL to become feasible. A QCL is a unipolar solid-state laser, which uses electrons as its only charge carrier. It consists of alternating layers of two different semiconductors (typically GaAs/AlGaAs) that form a series of quantum wells. The layers are grown on a substrate using MBE and the thickness is in the order of nanometers. The electrons are constrained within the quantum wells and they move through the structure by jumping between the levels and tunneling between the wells. The quantum well structures form a series of equal energy steps and each electron emits one photon at each of these steps, unlike diode lasers that only emit one photon per electron transmitted. This structure forms a series of injection and active regions along the structure and each region has slightly lower energy than the previous one to produce the cascade effect. In the active region, a population inversion exists, and the electron transits to a lower energy level and emits a photon. Then this level acts as the injector for the active region of the next quantum well and the process is repeated again [72]. Typically, the number of active regions increases from 25 to 75, which allows producing 25 to 75 photons per electron journey. The laser process usually emits in many frequencies. To select a specific frequency, a structure called distributed feedback (DFB) resonator is built on top of the laser crystal to prevent it from emitting at any wavelength other than the desired wavelength. One of the attractions of such systems is that the emission wavelength is determined only by the thickness of the crystal layers and the layer materials itself, unlike diode laser, whose wavelength depends on the bandgap. They are also small (millimeter size), compact, and robust, but the main problems are that they must operate under a cryogenic environment (liquid helium or liquid nitrogen) and they are difficult to tune. Recent results report powers in the order of tens of milliwatts and with easier construction techniques, although the low temperature environment is still a strong limitation [73,74].

Optical techniques have also been pursued for generating narrowband THz radiation. Original efforts that began in the 1970s used nonlinear photomixing of two laser sources, but struggled with low conversion efficiencies [75]. In this technique, two CW lasers with slightly differing center frequencies are combined in a material exhibiting a high second-order optical nonlinearity such as DAST. The two laser frequencies mutually interfere in the material to result in output oscillations at the sum and difference of the laser

**TABLE 10.5**
**Generation Performance for CW Systems**

| | Gunn Oscillators | Gas Laser | BWO | FEL | QCL |
|---|---|---|---|---|---|
| Power [mW] | 0.1–100 | 1–180 | 1–50 | $10^8$ | 10 |
| Bandwidth [GHz] | $10^{-2}$ | $10^{-2}$ | $10^{-2}$ | $10^4$ | $10^3$ |
| Tunability | Changing the length of the resonant cavity | Changing the excitation gas | Continuously tunable | Changing magnetic field and electrons' speed | Nontunable |
| Range min–max [THz] | 0.1–0.8 | 1–6 | 0.2–1 | 10–100 | 1–3 |

frequencies. These systems can be designed such that the difference term is in the THz range. Tunable CW THz radiation has been demonstrated by mixing two frequency-offset lasers in LTG GaAs [76] and by mixing two-frequency modes from a single multimode laser. Further techniques utilize optical parametric generators and oscillators where a Q-switched Nd:YAG laser pump beam generates a second idler beam in a nonlinear crystal and the pump and idler signals beat to emit THz radiation [77–79]. Optical techniques provide broadly tunable THz radiation, are relatively compact because of the availability of solid-state laser sources, and demonstrate output powers in excess of 100 mW (pulsed) [80]. Optical down-conversion is a rich area for materials research as MBE and other materials advances allow engineered materials with improved photomixing properties [81] (Table 10.5).

### 10.3.2 Detection

The detector for CW THz can be classified into two groups: narrowband and broadband. Bolometers, pyroelectric, and Golay cell detectors are broadband, while Schottky diodes and superconducting structures are narrowband. To extract spectral information with CW THz sources and detectors, interferometry techniques may be used. Beam modulation, direct detection, and heterodyne detection techniques are used when narrowband detection is required.

Broadband detectors are usually based on thermal absorption, but the low output power of THz sources along relatively high levels of thermal background radiation in this spectral range requires highly sensitive detection methods. The most common systems are helium-cooled silicon; germanium and InSb bolometers are also cooled to reduce the thermal background. Superconductor research has yielded extremely sensitive bolometers based on the change of state of a superconductor such as niobium, and a single-photon detector for THz photons has been demonstrated [82]. This detector offers unparalleled sensitivity using a single-electron transistor consisting of a quantum dot in a high magnetic field. Although detection speeds are currently limited to 1 ms, high-speed designs are proposed, and this has the potential to revolutionize the field of THz detection. Pyroelectric infrared detectors are also used at THz frequencies. These are commercially available and provide good performance, but speed is low.

The Golay cell is an acousto-optic detector that generates a signal based on the expansion and contraction of a gas inside a sealed cell resulting from the absorption of modulated radiation. A thin gold film inside is deformed by the expansion of the gas, which provides the mechanism for measuring the pressure. Changes in this pressure provide information about the power absorbed of the incident radiation with a proper calibration. The responsivity of the Golay is very high, but it has a poor time response and there is a maximum input power that usually is in the order of tens of microwatts. If a higher power is applied, the gas expands too much and the gold film is damaged. The beam must be modulated around 10 Hz for detection and to reduce noise effects. The SNR can be improved by employing a low-pass filter to eliminate MIR noise.

At room temperature, semiconductor structures may be used. The most common is the Schottky diode used as direct detector or with a heterodyne approach. For higher sensitivity, cryogenic cooling is used for heterodyne superconductor detectors. Several superconductor structures can be used and have been used for over 20 years. The most widely used is the superconductor–insulator–superconductor tunnel junction mixer [83]. High-temperature superconductors such as YBCO are under investigation for their potentially higher bandwidth operation. Alternative narrowband detectors such as electronic resonant detectors, based on the fundamental frequency of plasma waves in field effect transistors, have been demonstrated up to 0.6 THz [84].

The maximum frequency that commercially available Schottky diodes can achieve is around 0.8 THz using direct detection, but the frequency can be higher if the heterodyne technique is used and then the Schottky diode works as a mixer. A planar Schottky diode mixer has been operated successfully at 2.5 THz for space-sensing applications [85]. The requirement to implement a heterodyne approach is that another source with a difference is required and the difference between the two frequencies must be in the range of detection of the mixer, which can be at the megahertz or the gigahertz range. One frequency is used as a probe signal and carries the information of the sample and the other is called local oscillator (LO). The two signals are mixed onto the diode, which gives outputs at the difference and summation of the frequencies. Usually the difference of the frequencies (IF) is picked out, filtered, and amplified with a low-noise amplifier. The sensitivity of this technique can be very high. A Schottky diode is very sensitive to electric

**TABLE 10.6**
**Detection Performance for CW Systems**

| | Bolometers | Pyroelectric | Golay Cell | Schottky Diode |
|---|---|---|---|---|
| Dynamic range | ~40 dB | ~30 dB | ~50 dB | Direct detection: 35 dB<br>Heterodyne: >130 dB |
| SNR | 100 (FTIR) | 1–10 K | Vol 246 | 1–10 K |
| NEP<br>$W/Hz^{1/2}$ | $\sim 10^{-16}$ | $\sim 10^{-10}$ | $\sim 10^{-10}$ | Direct detection: $\sim 10^{-8}$ $W/Hz^{1/2}$<br>Heterodyne: $\sim 10^{-19}$ $W/Hz^{1/2}$ |
| Range [THz] | 1–15 THz | 0.5–100 THz | 0.05–15 | 0.001–0.8 |
| Bandwidth [THz] | 15 (w/filter) | 15 (w/filter) | 15 (w/filter) | $10^{-2}$ |
| Responsivity [V/W] | 100 | 100 | $\sim 10^5$ | 100–3000 |

static discharges, and special caution must be taken when manipulating and operating it; however, it is both compact and mechanically robust. Its sensitivity and dynamic range are high and, due to its narrowband nature, it is less susceptible to external noise than other detectors (Table 10.6).

## 10.4 SENSING TECHNIQUES AND APPLICATIONS

### 10.4.1 Time Domain Spectroscopy

This technique is the workhorse of most of the spectroscopic measurements performed in the THz range. It combines the pulse generation by using a femtosecond laser onto a nonlinear crystal or photoconductive antenna with a coherent detection by using an EO crystal or also a photoconductive antenna [86]. This is a time-resolved technique in which an ultrashort THz pulse is generated and sent to a sample. The pulse is detected coherently and provides both transmission and phase information. The general way in which it works is by splitting an ultrashort laser pulse into two beams: the pump and the probe beam. The pump beam is used to generate the THz pulse by shining a nonlinear crystal or photoconductive antenna. This THz is guided and focused onto the sample by mirrors or lenses. The transmitted or reflected pulse is again guided and focused onto an EO crystal together with the probe beam. To get the THz and laser beams together, an Indium Tin Oxide (ITO) glass or silicon pellet can be used. The ITO glass is transparent to optical radiation but reflects THz, while the silicon pellet is transparent to THz but reflects the optical radiation. The two polarizations of the probe beam (s and p) are split and each one is detected by a single photodiode (Figure 10.6). The photodiodes are configured in such a way that the current they generate subtracts each other. In this way,

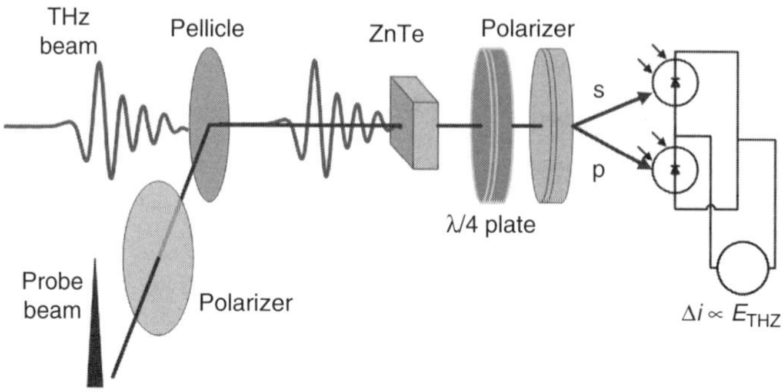


**FIGURE 10.6** Electro-optical sampling diagram. The polarization s and p of the probe beam are modulated by the THz electric field applied onto the EO crystal. The differential intensity between the two currents measured for each polarization is proportional to the THz electric field.

allows the net current to be adjusted to 0 when no THz pulse is applied onto the EO crystal. When THz is present onto the EO crystal, its birefringence changes and the balance between s and p polarizations is broken, resulting in a nonzero net current, which is proportional to the electric field. Either the probe or the pump beam is delayed with a linear delay stage so the modulation of the probe beam is different at each point, depending on the THz pulse's amplitude, thus allowing recording of the THz pulse profile (Figure 10.7).

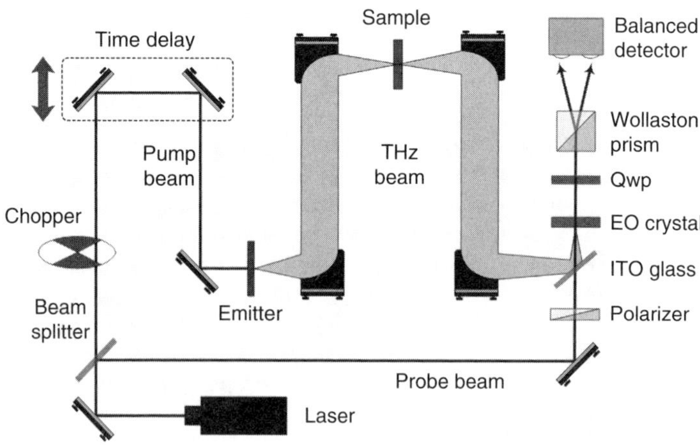


**FIGURE 10.7** THz-TDS diagram. The generation of a THz pulse is based on the use of a femtosecond laser pulse onto a nonlinear crystal or photoconductive antenna. Likewise, the detection is done by using an electro-optical crystal or also a photoconductive antenna.

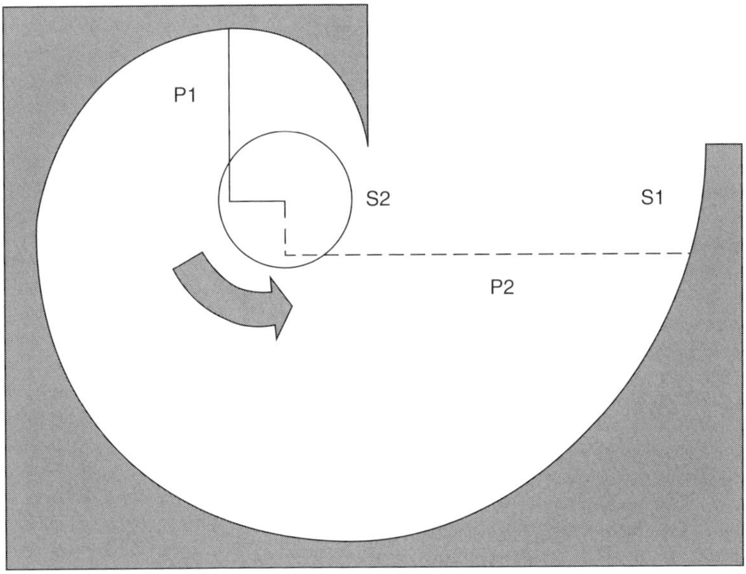


**FIGURE 10.8** Involute stage diagram. Surface S1 follows an involute curve and S2 is a circular surface. A beam perpendicular to S2 is retroreflected by S1, and by rotating S2, the beam path's distance changes (P1 and P2).

A variation to the linear stage is the involute stage (Figure 10.8), which is intended to improve the acquisition speed of the pulse and reduce the size. This stage is based on a reflecting surface that follows an involute curve. Another circular surface is placed in the center of the involute curve, and it can be demonstrated that a beam tangent to this circular surface will be retroreflected by the involute curve [87,88]. It can also be seen that by rotating the circular surface at a constant speed, the distance between the two surfaces increases linearly. This allows a laser beam properly aligned to be delayed linearly by rotating the circular surface at a constant speed, which could be at kilohertz or megahertz rate. The linearity of such a device is excellent and it is very useful in performing delay sweeps in which several waveforms must be taken quickly. A sweep frequency of a linear stage is limited to few tens or hundreds of hertz because of mechanical limitations. However, the involute stage is less indicated to perform step-by-step delay.

An alternative to THz-TDS is the spectroscopy with a chirped pulse. (Figure 10.9). Electro-optic detection of a THz pulse using a chirped probe pulse was first demonstrated by Jiang and Zhang [89,90]. This technique allows the full THz waveforms to be measured simultaneously rather than scanning the temporal pulse with a stepped motion stage. This provides a

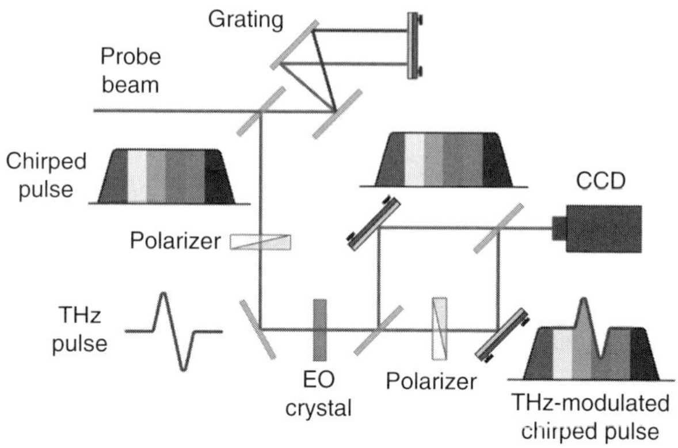


**FIGURE 10.9** Chirped pulse diagram. The chirped pulse is modulated by the THz pulse.

significant reduction in the acquisition time and greatly extends the applicability of THz systems in situations where the sample is dynamic or moving. In the standard THz-TDS, the femtosecond laser pulse is used to probe the instantaneous THz field at a certain time delay. The relative delay between the probe pulse and the THz pulse is then adjusted and the measurement is repeated. In this way, the full temporal profile of the THz pulse is measured, but it requires some time to scan the full duration of the pulse. This process can be greatly accelerated by applying a linear chirp to the probe pulse by using a diffraction grating. This technique derives from real-time picosecond optical oscilloscopes [91] and the geometry is similar to the standard THz-TDS. The optical probe pulse is frequency chirped and time stretched with a grating pair from subpicosecond to few tens of picosecond. The negative chirp of the grating makes the blue component lead the red component. The output from the grating is a pulse with longer pulse duration and a wavelength that varies linearly with time. The chirped probe pulse is modulated by the THz pulse when they are mixed onto the EO crystal. Conceptually, the chirped probe pulse can be seen as a succession of short pulses each with a different wavelength. Each of these wavelength components encodes a different portion of the THz pulse. A spectrometer spatially separates the different wavelength components and thus reveals the temporal THz pulse. The spatial signal output from the spectrometer is measured using a CCD. For maximum image acquisition speed, the THz pulse and probe pulse may be expanded in the vertical dimension using cylindrical lenses. The CCD is then able to capture both the THz temporal waveforms and several hundred vertical pixels simultaneously and only a single translation stage is required for spectroscopic image acquisition.

### 10.4.2 Explosive Identification

Explosives and explosives-related compounds (ERCs) sensing and identification have been a very important topic for most security and defense organizations. In particular, a noncontact, fast technique is required for use in real applications. The most recent approach for such sensing and identification has been the use of Raman and Fourier transform infrared (FTIR) spectroscopy for near- and mid-infrared and is intended for use in the FIR range. In that range, THz time domain spectroscopy has proved to be the best approach. THz wave offers several advantages other than identifying the fingerprints of the explosives. THz waves can penetrate cloth, plastic, paper, and other nonmetallic and dry materials and thus provide the possibility of identifying the explosives behind nonoptically transparent covers. The identification can be performed either in transmission or reflection geometry. For a real application, reflection geometry is the most realistic approach and, in particular, the diffuse reflection, which is the reflection that objects reemit due to scattering and not due to specular reflection.

The spectra of ERCs' samples such as TNT, RDX, HMX, PETN, tetryl (2,4,6-trinitrophenyl-*n*-methylnitrami 2-amino-4,6-DNT (2-amino-4,6-dinitrotoluene), 4-amino-2,6-DNT, 4-nitrotoluene, 1,3,5-TNB (1,3,5-trinitrobenzene), 1,3-DNB (1,3-dinitrobenzene), 1,4-DNB, 2,4-DNT, 2,6-DNT, 3,5-dinitro aniline, and 2-nitro-diphenyl-aniline can be measured with THz-TDS and FTIR to compare the two techniques. The THz spectra of the ERCs are measured with a standard TDS setup, whereas the FTIR spectra are obtained by using a commercial FTIR spectrometer Bruker IFS 66V/S in the 0.1–20 THz (4–670 $cm^{-1}$) range. Both spectral information are measured either in the vacuum or nitrogen atmosphere. Table 10.7 shows a summary of the absorption peak positions. Most of the data agree with the literature results [92–94], and some others are new observations. The spectra features can be simulated by using specific software (such as Gaussian 03) based on the density function theory. Figure 10.10 shows the comparison between the experimental results and the calculated results obtained using Gaussian 03. Most of these absorption features result from the normal vibrational modes of ERC's molecules. The experimental results include transmission spectra and diffuse reflectance spectra using an FTIR spectrometer in the 1.5–21 THz range. Good agreements are achieved among calculated results, transmission, and diffuse reflection spectra. THz-TDS shows a higher SNR from 0.1 to 3.0 THz compared to the FTIR spectrometer. Therefore, the absorption spectra of common ERCs in the 0.1–3 THz range were measured using THz-TDS in a nitrogen environment. Most of these low frequency absorption fingerprints come from lattice vibrations or phonon modes of crystalline ERCs. Few ERCs are featureless in the range of 0.1–3 THz, such as tetryl, 1,3,5-TNB, 1,4-DNB, and ammonium nitrate. The detection of these ERCs requires special considerations because THz waves at a higher frequency have lower penetration

**TABLE 10.7**
**THz Absorption Peaks in THz of Explosives and Related Compounds (ERCs)**

| ERCs | Measured Absorption Peak Position (THz) |
|---|---|
| TNT | 1.66, 2.20, 3.69, 4.71, 5.58, 8.16, 9.15, 9.75, 10.65, 13.89, 15.09, 19.17 |
| RDX | 0.82, 1.05, 1.50, 1.96, 2.20, 3.08, 6.73, 10.34, 12.33, 13.86, 14.54, 17.74, 18.12, 20.13 |
| HMX | 1.78, 2.51, 2.82, 6.06, 11.10, 11.97, 13.56, 14.52, 18.18, 18.51 |
| PETN | 2.0, 2.84 |
| Tetryl | 5.97, 10.11, 11.28, 14.67, 16.14, 18.36 |
| 2-Amino-4,6-DNT | 0.96, 1.43, 1.87, 3.96, 5.07, 6.27, 8.49, 9.87, 10.77, 12.15, 13.44, 16.68 |
| 4-Amino-2,6-DNT | 0.52, 1.24, 2.64, 3.96, 5.04, 5.82, 7.53, 9.30, 10.20, 11.13, 13.86, 14.97, 17.70 |
| 4-Nitrotoluene | 1.20, 1.37, 1.86, 6.75, 8.85, 10.83, 14.04, 15.66, 18.51 |
| 1,3,5-TNB | 4.17, 4.62, 10.05, 11.19, 13.80, 15.75, 19.05 |
| 1,3-DNB | 0.94, 1.19, 2.37, 10.56, 12.18, 15.33, 17.13 |
| 1,4-DNB | 3.24, 3.96, 5.55, 10.38, 12.45, 13.29, 15.21, 15.54 |
| 2,4-DNT | 0.45, 0.66, 1.08, 2.52, 5.01, 8.88, 10.56, 11.58, 12.81, 14.34, 15.81, 19.05 |
| 2,6-DNT | 1.10, 1.35, 1.56, 2.50, 5.61, 6.75, 9.78, 11.43, 13.32, 13.89, 15.39, 17.25 |
| 3,5-Dinitro-aniline | 0.96, 1.20, 3.18, 4.62, 5.04, 5.91, 7.44, 10.62, 10.98, 14.46, 16.41, 18.18 |
| 2-Nitro-diphenyl-aniline | 2.19, 2.58, 2.88, 3.45, 5.13, 6.18, 7.56, 10.08, 12.33, 13.05, 15.00, 15.60, 16.29, 17.34, 18.51, 19.32 |

depths for many daily-used materials and their atmospheric attenuation is also different.

If a real-world scenario is considered, both atmospheric attenuation and scattering effect will affect spectroscopic results. For instance, THz sensing in the transmission windows is necessary for real-world applications. ERCs have many distinguishable THz fingerprints in the range of 10–21 THz, where THz waves show low atmospheric attenuation. However, THz waves in this range do not have high penetrations for many daily-used materials compared with those at lower frequency ranges. Therefore, standoff detection is limited to exposed explosives in the range of 10–21 THz. ERCs also have abundant THz fingerprints in the 0.1–3 THz range, where daily-used materials become much more transparent, and standoff sensing hidden explosives under cover in the atmosphere are promising. Sensing with narrowband CW THz waves is another promising direction. CW sources and a detector array covering a range wherein ERCs have THz fingerprints and the atmosphere has low attenuation are applicable for standoff sensing. The relative atmospheric

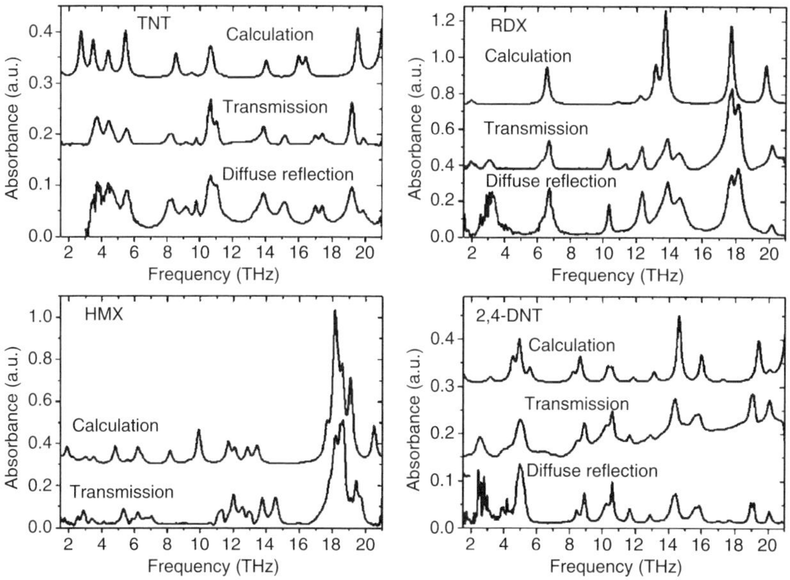


**FIGURE 10.10** Comparison of calculated results, transmission spectra, and diffuse reflectance spectra of TNT, RDX, HMX, and 2,4-DNT measured with FTIR. The spectra have been vertically shifted for clarity.

attenuation up to 7 THz is shown in Figure 10.11. Several windows can be used in this range to perform standoff detection. The reference is taken at low humidity and the signal at 17% RH. The position of the peaks is clear.

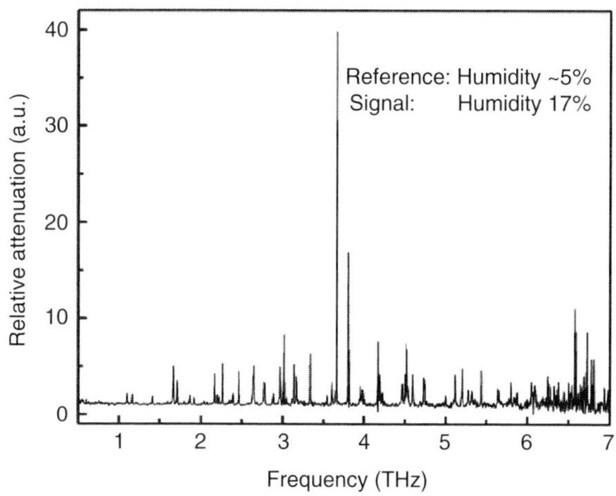


**FIGURE 10.11** THz transmission windows up to 7 THz. Several windows can be used to perform standoff sensing and identification.

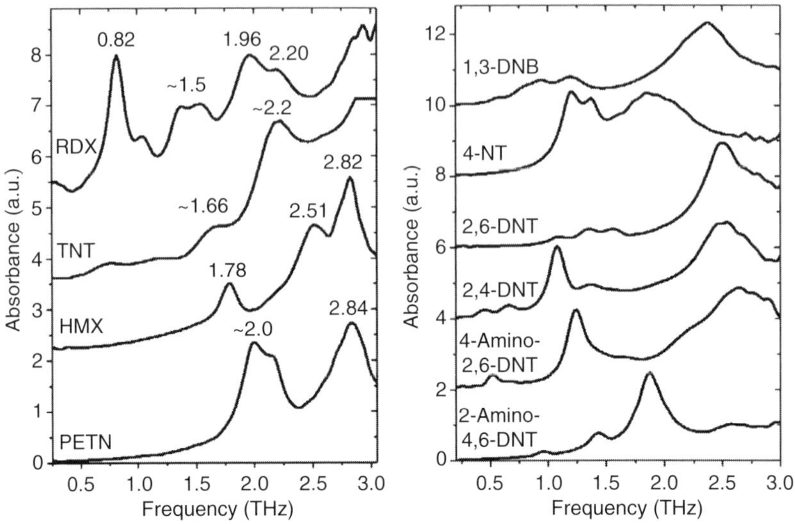


**FIGURE 10.12** (a) THz absorption spectra of RDX, TNT, HMX, and PETN obtained with THz-TDS; (b) THz absorption spectra of explosive related compounds obtained using THz-pulsed spectroscopy: (1) 1,3-DNB; (2) 4-NT; (3) 2,6-DNT; (4) 2,4-DNT; (5) 4-amino-2,6-DNT; (6) 2-amino-4,6-DNT. The spectra have been vertically shifted for clarity.

Results shown in Figure 10.12 are obtained by performing transmission. However, for real-field applications, reflection measurements are preferred because most bulky targets are impossible to measure in a transmission mode. In particular, diffuse reflection spectroscopy is more applicable for standoff detection because it does not require the sample to be perpendicular to the line of sight. Detection of RDX by THz spectroscopic imaging based on specular reflection has been demonstrated [95]; and it can also be done in diffuse reflection. The diffuse reflection spectrum of an RDX pellet was acquired from a diffuse reflection measurement using a THz-pulsed spectroscopy system in combination with a diffuse reflectance accessory. The measurements were conducted in a chamber purged with nitrogen to avoid the influence of water vapor absorptions in ambient air. The obtained absorption spectrum (0.2–1.8 THz) by applying the Kramers–Kronig transform to the reflection spectrum agrees with that from the transmission measurement.

### 10.4.3 Other Spectroscopic Applications

Petroleum products analysis has been performed with several spectroscopic techniques such as ultraviolet absorption, infrared absorption, Raman scattering, mass spectroscopy, emission spectroscopy, nuclear magnetic resonance, and chromatography. The closest techniques to THz range are FTIR and

Raman spectroscopy [96], and both have proved to be major contributors to near-infrared (NIR) and mid-infrared (MIR) [97] ranges and are able to provide reliable spectral data from many kinds of liquid and solid samples. THz-TDS can also be used in this field to analyze gasoline, diesel, and lubricating oil. For gasoline and diesel, Co-operative Fuel Research (CFR) developed a standard variable in the 1930s to measure the performance of the fuels based on the octane number [98], but alternative spectroscopic techniques have been developed because of the complexity of CFR methods [99]. The characterization of lubricating oil depends strongly on the product complexity and the nature of the application. This characterization is usually based on a statistical basis that generates a particular fingerprint associated for a particular blend of components and additives. The comparison of the fingerprint with a known standard is used for monitoring oil degradation. This degradation has been studied in the infrared region and includes processes such as oxidation, nitration, sulfate formation, and others [100]. The absorption of THz waves in liquid phase is mainly due to vibrational modes that can be torsional or bending. The real and imaginary parts of the dielectric permittivity, refraction index, and absorption coefficient can be obtained with THz-TDS because of its coherency. These parameters can be used to identify the different grades and are providers of gasoline and diesel. Aromatic compounds such as *o*-xylene, *m*-xylene, and *p*-xylene can also be identified with THz-TDS.

THz spectroscopy also finds applications in the identification and chemical dynamics of organic and biological molecules. The spectra can give information about the evolution of a particular molecule and how the interaction with other molecules evolves in time. For instance, sulfamethoxazole (SMZ) is used worldwide for the treatment of bacterial infections. The interactions between SMZ and other drugs, like caffeine and trimethoprim (TMP), may eventually be very problematic for a patient. Therefore, a technique that can monitor such interactions is of great interest. Figure 10.13 shows the changes in the THz absorption spectra after SMZ bonds to caffeine or

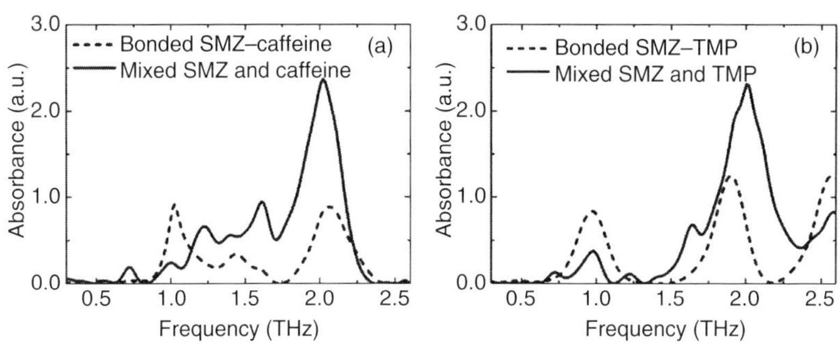


**FIGURE 10.13** The spectra of SMZ–caffeine and SMZ–TMP are different, depending on the bonding condition of the drugs.

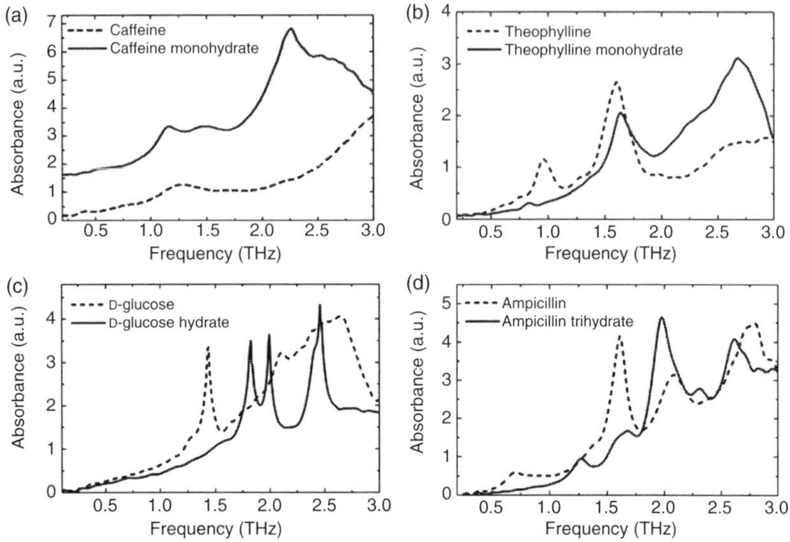


**FIGURE 10.14** The spectra of these drugs change, depending on the hydration state of the molecule.

TMP: absorption peaks at ~2.02 and ~1.62 THz decrease after bonding, absorption peaks at ~1.0 THz increase, and absorption peaks at ~0.72 and ~1.22 THz disappear. The changes in the spectra can be monitored in time and the chemical dynamics can be studied. Figure 10.14 shows examples of caffeine, theophylline, D-glucose, and ampicillin, which are very important molecules for the pharmaceutical industry. In this case, the spectra are different, depending on the hydration state of the molecules. THz spectroscopy is well suited for distinguishing anhydrous and hydrated pharmaceutical materials, which have different dissolution rates, stability, and bioavailability, therefore their study is of significance. D-glucose monohydrate dehydrates on heating (>40°C). The absorption peaks at 1.80 and 1.95 THz decrease and those at 1.28, 1.44, and 2.07 THz increase over time due to the reduction of D-glucose monohydrate and augmentation of anhydrous D-glucose.

## 10.5 IMAGING TECHNIQUES AND APPLICATIONS

### 10.5.1 Passive and Active Imaging

Passive imaging is based on the detection of the thermal radiation that all objects radiate because of its temperature. The distribution of the radiation is governed by Planck's law and it describes the radiation intensity emitted by a so-called blackbody at a given temperature. Most of the background cosmic radiation is within the THz range, radiated by cool (30 K) stellar dust. The oldest form of THz imaging is passive submillimeter sensing, which has been used for many decades for space imaging applications. In these systems, a heterodyne detector (on Earth or aboard a satellite) is used to sense the amount of THz radiation emitted by distant galaxies. By tuning the frequency of the detector, a spectrum can be obtained. This spectrum contains vital information regarding the presence of certain molecules in that distant galaxy. For instance, water molecules have strong characteristic absorption resonances at 0.557, 0.752, 1.097, 1.113, 1.163, and 1.207 THz [101–103]. By comparing the amplitude of the received THz power at these frequencies relative to the background radiation, astronomers can determine whether water is likely to exist on distant planets. This is a vital tool in the search for extraterrestrial life. Other molecules that can be easily identified using this technique include oxygen, carbon monoxide, and nitrogen [104].

Similarly, passive THz imaging principles have recently been employed in terrestrial applications. This type of imaging system is aided by the fact that a wide variety of common materials have very low absorption coefficients at THz frequencies and thus appear transparent to THz imaging systems. Materials such as plastics, cloth, paper, and cardboard are transparent at THz frequencies and are totally opaque in the optical spectrum. This led groups such as the European Space Agency (ESA) [105] to invest in the development of a passive CCD camera operating at THz frequencies. This project focused on combining micromachined THz antenna with a silicon photonic bandgap backing plane to form an imaging array. THz emission by the object because of its temperature enables passive imaging, but it is also a source of severe noise. For this reason, passive THz imaging methods have had most success in space, where the detector can be mounted on a satellite, away from the strong thermal background that exists on Earth and directed solely at the target of interest.

In active imaging, the target is illuminated with the radiation of interest and then the reflected or transmitted radiation is measured. A well-known example is the radar, which emits pulses of radiation at a particular frequency and often with a particular modulation and then the receiver detects the reflected radiation and looks for the same frequency and modulation. This allows the radar to detect a weak signal in the presence of strong background noise. The location of the target can be accurately determined [106] on the basis of the time delay of the received pulse and its direction.

Active imaging systems can use a pulsed or CW illumination. Early THz imaging systems used CW gas THz lasers to illuminate the target and thermal detectors [107,108] or pyroelectric cameras [109], but thermal background noise was and still is a common problem. The background radiation emitted by the target or the surroundings is generally indistinguishable from the active reflected or transmitted radiation, resulting in noise in the image. One way to overcome this problem is by using a source more powerful than the background. For pulsed systems, the illumination power is compressed into a short pulse width (a few picoseconds), causing a very high peak illumination power. This high peak power together with coherent detection techniques allows the use of much lower average power sources while providing the same SNR.

Pulsed THz imaging (T-ray imaging) was first demonstrated by Hu and Nuss at Bell Laboratories in 1995 [110] and many applications have been developed including microchips imaging [111], skin burn severity evaluation [112], leaf moisture content [113], tooth cavities inspection [114], and skin cancer [115]. T-ray imaging provides many desirable characteristics such as the following:

1. T-rays have low phonon energies (4 meV @ 1 THz) and, therefore, are not ionizing for biological tissues. In comparison, a typical x-ray photon has an energy in the KeV range and causes ionization and other potentially harmful effects on biological tissues [116,117].
2. T-ray imaging provides spectroscopic information within the THz range as well as the amplitude and phase information. This allows identifying uniquely many different materials within an image because its unique rotational, vibration, and translational modes are within the THz range. This is not obtained with optical, x-ray, or NMR images.
3. T-rays can easily penetrate most dielectric and dry materials, allowing the imaging of the internal content and structure.

Although CW systems do not provide spectroscopic and fingerprint identification, the radiation is also not ionizing and can penetrate most of nonmetal and dry materials. The use of CW systems for imaging has recently gained more popularity because they can be compact and portable and are more suitable for field applications compared to actual pulsed systems. Similar to the pulsed system, different kinds of mirrors and lenses can be used to focus and guide the beam, resulting in an improvement in the collection efficiency. The beam can be modulated and very narrowband detectors and techniques can be used to reduce the amount of noise. Regarding the applications, security inspection was first proposed using FIR radiation in the 1970s [118] and it has been extended to the THz, especially at frequencies below 1 THz. More recently, CW THZ imaging has proved to be a very valuable NDE tool, and found some key applications such as the inspection of the insulation foam used in the space shuttle.

THz imaging has been used for target screening in plastic and low-refractive-index materials, biomedical imaging, security scanning, and microscopic imaging. THz imaging has the advantage of being noninvasive and nonionizing, thus representing little harm to biological tissues. The wavelength is substantially smaller than that offered by microwave radiation and can provide much higher spatial resolution. When used in tomography application, THz radiation is also significantly less susceptible to scattering than infrared light, thus allowing improved reconstruction fidelity. THz imaging systems are a recent addition to the wide array of available imaging modalities, but the unique properties of THz radiation allow THz imaging to fill niches that are unreachable using other techniques.

### 10.5.2 CCD-Based Two-Dimensional Imaging

This technique was developed for pulsed systems and represented a dramatic improvement in the acquisition speed and made real-time imaging feasible [119]. This technique provides a parallel detection capability and removes the need to scan the target. This method is based on electro-optic sampling and uses a large EO crystal as a detector. It implements a quasiplane wave illumination instead of focusing the THz pulse on the sample. Both the probe beam and the THz beam are expanded and directed onto the EO detector crystal. The THz pulse acts as a transient bias on the EO crystal (usually a ZnTe crystal as discussed in Ref. [110]) and induces a birefringence in the crystal that modulates the probe beam. The two-dimensional (2D) THz field distribution is then converted to a 2D intensity modulation of the optical probe beam after it passes through a crossed polarizer analyzer and finally is recorded by a CCD camera. The use of a CCD does not make the use of balanced detection possible as in the standard EO sampling. However, a phase-sensitive detection principle similar to that used in the lock-in amplifier can be implemented by dynamic subtraction. The THz beam is modulated by an optical chopper synchronized with the CCD camera with a modulation frequency equal to half of the CCD frame rate. Thus, the THz signal appears in the probe beam in an alternate frame and the difference between two consecutives frames gives the signal. SNR can be improved by accumulating several frames.

### 10.5.3 Imaging Geometries and Beam Guidance

Both pulsed and CW scanning THz systems are based on a point emitter–receiver; therefore, a raster scan is performed to obtain the image. Each pixel is the result of scanning a particular point (transmitted or reflected spectral information for a pulsed system and transmitted or reflected energy for a CW). Transmission or reflection geometry can be implemented to perform imaging with a pulsed or CW system (Figure 10.15). Either the sample or the imaging system is moved with an XY stage and data are acquired point by point. The main drawback of this technique is the speed; although very fast stages are available, it is very difficult to achieve real-time imaging of samples of medium to big size (hundreds of centimeters). To perform real-time imaging, the development of detector arrays is required.

A variation of a normal geometry is the use of two different frequencies to illuminate the sample by using the chopper as a modulator and switch at the

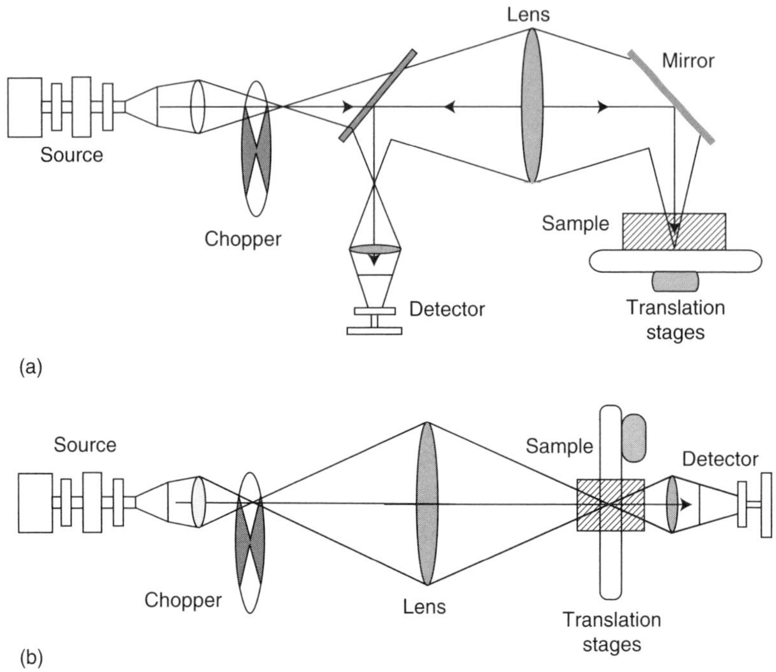


**FIGURE 10.15** Normal reflection geometry (a) and transmission geometry (b) setups.

same time (Figure 10.16). The beam of frequency $\omega_1$ is modulated by the chopper in the usual way, but when $\omega_1$ is blocked then the beam of frequency $\omega_2$ is reflected by the chopper's blade and sent to the sample.

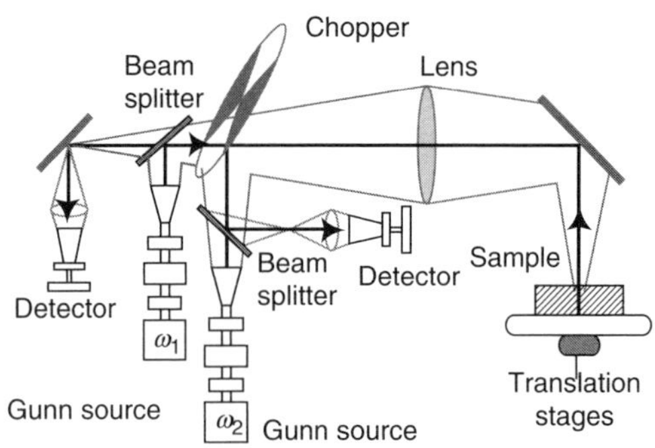


**FIGURE 10.16** Two-frequency geometry. The chopper is used as a modulator and switch at the same time.

When the beam $\omega_1$ is not blocked, beam $\omega_2$ is not reflected and goes out of the optical path through the chopper's blade. As a result, there are two alternating frequencies imaging the sample, which is modulated at the same frequency, but delayed by a half-period of this modulation frequency. With this geometry, the pictures at the two frequencies can be obtained at once. Therefore, the first clear advantage is that the time to obtain the two pictures at different frequencies is reduced. However, the purpose of this geometry is that, for each point, the intensity of the reflected signal at both frequencies is recorded; hence, theoretically, the common background can be removed and one can retrieve the difference in the absorption of the sample between the frequencies.

Likewise, the lens is a critical component in most optical imaging systems; it is also critical for THz systems. In optical systems, the resolution is fundamentally limited by the wavelength and it also happens in a THz system. The resolution is approximately given by the Rayleigh criterion for far-field imaging in the absence of lens aberrations. The same principles and equations used to design a lens in the optical range can be used in the THz range, taking precaution while using thin lenses and paraxial approximations. Lenses can be fabricated for THz range with materials like polyethylene or Teflon. Several types of lenses can be designed and fabricated, the most common are the spherical, Fresnel [120–122], and hyperbolic. The design of a spherical lens is simple, but it results in a thicker lens with prominent aberrations at large apertures. A Fresnel lens is thinner, but the focal length depends on the frequency and still shows prominent aberrations at large apertures. A hyperbolic lens corrects for the spherical aberration providing a better resolution, but the thickness is similar to the spherical lens.

### 10.5.4 Compact CW THz Imaging Systems

Recently, a CW compact imaging system has been designed and built to perform the scanning of the space shuttle foam. The system consists of two main parts: the THz unit and the translation stages. The system implements normal reflection geometry and the lenses are hyperbolic and designed to focus the beam at a distance of 225 mm from the front surface to allow enough clearance to scan a sample. The detection is performed by a Schottky diode in direct detection mode. The unit incorporating a laser diode to indicate the position of the THz beam's focus by splitting the laser beam in two spots that are coincident on the focal plane. The unit can be moved along the $z$-axis until the two spots appear together onto the sample's substrate. If the sample's substrate is not reachable, the separation between the two spots is related to the distance between the testing surface and the focal plane.

The structure of the system is shown in Figure 10.17. The THz wave is generated by a 0.1 THz Gunn diode followed by a doubler, resulting in an

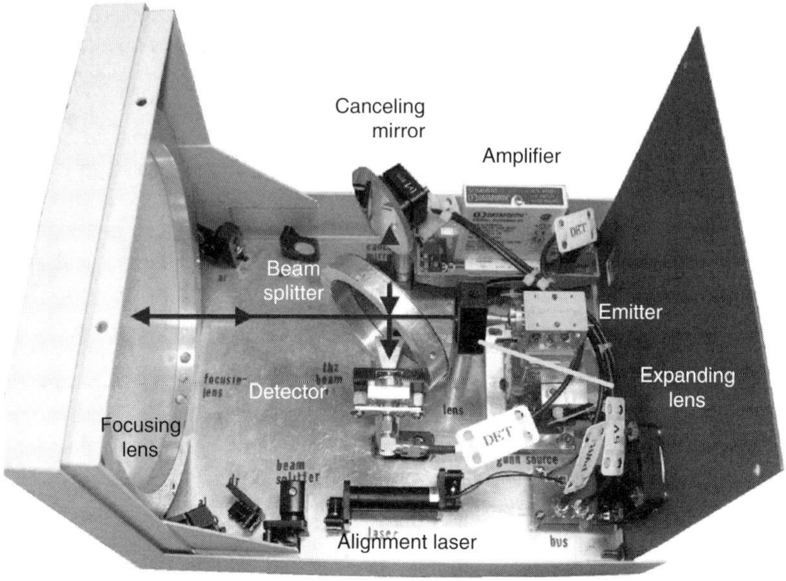


**FIGURE 10.17** THz unit components and geometry. Part of the outgoing beam is reflected back to the detector by the canceling mirror to cancel the waves reflected by the focusing lens.

output frequency of 0.2 THz. The Gunn diode and the doubler form the emitter. The THz wave is coupled into air by a horn antenna and expanded by a 4 mm focal length polyethylene hyperbolic lens (expanding lens). The expanded beam is focused at 225 mm from the front side by another polyethylene lens of 66 mm focal length. The separation between the focusing lens and the expanding beam is 93 mm. After reflection from the sample's substrate, the THz wave is collected and refocused by the focusing lens into the detector, which is an unbiased Schottky diode. The incoming beam is separated from the outgoing beam by a high resistivity silicon beam splitter. The signal is amplified by a low-noise amplifier and sent to the computer (Table 10.8).

An important problem in CW systems that work in reflection is the generation of standing waves, which is caused by the interference between the outgoing and reflected beams. There are two possible sources: the standing waves generated by the internal reflections within the system and the standing waves generated by the reflections coming from the sample. These standing waves introduce a waved pattern into the picture that results in a succession between clearer and darker areas. A maximum acceptable modulation percentage introduced by this kind of pattern is around 20% to

**TABLE 10.8**
**Performance and Specifications of the THz Unit**

| | |
|---|---|
| Dimensions | 8.25″ × 8.25″ × 11″ |
| Weight | 9 lbs |
| Maximum signal (at focal distance, with amplifier) | 2 V |
| Background signal (at focal distance, with amplifier) | 1 mV |
| Dynamic range | 33 dB |
| Resolution | 2.6 mm |
| Depth of focus | 25 mm |
| Focal distance | 225 mm |

30%. Higher modulation percentage may make the images more difficult to interpret the pictures because of the reduced contrast of the dark areas.

Although canceling the standing waves generated by the sample is difficult because it depends strongly on the particular geometry, it is possible to reduce the standing waves generated by the internal reflection of the system. For the 0.2 system, it has been observed that the reflections coming from the focusing lens are relatively strong. To cancel the standing waves generated by this reflection, a mirror is placed at the opposite side of the detector. This mirror (canceling mirror) reflects part of the outgoing radiation deviated by the beam splitter back into the detector. After optimizing the position of all components by using a test mirror at the focal plane, the canceling mirror can be adjusted to reduce the background signal of the detector. To do this, only the reflection coming from the lens must be present, so it is particularly important to avoid placing anything in front of the lens. When the canceling mirror is adjusted, then the test mirror can be placed again on the focal plane. If this test mirror is moved along the $z$-axis, the modulation should be less than 30%.

The system does not use modulation to reduce the noise and relies on the narrowband of the detector and a digital filtering. To extract the resolution of the system, a fan pattern is imaged and then the separation between the blades is measured, thus providing a value for the resolution. The stages are linearly guided and belt-driven with servomotors. They are able to provide an accuracy of less than 30 $\mu$m, a maximum speed of 1000 mm/s, and a maximum acceleration of 3500 mm/s$^2$. The maximum load is 15 lbs for the vertical stage and 10 lbs for the horizontal stage (Figure 10.18).

### 10.5.5 Space Shuttle Foam Inspection

The tragedy of Columbia took place on February 1, 2003, because of the detachment of a foam panel during lift off. This drove NASA to study and analyze possible solutions and technologies that could be used to inspect

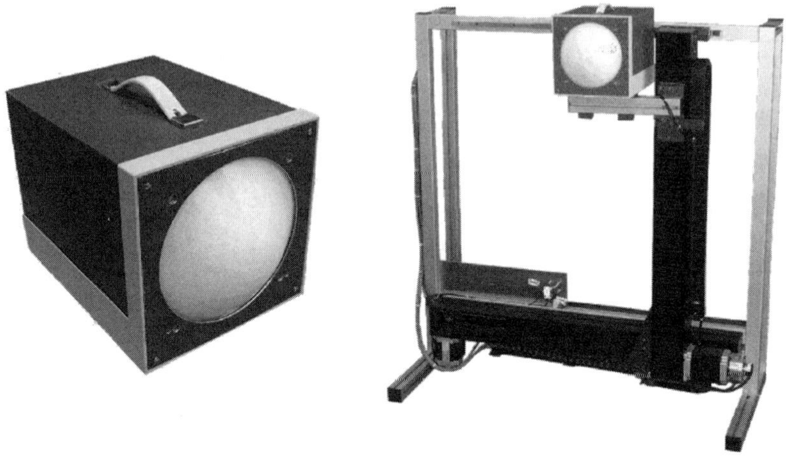

**FIGURE 10.18** THz unit (left) and linear stages to scan the samples (right).

those panels and avoid the detachment. The detachment was mainly due to the presence of defects (voids and delaminations) that may have appeared during its fabrication. During the first phase, several technologies and techniques were tested, but the most promising were back-scattered x-ray and THz imaging. Actually, THz imaging ranks among the best techniques studied [123], and it turns out that the inspection of the insulation foam is a key application of THz technology as an NDE tool.

The sprayed-on foam insulation (SOFI) is a good subject for THz imaging because it has a low absorption coefficient and index of refraction [124] at frequencies lower than 1 THz. The samples, provided by Lockheed Martin Space Systems and NASA Marshall Flight Center, are sprayed layer by layer onto an aluminum substrate. In the inspections, the defects appear as dark circles with light interiors, corresponding to the scattering and interference at the edge of the feature and enhanced transmission due to the lack of material in the interior. The inspection can be carried with a pulsed or CW system. The pulsed system requires working at a central frequency of around 0.5 THz, and the substrate and surface must be perfectly perpendicular to the beam to obtain the pulse information properly. CW system is more tolerant to nonperpendicular conditions of the substrate and the roughness of the surface, but some standing pattern may affect the pictures. However, this standing wave pattern can be reduced resulting in a picture where the defects and the general structure of the foam show up clearly [125].

Several samples provided by NASA were scanned with a CW system at 0.2 THz and 0.4 THz. The samples have a typical size of $2 \times 2$ feet with a foam thickness between 2″ and 4″ (Figure 10.19). The defects were built by placing a circular polyurethane slice on the substrate or solid foam and by

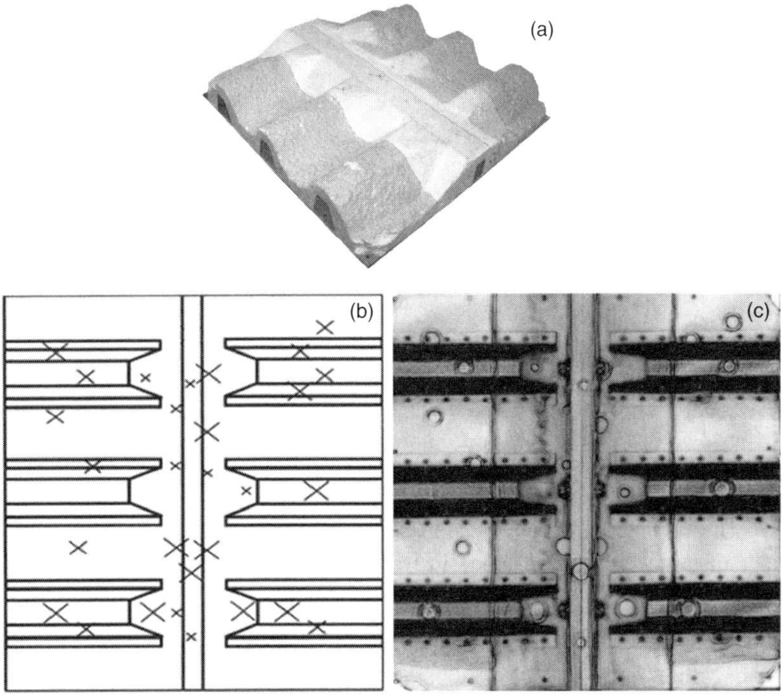


**FIGURE 10.19** THz image of the 2 × 2 feet panel (c) at 0.2 THz taken with the THz system. (a) Optical image of the sample and defect map (b).

spraying foam on the slice. The samples are provided with a defect map. Sample in Figure 10.19a has a flat substrate and a stringer on the middle. The stringer is a U-shaped inverted piece, which simulates the reinforcements of the tank structure. It has been observed that most of the defects appear in the vicinities of those stringer structures. The sample shown in Figure 10.19b has six stringers and the foam is sprayed following the resulting geometry with an average thickness of 2″. The THz images (Figure 10.19c) show the position of the big (>0.5″) and medium (0.25″ to 0.5″) defects very clearly. The pictures can be processed to enhance the defects. In particular, the representation of the pictures on a log scale is very useful because it reduces the noise and enhances the contrast of the boundaries generated by the presence of the defects. Other types of processing such as deconvolution can be used if the point-spread function is known. The point-spread function can be obtained by measuring the beam spot with a pyroelectric camera and fitting the data to an Airy disc pattern. Basically, the deconvolution technique enhances the sharpness of the image and improves the definition of the defects. Additionally, local contrast enhancement can be performed on the image as well, which serves to emphasize the structure through more effective use of the dynamic range of the display. As a result, defects as small as 0.25″ are identified and defects greater than 0.5″ are easily distinguished from background fluctuation. The system also detects natural defects such as rollovers. The system is not very sensitive to the surface conditions, and it is very tolerant to the depth of the defect. The system complies with the most desired characteristic criteria given by NASA.

### 10.5.6 Security Scan and Content Inspection

It is beneficial to use THz for security screening because it can perform functions similar to x-ray screening, but with a lower cost and health risk. Some examples are given in this section to show the feasibility. The systems used are a Gunn diode at 0.2 THz with a nominal power of 20 mW and another Gunn diode at 0.6 THz with a nominal power of 350 μW. For detection, a Schottky diode is used at 0.2 THz and a Golay cell at 0.6 THz. The typical resolution with hyperbolic lenses at 0.2 THz is 2.7 mm and 1.3 mm at 0.6 THz.

The screening examples include a lady's purse, containing some objects of daily use, which is scanned in transmission geometry at 0.2 THz and a laptop computer is scanned in reflection geometry at 0.6 THz (Figure 10.20). The step size is 1 mm at 0.2 THz and 0.5 mm at 0.6 THz. The objects in the lady's purse include a perfume bottle, a book, an MP3 player, a bottle of pills, a chewing gum, a hand mirror, a Swiss army knife, and a hand care set. The perfume and pills bottle, book, Swiss army knife, and hand care set can be identified, but the others are mixed on the bottom-right of the bag and no clear shape is recognized. The suspicious objects to be detected are the knife and the hand care set, which are seen in the center and on the bottom-center. The results of the computer imaging show that the internal structure of both top and bottom sides can be seen: the circuitry of the display and the screws, the fan, the memory slot, the battery, the hard disk, and some cabling and connectors.

Another possible use of CW THz imaging could be in the quality control inspection of some manufacturing processes. The examples in Figure 10.21 show the feasibility of THz imaging to check the quantity, position, and quality of the contents of nontransparent recipients. A ceramic teapot is inspected empty and half full of water. The two pictures are taken in a transmission at 0.2 THz with a step size of 1 mm and a resolution of 2.7 mm. The contrast between the volume with water and the volume without water is high because THz radiation is strongly absorbed by the water, which allows the amount of water (volume) to go inside containers that are not transparent in the optical range such as opaque plastic or ceramic bottles. In transparent containers, the amount of liquid is usually checked by using intelligent optical cameras. Another example is the inspection of a racquetball inside its bag. The THz image was taken at 0.6 THz also in transmission geometry with a

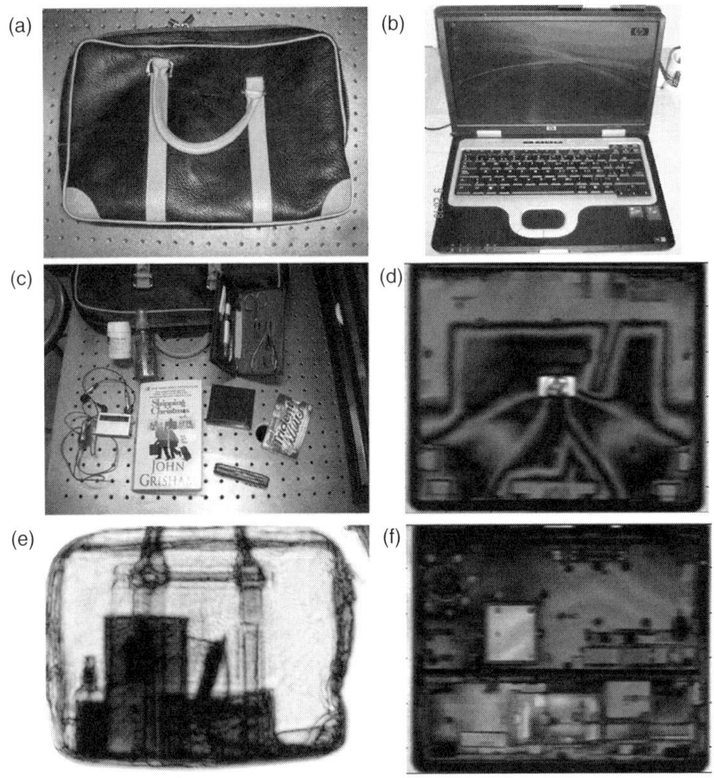


**FIGURE 10.20** Lady handbag (a) with some harmless objects and other potentially dangerous (c) at 0.2 THz in transmission geometry (e). Inspection of a laptop (b) in normal reflection geometry at 0.6 THz. Image of the top (d) and image of the bottom (f).

step size of 1 mm and a resolution of 1 mm. The shape, the wires (with a diameter of 1.27 mm), and the zipper are clearly recognized.

### 10.5.7 Tomography

The use of THz to perform tomography has always been in the mind of the researchers because, besides target reconstruction, it can also provide spectroscopic information of that target, whereas microwave and x-ray imaging modalities produce only density pictures. A THz wave can easily penetrate and inspect the insides of most dielectric materials, which are opaque to visible light and low in contrast to x-rays, making THz wave imaging a useful complementary tool source in this context. Most of the work has been done with pulsed THz systems, with a typical range from 0.1 to 3 THz. The broadband frequency characteristic is used to get the spectroscopic data of

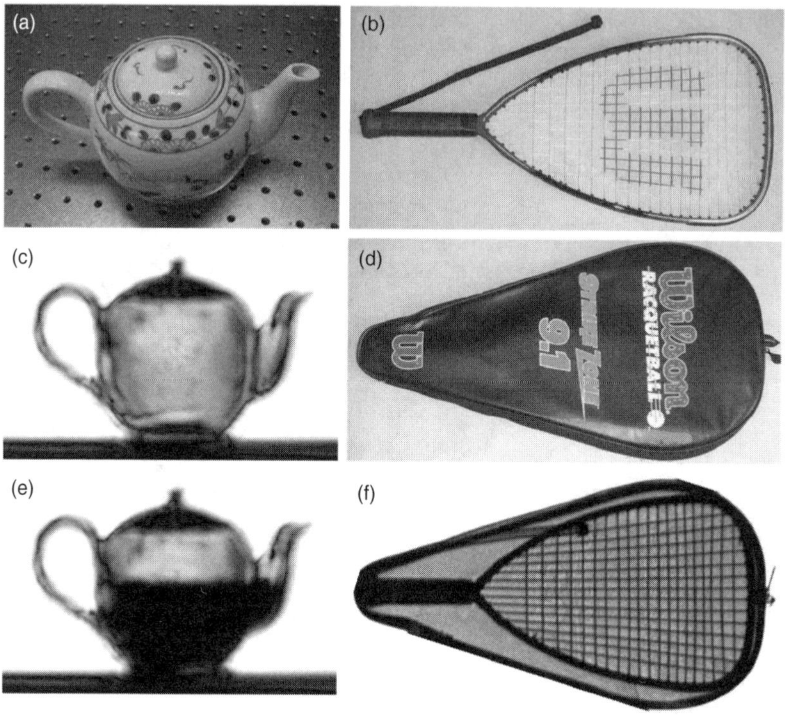


**FIGURE 10.21** Content inspection with transmission geometry. Imaging of a teapot (a) at 0.2 THz with no water (c) and half full of water (e). Imaging of a racket (b) inside the bag (d) taken at 0.6 THz (f). The racket and wires can be seen clearly.

the sample, whereas the time domain data can be used to get information on the time of the flight, which provides depth information. The resolution achieved with THz wave imaging is reasonably good spatial resolution (millimeter or submillimeter) compared to microwave resolution (centimeter).

A tomography imaging system setup can be implemented by using a Fresnel lens. The focal length of such a lens is frequency-dependent and this dependency can be used for tomography imaging of a target. Objects at various positions along the beam propagation path can be uniquely imaged on the same imaging plane using a Fresnel lens at different frequencies. This procedure allows the reconstruction of an object's tomography contrast image by assembling the frequency-dependent images, providing a new tomography imaging modality. A Fresnel lens is a Fresnel zone plate with phase or amplitude patterns formed by a series of concentric ring structures. The main focal length of this lens is proportional to the frequency of the radiation [126]. The experimental setup of such an imaging system using a CCD camera is similar to the one used for the characterization of a THz wave reported elsewhere [127]. By scanning the time delay between the THz and optical probe beam, a temporal waveform of the THz wave at each pixel on the image plane was measured using the CCD camera. Fourier transformation of the temporal waveforms provides the THz field amplitude (or intensity) distribution on the image plane at each frequency. The measured 2D THz field distribution at each frequency provides images of the THz field transmission of a target at each corresponding position along the $z$-axis. Figure 10.22 schematically illustrates the tomography imaging arrangement. Three plastic sheets with different patterns are placed along the THz beam path at 3, 7, and 14 cm respectively, away from the Fresnel lens (in the $z$ direction). The sensor is located at 6 cm from the Fresnel lens (in the $z'$ direction). The frequencies at which the patterns are well imaged are 0.74, 1.24, and 1.57 THz, respectively. At each frequency, the Fresnel lens images a different plane section of a target object, while images from other depths remain blurred. Each point in the different object planes along the $z$-axis is mapped onto a corresponding point on the $z'$ plane (sensor plane) with the magnification factor $-z'/z$. The depth of focus of the THz wave introduces a depth uncertainty equal to the depth of focus divided by the square of the magnification factor [128]; thus, the uncertainty of the target position is also a function of $z$ and, for a large value of $z$, the depth resolution decreases. Although this concept has been exposed using THz wave pulses, this concept is also applicable to any tunable narrowband imaging beam. Unlike other transmitted THz tomography techniques, which require the rotation of the

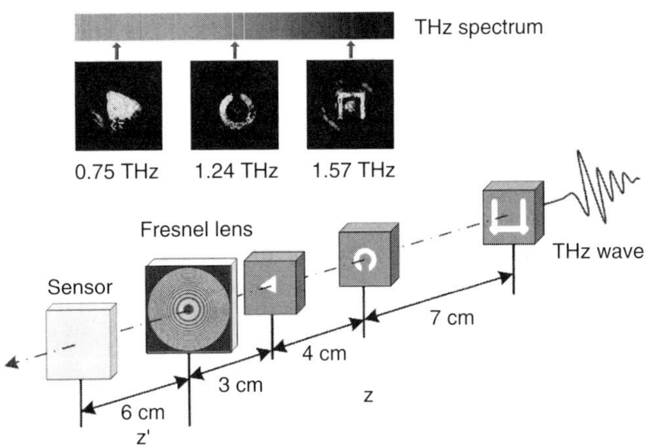


**FIGURE 10.22** Schematic illustration of tomographic imaging with a Fresnel lens. Targets at various locations along the propagation path are uniquely imaged on the same imaging sensor plane with different frequencies.

target [129], this tomography image is achieved without rotating or moving the target.

An alternative method of performing tomography without using a Fresnel lens is the THz computed tomography (T-ray CT). T-ray CT allows pulsed THz radiation to probe the optical properties of 3D structures providing sectional images of objects in a similar way to conventional computed tomography techniques such as x-ray CT. T-ray CT systems directly measure the transmitted amplitude and phase of broadband THz pulses at multiple projection angles through a target, which allows a wealth of information to be extracted from the target object, including both its 3D structure and frequency-dependent FIR optical properties. It extends the capabilities of previous T-ray tomography systems by allowing more general 3D targets to be imaged.

A typical T-ray CT setup is inspired on a regular x-ray CT system. The hardware is a relatively simple extension of modern transmission-mode THz imaging systems. The target sample is mounted on a rotation stage, and a 2D THz image is obtained at each projection angle. The data can then be processed using the filtered back-projection algorithm to reconstruct the refractive index and absorption coefficient at every position within the sample volume. The data acquisition speed is an important concern in all THz imaging systems and it is of particular concern for T-ray CT, in which multiple images of the object must be obtained. Hence, a linearly chirped optical probe beam is used for EO detection of the THz pulses. Using this technique, the full THz waveform is measured simultaneously, dramatically accelerating the imaging speed. The target is then raster scanned in $x$- and $y$-dimensions to form a 2D image.

The reconstruction of the 3D object from the measured projection data is performed using mathematical inverse algorithms. T-ray CT borrows algorithms from x-ray CT such as the filtered back-projection algorithm. This algorithm has long been the workhorse in this domain and it is used to invert the Radon transform to reconstruct the object of interest. Unlike traditional x-ray CT, which only measures the amplitude of the transmitted radiation, T-ray CT measures the transmitted pulse shape and phase information. The reconstruction algorithm can be performed using a number of parameters extracted from the measured data such as the THz pulse amplitude and the timing of the peak of the pulse.

If the amplitude is used, the reconstructed image provides a 3D image dependent on the bulk absorption of the sample in the FIR. Using the timing will reconstruct a refractive-index map of the sample in 3D. Full reconstruction algorithms may use the Fourier transform of the measured THz pulses to reconstruct the frequency-dependent refractive index and absorption of the sample and it may allow different materials to be identified. Figure 10.23 shows the result of the reconstruction for a couple of samples: a turkey bone and a hollow ball. The basic shape of the sphere and the affixed plastic tube is

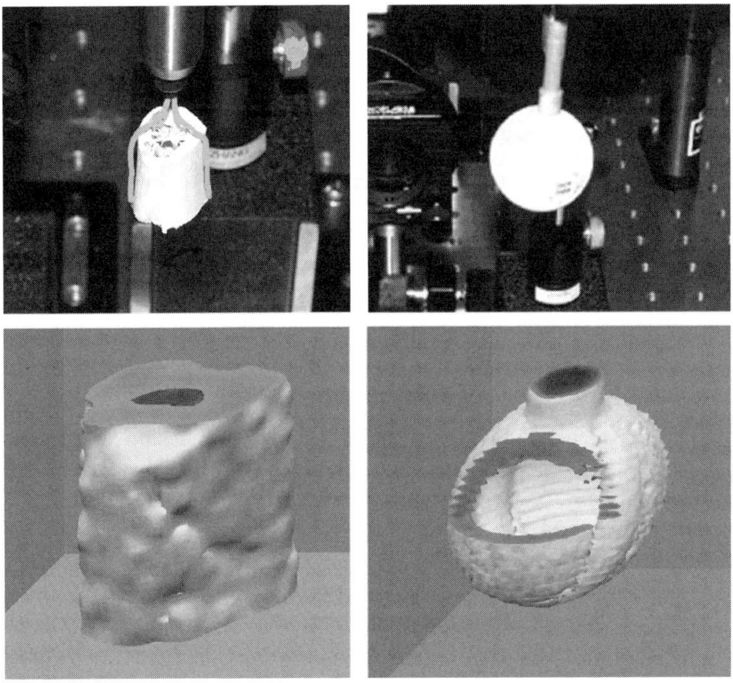

FIGURE 10.23 3D reconstruction of a turkey bone and a hollow ball by using pulsed THz imaging.

clearly visible. A turkey bone was also imaged and reconstructed. In this case, the fine internal structure was not well reconstructed because of its similar size with respect to the THZ wavelength.

A similar task can be performed by using a CW source. The inconvenience is that no phase information is obtained; hence, the reconstruction processes are very similar to those used in regular x-ray CT. The filtered back-propagation and the inversion of the radon transform can be used in a similar way as the T-ray CT with the amplitude information. Figure 10.24 shows the projection and the reconstruction of a cross section of an air freshener. The source is a CW gas laser at 1.63 THz with a power of 180 mW. The detection is performed with a pyroelectric detector. An optical chopper is used to modulate the beam at 80 Hz and a lock-in amplifier is used to register the signal. The projections are taken in transmission geometry in a raster fashion with a step size of 0.5 mm, which is close to the resolution of the system. After all cross sections are reconstructed, a 3D reconstruction is attempted (Figure 10.25) with a reasonable result. The internal structure is well reconstructed and the external pattern of the enclosure is also distinguished.

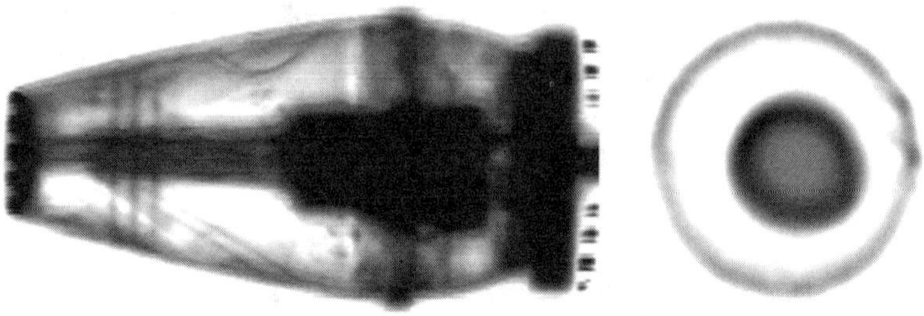

**FIGURE 10.24** Air freshener's 2D projection image (left) by using CW THz gas laser in transmission and cross-section reconstruction (right). The outside ring is the enclosure and the inner structure is the bulk freshener.

The filtered back-projection algorithm and the inversion of the radon transform assume that diffraction effects and Fresnel losses can be neglected, which works well for targets with features that are large relative to the wavelength of the THz radiation (0.3 mm at 1 THz). However, for more complex targets with fine structures, the filtered back-projection algorithm is unable to accurately reconstruct the target because diffraction effects dominate the measurements. The applicability of the T-ray and CW THz CT techniques is limited by two important restrictions: the limited THz power available and the approximations made by the reconstruction algorithm. As T-ray and CW THz CT operate in transmission mode, they are only suitable

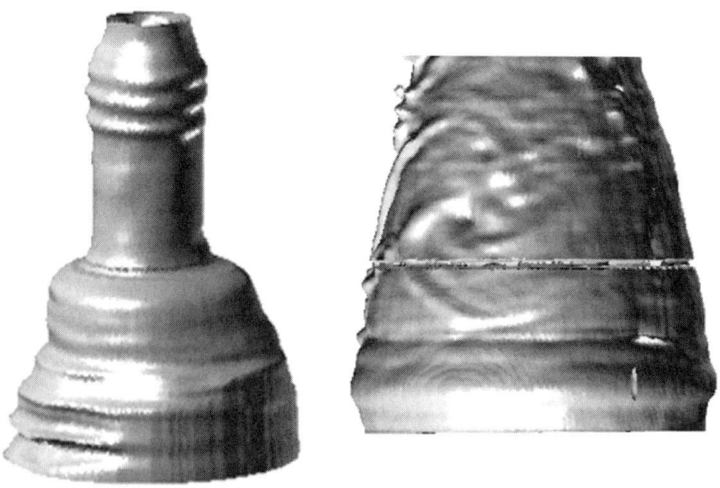

**FIGURE 10.25** Air freshener's internal 3D reconstruction (bottom-left) and external 3D reconstruction (bottom-right). The reconstructed pattern agrees well with the real one.

for objects that do not attenuate or scatter the THz radiation too severely. This is a particular limitation for biomedical applications in which the absorption of moist tissue is prohibitive. In addition, the current simple reconstruction algorithm does not describe the full interaction of THz radiation with complex structures and more sophisticated methods are required before strongly diffracting objects can be imaged accurately. Fortunately, recent research is making progress on both of these fronts. Applications of this technique are foreseen in nondestructive mail and packaging inspection, semiconductor testing, manufacturing quality control, and even potentially in some biomedical applications where the absorption of THz is not too extreme.

## 10.6 CONCLUSION

Pulsed systems based on optical rectification and photoconducting antennas have been well developed in recent years and have revealed a large amount of scientific information in the THz region. It has triggered many academic and industrial applications, such as security scanning, defect characterization, label-free gene and protein analysis, biochemicals identification, gas and liquid measurement, dynamics study of semiconductors, etc. In particular, THz-TDS has been very successful in the identification of small molecules such as explosives, medicines, and biomolecules, because many of the rotational and vibrational peaks of these small molecules are located between 0.3 and 3.0 THz. However, there are still some issues in these applications which may make further developments difficult:

1. Current pulsed THz wave emitters offer extremely low conversion efficiency from the input laser power into THz wave power and a lock-in amplifier has to be used to increase the SNR and time constants around milliseconds are usually set.
2. The high water vapor absorption significantly weakens the THz wave signal during its propagation in the air and, therefore, it is challenging to accomplish remote sensing with THz wave in the air for several meters.
3. Thick samples or with a high absorption coefficient in THz range greatly attenuate the transmitted THz wave and in some cases only the reflected or scattered THz wave signal can be collected and analyzed, although the signal is very weak.

All these issues require a common solution of improving the systems with large SNR. A real-world application utilizing a THz-TDS or THz imaging system with a large SNR provides quick scanning or even snapshot data acquisition, as such a system with a large SNR needs less averaging time for the lock-in amplifier. A powerful source will provide enough THz wave power to compensate for both the water vapor absorption during the propagation and the attenuation through the transmitted object and a very sensitive detector will enable the detection of a severely attenuated or scattered signal from the noise. To get a large SNR, more powerful THz wave sources and more sensitive THz wave detectors are necessary. For some applications, semiconductors such as GaAs and ZnTe are widely used as THz wave emitters and detectors. However, these materials cannot satisfy some of the applications that have critical high SNR requirements. A better THz wave emitter and detector would increase the SNR, reduce scanning time significantly, and would make it possible to perform long-distance and scattering spectroscopy or imaging. The new devices would provide better measurement precision, higher sensitivity, and larger suitability than the ones that are currently in use. With CW applications, the water vapor absorption effects can be reduced by selecting frequencies within the transmission windows available. Nevertheless, in both cases there are regions of the spectra that are useless because of their high absorption.

The nontransperancy of metals could represent an issue for THz imaging in security screening. Also it cannot be used to scan metal packages or regular packages with a high content of metal, but it can still be used for scanning envelopes and other small mail packages to look for explosives or biological agents, which cannot be done with x-rays. However, it is still an attractive technology because it is safer and more reliable than the x-ray technology widely used today. The radiation can be used to screen people for weapons and explosives without endangering their health, which is also a key application compared to x-ray technology.

Acquisition speed is a very important technological problem especially for those applications involving imaging. This problem will be solved when 2D detector arrays become available. Actually, most of the detection is done by a single device, so raster scan or other techniques implying moving the sample, the transceiver, or the beam are required to get a 2D picture of the target. Even though the 2D-CCD-based imaging is available for pulsed systems, its use is still limited to lab environment and some engineering is required to come up with a compact and field use device. There is a lot of work being done to create an array of detectors and invent a THz camera. Recently, research on microbolometers carried by Lee and Hu [130] showed a system that can perform real-time CW THz imaging.

Another problem in reality is that the cost of THz is relatively high compared to other such technologies. Figure 10.26 shows the number of photons generated by each femtodollar for microwave, infrared, and optical technologies compared to THz. It can be seen that the THz technology is the most expensive of all, and, in some cases, with several orders of magnitude. However, the cost is reducing as research and interest in the industry is increasing, and it could reach a cost similar to the optical technologies.

The future applications of THz technology can be divided into three time periods: short term, medium term, and long term. Short-term period implies

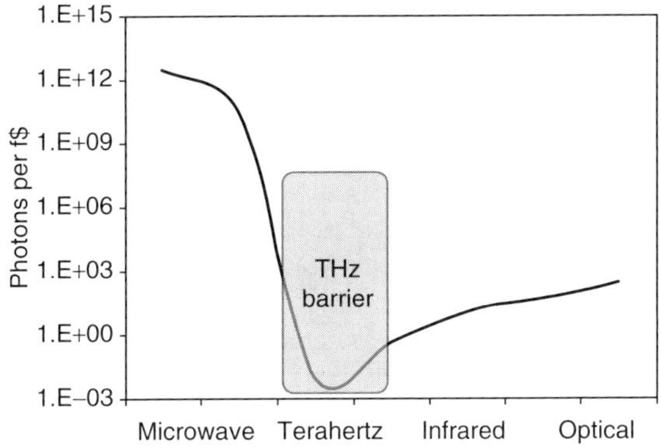


**FIGURE 10.26** Photons per femtodollar of THz technology compared with its neighbor's technologies. Vertical scale is in log scale.

applications that are already using THz technology as a tool and are in the process of optimization and perfection. Medium-term applications are those in which the proof of principle has been demonstrated and are being developed from the laboratory environment to a more realistic environment within 3 to 5 years. Long-term applications are beyond 5 years and are applications that are envisioned as feasible, but they still show important constraints and issues.

Examples of short-term applications are those related to NDE and spectroscopy. Medium-term applications are long-distance sensing and identification together with imaging. Long-term applications involve biological and medical fields. The momentum that the THz technology has gained recently will bring it to a level of maturity of its own similar to optics and radio technologies in a few years. Our estimation of the most relevant projects in future is indicated in Figure 10.27:

1. Complete spectroscopic database of ERC
2. THz sensing and imaging instrument development
3. THz camera and spectrometer
4. THz real time imaging and sensing
5. Biomedical applications: skin cancer and superficial diagnosis
6. Biomedical applications: *in vivo* analysis and diagnosis

The main problem in applying THz in a biomedical field is the high water content of samples. This makes it difficult to use THz for *in vivo* inspection. However, it may be possible in the future to introduce THz radiation with some kind of THz fiber and perform imaging or sensing by introducing the probe inside the body.

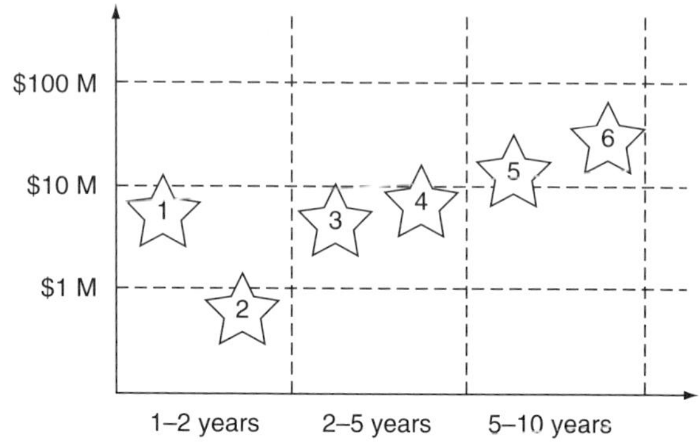


**FIGURE 10.27** Expected funding and time frame for the most relevant future projects.

## ACKNOWLEDGMENTS

The authors would like to thank Bradley Ferguson, Kai Liu, Fatemah al Douseri, Zhiping Jiang, and Shaoghong Wang for providing the data and content. The authors would also like to thank Warren Ussery, Ron Reightler, Eric Madras, and James Walker for the foam samples and critical discussions, and Peter Siegel, Michael Shur, and Nezih Pala for the technical support, Jingzhou Xu for his suggestions and proofreading, Yunqing Chen and Haibo Liu for the explosives' samples preparation, Nick Karpowicz and Hua Zhong for their support on THz imaging, and Xia Li and Kevin Mistry for help in conducting the experiments. Part of the work is supported by the U.S. National Aeronautics and Space Administration and Lockheed Martin Space Systems, the National Science Foundation, the Army Research Office, and IGERT.

## REFERENCES

1. C.H. Lee, "Picosecond optoelectronic switching in GaAs," *Appl. Phys. Lett.*, **30**, 2, pp. 84–86 (1977).
2. C.H. Lee and V.K. Mathur, "Picosecond photoconductivity and its applications," *IEEE J. Quantum Electron.*, **17**, 10, pp. 2098–2112 (1981).
3. G.A. Mourou, C. Stancampiano, A. Antonetti, and A. Orszag, "Picosecond microwave pulses generated with a subpicosecond laser driven semiconductor switch," *Appl. Phys. Lett.*, **39**, 4, pp. 295–365 (1981).
4. J.A. Valdmanis, G.A. Mourou, and C.W. Gabel, "Subpicosecond electrical sampling," *IEEE J. Quantum Electron.*, **19**, 4, pp. 664–667 (1983).
5. C. Fattinger and D. Grischkowsky, "Terahertz beams," *Appl. Phys. Lett.*, **54**, 6, pp. 490–494 (1989).
6. C. Fattinger and D. Grischkowsky, "Point source terahertz optics," *Appl. Phys. Lett.*, **53**, 16, pp. 1483–1485 (1988).
7. M. Bass, P.A. Franken, J.F. Ward, and G. Weinreich, "Optical rectification," *Phys. Rev. Lett.*, **9**, 11, pp. 446–448 (1962).

8. K.H. Yang, P.L. Richards, and Y.R. Shen, "Generation of far-infrared radiation by picosecond light pulses in $LiNbO_3$," *Appl. Phys. Lett.*, **19**, 9, pp. 320–323 (1971).
9. X.-C. Zhang, X.F. Ma, Y. Jin, T.-M. Lu, E.P. Boden, P.D. Phelps, K.R. Stewart, and C.P. Yakymyshyn, "Terahertz optical rectification from a nonlinear organic crystal," *Appl. Phys. Lett.*, **61**, 26, pp. 3080–3082 (1992).
10. X.-C. Zhang, Y. Jin, K. Yang, and L.J. Schowalter, "Resonant nonlinear susceptibility near the GaAs band gap," *Phys. Rev. Lett.*, **69**, 15, pp. 2303–2310 (1992).
11. A. Rice, Y. Jin, X.F. Ma, X.-C. Zhang, D. Bliss, J. Larkin, and M. Alexander, "Terahertz optical rectification from <110> zinc-blende crystals," *Appl. Phys. Lett.*, **64**, 11, pp.1324–1326 (1994).
12. P.Y. Han, M. Tani, F. Pan, and X.-C. Zhang, "Use of the organic crystal DAST for terahertz beam applications," *Opt. Lett.*, **25**, 9, pp. 675–677 (2000).
13. A.M. Sinyukov and L.M. Hayden, "Generation and detection of terahertz radiation with multilayered electro-optic polymer films," *Opt. Lett.*, **27**, 1, pp. 55–57 (2002).
14. A.M. Sinyukov, M.R. Leahy, M. Hayden, M. Haller, J. Luo, A.K-Y. Jen, and L.R. Dalton, "Resonance enhanced THz generation in electro-optic polymers near the absorption maximum," *Appl. Phys. Lett.*, **85**, 24, pp. 5827–5829 (2004).
15. A. Leitenstorfer, S. Hunsche, J. Shah, M.C. Nuss, and W.H. Knox, "Detectors and sources for ultrabroadband electro-optic sampling: Experiment and theory," *Appl. Phys. Lett.*, **74**, 11, pp. 1516–1518 (1999).
16. Imaging technology and systems section, in *The encyclopedia of imaging science and technology*, vol. 2, pp. 1393–1404, ed. L. Hornak, John Wiley & Sons, NY (2002).
17. P.R. Smith, D.H. Auston, and M.C. Nuss, "Subpicosecond photoconducting dipole antennas," *IEEE J. Quantum Electron.*, **24**, 2, pp. 255–260 (1988).
18. M. van Exter, C. Fattinger, and D. Grischkowsky, "High-brightness terahertz beams characterized with an ultrafast detector," *Appl. Phys. Lett.*, **55**, 4, pp. 337–339 (1989).
19. G. Zhao, R.N. Schouten, N. Van-der-Valk, W.T. Wenckebach, and P.C.M. Planken, "Design and performance of a THz emission and detection setup based on a semi-insulating GaAs emitter," *Rev. Sci. Instrum.*, **73**, 4, pp. 1715–1719 (2002).
20. M. Tani, M. Nakajima, S. Kono, and K. Sakai, "Generation of ultrabroadband terahertz radiation with a photoconductive emitter," *IEEE/LEOS Annu. Meeting Conf. Proc.*, vol. 2, p. 532, Glasgow, UK (2002).
21. M. Nakajima, M. Takahashi, and M. Hangyo, "Strong enhancement of THz radiation intensity from semi-insulating GaAs surfaces at high temperatures," *Appl. Phys. Lett.*, **81**, 8, pp. 1462–1464 (2002).
22. H.M. Heiliger, M. Vosseburger, H.G. Roskos, H. Kurz, R. Hey, and K. Ploog, "Applications of liftoff low-temperature-grown GaAs on transparent substrates for THz signal generation," *Appl. Phys. Lett.*, **69**, 19, pp. 2903–2905 (1996).
23. M. Tani, S. Matsuura, K. Sakai, and S. Nakashima, "Emission characteristics of photoconductive antennas based on low-temperature-grown GaAs and semi-insulating GaAs," *Appl. Opt.*, **36**, 30, pp. 7853–7859 (1997).
24. P.Y. Han and X.-C. Zhang, "Coherent, broadband midinfrared terahertz beam sensors," *Appl. Phys. Lett.*, **73**, 21, pp. 3049–3051 (1998).

25. K. Reimann, R.P. Smith, A.M. Weiner, T. Elsaesser, and M. Woerner, "Direct field-resolved detection of terahertz transients with amplitudes of megavolts per centimeter," *Opt. Lett.*, **28**, 6, pp. 471–473 (2003).
26. X.-C. Zhang and D.H. Auston, "Optoelectronic measurement of semiconductor surfaces and interfaces with femtoseconds optics," *J. Appl. Phys.*, **71**, 1, pp. 326–338 (1992).
27. T. Dekorsy, H. Auer, H.J. Bakker, H.G. Roskos, and H. Kurtz, "THz electromagnetic emission by coherent infrared-active phonons," *Phys. Rev. B*, **53**, 7, pp. 4005–4014 (1996).
28. C.A. Schumttenmaer, "Exploring dynamics in the far-infrared with terahertz spectroscopy," *Chem. Rev.*, **104**, 4, pp. 1759–1780 (2004).
29. E. Gornik and R. Kersting, "Coherent THz emission in semiconductors," *Ultrafast Physical Processes in Semiconductors*, vol. 67 of semiconductors and semimetals, Chapter 8, Academic Press, NY (2001).
30. A. Leitenstorfer, S. Hunsche, J. Shah, M.C. Nuss, and W.H. Knox, "Femtosecond high-field transport in compound semiconductor," *Phys. Rev. B*, **61**, 24, pp. 16642–16652 (1999).
31. R. Huber, A. Brodshelm, F. Tauser, and A. Leitenstorfer, "Generation and field-resolved detection of femtosecond electromagnetic pulses tunable up to 41 THz," *Appl. Phys. Lett.*, **76**, 22, pp. 3191–3193 (2000).
32. J.T. Darrow, X.-C. Zhang, and D.H. Auston, "Power scaling of large-apertures photoconducting antennas," *Appl. Phys. Lett.*, **58**, 1, pp. 25–27 (1991).
33. J.T. Darrow, Z.-C. Zhang, D.H. Auston, and J.D. Morse, "Saturation properties of large-apertures photoconduction antennas," *IEEE J. Quantum Electron.*, **28**, 6, pp. 1607–1616 (1992).
34. B. Ferguson and D. Abbott, "De-noising techniques for terahertz responses of biological samples," *Microelec. J.*, **32**, 12, pp. 943–953 (2001).
35. P.C.M. Planken, M.C. Nuss, I. Brener, K.W. Goossen, M.S.C. Luo, S.L. Chuang, and L. Pfeiffer, "Terahertz emission in single quantum wells after coherent optical excitation of light hole and heavy hole excitons," *Phys. Rev. Lett.*, **69**, 26, pp. 3800–3803 (1992).
36. H.G. Roskos, M.C. Nuss, J. Shah, K. Leo, D.A. Miller, A.M. Fox, S. Schmitt-Rink, and K. Kohler, "Coherent Submillimeter-wave emission from charge oscillations in a double-well potential," *Phys. Rev. Lett.*, **68**, 14, pp. 2216–2219 (1992).
37. N. Sekine, K. Yamanaka, K. Hirakawa, M. Vosseburger, P. Haring-Bolivar, and H. Kurz, "Observation of terahertz radiation from higher-order two-dimensional Plasmon modes in GaAs/AlGaAs single quantum wells," *Appl. Phys. Lett.*, **74**, 7, pp. 1006–1008 (1999).
38. X.-C. Zhang, J.T. Darrow, B.B. Hu, D.H. Auton, M.T. Schmidt, P. Tham, and E.S. Yang, "Optically induced electromagnetic radiation from semiconductor surfaces," *Appl. Phys. Lett.*, **56**, 22, pp. 2228–2230 (1990).
39. Y. Shimada, K. Hirakawa, and S.-W. Lee, "Time-resolved terahertz emission spectroscopy of wide miniband GaAs/AlGaAs superlattices," *Appl. Phys. Lett.*, **81**, 9, pp. 1642–1644 (2002).
40. M. Hangyo, S. Tomozawa, Y. Murakami, M. Tonouchi, M. Tani, Z. Wang, K. Sakai, and S. Nakashima, "Terahertz radiation from superconducting $YBa_2Cu_3O_{7-\delta}$ thin films excited by femtosecond optical pulses," *Appl. Phys. Lett.*, **69**, 14, pp. 2122–2124 (1996).

41. C. Jaekel, H.G. Roskos, and H. Kurz, "Emission of picosecond electromagnetic pulses from optically excited superconducting bridges," *Phys. Rev. B*, **54**, 10, pp. 6889–6892 (1996).
42. H. Wald, C. Steigmeier, P. Seidel, S. Nashima, M. Tonouchi, and M. Hangyo, "Comparison of the electromagnetic pulse emission from $YBa_2Cu_3O_{7-\delta}$ and $Y_{0.7}Pr_{0.3}Ba_2Cu_3O_7$ excited by femtosecond laser pulses," *Physica C*, **341**, 3, pp. 1899–1900 (2000).
43. M.C. Beard, G.M. Turner, and C.A. Schmuttenmaer, "Measuring intramolecular charge transfer via coherent generation of THz radiation," *J. Phys. Chem. A*, **106**, 6, pp. 878–883 (2002).
44. N. Kida and M. Tonouchi, "Terahertz radiation from magnetoresistive $Pr_{0.3}Ca_{0.3}MnO_3$ thin films," *Appl. Phys. Lett.*, **78**, 26, pp. 4115–4117 (2001).
45. N. Kida and M. Tonouchi, "Reversible and bistable terahertz radiation from magnetoresistive $Pr_{0.3}Ca_{0.3}MnO_3$ thin films," *Appl. Phys. Lett.*, **82**, 20, pp. 3412–3414 (2003).
46. G.L. Carr, M.C. Martin, W.R. McKinney, K. Jordan, G.R. Neil, and G.P. Williams, "High-power terahertz radiation from relativistic electrons," *Nature*, **420**, pp. 153–156 (2002).
47. G.L. Carr, M.C. Marting, W.R. McKinney, K. Jordan, G.R. Neil, and G.P. Williams, "Very high power THz radiation sources," *J. Bio. Phys.*, **29**, 3, pp. 319–325 (2003).
48. I. Wilke, A.M. MacLeod, W.A. Gillespie, G. Berden, G.M.H. Knippels, and A.F.G. van-der-Meer, "Single-shot electron-beam bunch length measurements," *Phys. Rev. Lett.*, **88**, 12, p. 124801 (2002).
49. X. Xie, J. Dai, and X.-C. Zhang, "Coherent control of THz wave generation in ambient air," *Phys. Rev. Lett.*, **96**, 7, p. 075005 (2006).
50. E.L. Dereniak and G.D. Boreman, "*Infrared detectors and systems*," John Wiley & Sons, Inc., ISBN-0471122092, NY (1996).
51. M.R. Stone, M. Naftaly, R.E. Miles, and J.R.F.D.P. Steenson, "Electrical and radiation characteristics of semi-large photoconductive terahertz emitters," *IEEE Trans. Microw. Theory Tech.*, **52**, 10, pp. 2420–2429 (2004).
52. D.H. Auston and P.R. Smith, "Generation and detection of millimeter waves by picosecond photoconductivity," *Appl. Phys. Lett.*, **43**, 7, pp. 631–633 (1983).
53. S. Kono, M. Tani, and K. Sakai, "Ultra-broadband photoconductive detection: Comparison with free-space electro-optic sampling," *Appl. Phys. Lett.*, **79**, 7, pp. 898–900 (2001).
54. S. Kono, M. Tani, and K. Sakai, "Coherent detection of mid-infrared radiation up to 60 THz with an LT-GaAs photoconductive antenna," *IEE Proc. Optoelectron.*, **149**, 3, pp. 105–109 (2002).
55. Q. Wu and X.-C. Zhang, "Free-space electro-optic sampling of terahertz beams," *Appl. Phys. Lett.*, **67**, 24, pp. 3523–3525 (1995).
56. A. Nahata, A.S. Weling, and T.F. Heinz, "A wideband coherent terahertz spectroscopy system using optical rectification and electro-optic sampling," *Appl. Phys. Lett.*, **69**, 16, pp. 2321–2323 (1996).
57. P.U. Jepsen, C. Winnewisser, M. Schall, V. Schyja, S.R. Keiding, and H. Helm, "THz electromagnetic emission by coherent infrared-active phonons," *Phys. Rev. E*, **53**, 4, pp. 3052–3054 (1996).
58. Q. Wu and X.-C. Zhang, "7 Terahertz broadband GaP electro-optic sensor," *Appl. Phys. Lett.*, **70**, 14, pp. 1784–1786 (1997).

59. M. Vossebrger, M. Brucherseifer, G.C. Cho, H.G. Roskos, and H. Kurz, "Propagation effects in electro-optic sampling of terahertz pulses in GaAs," *Appl. Opt.*, **37**, 15, pp. 3368–3371 (1998).
60. Q. Wu and X.-C. Zhang, "Ultrafast electro-optic field sensor," *Appl. Phys. Lett.*, **68**, 12, pp. 1604–1606 (1996).
61. A. Brodschelm, F. Tauser, R. Huber, J.Y. Sohn, and A. Leitenstorfer, "Amplitude and phase resolved detection of tunable femtosecond pulses with frequency components beyond 100 THz," *Springer Ser. Chem. Phys.*, **66**, pp. 215–217 (2000).
62. Robert S. Elliott, "*An introduction to guided waves and microwave circuits*", 1st Ed., Prentice-Hall, ISBN 0–13–481052-X (1993).
63. F. Maiwald, S. Martin, J. Bruston, A. Maestrini, T. Crawford, and P.H. Siegel, "2.7 THz waveguide tripler using monolithic membrane diodes," *Microw. Symp. Dig., IEEE MTT-S International*, vol. 3, pp. 1637–1640 (2001).
64. V. Ryzhii, I. Khmyrova, and M. Shur, "Terahertz photo-mixing in quantum well structures using resonant excitations of plasma oscillations," *J. Appl. Phys.*, **91**, 4, pp. 1875–1881 (2002).
65. E. Mueller, "Submillimeter wave lasers," *Wiley encyclopedia of electrical and electronics engineering*, vol. 20, Ed. J.G. Webster, John Wiley & Sons, Inc., pp. 597–615 (1999).
66. R. Densing, A. Erstling, M. Gogolewski, H.P. Gemund, G. Lundershausen, and A. Gatesman, "Effective far infrared laser operation with mesh couplers," *Infrared Phys.*, **33**, 3, pp. 219–226 (1992).
67. A. Dobroiu, M. Yamashita, Y.N. Ohshima, Y. Morita, C. Otani, and K. Kawase, "Terahertz imaging system based on a backward-wave oscillator," *Appl. Opt.*, **43**, 30, pp. 5637–5646 (2004).
68. G.P. Williams, "Far-IR/THz radiation from the Jefferson Laboratory, energy recovered linac, free electron laser," *Rev. Sci. Instrum.*, **73**, 3, pp. 1461–1463 (2002).
69. J. Faist, F. Capasso, D.L. Sivco, C. Sirtori, A.L. Hutchinson, and A.Y. Cho, "Quantum cascade laser," *Science*, **264**, 5158, pp. 553–556 (1994).
70. S. Komiyama, "Far-infrared emission from population-inverted hot-carrier system in p-Ge," *Phys. Rev. Lett.*, **48**, 4, pp. 271–274 (1982).
71. Y.P. Gousev, I.V. Altukhov, K.A. Korolev, V.P. Sinis, M.S. Kagan, E.E. Haller, M.A. Odnoblyudov, I.N. Yassievich, and K.A. Chao, "Widely tunable continuous wave THz laser," *Appl. Phys. Lett.*, **75**, 6, pp. 757–759 (1999).
72. R. Kohler, A. Tredicucci, F. Beltram, H.E. Beere, E.H. Linfield, A.G. Davies, D.A. Ritchie, R.C. Iotti, and F. Rossi, "Terahertz semiconductor-heterostructure laser," *Nature*, 417, pp. 156–159 (2002).
73. L. Mahler, A. Tredicucci, R. Kohler, F. Beltram, H.E. Beere, E.H. Linfield, and D.A. Ritchie, "High-performance operation of single-mode terahertz quantum cascade lasers with metallic gratings," *Appl. Phys. Lett.*, **87**, p. 181101 (2005).
74. L. Mahler, R. Kohler, A. Tredicucci, F. Beltram, H.E. Beere, E.H. Linfield, D.A. Ritchie, and A.G. Davies, "Single-mode terahertz quantum cascade lasers with distributed feedback resonators," *Appl. Phys. Lett.*, **84**, 26, pp. 5446–5448 (2004).

75. J.R. Morris and Y.R. Shen, "Theory of far-infrared generation by optical mixing," *Phys. Rev. A*, **15**, 3, pp. 1143–1156 (1977).
76. E.R. Brown, K.A. McIntosh, K.B. Nichols, and C.L. Dennis, "Photo-mixing up to 3.8 THz in low-temperature-grown GaAs," *Appl. Phys. Lett.*, **66**, 3, pp. 285–287 (1995).
77. K. Kawase, M. Sato, T. Taniuchi, and H. Ito, "Coherent tunable THz wave generation from $LiNbO_3$ with monolithic grating coupler," *Appl. Phys. Lett.*, **68**, 18, pp. 2483–2485 (1996).
78. J. Shikata, K. Kawase, M. Sato, T. Taniuchi, and H. Ito, "Enhancement of THz-wave output from $LiNbO_3$ optical parametric oscillators by cryogenic cooling," *Opt. Lett.*, **24**, 4, pp. 202–204 (1999).
79. K. Imai and K. Kawase, "A frequency-agile terahertz-wave parametric oscillator," *Opt. Express*, **8**, 13, pp. 699–704 (2001).
80. K. Kawase, J. Shikata, K. Imai, and H. Ito, "Transform-limited, narrow-linewidth, terahertz-wave parametric generator," *Appl. Phys. Lett.*, **78**, 19, pp. 2819–2821 (2001).
81. C. Kadow, A.W. Jackson, A.C. Gossard, S. Matsuura, and G.A. Blake, "Self assembled ErAs islands in GaAs for optical-heterodyne terahertz generation," *Appl. Phys. Lett.*, **76**, 24, pp. 3510–3512 (2000).
82. S. Komiyama, O. Astaflev, V. Antonov, T. Kutsuwa, and H. Hirai, "A single-photon detector in the far-infrared range," *Nature*, **403**, pp. 405–407 (2000).
83. G.J. Dolan, T.G. Phillips, and D.P. Woody, "Low noise 115 GHz mixing in superconductor oxide barrier tunnel junctions," *Appl. Phys. Lett.*, **34**, 5, pp 347–349 (1979).
84. W. Knap, Y. Deng, S. Rumyantsev, J.Q. Lu, M.S. Shur, C.A. Saylor, and L.C. Brunel, "Resonant detection of sub-terahertz radiation by plasma waves in the submicron field effect transistor," *Appl. Phys. Lett.*, **80**, 18, pp. 3433–3435 (2002).
85. M.C. Gaidis, H.M. Pickett, C.D. Smith, S.C. Martin, R.P Smith, and P.H Siegel, "A 2.5 THz receiver front-end for space borne applications," *IEEE Trans. Microw. Theory Tech.*, **48**, 4, pp. 733–739 (2000).
86. D. Grischkowsky, S.R. Keiding, M.P van-Exter, and C. Fattinger, "Far-infrared time-domain spectroscopy with terahertz beams of dielectrics and semiconductors," *J. Opt. Society Am. B*, **7**, 10, pp. 2006–2015 (1990).
87. J. Xu and X.-C. Zhang, "Circular involute stage," *Opt. Lett.*, **29**, pp. 2082–2084 (2004).
88. J. Xu, Z. Lu, and X.-C. Zhang, "Compact involute optical delay line," *Electron. Lett.*, **40**, p. 1218 (2004).
89. Z. Jiang and X.-C. Zhang, "Electro-optic measurement of THz field pulses with a chirped optical beam," *Appl. Phys. Lett.*, **72**, 16, pp. 1945–1947 (1998).
90. Z. Jiang and X.-C. Zhang, "Single-shot spatiotemporal terahertz field imaging," *Opt. Lett.*, **23**, 14, pp. 1117–1119 (1998).
91. A. Galvanauskas, J.A. Tellefsen, A. Krotkus, M. Oberg, and B. Borberg, "Real-time picosecond electro-optic oscilloscope technique using a tunable semiconductor laser," *Appl. Phys. Lett.*, **60**, 2, pp. 145–147 (1992).

92. M.C. Kemp, P.F. Taday, B.E. Cole, J.A. Cluff, A.J. Fitzgerald, and W.R. Tribe, "Security applications of terahertz technology," *Proc. SPIE*, **5070**, pp. 44–52 (2003).
93. Y. Chen, H. Liu, Y. Deng, D. Veksler, M. Shur, X.-C. Zhang, D. Schauki, M.J. Fitch, and R. Osiander, "Spectroscopic characterization of explosives in the far infrared region," *Proc. SPIE*, **5411**, pp. 1–8 (2004).
94. K. Yamamoto, M. Yamaguchi, F. Miyamaru, M. Tani, M. Hangyo, T. Ikeda, A. Matsushita, K. Koide, M. Tatsuno, and Y. Minami, "Noninvasive inspection of C-4 explosive in mails by terahertz time-domain spectroscopy," *Jpn. J. Appl. Phys.*, **43**, pp. L414–417 (2004).
95. Y.C. Shen, T. Lo, P.F. Taday, B.E. Cole, W.R. Tribe, and M.C. Kemp, "Detection and identification of explosives using terahertz pulsed spectroscopic imaging," *Appl. Phys. Lett.*, **86**, p. 241116 (2005).
96. A. Fadini and F.M. Schnepel, "*Vibrational spectroscopy methods and applications*," John Wiley & Sons, NY (1989).
97. M.D. Fayer, "*Ultrafast infrared and Raman spectroscopy*," Marcel Dekker Inc, NY, Basel (2001).
98. American Society for Testing and Materials, Philadelphia, "Standard testing method for knock characteristics of motor fuels by the research method," ASTM, D2699 (1985).
99. M.H. Brooker and R.W. Berg, "Non-invasive spectroscopic on-line methods to measure industrial processes, A review," Green Industrial Applications of Ionic Liquids, a NATO Advanced Research Workshop (2000).
100. A. Geach, "Infrared analysis as a tool for assessing degradation in used engine lubricants", Technical Report, Weracheck Canada Inc. (1996).
101. H.M. Pickett, R.L. Poynter, and E.A. Cohen, "Submillimeter, millimeter and microwave spectral line catalog" (2003). http://spec.jpl.nasa.gov.
102. H.M. Pickett, R.L. Poynter, E.A. Cohen, M.L. Delitsky, J.C. Pearson, and H.S.P. Muller, "Submillimeter, millimeter and microwave spectral line catalog," *J. Quant. Spectrosc. Radiat. Transf.*, **60**, 5, pp. 883–890 (1998).
103. R.L. Poynter and H.M. Pickett, "Submillimeter, millimeter and microwave spectral line catalog," *Appl. Opt.*, **24**, 14, p. 2235 (1985).
104. P.H. Siegel, "Terahertz technology," *IEEE Trans. Microw. Theory Tech.*, **50**, 3, pp. 910–928 (2002).
105. C.M. Mann, P. de Maagt, G. McBride, F. van-de-Water, D. Castiglione, A. McCalden, L. Deias, J. O'Neill, A. Laisne, J.T. Vallinas, I. Ederra, and D. Haskett, "Micro-fabrication of 3D terahertz circuitry," *IEEE MTT-S Int. Microw. Symp. Digest*, **2**, IEEE Piscataway, NJ, USA, 739 (2003).
106. G.W. Stimson, "*Introduction to airborne radar*," 2nd Ed, SciTech Publishing Inc Mendham, NJ, (1998).
107. N.I. Malykh, A.G. Nagornyi, and E.S. Yampolskii, "Submillimeter-wave imagery," *Instrum. Exp. Tech.*, **18**, p. 1 (1975).
108. T.S. Hartwick, D.T. Hodges, D.H. Barker, and F.B. Foote, "Far infrared imagery," *Appl. Opt.*, **15**, 8, pp. 1919–1922 (1976).
109. A.A. Lash and D.N. Yundev, "Visualization of sub-millimeter laser radiation with a pyroelectric television camera", *Instrum. Exp. Tech.*, **27**, p. 3 (1984).

110. B.B. Hu and M.C. Nuss, "Imaging with terahertz waves," *Opt. Lett.*, **20**, 16, pp. 1716–1718 (1995).
111. D.M. Mittleman, R.H. Jacobson, and M.C. Nuss, "T-ray imaging," *IEEE J. Sel. Top. Quantum Electron.*, **2**, 3, pp. 679–692 (1996).
112. D.M. Mittleman, M. Gupta, R. Neelamani, R.G. Baraniuk, J.V. Rudd, and M. Koch, "Recent advances in terahertz imaging," *Appl. Phys. B*, **68**, p. 1085 (1999).
113. S. Hadjiloucas, L.S. Haratzas, and J.W. Bowen, "Measurement of leaf water content using terahertz radiation," *IEEE Trans. Microw. Theory Tech.*, **47**, 2, pp. 142–149 (1999).
114. M. Knott, "See-through teeth," *New Scientist*, **162**, 2192, p. 22 (1999).
115. R.M. Woodward, B. Cole, V.P. Wallace, D.D. Arnone, R. Pye, E.H. Linfield, M. Pepper, and A.G. Davies, "Terahertz pulse imaging of *in vitro* basal cell carcinoma samples," *Proc. Conf. Lasers Electro-Optics*, Baltimore, p. 329 (2001).
116. W.S. Smye, J.M. Chamberlain, A.J. Fitzgerald, and E. Berry, "The interaction between terahertz radiation and biological tissue," *Phys. Med. Biol.*, **46**, 9, p. 101 (2001).
117. G.C. Walker, E. Berry, N.N. Zinov'ev, A.J. Fitzgerald, R.E. Miles, J. Chamberlain, and M.A. Smith, "Terahertz imaging and international safety guidelines," *Proc. SPIE Med. Imaging*, **4682**, San Diego, p. 683 (2002).
118. T.S. Hartwick, "Far infrared imaging techniques for law enforcement applications," *Optics Secur. Law Enforcement, SPIE*, **108**, p. 139 (1977).
119. Q. Wu, T.D. Hewitt, and X.-C. Zhang, "Two dimensional electro-optic imaging of terahertz beams," *Appl. Phys. Lett.*, **69**, 8, pp. 1026–1028 (1996).
120. V. Minin and O.V. Minin, "System of microwave radio-vision of three-dimensional objects in real time," *Proc. SPIE-Int. Soc. Opt. Eng.*, **4129**, p. 616 (2000).
121. E.D. Walsby, S. Wang, B. Ferguson, J. Xu, T. Yuan, R. Blaikie, S.M. Durbin, D.R.S. Cumming, and X.-C. Zhang, "Investigation of a THz Fresnel lenses," Ed. by R.D. Miller, M.M. Murnane, N.F. Scherer, A.M. Weiner, *Chemical Physics, Ultrafast Phenomena XIII*, p. 292, Springer, New York (2002).
122. S.H. Wang and X.-C. Zhang, "Terahertz tomographic imaging with a Fresnel lens," *Optics & Photonics News,* Special December issue Optics in 2002, 59 (2002).
123. "NASA's implementation plans for space shuttle return to flight and beyond," vol. 1, rev. 2, April 26 (2004). Available at http://www.spaceref.com/news/viewer.html?pid = 21093
124. J. Xu, H. Zhong, T. Yuan, X. Xu, X.-C. Zhang, R. Reightler, and E. Madras, "T-rays identify defects in insulating materials," *CLEO*, CMB2 (2004).
125. N. Karpowicz, H. Zhong, C. Zhang, K. Lin, J.S. Hwang, J. Xu, and X.-C. Zhang, "Compact continuous-wave sub-terahertz system for inspection applications," *Appl. Phys. Lett.*, **86**, p. 054105 (2005).
126. J. Jahns and S.J. Walker, "Two-dimensional array of diffractive micro-lenses fabricated by thin film deposition," *Appl. Opt.*, **29**, 7, pp. 931–936 (1990).
127. S.H. Wang and X.-C. Zhang, "Tomographic imaging with a terahertz binary lens," *Appl. Phys. Lett.*, **82**, 12, pp. 1821–1823 (2003).

128. S.H. Wang, B. Ferguson, D. Abbott, X.-C. Zhang, "THz imaging and tomography," *J. Bio. Phys.*, **29**, pp. 247–256 (2003).
129. X.-C. Zhang, "Three-dimensional terahertz wave imaging," *Phil. Trans. R. Soc. Lond. A*, **362**, pp. 283–299 (2004).
130. A.W.M. Lee and Q. Hu, "Real-time, continuous-wave terahertz imaging by use of a micro-bolometer focal plane array," *Opt. Lett.*, **30**, 19, pp. 2563–2565 (2005).

# Index


2D CCD based imaging, 406
3R function, 35

## A

Absorption recovery, 38, 54–55, 58
Absorption saturation measurement, 55, 58
Absorption spectra, 383, 387–388
Active imaging, 388–391
Airy disk, 397
All-optical switches, 33, 41–72
Ambipolar diffusion constant, 274–275
Amplified spontaneous emission (ASE), 78, 80–81, 88–89
Analog-to-digital converter (ADC), 308–310, 313, 316–318, 331, 336, 346, 349, 350–351, 357
Antenna base stations (BS), 159, 162–169, 172, 174, 176–181
Antenna dispersion, 248
Antenna measurement, 140
Atmospheric attenuation, 383–385
Average power, 364, 368, 390

## B

Backward-wave oscillator, 373–374
BAMH-PPV, 2, 5-bis[N-methyl-N-hexylamino] phenylene vinylene, 262, 290–292, 294, 297, 299, 302
Bandgap, 370, 376, 389
Baseband-over-fiber, 164, 167–168, 173
Broadband radio networks, 158–160
Biological molecules inspection, 387
Bit error ratio (BER), 91, 198–199, 208
Bolometer, 371–372, 377–378
Boltzman constant, 316
Broadband pulse generation, 368–369

## C

Carbon nanotube, 42, 67
Carrier heating effect, 63
Carrier lifetime, 262, 273–275, 281, 289–290, 292–294, 297, 301
Carrier-to-interference-and-noise ratio (CINR) in HFR systems, 170, 172
Carrier-to-noise ratio (CNR) in HFR systems, 164, 168, 170
Central Office (CO) in HFR systems, 160, 162–163, 165–168, 174–175, 178
Commercial HFR systems, 165–167
CCD camera, 389, 391, 401
Central frequency, 364, 365, 396
Characteristic switching time, 261, 276, 286–287, 289, 302
Chirp, 78, 83, 91
Chirped modulation, 322, 328
Chirped optical pulse, 310–311, 314, 340
Chirped pulse, 311, 313, 342, 345, 352–353, 357, 381
$CO_2$ gas laser, 375
Coherent detection, 82–83
Colliding pulse mode-locked (CPM) laser, 37–40
Collision frequency, 268
Continuous-wave systems, 365, 373–379
Conversion efficiency, 374, 405
Cosmic radiation, 388
Coupling coefficient, 264, 270
Cross phase modulation, 60

## D

Deconvolution, 397
Demultiplexing (DEMUX), 35, 43–47, 51, 57, 69–70

Density function theory, 383
Dielectric constant, 268
Diffuse reflectance, 383, 386
Diffusion length, 273
Digital signal processing (DSP), 308, 351
Direct modulation, 79, 82, 104
Dispersion panelty, 319, 335
Dispersion tolerant signal transport in HFR systems, 168–169
Dispersion, 193–195, 202–203, 205, 308, 312–315, 318–324, 326–331, 333–340, 344, 346, 348–349, 352–353, 357–358
Distribute feedback resonator, 376
Distributed Bragg reflector (DBR), 48
Distributed feedback (DFB) laser, 60–61
Domain switching, 276, 287, 289
Duty cycle, 364
Dynamic range, 365, 379, 398

## E

Einstein relation, 273–274
Electroabsorption modulator (EAM), 3, 6–15, 18, 21–22, 24, 26
Electroabsorption transceiver (EAT), 10–12
Electromagnetic waveform synthesis, 215–216, 218, 227, 255
  frequency limits, 217
  millimeter waveforms, 218–219, 221–222, 225–227
  ultrabroadband microwave waveforms, 227, 229, 234
Electro-optic (EO) effect, 115
Electro-optic (EO) material, 117–120, 122, 136
Electro-optic (EO) probing, 116
Electro-optic sampling (EOS), 22–24, 116, 133, 138–139, 150–151, 371, 373
Electro-optics crystal, 368
EMC sensor, 20–22
Erbium doped fiber amplifier (EDFA), 130, 202–207
Error-vector magnitude (EVM), 198, 208
Etalon, 48–49, 51
Evanescent field, 143–146
Explosives and related compounds, 383

## F

Femtosecond laser, 368, 379, 382
Ferroelectric capacitor, 261, 275–284
Ferroelectric nonvolatile random access memories, FERAM, 275
Fiber radio systems, 23–24
Fiber-chip coupling, 3, 14–15
Fingerprints, 383–384
Focused ion beam (FIB), 135
Fourier transform infrared spectroscopy (FTIR), 383, 386–387
Four-wave mixing, 60–61
Free carrier injection, 261–263
Free carriers, 261–262, 266, 268, 275, 289, 302
Free electron laser, 371, 374–375
Fresnel lens, 393, 400–402

## G

GaAs, 261, 266, 268, 275
Gain modulation, 60, 62
Gain saturation, 62
Gain-lever laser, injection locking of, 97–98
Gaussian beam, 367
Golay cell, 371–372, 377–378, 398
Graded-index separate confinement heterostructure (GRIN–SCH), 38
Group velocity dispersion, GVD, 41, 312–315, 319, 355
Gunn diode, 374, 393–394, 398

## H

HBT-EAM, 13–14
Header recognition circuit, 35–36
Heterodyne detection, 377
Heterodyne mixing, 9–10

HFR network architectures, 163–164, 178
HFR signal transport schemes, 171
Hybrid integration in antenna BSs, 179, 181
High impedance probe, 136
Hole burning effect, 62–63
Hyperbolic lens, 393–394, 398

## I

IF-over-fiber, 164, 168, 171–173
Imaging, 146–150
IMPATT diode, 374
Injection locking, 77–105
Intersubband transition (ISBT), 37, 42, 51–60
Intersymbol interference (ISI), 91
Inversion symmetry, 368
Involute stage, 381
Ishibashi–Merz model, 276, 280, 286–287, 289

## J

J-aggregate, 67, 70
Johnson noise, 366

## K

Kerr effect, 115, 368
Kramers–Kronig transform, 386

## L

Laplace transform, 278
Laser rate equations, injection-locked, 85
Laser, DBR, 91, 97–98, 100
Laser, DFB, 86–87, 90–92, 96, 100, 103
Laser, Fabry–Perot, 78, 86, 91
Link gain, 78, 93, 96–98
Local multipoint distribution system (LMDS), 159, 170, 173
Locking range, 79–81, 84–88, 92–95, 100, 103
Low-temperature growth, 51

## M

Mach–Zehnder (MZ) modulator, 322, 329, 337, 342–346, 352–353
Mach–Zehnder interferometer (MZM), 37, 42, 46, 187–189, 192–194, 200, 202, 204–206, 210, 270
Magneto-optic (MO) effect, 115–116
Magneto-optic (MO) material, 122–123
Magneto-optic (MO) probing, 120–122, 130, 134, 150
Mapping, 143–146
Metal–polymer–metal switch, 291, 301
Metropolitan area networks (MAN), 162–163, 175–176
Microcellular radio networks, 158
Millimeter-wave radio, 158–160, 162–163, 167–169, 174–175, 177, 179, 181
Monolithic integration in antenna BSs, 179, 181
Microring resonant wavelength, 261, 266, 302
Microring, 261–266, 268, 270, 275, 302
Microwave mode locking, 22–23
Microwave phase shifter, 15–16
Microwave photonic switching, 261
Millimeter wave communication, 23
Millimeter-wave generation, 93, 98–100
Mode-locked laser diode (MLLD), 128
Mode-locked laser, injection locking of, 101–102
Modulation bandwidth, 84, 91, 92–94, 96–98
Modulation error ratio (MER), 198–200, 204, 208
Modulation frequency, 365–366, 391, 393
Multimode interferometer (MMI), 329
Multiple QW (MQW), 38, 47–51, 54–55, 58

## N

NASA, 395–396, 398
Near-field, 114, 124, 140–143, 146
Network analyzer (NA), 127, 130, 137–139

Noise equivalent power, 365
Noise factor, 190
Noise figure, 187–191, 198, 200–202, 204, 207
Noise floor, 365–366
Noise, 78, 83–86, 96, 98, 100, 103–104
Nondegenerate four wave mixing (NDFWM), 60
Nyquist rate, 308, 310, 350, 357

## O

O-E converter, 127–128
Optical (electronic) time division multiplexing (O(E)TDM), 32
Optical frequency-to-time conversion, 227, 234
Optical Kerr gate, 64
Optical links, 2–3, 17–19, 26
Optical pulse shaping, 215, 227, 229, 238, 255
  direct space-to-time, 215–216
  Fourier transform, 215, 367–368, 383, 401–402
  general, 215–218
Optical rectification, 368–370, 405
Optical single sideband with carrier (OSSB+C) modulation, 168–170, 174, 178
Optical spectral efficiency, 174–175
Optoelectronic switching, 260–261, 264, 302

## P

Passive imaging, 388–391
Performance of HFR systems, 166, 168–174
Picocellular radio networks, 158, 162
  Radio spectrum, 159
Petroleum analysis, 386
Phase diversity, 308, 314, 330, 336–339, 346, 349
Phase switching, 270–272
Phase-array antennas, 260
Phase-match conditions, 369, 373
Phase-shift, 262–263, 270
Phasor model, 79, 89–91
Phonon modes, 383
Photocarrier acceleration, 368
Photoconductance, 262, 293–294, 296–297, 299
Photoconduction, 368
Photoconductive antennas, 367, 369–373
Photoconductive switch, 261, 276, 281, 284, 288–291, 296–299, 302
Photoconductivity effect, 260–261
Photoconductor, 261–262
Photo-Dember effect, 370
Photodetector, 2–5, 7–10, 12–13, 20, 26
Photo-mixing, 374, 376–377
Photon energy, 260, 262–263, 281, 298
Photonic crystals (PC), 41, 71
Photonic integration, 13–14
Photonic local oscillator, 9, 20–21
Photonic microwave signal processing, 3, 14–17
Photonic mixing, 11
Picosecond photoconductor, 261
Pink noise, 366
Planar lightwave circuit (PLC), 43, 51
Planck's law, 388
Plasma frequency, 268
Plasma oscillations, 368
Pockels effect, 115, 368
Point-spread function, 397
Polarization insensitivity, 63
Polarization switching process, 261, 275–281, 285-288
Polymer, 262, 289–292, 294, 296–302
Probe beam, 368, 373, 379–380, 391, 401–402
Pulse duration, 364–365, 369, 382
Pulse energy, 364
Pulsed systems, 364, 367–373, 389–391, 405–406
Pump beam, 368, 377–380
Pump-(and-)probe measurement, 49, 54–55
Pyroelectric detectors, 377–378, 389, 397, 403

## Q

Quadrature amplitude modulation (QAM), 185–186, 198–202, 204, 206–208, 210–211
Quantum cascade laser, 374
Quantum dots (QD), 37, 42, 60–66, 71
Quantum efficiency, 262, 300–301
Quantum wells (QW), 37–38, 42, 46, 51–52, 54, 57–60

## R

Radar, 364, 389
Raman (amplification, amplified, amplifier), 202, 205–208
Raman spectroscopy, 386–387
Raster scan, 391, 402, 406
Rayleigh range, 367
Rayleigh scattering, 195–196
Refractive index, 261–268, 302
Relative intensity noise (RIN), 188–191
Relaxation oscillation, See Resonance frequency
Remote nodes (RN), 163, 174–176, 178
RF-over-fiber, 164–165, 168, 170–173, 177–178
Repetition rate, 364
Resolution, 367, 373, 390, 393, 395, 398–401, 403
Resonance frequency, 84, 86–89, 91–96, 104–105
Responsivity, 365–366, 378
RF bandwidth, 314, 322, 325–329, 333
RF fiber-optic link, 317, 321–323
RIN, 78
Runga–Kutta method, 278–279

## S

Saturable absorber, 37–39, 67
Saturation energy, 55–56, 58, 71
Schottky diode, 374, 377–379, 393–398
Second harmonic generation (SHG), 38, 60
Security screening, 398, 406
SEED elements, 12
Semiconductor optical amplifier (SOA), 42–44, 46, 60–63
SFDR, 91, 96–97
Shot noise, 188–189, 366
Signal-to-noise ratio (SNR), 308, 315–318, 329, 337–340, 346–348, 352–353, 357, 365
Single shot measurement, 381
Single-photon absorption, 260, 263, 270, 275, 302
Single-sideband (SSB) modulation, 308, 330–333
Specific absorption rate (SAR), 141–142
Spectral Engineering, 216, 236–243
Spectrometer, 382–383, 407
Spherical lens, 393
Spin polarization switch, 46–51, 64
Spin relaxation, 37, 42, 49–50
Spontaneous emission, 189–190
Sprayed-on foam insulation inspection, 396
Spur free dynamic range (SFDR), 191–193, 198, 202–204, 206–207, 210, 329, 335, 353
Spurious free dynamic range (SFDR), 165–166, 172–173
Standing wave pattern, 396
Stand-off detection, 384–386
Stimulated Brillouin scattering (SBS), 195–197, 201–203, 205–207
Sub-harmonic optical synchronous mode-locking (SSML), 39
Surface emission, 370
Switching time, 70, 261, 275–277, 281, 284–289, 302

## T

Terahertz computed tomography, 402
Terahertz emission, 289, 300
Terahertz gas laser, 375
Terahertz imaging, 368, 388
Terahertz time domain spectroscopy, 367, 379

Terahertz, 312
Thermal noise, 188–189, 191
Third order intercept (OIP3), 191–193, 201, 203–204, 207
Third order non-linearity, 371
Third-order nonlinear susceptibility, 67
THz gap, 364
Ti:Sapphire laser, 265, 282, 291, 298
Time delay, 368, 382, 389, 401
Time division multiplexing (TDM), 32–33
Time reversal, 310
Time-bandwidth product (TBP), 327–330
Time-stretch, 308, 310–319, 321–333, 337, 339–344, 346, 348–349, 352, 356–357
Time-to-wavelength mapping, 307, 311
Tomography with Fresnel lenses, 399
Transform-limited pulse, 39
Transient digitizer, 330, 339–342
Transient mobility, 262, 290, 302
Transient photoconductance, 293–296, 299–300
Transmission losses, 19
Traveling wave devices, 3, 5, 26
Two-photon absorption, 55–57, 64, 262–264, 268, 270, 275

## U

Ultrafast optoelectronic switching, 260–261, 302
Ultrafast photoconductivity, 260–261
Ultra-wideband (UWB), 214–216, 227–229, 234–240, 242–244, 247–248, 255
Uni-traveling-carrier photodiode (UTC–PD), 128–130, 133, 139
UWB pulse generation, 16–17

## V

VCSELS, injection locking of, 95–96
Vector signal analyzer (VSA), 204
Vector signal generator (VSG), 204, 208
Virtual time-gating, 351–352, 354

## W

Waveguide, 261–262, 264–266, 268, 270, 273, 275, 290, 302
Wavelength conversion, 34–35, 46, 60–63, 72
Wavelength division multiplexing (WDM) in HFR networks, 161–163, 174–177
Wavelength interleaving, 174–176
Wavelength interleaved optical add drop multiplexer (WI–OADM), 174–175
WDM HFR and WDM MAN interface, 163, 175–176
Wireless personal area networks (WPAN), 159
Wireless local area networks (WLAN), 158–159, 161, 166–167
Wavelength division multiplexing (WDM), 32–35, 71–72, 210
Wavelength-to-time mapping, 311, 325, 345

## X

X-ray computed tomography, 402